WILEY SERIES IN PROBABILITY
AND MATHEMATICAL STATISTICS

ESTABLISHED BY WALTER A. SHEWHART AND SAMUEL S. WILKS

Editors
*Vic Barnett, Ralph A. Bradley, J. Stuart Hunter,
Joseph B. Kadane, David G. Kendall, Adrian F. M. Smith,
Stephen M. Stigler, Jozef L. Teugels, Geoffrey S. Watson*

Probability and Mathematical Statistics

ADLER · The Geometry of Random Fields
ANDERSON · The Statistical Analysis of Time Series
ANDERSON · An Introduction to Multivariate Statistical Analysis, *Second Edition*
ARNOLD · The Theory of Linear Models and Multivariate Analysis
BARNETT · Comparative Statistical Inference, *Second Edition*
BERNARDO and SMITH · Bayesian Statistical Concepts and Theory
BHATTACHARYYA and JOHNSON · Statistical Concepts and Methods
BILLINGSLEY · Probability and Measure, *Second Edition*
BOROVKOV · Asymptotic Methods in Queuing Theory
BOSE and MANVEL · Introduction to Combinatorial Theory
CAINES · Linear Stochastic Systems
CHEN · Recursive Estimation and Control for Stochastic Systems
COCHRAN · Contributions to Statistics
COCHRAN · Planning and Analysis of Observational Studies
CONSTANTINE · Combinatorial Theory and Statistical Design
*DOOB · Stochastic Processes
DUDEWICZ and MISHRA · Modern Mathematical Statistics
EATON · Multivariate Statistics: A Vector Space Approach
ETHIER and KURTZ · Markov Processes: Characterization and Convergence
FELLER · An Introduction to Probability Theory and Its Applications, Volume I, *Third Edition*, Revised; Volume II, *Second Edition*
FULLER · Introduction to Statistical Time Series
FULLER · Measurement Error Models
GIFI · Nonlinear Multivariate Analysis
GRENANDER · Abstract Inference
HALD · A History of Probability and Statistics and Their Applications before 1750
HALL · Introduction to the Theory of Coverage Processes
HAMPEL, RONCHETTI, ROUSSEEUW, and STAHEL · Robust Statistics: The Approach Based on Influence Functions
HANNAN · Multiple Time Series
HANNAN and DEISTLER · The Statistical Theory of Linear Systems
HOEL · Introduction to Mathematical Statistics, *Fifth Edition*
HUBER · Robust Statistics
IMAN and CONOVER · A Modern Approach to Statistics
IOSIFESCU · Finite Markov Processes and Applications
JOHNSON and BHATTACHARYYA · Statistics: Principles and Methods, *Revised Printing*
LAHA and ROHATGI · Probability Theory
LARSON · Introduction to Probability Theory and Statistical Inference, *Third Edition*
LEHMANN · Testing Statistical Hypotheses, *Second Edition*
LEHMANN · Theory of Point Estimation
MATTHES, KERSTAN, and MECKE · Infinitely Divisible Point Processes
MUIRHEAD · Aspects of Multivariate Statistical Theory
OLIVER and SMITH · Influence Diagrams, Belief Nets and Decision Analysis
PRESS · Bayesian Statistics: Principles, Models, and Applications
PURI and SEN · Nonparametric Methods in General Linear Models
PURI and SEN · Nonparametric Methods in Multivariate Analysis
PURI, VILAPLANA, and WERTZ · New Perspectives in Theoretical and Applied Statistics
RANDLES and WOLFE · Introduction to the Theory of Nonparametric Statistics
RAO · Asymptotic Theory of Statistical Inference

*Now available in a lower priced paperback edition in the Wiley Classics Library.

*Now available in a lower priced paperback edition in the Wiley Classics Library.

Applied Probability and Statistics (Continued)

Continued on back end papers

*Now available in a lower priced paperback edition in the Wiley Classics Library.

A User's Guide To
Principal Components

A User's Guide To Principal Components

J. EDWARD JACKSON

A Wiley-Interscience Publication
JOHN WILEY & SONS, INC.
New York · Chichester · Brisbane · Toronto · Singapore

BMDP is a registered trademark of BMDP Statistical Software, Inc., Los Angeles, CA.
LISREL is a registered trademark of Scientific Software, Inc., Mooresville, IN.
SAS and SAS Views are registered trademarks of SAS Institute, Inc., Cary, NC.
SPSS is a registered trademark of SPSS Inc., Chicago, IL.

In recognition of the importance of preserving what has been
written, it is a policy of John Wiley & Sons, Inc., to have books
of enduring value published in the United States printed on
acid-free paper, and we exert our best efforts to that end.

Library of Congress Cataloging in Publication Data:
Jackson, J. Edward.
   A user's guide to principal components / J. Edward Jackson.
       p.   cm. — (Wiley series in probability and mathematical
statistics.   Applied probability and statistics)
   Includes bibliographical references and index.
   1. Principal components analysis.   I. Title.   II. Series.
QA278.5.J27   1991
519.5'354—dc20
     ISBN 0-471-62267-2   90-28108
                                                          CIP

Printed in the United States of America

10 9 8 7 6 5 4 3 2

To my wife,
Suzanne

Contents

Preface

Principal Component Analysis (PCA) is a multivariate technique in which a number of related variables are transformed to (hopefully, a smaller) set of uncorrelated variables. This book is designed for practitioners of PCA. It is, primarily, a "how-to-do-it" and secondarily a "why-it-works" book. The theoretical aspects of this technique have been adequately dealt with elsewhere and it will suffice to refer to these works where relevant. Similarly, this book will not become overinvolved in computational techniques. These techniques have also been dealt with adequately elsewhere. The user is focusing, primarily, on data reduction and interpretation. Lest one considers the computational aspects of PCA to be a "black box," enough detail will be included in one of the appendices to leave the user with the feeling of being in control of his or her own destiny.

The method of principal components dates back to Karl Pearson in 1901, although the general procedure as we know it today had to wait for Harold Hotelling whose pioneering paper appeared in 1933. The development of the technique has been rather uneven in the ensuing years. There was a great deal of activity in the late 1930s and early 1940s. Things then subsided for a while until computers had been designed that made it possible to apply these techniques to reasonably sized problems. That done, the development activities surged ahead once more. However, this activity has been rather fragmented and it is the purpose of this book to draw all of this information together into a usable guide for practitioners of multivariate data analysis. This book is also designed to be a sourcebook for principal components. Many times a specific technique may be described in detail with references being given to alternate or competing methods. Space considerations preclude describing them all and, in this way, those wishing to investigate a procedure in more detail will know where to find more information. Occasionally, a topic may be presented in what may seem to be less than favorable light. It will be included because it relates to a procedure which is widely used—for better or for worse. In these instances, it would seem better to include the topic with a discussion of the relative pros and cons rather than to ignore it completely.

As PCA forms only one part of multivariate analysis, there are probably few college courses devoted exclusively to this topic. However, if someone did teach a course about PCA, this book could be used because of the detailed development of methodology as well as the many numerical examples. Except for universities with large statistics departments, this book might more likely find use as a supplementary text for multivariate courses. It may also be useful for departments of education, psychology, and business because of the supplementary material dealing with multidimensional scaling and factor analysis. There are no class problems included. Class problems generally consist of either theoretical proofs and identities, which is not a concern of this book, or problems involving data analysis. In the latter case, the instructor would be better off using data sets of his or her own choosing because it would facilitate interpretation and discussion of the problem.

This book had its genesis at the 1973 Fall Technical Conference in Milwaukee, a conference jointly sponsored by the Physical and Engineering Sciences Section of the American Statistical Association and the Chemistry Division of the American Society for Quality Control. That year the program committee wanted two tutorial sessions, one on principal components and the other on factor analysis. When approached to do one of these sessions, I agreed to do either one depending on who else they obtained. Apparently, they ran out of luck at that point because I ended up doing both of them. The end result was a series of papers published in the *Journal of Quality Technology* (Jackson, 1980, 1981a,b). A few years later, my employer offered an early retirement. When I mentioned to Fred Leone that I was considering taking it, he said, "Retire? What are you going to do, write a book?" I ended up not taking it but from that point on, writing a book seemed like a natural thing to do and the topic was obvious.

When I began my career with the Eastman Kodak Company in the late 1940s, most practitioners of multivariate techniques had the dual problem of performing the analysis on the limited computational facilities available at that time and of persuading their clients that multivariate techniques should be given any consideration at all. At Kodak, we were not immune to the first problem but we did have a more sympathetic audience with regard to the second, much of this due to some pioneering efforts on the part of Bob Morris, a chemist with great natural ability in both mathematics and statistics. It was my pleasure to have collaborated with Bob in some of the early development of operational techniques for principal components. Another chemist, Grant Wernimont, and I had adjoining offices when he was advocating the use of principal components in analytical chemistry in the late 1950s and I appreciated his enthusiasm and steady stream of operational "one-liners." Terry Hearne and I worked together for nearly 15 years and collaborated on a number of projects that involved the use of PCA. Often these assignments required some special procedures that called for some ingenuity on our part; Chapter 9 is a typical example of our collaboration.

A large number of people have given me encouragement and assistance in the preparation of this book. In particular, I wish to thank Eastman Kodak's Multivariate Development Committee, including Nancy Farden, Chuck Heckler, Maggie Krier, and John Huber, for their critical appraisal of much of the material in this book as well as some mainframe computational support for some of the multidimensional scaling and factor analysis procedures. Other people from Kodak who performed similar favors include Terry Hearne, Peter Franchuk, Peter Castro, Bill Novik, and John Twist. The format for Chapter 12 was largely the result of some suggestions by Gary Brauer. I received encouragement and assistance with some of the inferential aspects from Govind Mudholkar of the University of Rochester. One of the reviewers provided a number of helpful comments. Any errors that remain are my responsibility.

I also wish to acknowledge the support of my family. My wife Suzanne and my daughter Janice helped me with proofreading. (Our other daughter, Judy, managed to escape by living in Indiana.) My son, Jim, advised me on some of the finer aspects of computing and provided the book from which Table 10.7 was obtained (Leffingwell was a distant cousin.)

I wish to thank the authors, editors, and owners of copyright for permission to reproduce the following figures and tables: Figure 2.4 (Academic Press); Figures 1.1, 1.4, 1.5, 1.6, and 6.1 (American Society for Quality Control and Marcel Dekker); Figure 8.1 and Table 5.9 (American Society for Quality Control); Figures 6.3, 6.4, 6.5, and Table 7.4 (American Statistical Association); Figures 9.1, 9.2, 9.3, and 9.4 (Biometrie-Praximetrie); Figures 18.1 and 18.2 (Marcel Dekker); Figure 11.7 (Psychometrika and D. A. Klahr); Table 8.1 (University of Chicago Press); Table 12.1 (SAS Institute); Appendix G.1 (John Wiley and Sons, Inc.); Appendix G.2 (Biometrika Trustees, the Longman Group Ltd, the Literary Executor of the late Sir Ronald A. Fisher, F.R.S. and Dr. Frank Yates, F.R.S.); Appendices G.3, G.4, and G.6 (Biometrika Trustees); and Appendix G.5 (John Wiley and Sons, Inc., Biometrika Trustees and Marcel Dekker).

Rochester, New York   J. EDWARD JACKSON
January 1991

A User's Guide To
Principal Components

Introduction

The method of *principal components* is primarily a data-analytic technique that obtains linear transformations of a group of correlated variables such that certain optimal conditions are achieved. The most important of these conditions is that the transformed variables are uncorrelated. It will be the purpose of this book to show why this technique is useful in statistical analysis and how it is carried out.

The first three chapters establish the properties and mechanics of principal component analysis (PCA). Chapter 4 considers the various inferential techniques required to conduct PCA and all of this is then put to work in Chapter 5, an example dealing with audiometric testing.

The next three chapters deal with grouped data and with various methods of interpreting the principal components. These tools are then employed in a case history, also dealing with audiometric examinations.

Multidimensional scaling is closely related to PCA, some techniques being common to both. Chapter 10 considers these with relation to *preference*, or *dominance*, scaling and, in so doing, introduces the concept of *singular value decomposition*. Chapter 11 deals with *similarity* scaling.

The application of PCA to linear models is examined in the next two chapters. Chapter 12 considers, primarily, the relationships among the predictor variables and introduces principal component regression along with some competitors. Principal component ANOVA is considered in Chapter 13.

Chapter 14 discusses a number of other applications of PCA, including missing data, data editing, tests for multivariate normality, discriminant and cluster analysis, and time series analysis. There are enough special procedures for the two-dimensional case that it merits Chapter 15 all to itself. Chapter 16 is a "catch-all" that contains a number of extensions of PCA including cross-validation, procedures for two or more samples, and robust estimation.

The reader will notice that several chapters deal with subgrouped data or situations dealing with two or more populations. Rather than devote a separate chapter to this, it seemed better to include these techniques where relevant. Chapter 6 considers the situation where data are subgrouped as one might find

1

in quality control operations. The application of PCA in the analysis of variance is taken up in Chapter 13 where, again, the data may be divided into groups. In both of these chapters, the underlying assumption for these operations is that the variability is homogeneous among groups, as is customary in most ANOVA operations. To the extent that this is not the case, other procedures are called for. In Section 16.6, we will deal with the problem of testing whether or not the characteristic roots and vectors representing two or more populations are, in fact, the same. A similar problem is considered in a case study in Chapter 9 where some ad hoc techniques will be used to functionally relate these quantities to the various populations for which data are available.

There are some competitors for principal component analysis and these are discussed in the last two chapters. The most important of these competitors is *factor analysis*, which is sometimes confused with PCA. Factor analysis will be presented in Chapter 17, which will also contain a comparison of the two methods and a discussion about the confusion existing between them. A number of other techniques that may relevant for particular situations will be given in Chapter 18.

A basic knowledge of matrix algebra is essential for the understanding of this book. The operations commonly employed are given in Appendix A and a brief discussion of computing methods is found in Appendix C. You will find very few theorems in this book and only one proof. Most theorems will appear as statements presented where relevant. It seemed worthwhile, however, to list a number of basic properties of PCA in one place and this will be found in Appendix B. Appendix D deals with symbols and terminology—there being no standards for either in PCA. Appendix E describes a few classic data sets, located elsewhere, that one might wish to use in experimenting with some of the techniques described in this book. For the most part, the original sources contain the raw data. Appendix F summarizes all of the data sets employed in this book and the uses to which they were put. Appendix G contains tables related to the following distributions: normal, t, chi-square, F, the Lawley–Hotelling trace statistic and the extreme characteristic roots of a covariance matrix.

While the bibliography is quite extensive, it is by no means complete. Most of the citations relate to methodology and operations since that is the primary emphasis of the book. References pertaining to the theoretical aspects of PCA form a very small minority. As will be pointed out in Chapter 4, considerable effort has been expended elsewhere on studying the distributions associated with characteristic roots. We shall be content to summarize the results of this work and give some general references to which those interested may turn for more details. A similar policy holds with regard to computational techniques. The references dealing with applications are but a small sample of the large number of uses to which PCA has been put.

This book will follow the general custom of using Greek letters to denote population parameters and Latin letters for their sample estimates. Principal component analysis is employed, for the most part, as an exploratory data analysis technique, so that applications involve sample data sets and sample estimates obtained from them. Most of the presentation in this book will be within that context and for that reason population parameters will appear primarily in connection with inferential techniques, in particular in Chapter 4. It is comforting to know that the general PCA methodology is the same for populations as for samples.

Fortunately, many of the operations associated with PCA estimation are distribution free. When inferential procedures are employed, we shall generally assume that the population or populations from which the data were obtained have multivariate normal distributions. The problems associated with non-normality will be discussed where relevant.

Widespread development and application of PCA techniques had to wait for the advent of the high-speed electronic computer and hence one usually thinks of PCA and other multivariate techniques in this vein. It is worth pointing out, however, that with the exception of a few examples where specific mainframe programs were used, the computations in this book were all performed on a 128K microcomputer. No one should be intimidated by PCA computations.

Many statistical computer packages contain a PCA procedure. However, these procedures, in general, cover some, but not all, of the first three chapters, in addition to some parts of Chapters 8 and 17 and in some cases parts of Chapters 10, 11, and 12. For the remaining techniques, the user will have to provide his or her own software. Generally, these techniques are relatively easy to program and one of the reasons for the many examples is to provide the reader some sample data with which to work. Do not be surprised if your answers do not agree to the last digit with those in the book. In addition to the usual problems of computational accuracy, the number of digits has often been reduced in presentation, either in this book or the original sources, to two or three digits for reason of space of clarity. If these results are then used in other computations, an additional amount of precision may be lost. The signs for the characteristic vectors may be reversed from the ones you obtain. This is either because of the algorithm employed or because someone reversed the signs deliberately for presentation. The interpretation will be the same either way.

CHAPTER 1

Getting Started

1.1 INTRODUCTION

The field of *multivariate analysis* consists of those statistical techniques that consider two or more related random variables as a single entity and attempts to produce an overall result taking the relationship among the variables into account. A simple example of this is the correlation coefficient. Most inferential multivariate techniques are generalizations of classical univariate procedures. Corresponding to the univariate t-test is the multivariate T^2-test and there are multivariate analogs of such techniques as regression and the analysis of variance. The majority of most multivariate texts are devoted to such techniques and the multivariate distributions that support them.

There is, however, another class of techniques that is unique to the multivariate arena. The correlation coefficient is a case in point. Although these techniques may also be employed in statistical inference, the majority of their applications are as data-analytic techniques, in particular, techniques that seek to describe the multivariate *structure* of the data. *Principal Component Analysis* or PCA, the topic of this book, is just such a technique and while its main use is as a descriptive technique, we shall see that it may also be used in many inferential procedures as well.

In this chapter, the method of principal components will be illustrated by means of a small hypothetical two-variable example, allowing us to introduce the mechanics of PCA. In subsequent chapters, the method will be extended to the general case of p variables, some larger examples will be introduced, and we shall see where PCA fits into the realm of multivariate analysis.

1.2 A HYPOTHETICAL EXAMPLE

Suppose, for instance, one had a process in which a quality control test for the concentration of a chemical component in a solution was carried out by two different methods. It may be that one of the methods, say Method 1, was the

4

standard procedure and that Method 2 was a proposed alternative, a procedure that was used as a back-up test or was employed for some other reason. It was assumed that the two methods were interchangeable and in order to check that assumption a series of 15 production samples was obtained, each of which was measured by both methods. These 15 pairs of observations are displayed in Table 1.1. (The choice of $n = 15$ pairs is merely for convenience in keeping the size of this example small; most quality control techniques would require more than this.)

What can one do with these data? The choices are almost endless. One possibility would be to compute the differences in the observed concentrations and test that the mean difference was zero, using the paired difference t-test based on the variability of the 15 differences. The analysis of variance technique would treat these data as a two-way ANOVA with methods and runs as factors. This would probably be a mixed model with *methods* being a fixed factor and *runs* generally assumed to be random. One would get the by-product of a run component of variability as well as an overall measure of inherent variability if the inherent variability of the two methods were the same. This assumption could be checked by a techniques such as the one due to Grubbs (1948, 1973) or that of Russell and Bradley (1958), which deal with heterogeneity of variance in two-way data arrays. Another complication could arise if the variability of the analyses was a function of level but a glance at the scattergram of the data shown in Figure 1.1 would seem to indicate that this is not the case.

Certainly, the preparation of Figure 1.1 is one of the first things to be considered, because in an example this small it would easily indicate any outliers or other aberrations in the data as well as provide a quick indication of the relationship between the two methods. Second, it would suggest the use of

Table 1.1. Data for Chemical Example

Obs. No.	Method 1	Method 2
1	10.0	10.7
2	10.4	9.8
3	9.7	10.0
4	9.7	10.1
5	11.7	11.5
6	11.0	10.8
7	8.7	8.8
8	9.5	9.3
9	10.1	9.4
10	9.6	9.6
11	10.5	10.4
12	9.2	9.0
13	11.3	11.6
14	10.1	9.8
15	8.5	9.2

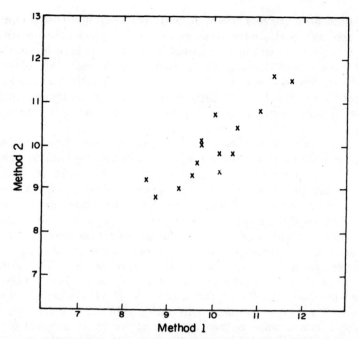

FIGURE 1.1. Chemical example: original data. Reproduced from Jackson (1980) with permission of the American Society for Quality Control and Jackson (1985) with permission of Marcel Dekker.

regression to determine to what extent it is possible to predict the results of one method from the other. However, the requirement that these two methods should be interchangeable means being able to predict in either direction, which (by using ordinary least-squares) would result in two different equations. The least-squares equation for predicting Method 1 from Method 2 minimizes the variability in Method 1 given a specific level of Method 2, while the equation for predicting Method 2 from Method 1 minimizes the variability in Method 2 given a specific level of Method 1.

A single prediction equation is required that could be used in either direction. One could invert either of the two regression equations, but which one and what about the theoretical consequences of doing this? The line that will perform this role directly is called the *orthogonal* regression line which minimizes the deviations *perpendicular to the line itself.* This line is obtained by the method of principal components and, in fact, was the first application of PCA, going back to Karl Pearson (1901). We shall obtain this line in the next section and in so doing will find that PCA will furnish us with a great deal of other information as well. Although many of these properties may seem superfluous for this small two-variable example, its size will allow us to easily understand these properties and the operations required to use PCA. This will be helpful when we then go on to larger problems.

In order to illustrate the method of PCA, we shall need to obtain the sample means, variances and the covariance between the two methods for the data in Table 1.1. Let x_{1k} be the test result for Method 1 for the kth run and the corresponding result for Method 2 be denoted by x_{2k}. The vector of sample means is

$$\bar{\mathbf{x}} = \begin{bmatrix} \bar{x}_1 \\ \bar{x}_2 \end{bmatrix} = \begin{bmatrix} 10.00 \\ 10.00 \end{bmatrix}$$

and the sample covariance matrix is

$$\mathbf{S} = \begin{bmatrix} s_1^2 & s_{12} \\ s_{12} & s_2^2 \end{bmatrix} = \begin{bmatrix} .7986 & .6793 \\ .6793 & .7343 \end{bmatrix}$$

where s_i^2 is the variance and the covariance is

$$s_{ij} = \frac{n \sum x_{ik} x_{jk} - \sum x_{ik} \sum x_{jk}}{[n(n-1)]}$$

with the index of summation, k, going over the entire sample of $n = 15$. Although the correlation between x_1 and x_2 is not required, it may be of interest to estimate this quantity, which is

$$r = \frac{s_{12}}{(s_1 s_2)} = .887$$

1.3 CHARACTERISTIC ROOTS AND VECTORS

The method of principal components is based on a key result from matrix algebra: A $p \times p$ symmetric, nonsingular matrix, such as the covariance matrix $\mathbf{S}$, may be reduced to a diagonal matrix $\mathbf{L}$ by premultiplying and postmultiplying it by a particular orthonormal matrix $\mathbf{U}$ such that

$$\mathbf{U}'\mathbf{S}\mathbf{U} = \mathbf{L} \tag{1.3.1}$$

The diagonal elements of $\mathbf{L}$, $l_1, l_2, \ldots, l_p$ are called the *characteristic roots, latent roots* or *eigenvalues* of $\mathbf{S}$. The columns of $\mathbf{U}$, $\mathbf{u}_1, \mathbf{u}_2, \ldots, \mathbf{u}_p$ are called the *characteristic vectors* or *eigenvectors* of $\mathbf{S}$. (Although the term *latent* vector is also correct, it often has a specialized meaning and it will not be used in this book except in that context.) The characteristic roots may be obtained from the solution of the following determinental equation, called the *characteristic equation*:

$$|\mathbf{S} - l\mathbf{I}| = 0 \tag{1.3.2}$$

where **I** is the identity matrix. This equation produces a pth degree polynomial in l from which the values $l_1, l_2, \ldots, l_p$ are obtained.

For this example, there are $p = 2$ variables and hence,

$$|\mathbf{S} - l\mathbf{I}| = \begin{vmatrix} .7986 - l & .6793 \\ .6793 & .7343 - l \end{vmatrix} = .124963 - 1.53291 + l^2 = 0$$

The values of l that satisfy this equation are $l_1 = 1.4465$ and $l_2 = .0864$.

The characteristic vectors may then be obtained by the solution of the equations

$$[\mathbf{S} - l\mathbf{I}]\mathbf{t}_i = 0 \tag{1.3.3}$$

and

$$\mathbf{u}_i = \frac{\mathbf{t}_i}{\sqrt{\mathbf{t}_i' \mathbf{t}_i}} \tag{1.3.4}$$

for $i = 1, 2, \ldots, p$. For this example, for $i = 1$,

$$[\mathbf{S} - l_1\mathbf{I}]\mathbf{t}_1 = \begin{bmatrix} .7986 - 1.4465 & .6793 \\ .6793 & .7343 - 1.4465 \end{bmatrix}\begin{bmatrix} t_{11} \\ t_{21} \end{bmatrix} = \begin{bmatrix} 0 \\ 0 \end{bmatrix}$$

These are two homogeneous linear equations in two unknowns. To solve, let $t_{11} = 1$ and use just the first equation:

$$-.6478 + .6793 t_{21} = 0$$

The solution is $t_{21} = .9538$. These values are then placed in the normalizing equation (1.3.4) to obtain the first characteristic vector:

$$\mathbf{u}_1 = \frac{\mathbf{t}_1}{\sqrt{\mathbf{t}_1' \mathbf{t}_2}} = \frac{1}{\sqrt{1.9097}}\begin{bmatrix} 1.0 \\ .9538 \end{bmatrix} = \begin{bmatrix} .7236 \\ .6902 \end{bmatrix}$$

Similarly, using $l_2 = .0864$ and letting $t_{22} = 1$, the second characteristic vector is

$$\mathbf{u}_2 = \begin{bmatrix} -.6902 \\ .7236 \end{bmatrix}$$

These characteristic vectors make up the matrix

$$\mathbf{U} = [\mathbf{u}_1 \mid \mathbf{u}_2] = \begin{bmatrix} .7236 & -.6902 \\ .6902 & .7236 \end{bmatrix}$$

which is *orthonormal*, that is,

$$\mathbf{u}_1'\mathbf{u}_1 = 1 \qquad \mathbf{u}_2'\mathbf{u}_2 = 1 \qquad \mathbf{u}_1'\mathbf{u}_2 = 0$$

Furthermore,

$$\mathbf{U'SU} = \begin{bmatrix} .7236 & 0.6902 \\ -.6902 & .7236 \end{bmatrix} \begin{bmatrix} .7986 & .6793 \\ .6793 & .7343 \end{bmatrix} \begin{bmatrix} .7236 & -.6902 \\ .6902 & .7236 \end{bmatrix}$$

$$= \begin{bmatrix} 1.4465 & 0 \\ 0 & .0864 \end{bmatrix} = \mathbf{L}$$

verifying equation (1.3.1).

Geometrically, the procedure just described is nothing more than a *principal axis* rotation of the original coordinate axes x_1 and x_2 about their means as seen in Figures 1.2 and 1.3. The elements of the characteristic vectors are the direction cosines of the new axes related to the old. In Figure 1.2, $u_{11} = .7236$ is the cosine of the angle between the x_1-axis and the first new axis; $u_{21} = .6902$ is the cosine of the angle between this new axis and the x_2-axis. The new axis related to $\mathbf{u}_1$ is the orthogonal regression line we were looking for. Figure 1.3 contains the same relationships for $\mathbf{u}_2$.

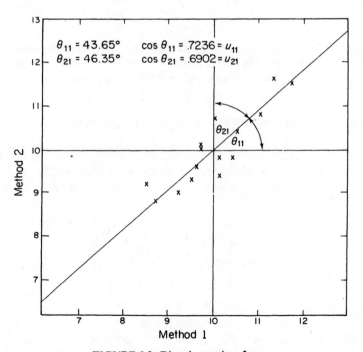

FIGURE 1.2. Direction cosines for $\mathbf{u}_1$.

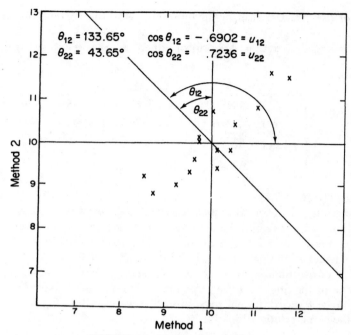

FIGURE 1.3. Direction cosines for $\mathbf{u}_2$.

Except for $p = 2$ or $p = 3$, equation (1.3.2) is not used in practice as the determinental equations become unwieldy. Iterative procedures, described in Appendix C, are available for obtaining both the characteristic roots and vectors.

1.4   THE METHOD OF PRINCIPAL COMPONENTS

Now that all of the preliminaries are out of the way, we are now ready to discuss the method of principal components (Hotelling, 1933). The starting point for PCA is the sample covariance matrix $\mathbf{S}$ (or the correlation matrix as we shall see in Chapter 3). For a p-variable problem,

$$
\mathbf{S} = \begin{bmatrix} s_1^2 & s_{12} & \cdots & s_{1p} \\ s_{12} & s_2^2 & \cdots & s_{2p} \\ \vdots & \vdots & & \vdots \\ s_{1p} & s_{2p} & \cdots & s_p^2 \end{bmatrix}
$$

where s_i^2 is the variance of the ith variable, x_i, and s_{ij} is the covariance between the ith and jth variables. If the covariances are not equal to zero, it indicates that a linear relationship exists between these two variables, the strength of that relationship being represented by the correlation coefficient, $r_{ij} = s_{ij}/(s_i s_j)$.

The principal axis transformation obtained in Section 1.3 will transform p correlated variables $x_1, x_2, \ldots, x_p$ into p new *uncorrelated* variables $z_1, z_2, \ldots, z_p$. The coordinate axes of these new variables are described by the characteristic vectors $\mathbf{u}_i$ which make up the matrix $\mathbf{U}$ of direction cosines used in the transformation:

$$\mathbf{z} = \mathbf{U}'[\mathbf{x} - \bar{\mathbf{x}}] \tag{1.4.1}$$

Here $\mathbf{x}$ and $\bar{\mathbf{x}}$ are $p \times 1$ vectors of observations on the original variables and their means.

The transformed variables are called the *principal components* of $\mathbf{x}$ or *pc*'s for short. The ith principal component is

$$z_i = \mathbf{u}_i'[\mathbf{x} - \bar{\mathbf{x}}] \tag{1.4.2}$$

and will have mean zero and variance l_i, the ith characteristic root. To distinguish between the transformed variables and the transformed observations, the transformed *variables* will be called *principal components* and the individual transformed observations will be called *z-scores*. The use of the word *score* has its genesis in psychology and education, particularly in connection with factor analysis, the topic of Chapter 17. However, this term is now quite prevalent with regard to PCA as well, particularly with the advent of the mainframe computer statistical packages, so we will employ it here also. The distinction is made here with regard to z-scores because another normalization of these scores will be introduced in Section 1.6.

The first observation from the chemical data is

$$\mathbf{x} = \begin{bmatrix} 10.0 \\ 10.7 \end{bmatrix}$$

Substituting in (1.4.1) produces

$$\mathbf{z} = \begin{bmatrix} .7236 & .6902 \\ -.6902 & .7236 \end{bmatrix} \begin{bmatrix} 10.0 - 10.00 \\ 10.7 - 10.00 \end{bmatrix} = \begin{bmatrix} .48 \\ .51 \end{bmatrix}$$

so the z-scores for the first observation are $z_1 = .48$ and $z_2 = .51$. The variance of z_1 is equal to $l_1 = 1.4465$ and the variance of z_2 is equal to $l_2 = .0864$. As we shall see in Section 1.5.3, $l_1 + l_2$ are equal to the sum of the variances of the original variables. Table 1.2 includes, for the original 15 observations, the deviations from their means and their corresponding pc's, z_1 and z_2 along with some other observations and quantities that will be described in Sections 1.6 and 1.7.

Table 1.2. Chemical Example. Principal Components Calculations

Obs. No.	$x_1 - \bar{x}_1$	$x_2 - \bar{x}_2$	z_1	z_2	$\sqrt{l_1}\,z_1$	$\sqrt{l_2}\,z_2$	y_1	y_2	T^2
1	.0	.7	.48	.51	.58	.15	.40	1.72	3.12
2	.4	-.2	.15	-.42	.18	-.12	.13	-1.43	2.06
3	-.3	.0	-.22	.21	-.26	.06	-.18	.70	.52
4	-.3	.1	-.15	.28	-.18	.08	-.12	.95	.92
5	1.7	1.5	2.27	-.09	2.72	-.03	1.88	-.30	3.62
6	1.0	.8	1.28	-.11	1.53	-.03	1.06	-.38	1.27
7	-1.3	-1.2	-1.77	.03	-2.13	.01	-1.47	.10	2.17
8	-.5	-.7	-.84	-.16	-1.02	-.05	-.70	-.55	.79
9	.1	-.6	-.34	-.50	-.41	-.15	-.28	-1.71	3.00
10	-.4	-.4	-.57	-.01	-.68	-.00	-.47	-.05	.22
11	.5	.4	.64	-.06	.77	-.02	.53	-.19	.32
12	-.8	-1.0	-1.27	-.17	-1.53	-.05	-1.06	-.58	1.46
13	1.3	1.6	2.04	.26	2.46	.08	1.70	.89	3.68
14	.1	-.2	-.07	-.21	-.08	-.06	-.05	-.73	.54
15	-1.5	-.8	-1.64	.46	-1.97	.13	-1.36	1.55	4.25
A	2.3	2.5	3.39	.22	4.08	.07	2.82	.75	8.51
B	-3.0	-2.7	-4.03	.12	-4.85	.03	-3.35	.40	11.38
C	1.0	-1.0	.03	-1.41	.04	-.42	.03	-4.81	23.14
D	-2.7	-.9	-2.57	1.21	-3.10	.36	-2.14	4.12	21.55
Variance for observations 1–15	.7986 $= s_1^2$	.7343 $= s_2^2$	1.4465 $= l_1$	.0864 $= l_2$	2.0924 $= l_1^2$	.0075 $= l_2^2$	1.0	1.0	

1.5 SOME PROPERTIES OF PRINCIPAL COMPONENTS

1.5.1 Transformations

If one wishes to transform a set of variables $\mathbf{x}$ by a linear transformation $\mathbf{z} = \mathbf{U}'[\mathbf{x} - \bar{\mathbf{x}}]$ whether $\mathbf{U}$ is orthonormal or not, the covariance matrix of the new variables, $\mathbf{S}_z$, can be determined directly from the covariance matrix of the original observations, $\mathbf{S}$, by the relationship

$$\mathbf{S}_z = \mathbf{U}'\mathbf{S}\mathbf{U} \qquad (1.5.1)$$

However, the fact that $\mathbf{U}$ is orthonormal is not a sufficient condition for the transformed variables to be uncorrelated. Only this characteristic vector solution will produce an $\mathbf{S}_z$ that is a diagonal matrix like $\mathbf{L}$ producing new variables that are uncorrelated.

1.5.2 Interpretation of Principal Components

The coefficients of the first vector, .7236 and .6902, are nearly equal and both positive, indicating that the first pc, z_1, is a weighted average of both variables. This is related to variability that x_1 and x_2 have in common; in the absence of correlated errors of measurement, this would be assumed to represent process variability. We have already seen that $\mathbf{u}_1$ defines the orthogonal regression line that Pearson (1901) referred to as the "line of best fit." The coefficients for the second vector, $-.6902$ and .7236 are also nearly equal except for sign and hence the second pc, z_2, represents differences in the measurements for the two methods that would probably represent testing and measurement variability. (The axis defined by $\mathbf{u}_2$ was referred to by Pearson as the "line of worst fit." However, this term is appropriate for the characteristic vector corresponding to the smallest characteristic root, not the second unless there are only two as is the case here.)

   While interpretation of two-variable examples is quite straightforward, this will not necessarily be the case for a larger number of variables. We will have many examples in this book, some dealing with over a dozen variables. Special problems of interpretation will be taken up in Chapters 7 and 8.

1.5.3 Generalized Measures and Components of Variability

In keeping with the goal of multivariate analysis of summarizing results with as few numbers as possible, there are two single-number quantities for measuring the overall variability of a set of multivariate data. These are

1. The determinant of the covariance matrix, $|\mathbf{S}|$. This is called the *generalized variance*. The square root of this quantity is proportional to the area or volume generated by a set of data.

2. The sum of the variances of the variables:

$$s_1^2 + s_2^2 + \cdots + s_p^2 = \text{Tr}(\mathbf{S}) \quad (\text{The } \textit{trace} \text{ of } \mathbf{S})$$

Conceivably, there are other measures of generalized variability that may have certain desirable properties but these two are the ones that have found general acceptance among practitioners.

A useful property of PCA is that the variability as specified by either measure is preserved.

1.
$$|\mathbf{S}| = |\mathbf{L}| = l_1 l_2 \ldots l_p \tag{1.5.2}$$

that is, the determinant of the original covariance matrix is equal to the *product* of the characteristic roots. For this example,

$$|\mathbf{S}| = .1250 = (1.4465)(.0864) = l_1 l_2$$

2.
$$\text{Tr}(\mathbf{S}) = \text{Tr}(\mathbf{L}) \tag{1.5.3}$$

that is, the sum of the original variances is equal to the *sum* of the characteristic roots. For this example,

$$s_1^2 + s_2^2 = .7986 + .7343 = 1.5329$$
$$= 1.4465 + .0864 = l_1 + l_2$$

The second identity is particularly useful because it shows that the characteristic roots, which are the variances of the principal components, may be treated as variance components. The ratio of each characteristic root to the total will indicate the proportion of the total variability accounted for by each pc. For z_1, $1.4465/1.5329 = .944$ and for z_2, $.0864/1.5329 = .056$. This says that roughly 94% of the total variability of these chemical data (as represented by $\text{Tr}(\mathbf{S})$ is associated with, accounted for or "explained by" the variability of the process and 6% is due to the variability related to testing and measurement. Since the characteristic roots are sample estimates, these proportions are also sample estimates.

1.5.4 Correlation of Principal Components and Original Variables

It is also possible to determine the correlation of each pc with each of the original variables, which may be useful for diagnostic purposes. The correlation of the ith pc, z_i, and the jth original variable, x_j, is equal to

$$r_{zx} = \frac{u_{ji}\sqrt{l_i}}{s_j} \tag{1.5.4}$$

For instance, the correlation between z_1 and x_1 is

$$\frac{u_{11}\sqrt{l_1}}{s_1} = \frac{.7236\sqrt{1.4465}}{\sqrt{.7986}} = .974$$

and the correlations for this example become

$$
\begin{array}{c c}
 & \begin{array}{cc} z_1 & z_2 \end{array} \\
\begin{array}{c} x_1 \\ x_2 \end{array} &
\left[\begin{array}{cc} .974 & -.227 \\ .969 & .248 \end{array} \right]
\end{array}
$$

The first pc is more highly correlated with the original variables than the second. This is to be expected because the first pc accounts for more variability than the second. Note that the sum of squares of each row is equal to 1.0.

1.5.5 Inversion of the Principal Component Model

Another interesting property of PCA is the fact that equation (1.4.1),

$$\mathbf{z} = \mathbf{U}'[\mathbf{x} - \bar{\mathbf{x}}]$$

may be inverted so that the original variables may be stated as a function of the principal components, viz.,

$$\mathbf{x} = \bar{\mathbf{x}} + \mathbf{U}\mathbf{z} \qquad (1.5.5)$$

because $\mathbf{U}$ is orthonormal and hence $\mathbf{U}^{-1} = \mathbf{U}'$. This means that, given the z-scores, the values of the original variables may be uniquely determined.

Corresponding to the first observation, the z-scores, .48 and .51, when substituted into (1.5.5), produce the following:

$$
\begin{bmatrix} 10.0 \\ 10.7 \end{bmatrix} =
\begin{bmatrix} 10.0 \\ 10.0 \end{bmatrix} +
\begin{bmatrix} .7236 & -.6902 \\ .6902 & .7236 \end{bmatrix}
\begin{bmatrix} .48 \\ .51 \end{bmatrix}
$$

Said another way, each variable is made up of a linear combination of the pc's. In the case of x_1 for the first observation,

$$x_1 = \bar{x}_1 + u_{11}z_1 + u_{12}z_2$$
$$= 10.0 + (.7236)(.48) + (-.6902)(.51) = 10.0$$

This property might seem to be of mild interest; its true worth will become apparent in Chapter 2.

1.5.6   Operations with Population Values

As mentioned in the Introduction, nearly all of the operations in this book deal with sample estimates. If the population covariance matrix, Σ, were known, the operations described in this chapter would be exactly the same. The characteristic roots of Σ would be noted by $\lambda_1, \lambda_2, \ldots, \lambda_p$ and would be population values. The associated vectors would also be population values. This situation is unlikely in practice but it is comforting to know that the basic PCA procedure would be the same.

1.6   SCALING OF CHARACTERISTIC VECTORS

There are two ways of scaling principal components, one by rescaling the original variables, which will be discussed in Chapter 3, and the other by rescaling the characteristic vectors, the subject of this section.

The characteristic vectors employed so far, the U-vectors, are orthonormal; they are orthogonal and have unit length. These vectors are *scaled to unity*. Using these vectors to obtain principal components will produce pc's that are uncorrelated and have variances equal to the corresponding characteristic roots.

There are a number of alternative ways to scale these vectors. Two that have found widespread use are

$$\mathbf{v}_i = \sqrt{l_i}\mathbf{u}_i \qquad \text{(i.e., } \mathbf{V} = \mathbf{U}\mathbf{L}^{1/2}) \tag{1.6.1}$$

$$\mathbf{w}_i = \mathbf{u}_i/\sqrt{l_i} \qquad \text{(i.e., } \mathbf{W} = \mathbf{U}\mathbf{L}^{-1/2}) \tag{1.6.2}$$

Recalling, for the chemical example, that

$$\mathbf{U} = \begin{bmatrix} .7236 & -.6902 \\ .6902 & .7236 \end{bmatrix}$$

with $l_1 = 1.4465$ and $l_2 = .0864$, these transformations will produce:

$$\mathbf{V} = \begin{bmatrix} .8703 & -.2029 \\ .8301 & .2127 \end{bmatrix}$$

$$\mathbf{W} = \begin{bmatrix} .6016 & -2.3481 \\ .5739 & 2.4617 \end{bmatrix}$$

For transformation (1.6.1), the following identities hold:

$$\mathbf{V}'\mathbf{V} = \mathbf{L} \tag{1.6.3}$$

$$\begin{bmatrix} .8703 & .8301 \\ -.2029 & .2127 \end{bmatrix} \begin{bmatrix} .8703 & -.2029 \\ .8301 & .2127 \end{bmatrix} = \begin{bmatrix} 1.4465 & 0 \\ 0 & .0864 \end{bmatrix}$$

$$VV' = S \qquad (1.6.4)$$

$$\begin{bmatrix} .8703 & -.2029 \\ .8301 & .2127 \end{bmatrix} \begin{bmatrix} .8703 & .8301 \\ -.2029 & .2127 \end{bmatrix} = \begin{bmatrix} .7986 & .6793 \\ .6793 & .7343 \end{bmatrix}$$

Although V-vectors are quite commonly employed in PCA, this use is usually related to model building and specification. The scores related to (1.6.6) are rarely used and hence we will not waste one of our precious symbols on it.

Corresponding to (1.3.1), there is

$$V'SV = L^2 \qquad (1.6.5)$$

$$\begin{bmatrix} .8703 & .8301 \\ -.2029 & .2127 \end{bmatrix} \begin{bmatrix} .7986 & .6793 \\ .6793 & .7343 \end{bmatrix} \begin{bmatrix} .8703 & -.2029 \\ .8301 & .2127 \end{bmatrix} = \begin{bmatrix} 2.0924 & 0 \\ 0 & .0075 \end{bmatrix}$$

Recalling that L is a diagonal matrix, equation (1.6.3) indicates that the V-vectors are still orthogonal but no longer of unit length. These vectors are *scaled to their roots*. Equation (1.6.4) shows that the covariance matrix can be obtained directly from its characteristic vectors. Scaling principal components using V-vectors, viz.,

$$\sqrt{l_i} z_i = v_i'[x - \bar{x}] \qquad (1.6.6)$$

may be useful because the principal components will be in the same units as the original variables. If, for instance, our chemical data were in grams per liter, both of these pc's and the coefficients of the V-vectors themselves would be in grams per liter. The variances of these components, as seen from (1.6.5) are equal to the *squares* of the characteristic roots. The pc's using (1.6.6) for the chemical data are also shown in Table 1.2.

Regarding now the W-vectors as defined in (1.6.2),

$$W'W = L^{-1} \qquad (1.6.7)$$

$$\begin{bmatrix} .6016 & .5739 \\ -2.3481 & 2.4617 \end{bmatrix} \begin{bmatrix} .6016 & -2.3481 \\ .5739 & 2.4617 \end{bmatrix} = \begin{bmatrix} .6913 & 0 \\ 0 & 11.5741 \end{bmatrix}$$

$$W'SW = I \qquad (1.6.8)$$

$$\begin{bmatrix} .6016 & .5739 \\ -2.3481 & 2.4617 \end{bmatrix} \begin{bmatrix} .7986 & .6793 \\ .6793 & .7343 \end{bmatrix} \begin{bmatrix} .6016 & -2.3481 \\ .5739 & 2.4617 \end{bmatrix} = \begin{bmatrix} 1 & 0 \\ 0 & 1 \end{bmatrix}$$

Equation (1.6.8) shows that principal components obtained by the transformation

$$y_i = w_i'[x - \bar{x}] \qquad (1.6.9)$$

will produce pc's that are still uncorrelated but now have variances equal to unity. Values of this quantity are called *y*-scores. Since pc's are generally regarded as "artificial" variables, scores having unit variances are quite popular for data analysis and quality control applications; *y*-scores will be employed a great deal in this book. The relation between *y*- and *z*-scores is

$$y_i = \frac{z_i}{\sqrt{l_i}} \qquad z_i = \sqrt{l_i}\, y_i \qquad (1.6.10)$$

The **W**-vectors, like **U** and **V**, are also orthogonal but are scaled to the *reciprocal* of their characteristic roots. The *y*-scores for the chemical data are also shown in Table 1.2.

Another useful property of **W**-vectors is

$$\mathbf{WW'} = \mathbf{S}^{-1} \qquad (1.6.11)$$

$$\begin{bmatrix} .6016 & -2.3481 \\ .5739 & 2.4617 \end{bmatrix} \begin{bmatrix} .6016 & .5739 \\ -2.3481 & 2.4617 \end{bmatrix} = \begin{bmatrix} 5.8755 & -5.4351 \\ -5.4351 & 6.3893 \end{bmatrix}$$

This means that if *all* of the characteristic vectors of a matrix have been obtained, it is possible to obtain the inverse of that matrix directly, although one would not ordinarily obtain it by that method. However, in the case of covariance matrices with highly correlated variables, one might obtain the inverse with better precision using (1.6.11) than with conventional inversion techniques.

There are other criteria for normalization. Jeffers (1967), for instance, divided each element within a vector by its largest element (like the t-vectors in Section 1.3) so that the maximum element in each vector would be 1.0 and all the rest would be relative to it.

One of the difficulties in reading the literature on PCA is that there is no uniformity in notation in general, and in scaling in particular. Appendix D includes a table of symbols and terms used by a number of authors for both the characteristic roots and vectors and the principal components resulting from them.

In summary, with regard to scaling of characteristic vectors, principal components can be expressed by (1.4.1), (1.6.6), or (1.6.9). The three differ only by a scale factor and hence the choice is purely a matter of taste. **U**-vectors are useful from a diagnostic point of view since the vectors are scaled to unity and hence the coefficients of these vectors will always be in the range of ± 1 regardless of the original units of the variables. Significance tests often involve vectors scaled in this manner. **V**-vectors have the advantage that they and their corresponding pc's are expressed in the units of the original variables. **W**-vectors produce pc's with unit variance.

1.7 USING PRINCIPAL COMPONENTS IN QUALITY CONTROL

1.7.1 Principal Components Control Charts

As will be seen elsewhere in this book, particularly in Chapters 6 and 9, PCA can be extremely useful in quality control applications because it allows one to transform a set of correlated variables to a new set of uncorrelated variables that may be easier to monitor with control charts. In this section, these same principles will be applied to the chemical data. These data have already been displayed, first in their original form in Table 1.1 and then in terms of deviations from their means in Table 1.2. Also included in Table 1.2 are the scores for the three scalings of the principal components discussed in the previous section and an overall measure of variability, T^2, which will be introduced in Section 1.7.4. Table 1.2 also includes four additional points that have not been included in the derivations of the characteristic vectors but have been included here to exhibit some abnormal behavior. These observations are:

Point	x_1	x_2
A	12.3	12.5
B	7.0	7.3
C	11.0	9.0
D	7.3	9.1

Figure 1.4 shows control charts both for the original observations, x_1 and x_2 and for the y-scores. All four of these charts have 95% limits based on the variability of the original 15 observations. These limits will all be equal to the standard deviations of the specific variables multiplied by 2.145, that value of the t-distribution for 14 degrees of freedom which cuts off .025 in each tail. (When controlling with individual observations rather than averages, 95% limits are often used rather than the more customary "three-sigma" limits in order to reduce the Type II error.)

Common practice is to place control limits about an established standard for each of the variables. In the absence of a standard, as is the case here, the sample mean for the base period is substituted. This implies that we wish to detect significant departures from the level of the base period based on the variability of that same period of time.

For these four control charts, these limits are:

$$x_1: \qquad 10.00 \pm (2.145)(.89) = 8.09, \ 11.91$$

$$x_2: \qquad 10.00 \pm (2.145)(.86) = 8.16, \ 11.84$$

$$y_1 \text{ and } y_2: \qquad 0 \pm (2.145)(1) = -2.145, \ 2.145$$

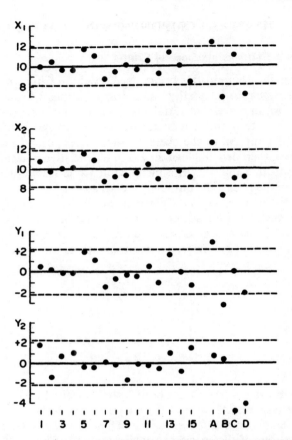

FIGURE 1.4. Chemical example: Control charts for original variables and principal components; 95% limits. Reproduced from Jackson (1980) with permission of the American Society for Quality Control and Jackson (1985) with permission of Marcel Dekker.

1.7.2 Type I Errors

When one uses two or more control charts simultaneously, some problems arise with the Type I error. This is the probability of a sample result being outside the control limits when the process level is at the mean or the standard established for that process. Consider first the two control charts for x_1 and x_2. The probability that each of them will be in control if the process is on standard is .95. If these two variables were uncorrelated (which they are not in this example), the probability that *both* of them would be in control is $(.95)^2 = .9025$ so the effective Type I error is roughly $\alpha = .10$, not .05. For 9 uncorrelated variables, the Type I error would be $1 - (.95)^9 = .37$. Thus if one was attempting to control 9 independent variables, at least one or more of these variables would indicate an out-of-control condition over one-third of the time.

The problem becomes more complicated when the variables are correlated as they are here. If they were perfectly correlated, the Type I error would remain at .05. However, anything less than that, such as in the present example, would leave one with some rather involved computations to find out what the Type I error really was. The use of principal component control charts resolved some of this problem because the pc's are uncorrelated; hence, the Type I error may be computed directly. This may still leave one with a sinking feeling about looking for trouble that does not exist.

One possible solution would be to use *Bonferroni* bounds (Seber, 1984, p. 12 and Table D.1), which is a method of opening up the limits to get the desired Type I error. The limits for each variable would have a significance level of α/p or, for this example, $.05/2 = .025$. These are conservative bounds yielding a Type I error of at most α. For the Chemical example, the limits would be increased from $\pm 2.145 s_i$ to $\pm 2.510 s_i$ for the original variables and would be ± 2.510 for the y-scores. These bounds would handle the Type I error problem for pc control charts but not for any situation where the variables are correlated.

1.7.3 Goals of Multivariate Quality Control

Any multivariate quality control procedure, whether or not PCA is employed, should fulfill four conditions:

1. A single answer should be available to answer the question: "Is the process in control?"
2. An overall Type I error should be specified.
3. The procedure should take into account the relationships among the variables.
4. Procedures should be available to answer the question: "If the process *is* out-of-control, what is the problem?"

Condition 4 is much more difficult than the other three, particularly as the number of variables increases. There usually is no easy answer to this, although the use of PCA may help. The other three conditions are much more straightforward. First, let us consider Condition 1.

1.7.4 An Overall Measure of Variability: T^2

The quantity shown in Figure 1.5

$$T^2 = \mathbf{y}'\mathbf{y} \qquad (1.7.1)$$

is a quantity indicating the overall conformance of an individual observation vector to its mean or an established standard. This quantity, due to Hotelling (1931), is a multivariate generalization of the Student t-test and does give a single answer to the question: "Is the process in control?"

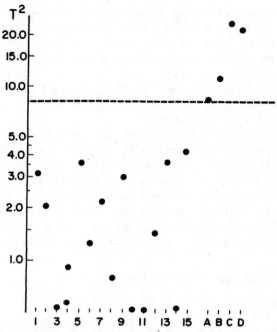

FIGURE 1.5. Chemical example: T^2-chart; 95% limits. Reproduced from Jackson (1980) with permission of the American Society for Quality Control and Jackson (1985) with permission of Marcel Dekker.

The original form of T^2 is

$$T^2 = [x - \bar{x}]'S^{-1}[x - \bar{x}] \tag{1.7.2}$$

which does not use PCA and is a statistic often used in multivariate quality control. From (1.6.11), $S^{-1} = WW'$. Substituting in (1.7.2) and using (1.6.9),

$$T^2 = [x - \bar{x}]'S^{-1}[x - \bar{x}]$$
$$= [x - \bar{x}]'WW'[x - \bar{x}] = y'y \tag{1.7.3}$$

so (1.7.1) and (1.7.2) are equivalent. The important thing about T^2 is that it not only fulfills Condition 1 for a proper multivariate quality control procedure but Conditions 2 and 3 as well. The only advantage of (1.7.1) over (1.7.2) is that if W has been obtained, the computations are considerably easier as there is no matrix to invert. In fact, $y'y$ is merely the sum of squares of the principal components scaled in this manner ($T^2 = y_1^2 + y_2^2$ for the two-variable case) and demonstrates another advantage in using W-vectors. If one uses U-vectors, the computations become, essentially, a weighted sum of squares:

$$T^2 = z'L^{-1}z \tag{1.7.4}$$

and the use of V-vectors would produce a similar expression.

Few books include tables for the distribution of T^2 because it is directly related to the F-distribution by the relationship

$$T^2_{p,n,\alpha} = \frac{p(n-1)}{n-p} F_{p,n-p,\alpha} \tag{1.7.5}$$

In this example, $p = 2$, $n = 15$, $F_{2,13,.05} = 3.8056$, so

$$T^2_{2,15,.05} = 8.187$$

An observation vector that produces a value of T^2 greater than 8.187 will be out of control on the chart shown in Figure 1.5. (This chart only has an upper limit because T^2 is a squared quantity, and for the same reason the ordinate scale is usually logarithmic.)

An alternative method of plotting T^2 is to represent it in histogram form, each value of T^2 being subdivided into the squares of the y-scores. This is sometimes referred to as a *stacked bar-graph* and indicates the nature of the cause of any out-of-control situations. However, the ordinate scale would have to be arithmetic rather than logarithmic. (This scheme was suggested to me by Ron Thomas of the Burroughs Corporation—a student in a Rochester Institute of Technology short course.)

1.7.5 Putting It All Together

Let us now examine, in detail, Figures 1.4 and 1.5. Note that the first 15 observations exhibit random fluctuations on all five control charts. This is as it should be since the limits were based on the variability generated by these 15 observations. Point A represents a process that is on the high side for both measurements and is out of control for x_1, x_2, y_1 (the component representing process) and T^2. Point B represents a similar situation when the process is on the low side. Point C is interesting in that it is out of control for y_2 (the testing and measurement component) and T^2 but not either x_1 or x_2. This point represents a mismatch between the two methods. (y_1, incidentally, is equal to zero.) This example shows that the use of T^2 and PCA adds some power to the control procedure that is lacking in the combination of the two original control charts. Point D is an outlier that is out of control on x_1, y_1, y_2, and T^2.

One advantage of a two-dimensional example is that the original data may be displayed graphically as is done in Figure 1.6. This is the same as Figure

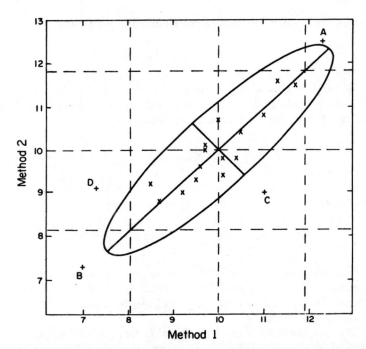

FIGURE 1.6. Chemical example: 95% control ellipse. Reproduced from Jackson (1980) with permission of the American Society for Quality Control and Jackson (1985) with permission of Marcel Dekker.

1.1 except that a number of things have been added, including the extra four observations. The original control limits for x_1 and x_2 have been superimposed and the "box" that they form represents the joint control region of the original control charts. There is also an ellipse constructed around the intersection of the means. This represents the T^2-limit and is a solution of (1.7.1) or (1.7.2) set equal to 8.187. Anything that is out of control on the T^2-chart will be outside this ellipse. This shows much more vividly the advantage of using a single measure, T^2, to indicate overall control. In particular, it shows that point C, while well within the box formed by the two sets of control limits, is well outside the ellipse. The implication is that a difference that large between the two test methods is highly unlikely when the methods, themselves, are that highly correlated. A procedure for constructing a control ellipse is given in Chapter 15, which deals with special applications of PCA for the two-dimensional case.

The notion of the ellipse goes back to Pearson (1901). It was recommended as a quality control device by Shewhart (1931) and, using small-sample statistics, by Jackson (1956). Figure 1.6 also serves to demonstrate that the principal components for the original 15 observations are uncorrelated since the axes of the ellipse represent their coordinate system.

1.7.6 Guideline for Multivariate Quality Control Using PCA

The procedure for monitoring a multivariate process using PCA is as follows:

1. For each observation vector, obtain the y-scores of the principal components and from these, compute T^2. If this is in control, continue processing.
2. If T^2 is out of control, examine the y-scores. As the pc's are uncorrelated, it would be hoped that they would provide some insight into the nature of the out-of-control condition and may then lead to the examination of particular original observations.

The important thing is that T^2 is examined first and the other information is examined *only* if T^2 is out of control. This will take care of the first three conditions listed in Section 1.7.3 and, hopefully, the second step will handle the fourth condition as well. Even if T^2 remains in control, the pc data may still be useful in detecting trends that will ultimately lead to an out-of-control condition. An example of this will be found in Chapter 6.

CHAPTER 2

PCA With More Than Two Variables

2.1 INTRODUCTION

In Chapter 1, the method of principal components was introduced using a two-variable example. The power of PCA is more apparent for a larger number of variables but the two-variable case has the advantage that most of the relationships and operations can be demonstrated more simply. In this chapter, we shall extend these methods to allow for any number of variables and will find that all of the properties and identities presented in Chapter 1 hold for more than two variables. One of the nice things about matrix notation is that most of the formulas in Chapter 1 will stay the same. As soon as more variables are added, however, some additional concepts and techniques will be required and they will comprise much of the subject material of this chapter.

The case of $p = 2$ variables is, as we have noted, a special case. So far it has been employed because of its simplicity but there are some special techniques that can be used only with the two-dimensional case and these will be given some space of their own in Chapter 15.

Now, on to the case $p > 2$. The covariance matrix will be $p \times p$, and there will be p characteristic roots and p characteristic vectors, each now containing p elements. The characteristic vectors will still be orthogonal or orthonormal depending on the scaling and the pc's will be uncorrelated pairwise. There will be p variances and $p(p-1)/2$ covariances in the covariance matrix. These contain all of the information that will be displayed by the characteristic roots and vectors but, in general, PCA will be a more expeditious method of summarizing this information than will an investigation of the elements of the covariance matrix.

2.2 SEQUENTIAL ESTIMATION OF PRINCIPAL COMPONENTS

Over the years, the most popular method of obtaining characteristic roots and vectors has been the *power method*, which is described in Appendix C. In this procedure, the roots and vectors are obtained sequentially starting with the largest characteristic root and its associated vector, then the second largest root, and so on. Although the power method has gradually been replaced by more efficient procedures in the larger statistical computer packages, it is more simple and easier to understand than the newer methods and will serve to illustrate some properties of PCA for the general case.

If the vectors are scaled to v-vectors, the variability explained by the first pc is $v_1 v_1'$. The variability unexplained by the first pc is $S - v_1 v_1'$. Using the chemical example from Chapter 1, the matrix of residual variances and covariances unexplained by the first principal component is

$$S - v_1 v_1' = \begin{bmatrix} .7986 & .6793 \\ .6793 & .7343 \end{bmatrix} - \begin{bmatrix} .7574 & .7224 \\ .7224 & .6891 \end{bmatrix}$$

$$= \begin{bmatrix} .0412 & -.0431 \\ -.0431 & .0452 \end{bmatrix}$$

This implies that $.0412/.7986 = .052$ or 5.2% of the variability in x_1 is unexplained by the first pc. Similarly, 6.6% of the variability in x_2 is unexplained. The off-diagonal element in the residual matrix is negative, which indicates that the residuals of x_1 and x_2 are negatively correlated. We already know from Section 1.5.2 that the second pc represents disagreements between x_1 and x_2. More will be said about residuals in Section 2.7.

It is worth noting that the determinant of this residual matrix is

$$(.0412)(.0452) - (-.0431)^2 = 0$$

The rank has been reduced from 2 to 1 because the effect of the first pc has been removed. The power method would approach this residual matrix as if it were a covariance matrix itself and look for its largest root and associated vector, which would be the second root and vector of S as we would expect. The variability unexplained by the first two pc's is

$$S - v_1 v_1' - v_2 v_2' = \begin{bmatrix} .00 & .00 \\ .00 & .00 \end{bmatrix}$$

for this two-dimensional example because the first two pc's have explained everything. (Recall from Section 1.6 that $S = VV'$.)

A four-variable example will be introduced in the next section. The operations for that example would be exactly the same as this one except that there will be more of them. The rank of the covariance matrix will be 4. After the effect of the first pc has been removed, the rank of the residual matrix will be 3; after the effect of the second pc has been removed, the rank will be reduced to 2, and so on.

Recall for the case $p = 2$ that the first characteristic vector minimized the sums of squares of the deviations of the observations perpendicular to the line it defined. Similarly, for $p = 3$ the first vector will minimize the deviations perpendicular to it in three-space, the first two vectors will define a plane that will minimize the deviations perpendicular to it and so on.

2.3 BALLISTIC MISSILE EXAMPLE

The material in this section represents some work carried out while the author was employed by the Hercules Powder Company at Radford Arsenal, Virginia (Jackson 1959, 1960). Radford Arsenal was a production facility and among their products at that time were a number of ballistic missiles used by the U.S. Army as artillery and anti-aircraft projectiles. Missiles ("rounds" in ordnance parlance) were produced in batches and a sample of each batch was subjected to testing in accordance with the quality assurance procedures in use at the time. One of these tests was called a *static* test (as contrasted with flight testing) where each round was securely fasted to prevent its flight during its firing. When a round was ignited it would push against one or more strain gauges, from which would be obtained a number of physical measures such as thrust, total impulse, and chamber pressure. This example will involve total impulse.

At the time a rocket is ignited it begins to produce thrust, this quantity increasing until a maximum thrust is obtained. This maximum thrust will be maintained until nearly all of the propellant has been burned and as the remaining propellant is exhausted the thrust drops back down to zero. A typical relation of thrust to time, $F(t)$, is shown in Figure 2.1. (The time interval for these products, typically, was just a few seconds.) Total impulse was defined as the area under this curve, that is,

$$\text{Total impulse} = \int_0^t F(t)\, dt$$

The method of estimating this quantity, which would seem crude in light of the technology of today, was as follows:

1. The thrust at a particular point in time would be represented as a single point on an oscilloscope.
2. A camera had been designed to continuously record this information to produce a curve similar to the one shown in Figure 2.1.
3. The area under the curve was obtained manually by means of a planimeter.

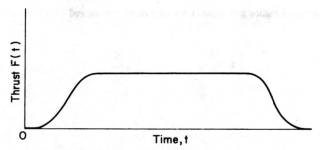

FIGURE 2.1. Ballistic Missile example: thrust curve.

Because of the cost associated with the manufacture and testing of these products, some redundancy in the testing procedure was desirable. For this test, two strain gauges were attached to the head of each rocket. Each gauge was connected to a separate oscilloscope. Later, an electronic device was developed to take the data from the strain gauges and perform the integration directly. Although considered to be much cheaper than the procedure described above as well as reducing the measurement error, it was not as reliable when first employed and so, rather than substitute this electronic integrator for oscilloscope–planimeter measurement, it was carried along in parallel. This example, involving booster rockets for the Nike-Ajax and Nike-Hercules systems, was carried out during this interim period and hence there were four measurements available:

x_1 = Gauge #1; integrator reading
x_2 = Gauge #1; planimeter measurement
x_3 = Gauge #2; integrator reading
x_4 = Gauge #2; planimeter measurement

The covariance matrix for a sample of 40 rounds from a production lot is displayed in Table 2.1. The U-vectors and the characteristic roots are given in Table 2.2. Later, use will be made of the V- and W-vectors, which are displayed in Table 2.3.

From Table 2.2, it is clear that the first pc, explaining 78.2%, represents the overall variability of the product. The second pc has different signs for each

Table 2.1. Ballistic Missile Example. Covariance Matrix

Gauge #1		Gauge #2	
Integrator x_1	Planimeter x_2	Integrator x_3	Planimeter x_4
102.74	88.67	67.04	54.06
88.67	142.74	86.56	80.03
67.04	86.56	84.57	69.42
54.06	80.03	69.42	99.06

Table 2.2. Ballistic Missile Example. Characteristic Roots and U-Vectors

	u_1	u_2	u_3	u_4
x_1	.468	−.622	.572	−.261
x_2	.608	−.179	−.760	−.147
x_3	.459	.139	.168	.861
x_4	.448	.750	.262	−.410
Characteristic root	335.34	48.03	29.33	16.41
% Explained	78.1	11.2	6.8	3.8

Table 2.3. Ballistic Missile Example. V- and W-vectors

	v_1	v_2	v_3	v_4	w_1	w_2	w_3	w_4
x_1	8.57	−4.31	3.10	−1.06	.0256	−.0897	.1055	−.0643
x_2	11.13	−1.24	−4.11	−.60	.0332	−.0258	−.1402	−.0364
x_3	8.41	.96	.91	3.49	.0251	.0200	.0310	.2126
x_4	8.20	5.20	1.42	−1.66	.0245	.1082	.0483	−.1013

gauge and hence represents gauge differences. The other two pc's are less clear. A case might be made for integrator–planimeter differences related to Gauge #1 for y_3 and for Gauge #2 for y_4 but the results may not prove overly convincing and more will be said about this in Section 2.6. If one is willing to accept, for the moment, that these last two pc's represent some sort of testing and measurement variability, then one can conclude that, on the basis of this sample, roughly 78% of the total variability is product and 22% is testing and measurement. The Army had required, with each released production lot, an estimate of the proportion of the total reported variability that could be attributed to testing and measurement and PCA was one of the methods proposed to produce this estimate (Jackson, 1960).

2.4   COVARIANCE MATRICES OF LESS THAN FULL RANK

Before going on to some new procedures required when one has more than two variables, it may be advisable to digress, briefly, to consider a special case of covariance matrix that is not of full rank. This situation will occur when one or more linear relationships exist among the original variables so that the knowledge of a subset of these variables would allow one to determine the remainder of the variables without error.

As an example, let us return to our chemical example and add a third variable, which will be the sum of the first two, that is, $x_3 = x_1 + x_2$. This is a case of a linear relationship because the sum of the first two variables uniquely determines the third. x_3 adds no information whatsoever. The covariance matrix now becomes

$$\begin{bmatrix} .7986 & .6793 & 1.4779 \\ .6793 & .7343 & 1.4136 \\ 1.4779 & 1.4136 & 2.8915 \end{bmatrix}$$

The third row and column are the result of the new variable, x_3. The other four quantities are the same as for the two-dimensional case. The characteristic roots of this matrix are 4.3880, .0864, and 0. These roots are directly related to the roots for the two-dimensional case. The first root, 4.3880, is equal to the first root for $p = 2$, 1.4465, plus the variance of x_3, 2.8915. The second root is the same as it was for $p = 2$. The third root is zero, indicating that the covariance matrix is not of full rank and there exists one linear relationship among the variables. In general the rank of a matrix will be reduced by 1 for each of these relationships.

The U-vectors corresponding to these roots are

$$\begin{bmatrix} .4174 & -.7017 & -.5774 \\ .3990 & .7124 & -.5774 \\ .8164 & .0107 & .5774 \end{bmatrix}$$

The coefficients in the first vector, not surprisingly, show that, $u_{31} = u_{11} + u_{21}$. All of the coefficients are still positive because all three values, generally, rise and fall together. The second vector is essentially $x_2 - x_1$ as it was before, but in this case, as with u_1, the third coefficient equals the sum of the first two. Since the third vector is associated with a zero root, do we need to bother with it? The answer is "yes" because u_3 explains a linear relationship. The coefficients are all equal except for sign and tell us that

$$-x_1 - x_2 + x_3 = 0$$

or

$$x_1 + x_2 = x_3$$

The existence of a zero root, $l_i = 0$ implies that $z_i = u_i'[x - \bar{x}] = 0$ for any x and hence, $u'[x - \bar{x}]'[x - \bar{x}]u/(n - 1) = l_i = 0$.

The V-vectors are:

$$\begin{bmatrix} .8695 & -.2063 & 0 \\ .8309 & .2094 & 0 \\ 1.7004 & .0031 & 0 \end{bmatrix}$$

The third vector is zero because it is normalized to its root, zero, and hence has no length. This means that the covariance matrix can be reconstructed from v_1 and v_2 alone. Another simple demonstration example may be found in Ramsey (1986). For the W-matrix, the corresponding third vector is undefined.

The practical implication of this is that an unsuspecting analyst may, from time to time, receive some multivariate data with one or more linear relationships (sums and differences being the most common) placed there by a well-intentioned client who did not realize that multivariate analysis abhors singular matrices. Many computer packages have, as an option, the ability to obtain all of the characteristic roots. Whenever the data are suspect, this should be done. The existence of one or more zero roots is prima facie evidence that trouble exists. An investigation of the vectors associated with them may give a clue as to what this trouble is. Even if one is not interested in performing a PCA on these data, in the first place, this technique can still be useful in such occurrences as singular $X'X$ matrices in multiple regression or its counterpart in MANOVA or discriminant analysis.

Examples of such constraints in a chemical problem along with the interpretation of the vectors associated with the zero roots may be found in Box et al. (1973). They pointed out that because of rounding, these roots may be positive rather than zero but would be so much smaller than the others that they should be detected anyhow. (See Section 2.8.5 on SCREE plots.)

Another cause of singular covariance matrices is the result of having more variables than observations. If $n < p$, then the maximum number of nonzero roots will be $n - 1$.

2.5 CHARACTERISTIC ROOTS ARE EQUAL OR NEARLY SO

Another anomaly of characteristic roots is the case where two or more of the roots are equal to each other. The simplest case of this would be for the case $p = 2$ where the two variances are equal and the variables are uncorrelated. The characteristic roots will be equal (and equal to the variances) and the ellipse defined by this will, in fact, be a circle. The major and minor axes will be of equal length and can be anywhere as long as they are orthogonal. This means that the vector orientation is undefined. In the more general case, any time there are two or more equal roots, that part of the orientation will be undefined even though there are distinct roots that are both larger and smaller.

Unless one is working with patterned matrices, the probability of the occurrence of identically equal roots in a real data set is remote, but what can occur is the existence of two or more roots that are nearly equal. As will be seen in Chapter 4, the standard errors of the coefficients of characteristic vectors are a function of the separation of these roots; these standard errors can become inflated by the occurrence of two or more roots close to each other. This means that even though the orientation of the axes is defined, it is not defined with very much precision and hence attempts to interpret these pc's might be unwise.

The most common occurrence, in practice, will for the first few roots to be fairly well separated and account for most of the variability; the remainder of the roots would all be small and of the same order of magnitude. This may imply that the last few population roots are equal and hence the vector subspace is undefined. This situation is generally assumed to represent inherent variation and, that being the case, there is little to be gained by using all of these last pc's since they explain very little of the variation and probably lack any realistic interpretation. Each of the last few pc's may often, in turn, relate primarily to a single variable, accounting for its residual variability. This will be the case if the inherent variability for the variables is uncorrelated. (We shall see that the factor analysis model in Chapter 17 will require this to be the case.)

If, for instance, one had a 20-variable problem and the first three pc's accounted for 95% of the variability, one might be tempted to use just those three and ignore the remaining 17 that account for the remaining 5%. This practice is sometimes referred to as *parsimony* defined as "economy in the use of means to an end." While PCA is certainly useful in transforming correlated variables into uncorrelated ones, its greater popularity probably stems from its ability, in many instances, to adequately represent a multivariate situation in a much-reduced dimensionality.

If one is to use less than a full set of pc's, two questions must be dealt with:

1. What criterion (called a *stopping rule*) should be used in deciding how many pc's to retain? In Section 2.6, one such procedure will be given for the purpose of illustration and a survey of proposed criteria will be given in Section 2.8.
2. What are the consequences of deleting one or more pc's? A procedure is required to monitor the residual variability not accounted for by the retained pc's. This will be discussed in Section 2.7.

2.6 A TEST FOR EQUALITY OF ROOTS

If the stopping rule is based on the assumption that the characteristic roots associated with the deleted pc's are not significantly different from each other, a procedure is required to test this hypothesis. A large-sample test for the hypothesis that the last $(p - k)$ roots are equal was developed by Bartlett (1950). There have been a number of modifications to this test since then and these will be dealt with in Chapter 4. For the moment, we shall employ the form found in Anderson (1963), which is

$$\chi^2 = -(v) \sum_{j=k+1}^{p} \ln(l_j) + (v)(p - k) \ln\left[\frac{\sum_{j=k+1}^{p} l_j}{(p - k)}\right] \qquad (2.6.1)$$

where χ^2 has $(1/2)(p - k - 1)(p - k + 2)$ degrees of freedom and v represents the number of degrees of freedom associated with the covariance matrix. This test can be performed after each stage, keeping in mind the effect on the Type I error by making successive tests. If the remaining roots are not significantly different from each other, the procedure is terminated at that point. (Some computer programs obtain all of the roots at once so this test could be incorporated to produce a series of tests before any of the vectors are obtained.) If the final number of pc's retained is k, formula (1.7.5) for T_α^2 must be modified by replacing p with k.

Turning to the Ballistics Missile example in Section 2.3, the first test that could be applied is for the hypothesis

$$H_0: \lambda_1 = \lambda_2 = \lambda_3 = \lambda_4$$

against the alternative that at least one root was different. For this hypothesis,

$$\chi^2 = -(39)[\ln(335.34) + \cdots + \ln(16.41)]$$
$$+ (39)(4)\ln\left[\frac{(335.34 + \cdots + 16.41)}{4}\right]$$
$$= 110.67$$

with 9 degrees of freedom, which is highly significant.

A test of the next hypothesis,

$$H_0: \lambda_2 = \lambda_3 = \lambda_4$$

which says, "Given that λ_1 is different from the others, are the others equal?", produces a value of $\chi^2 = 10.85$ which, with 5 degrees of freedom, is not quite significant at the 5% level. A test of $H_0: \lambda_3 = \lambda_4$ yields a value of $\chi^2 = 3.23$ which, with 2 degrees of freedom, is not even close. If the sample size were 100 instead of 40, for the same covariance matrix, all four roots would have tested out to be significantly different.

If the percentage of the trace unexplained by the significant pc's is appreciable, this is an indication that these pc's are not doing an acceptable job of spanning the space generated by the data. This is not necessarily a disaster. It may be that most of the variability is random variability associated with the original variables, and if this is large relative to the total trace it simply means that the original variables do not have that much in common. Parsimony is desirable but not always obtainable.

2.7   RESIDUAL ANALYSIS

2.7.1   Introduction

In Section 1.5.5, it was shown that if one used a full set of pc's, it was possible to invert the equation that produced the pc's from the data and, instead, determine the original data from the pc's. Since most of the applications in this book scale the pc's to have unit variances, we can rewrite equation (1.5.5)

$$x = \bar{x} + Uz$$

as

$$x = \bar{x} + Vy \tag{2.7.1}$$

However, x will be determined exactly only if all the pc's are used. If $k < p$ pc's are used, only an estimate $\hat{x}$ of x will be produced, viz.,

$$\hat{x} = \bar{x} + Vy \tag{2.7.2}$$

where V is now $p \times k$ and y is $k \times 1$. Equation (2.7.2) can be rewritten as

$$x = \bar{x} + Vy + (x - \hat{x}) \tag{2.7.3}$$

a type of expression similar to those often found in other linear models. In this case, the first term on the right-hand side of the equation represents the contribution of the multivariate mean, the second term represents the contribution due to the pc's, and the final term represents the amount that is unexplained by the pc model—the residual. Wherever any pc's are deleted, some provision should be made to check the residual.

Gnanadesikan and Kettenring (1972) divided multivariate analysis into

1. The analysis of internal structure.
2. The analysis of superimposed or extraneous structure.

There are outliers associated with each of these and it is important to keep their identities distinct. (Hawkins refers to these as Type A and Type B outliers.)

The "Type A" outlier refers to a general outlier from the distribution form one wishes to assume. Usually this assumption will be multivariate normal and these outliers will be detected by large values of T^2 and/or large absolute values of the y- or z-scores such as the example in Section 1.7. The important thing about this type of outlier is that it would be an outlier whether or not PCA has been employed and hence could be picked up by conventional multivariate techniques without using PCA. However, the use of PCA might well enhance the chance of detecting it as well as diagnosing what the problem might be.

In this section, we will be concerned with the "Type B" outlier, the third term in (2.7.3), which is an indication that a particular observation vector cannot be adequately characterized by the subset of pc's one chose to use. This result can occur either because too few pc's were retained to produce a good model or because the observation is, truly, an outlier from the model. It is also possible in repetitive operations, such as quality control, that the underlying covariance structure and its associated vector space may change with time. This would lead to a general lack-of-fit by the originally defined pc's.

2.7.2 The Q-Statistic

The residual term of (2.7.3) can be tested by means of the sum of squares of the residuals:

$$Q = (\mathbf{x} - \hat{\mathbf{x}})'(\mathbf{x} - \hat{\mathbf{x}}) \tag{2.7.4}$$

This represents the sum of squares of the distance of $\mathbf{x} - \hat{\mathbf{x}}$ from the k-dimensional space that the PCA model defines. [A form of this statistic was first proposed by Jackson and Morris, (1957), but the χ^2-approximation that they used for that statistic is incorrect and should not be used.]

To obtain an upper limit for Q, let:

$$\theta_1 = \sum_{i=k+1}^{p} l_i$$

$$\theta_2 = \sum_{i=k+1}^{p} l_i^2$$

$$\theta_3 = \sum_{i=k+1}^{p} l_i^3$$

and

$$h_0 = 1 - \frac{2\theta_1\theta_3}{3\theta_2^2}$$

Then the quantity

$$c = \theta_1 \frac{\left[\left(\dfrac{Q}{\theta_1}\right)^h - \dfrac{\theta_2 h_0(h_0 - 1)}{\theta_1^2} - 1\right]}{\sqrt{2\theta_2 h_0^2}} \tag{2.7.5}$$

is approximately normally distributed with zero mean and unit variance (Jackson and Mudholkar, 1979). Conversely, the critical value for Q is

$$Q_\alpha = \theta_1 \left[\frac{c_\alpha\sqrt{2\theta_2 h_0^2}}{\theta_1} + \frac{\theta_2 h_0(h_0 - 1)}{\theta_1^2} + 1\right]^{1/h_0} \tag{2.7.6}$$

where c_α is the normal deviate cutting off an area of α under the upper tail of the distribution if h_0 is positive and under the lower tail if h_0 is negative. This distribution holds whether or not all of the significant components are used or even if some nonsignificant ones are employed.

It should be noted that Q can also be written as

$$Q = \sum_{i=k+1}^{p} l_i y_i^2 = \sum_{i=k+1}^{p} z_i^2 \tag{2.7.7}$$

so that Q is a weighted sum of squares of the last $p - k$ components. Although this may be thought of as an alternative method of computing Q, it does not require the calculation of all the pc's and does not produce $[x - \hat{x}]$ directly. [In the form of equation (2.7.7), Q is sometimes referred to as the Rao-statistic. A number of texts state that it has an asymptotic χ^2-distribution but this, like the Jackson–Morris conjecture mentioned earlier, is incorrect.]

In Section 2.6, it was suggested that the last two characteristic roots in the Ballistic Missile example were not significantly different from each other and hence the last two pc's should be deleted. If only the first two pc's were retained, what would be the limit for Q? The last two roots were 29.33 and 16.41. From these, $\theta_1 = 45.74, \theta_2 = 1129.54, \theta_3 = 29\,650.12$, and from these $h_0 = .291$. Letting $\alpha = .05$, the limit for Q, using (2.7.6) is

$$Q_{.05} = 45.74 \left[\frac{(1.645)\sqrt{(2)(1129.54)(.291)^2}}{45.74} + \frac{(1129.54)(.291)(-.709)}{(45.74)^2} + 1 \right]^{1/.291}$$

$$= 140.45$$

Values of Q higher than this are an indication that a data vector cannot be adequately represented by a two-component model.

Now let us assume that a new round is tested with the results

$$x = \begin{bmatrix} 15 \\ 10 \\ 20 \\ -5 \end{bmatrix}$$

and assume that the mean is zero. Using the first two columns of W in Table 2.3, the y-scores are

$$y = W'(x - \bar{x}) = \begin{bmatrix} 1.094 \\ -1.744 \end{bmatrix}$$

The $\alpha = .05$ limits for the pc's for $n = 40$ are ± 2.20 so neither of these are significant. $T^2 = 4.2383$, which is also not significant when compared with its limit of 6.67. The predicted test values, given these pc's, are

$$\hat{x} = \bar{x} + Vy = \begin{bmatrix} 0 \\ 0 \\ 0 \\ 0 \end{bmatrix} + \begin{bmatrix} 16.9 \\ 14.3 \\ 7.5 \\ -.1 \end{bmatrix}$$

again, using the first two columns of **V** in Table 2.3. The residuals are

$$x - \hat{x} = \begin{bmatrix} -1.9 \\ -4.3 \\ 12.5 \\ -4.9 \end{bmatrix}$$

and their sum of squares is $Q = (x - \hat{x})'(x - \hat{x}) = 202.2$, which is significant. The conclusion is that the two-component model does not fit the data and the culprit appears to be a mismatch between the results on the second gauge. Note that if the other pc's had been computed, $y_3 = .559$ and $y_4 = 3.430$. Then, verifying (2.7.7),

$$(29.33)(.559)^2 + (16.41)(3.430)^2 = 202.2$$

It has been the experience of many practitioners, particularly but not restricted to the engineering and physical sciences, that the greatest utility of the Q-statistic (or the alternatives discussed below) is its ability to detect bad data, measurement errors, and the like. Every set of multivariate data that is to be subjected to PCA should be screened using one of these statistics. Some people use PCA, including the residual test, for screening multivariate data even though they have no intention of using the pc's afterward. We shall find a number of instances in this book where Q has also been used as an intermediate step in a particular procedure.

In obtaining the θ_i's, if one has a large number of deleted pc's, one is faced with having to obtain not only all of the characteristic roots associated with them but with their squares and cubes as well. There is a short cut for this because we are not interested in the individual roots but only their sums. Let **E** be the residual covariance matrix after k characteristic vectors have been extracted. Then

$$\theta_1 = \text{Tr}(E) \tag{2.7.8}$$

$$\theta_2 = \text{Tr}(E^2) \tag{2.7.9}$$

$$\theta_3 = \text{Tr}(E^3) \tag{2.7.10}$$

2.7.3 The Hawkins Statistic

In the previous section, it was shown (2.7.7) that Q could be expressed as

$$Q = \sum_{i=k+1}^{p} z_i^2$$

which is equal to the *weighted* sum of squares of the unretained pc's, the weights being their corresponding characteristic roots. The pc's with the larger roots have the greatest effect. Hawkins (1974, 1980), proposed using the *unweighted* sums of squares of the unretained pc's:

$$T_H^2 = \sum_{i=k+1}^{p} y_i^2 \tag{2.7.11}$$

T_H^2 has a T^2-distribution, the same as the sum of squares of the retained pc's except that p in (1.7.5) is replaced by $(p - k)$. (This is one statistic that *does* have an asymptotic χ^2-distribution and would have $p - k$ degrees of freedom.) Furthermore, the sum of these two sums of squares representing the "explained" and "unexplained" variability will be the same as the ordinary (non-pc) Hotelling's T^2 (1.7.2). For PCA, the Type A outliers refer to significant values of any of the first k pc's and Type B to any of the last $p - k$ pc's.

(One must be careful in discussing weights. In this context, the y_i appear to be unweighted since they all have unit variance. In fact, the sum of squares of all p y_i's is equal to $T^2 = (\mathbf{x} - \bar{\mathbf{x}})\mathbf{S}^{-1}(\mathbf{x} - \bar{\mathbf{x}})$, which is the Mahalinobis distance measure of the original variables weighted by $\mathbf{S}^{-1}$. As we shall see in Chapter 10, when discussing multidimensional scaling using interpoint distances among the pc's, it is the z_i that are considered *euclidean* or unweighted and the y_i that are considered weighted.)

Applying this to the same example from Section 2.7.2, $y_3 = .559$ and $y_4 = 3.430$, from whence the Hawkins-statistic, as given in (2.7.11) is 12.0774. By coincidence (since there are two retained and two deleted pc's in this model) the limit for T_H^2 is 6.67. This statistic, like the Q-statistic in the previous section, is significant. For diagnostics, one would then look at the individual pc's and here it is clear that the culprit is y_4, which from Table 2.3 would seem to represent, primarily, the difference in the results for the second gauge.

No formal comparison of these two methods has been carried out but some empirical work by Kane et al. (1977) and Spindler (1987) showed the results to be highly correlated as one would expect. (They will be perfectly correlated for the case $k = p - 1$, but that is an unlikely situation to occur in practice. If any pc's are to be deleted at all, there would be a minimum of two of them since it would be assumed that at least the last two roots were not significantly different.) Since these two statistics are related, the choice would reduce to one of personal preference. First is the conceptual question of whether or not to weight the unretained pc's. Second, Q does not require the calculation of the unretained pc's and T_H^2 does unless one wishes to compute, as an alternate form,

$$T_H^2 = [\mathbf{x} - \bar{\mathbf{x}}]'\mathbf{S}^{-1}[\mathbf{x} - \bar{\mathbf{x}}] - \sum_{i=1}^{k} y_i^2 \tag{2.7.12}$$

However, it is because of the unretained pc's that this statistic is advocated.

Finally, the Q-statistic is in terms of residuals and the diagnostics would also be in terms of residuals. The Hawkins statistic is in terms of the deleted pc's as are the diagnostics and it is on this latter point that its success would likely be measured. If the deleted pc's were readily interpretable, this might be preferred to Q; otherwise, it would not. The occasions when the deleted pc's are most likely to be interpretable are:

1. The smallest roots are related to near singularities, linear combinations of the variables.
2. Some of the deleted pc's are each related primarily to unexplained variability in a single variable. This is most likely to happen when there is a large number of variables rather than the 4-variable example used here.

The Hawkins statistic may be of less use if the number of variables is quite large. In some chemical applications, p is the order of 200 or so and k may be only 2 or 3. Suppose that $p = 200$ and $k = 3$. There would be 197 deleted pc's and one must, then, carry along 197 more characteristic vectors for the computations. Undoubtedly, some of these roots will not be significantly different, which would cast doubt on the precision with which the vectors are estimated. Even if the roots are all distinct, the chances of very many of them having physical significance seems remote. While the Q-statistic avoids all of these problems, significant values of Q would send the researcher to examine 200 individual residuals, which might not seem very appealing either. However, for a problem of this size, an algorithm could be imbedded to scan for large residuals, given that Q was significant.

The important thing is that some sort of residual analysis should be carried out for all PCA studies and, at this point in time, very few practitioners do this on any of them. On judgment day, in this author's view, the statistical equivalent of St. Peter will care not a whit about which method was employed but will come down with utmost ferocity upon those who have not used any of them at all.

2.7.4 Other Alternatives

There are some other possibilities for testing residuals. Hawkins (1974) and Fellegi (1975) suggested using the Max $|y_i|$ for $k + 1 < i < p$ as a test statistic comparing them with Bonferroni bounds based on $\alpha/(p - k)$. In the present example, the largest deleted pc was $y_4 = 3.430$. If $\alpha = .05$, the desired significance level for this test would be $(.05)/2 = .025$, which for 39 degrees of freedom would be $t_{39, .025} = 2.332$, so this test would also be significant. Chapter 8 takes up the subject of simple structure and rotation and, in that context, Hawkins and Fatti (1984) suggested performing a Varimax rotation on the deleted pc's, $y_{k+1}, \ldots, y_p$, and testing the maximum absolute value of these. The possible advantage of this procedure would lie in its diagnostic properties.

More of an adjunct than an alternative might be the use of an Andrews plot (Andrews, 1972) of $x - \hat{x}$ for a set of data. Curves that tend to cluster should exhibit the same sorts of residual behavior and might enhance the interpretation of this phenomenon. Limits for these curves, if desired, could be based on the procedure due to Kulkarni and Paranjape (1984) except that they would be based on $S - V_k V_k'$ rather than S itself. The Andrews technique will described in Section 18.6.

Before the distribution of Q had been derived, Gnanadesikan and Kettenring (1972) suggested making a gamma probability plot of Q, and this might still have merit in data analysis. This does require an estimate of the shape parameter of this distribution, which Gnanadesikan and Kettenring suggested obtaining on the basis of the unretained pc's. They also suggested:

1. Probability plots of the unretained pc's
2. Scatter plots of combinations of the unretained pc's
3. Plots of retained vs. unretained pc's

Related work on gamma plots was done by Wilk et al. (1962) and Wilk and Gnanadesikan (1964). Other plotting techniques that may be useful were given by Andrews (1979) in connection with robustness and by Daniel and Wood (1971), which, while designed for regression, have applications in PCA as well.

2.8 WHEN TO STOP?

2.8.1 Introduction

One of the greatest uses of PCA is its potential ability to adequately represent a p-variable data set in $k < p$ dimensions. The question becomes: "What is k?" Obviously, the larger k is, the better the fit of the PCA model; the smaller k is, the more simple the model will be. Somewhere, there is an optimal value of k; what is it?

To determine k, there must be a criterion for "optimality." This section will describe a large number of criteria that are in use. These criteria range all the way from significance tests to graphical procedures. Some of these criteria have serious weaknesses, but will be listed anyhow since they are commonly used.

2.8.2 When to Start?

Hotelling (1933) spoke of the "sand and cobblestone" theories of the mind with regard to test batteries. *Cobblestone* referred to the situation where a few pc's fairly well characterized the data. *Sand*, on the other hand, meant that there were many low correlations and the resultant major characteristic roots were relatively small with some possibly indistinguishable. Sections 2.8.3 through 2.8.11 will be concerned with a number of stopping procedures. However, it would be convenient to have a procedure that would tell us at the beginning whether or not we should even begin to embark on the PCA process. There are a number of guides available.

From a statistical point of view, the best way to determine whether more than a sandpile exists is to perform the significance test on the equality of the characteristic roots. A form of this was given in (2.6.1) and will be expanded in Section 2.8.3. This requires the computation of all of the roots and while this is generally not a problem these days it might be desirable to have some quick measures available that do not involve the roots.

In Section 4.2.3, some bounds are given on the maximum root. These bounds are simple functions of the elements of the covariance or correlation matrix. A "one-number" descriptor was proposed by Gleason and Staelin (1975) for use with correlation matrices:

$$\varphi = \sqrt{\frac{\|\mathbf{R}\|^2 - p}{p(p-1)}} \tag{2.8.1}$$

where $\|\mathbf{R}\|^2 = \sum\sum r_{ij}^2 = \sum l_i^2$ with all summations running from 1 to p. They called this a measure of *redundancy* but we shall not use this term in order to avoid confusion with another use of the word in Section 12.6. If there is no correlation at all among the variables, $\|\mathbf{R}\|^2 = p$ and $\varphi = 0$. If all of the variables are perfectly correlated, $l_1 = p$ and $\varphi = 1$. Although this coefficient has the same range as a multiple correlation coefficient, it is not unusual to obtain values of less than .5 for situations that appear to have fairly strong structure. The distribution of this quantity is unknown and it would appear that experience will be the best guide as how to interpret it. However, this quantity could be useful in comparing two or more data sets.

If one is working with covariance matrices, (2.8.1) becomes

$$\varphi = \sqrt{\frac{\|\mathbf{S}\|^2 - \sum_{i=1}^{p}(s_i^2)^2}{\sum_{i=1}^{p}\sum_{j \neq i}^{p}(s_i s_j)^2}} \tag{2.8.2}$$

For the Ballistic Missile example, which involved a covariance matrix,

$$\varphi = \sqrt{\frac{115\,887 - 47\,895}{136\,238}} = .706$$

Another single number descriptor is the *index* of a matrix, $\sqrt{l_1/l_p}$ which, for this example, would be $\sqrt{335.34/16.41} = 4.52$. Unlike the Gleason–Staelin statistic, this index does not have an upper bound and will generally increase with increasing p. The index of a singular covariance matrix is undefined.

A quantity that may be useful in connection with stopping rules is an extension of equation (1.5.4), the correlation of an original variable with a principal component. The multiple correlation of an original variable with all of the retained pc's is

$$R_{x_j(y_1 \ldots y_k)} = \sqrt{\sum_{i=1}^{k} \left(\frac{l_i u_{ji}^2}{s_j^2} \right)} = \sqrt{\sum_{i=1}^{k} \left(\frac{v_{ji}}{s_j} \right)^2} \tag{2.8.3}$$

2.8.3 Significance Tests

Before the advent of Bartlett's test, discussed in Section 2.6, the statistical stopping rules were, by necessity, ad hoc procedures such as the use of reliability coefficients (Hotelling, 1933) or tests for the rank of the covariance matrix using the generalized variance (Hoel, 1937). The various forms of Bartlett's test are large-sample procedures assuming that the sample size is large enough for the χ^2-approximation to hold. Section 4.4 will contain a number of small-sample procedures in detail. What is important here is to discuss the philosophy of significance tests in this context.

For significance tests in general, one should make the distinction between *statistical* and *physical* significance. Consider a simple t-test for a sample mean. If one obtained a sample of one million observations, nearly any departure of the sample mean from the hypothetical mean, no matter how small, would be judged significant. In more practical terms, if the precision with which the sample mean is estimated is small, a mean difference may be detected that is not considered important to the experimenter and in fact, the notion of "importance" or physical significance is the basis for sample size determinations. On the other hand, a true difference may exist between the true and hypothetical means but this difference may be obscured by a large variance in the sample. This analogy carries over to PCA. Bartlett's test is for the null hypothesis that the last $(p - k)$ characteristic roots are equal. It may be that this procedure will continue selecting pc's to be included in the model that explain very little of the total variance and, in addition, may not be readily interpretable. On the contrary, the precision of the estimates may be such that the PCA model may be oversimplified, that is, some pc's that one might expect to show up in the model are prevented from doing so by excessive variability. While either can happen, most practitioners would agree that the former is more prevalent, that is, Bartlett's test ends up retaining too many pc's.

Suppose one performs a significance test of this type on, say, a $p = 20$-variable problem and finds that only the last four roots are not significantly different so that 16 pc's should be retained. On reflection, this individual elects to delete 6 more and end up with 10 pc's in the model. Is this being hypocritical? Not necessarily. If the extra six discards were interpreted to be uncontrollable inherent variability, there might be a case for dropping them. Some of the procedures to be described below will address some of these issues but they should be considered as adjuncts to, not substitutes for, significance tests. This would be particularly true in the case of the situation where high variability suppresses some pc's; do NOT include pc's in the model that do not belong there statistically. The answer to this dilemma would be to investigate why this situation may have occurred. A possibility, not to be overlooked, is the existence of one or more outliers; this is an excellent place to use the residual analysis techniques of the previous section or some of the sensitivity techniques given in Section 16.4.

2.8.4 Proportion of Trace Explained

Over the years, a very popular stopping rule has been one related to the *proportion* of the trace of the covariance matrix that is explained by the pc's in the model. Its popularity lies in the fact that it is easy to understand and administrate. Some computer packages use this rule; the user will specify some particular proportion (the default is commonly .95) and when that much of the trace is explained, the process is terminated.

This procedure is NOT recommended. There is nothing sacred about any fixed proportion. Suppose, for $p = 20$, that the last 15 roots are not significantly different so that only five pc's would be retained, and further suppose that these five pc's explain only 50% of the trace. Should one keep adding pc's related to these other roots until the magic figure is reached? Definitely not. Conversely, the example to be presented in Section 2.9 has over 95% explained by the first two pc's and yet one more should be included in the model.

Having said that, it must be admitted that there are occasions when PCA is used as an exploratory tool when very little is known about the population from which the sample is obtained. In these instances, the proportion of the trace explained may be of some use in developing provisional PCA models and we shall not be above employing this procedure occasionally.

2.8.5 Individual Residual Variances

What would seem to be a better stopping rule would be one based on the *amount* of the explained and unexplained variability. In this procedure, one determines in advance the amount of residual variability that one is willing to tolerate. Characteristic roots and vectors are obtained until the residual has been reduced to that quantity. (Again, this should be carried out *after* the significance test so that one does not include pc's that the test would not permit, even if the desired residual variance were unobtainable.)

In some situations, prior information with regard to inherent variability may be available that might serve as the desired target residual. This method has been employed for some of the examples in this book—in particular, the audiometric examples in Chapters 5 and 9. While it is best to specify the desired residual variability for each variable separately, this may not be possible or practical. In this situation, an average residual could be specified that could then be expressed in form of the residual trace. For a chemical example, see Box et al. (1973).

2.8.6   The SCREE Test

This is a graphical technique. One plots all of the characteristic roots of the covariance matrix, the values of the roots themselves being the ordinate and the root number the abscissa. A typical SCREE plot is shown in Figure 2.2. [The name *SCREE* is due to Cattell (1966), scree being defined as the rubble at the bottom of a cliff, i.e. the retained roots are the cliff and the deleted ones are the rubble.] Note that the last few roots are much smaller than the rest and are nearly in a straight line. There is a break of sorts between the first three roots and these remaining five roots. (This break is sometimes called an *elbow*.) Cattell and Jaspers (1967) suggested using all of the pc's up to and including the first one in this latter group. In this example, the model would include four pc's.

   This technique has become quite popular although there can be some problems with it. First, it is only a graphical substitute for a significance test. Second, the plot may not have a break in it; it could appear for instance as a fairly smooth curve. Alternatively, it may have more than one break. In this case, it is customary to use the first break in determining the retained pc's. However, Wernimont suggested that for some analytical chemistry problems in which he found this, the pc's before the first break represented the components describing the system under study and the second set, between the breaks, was

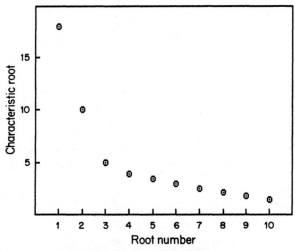

FIGURE 2.2. Typical SCREE plot.

made up of components representing instrumentation and/or other identifiable testing and measurement variability. An idealized representation of such a situation is shown in Figure 2.3. If these intermediate pc's are to be retained, their roots should be distinct. Cattell and Vogelmann (1977) also discussed procedures for multiple SCREEs.

Box et al. (1973) described a similar situation in which the third group of roots should have all been equal to zero because they represented linear constraints. However, due to rounding error, they were positive quantities, although much smaller than the roots representing inherent variability. [The opposite situation, $l_p = 0$ when $\lambda_p > 0$ is not likely. Yin et al. (1983) and Silverstein (1984) showed that l_p is bounded away from zero under fairly general conditions.]

A third situation that may occur is that the first few roots are so widely separated that it is difficult to plot them all without losing the detail about the rubble necessary to determine the break. This problem may be diminished by plotting the logs of the roots instead. (This is called an LEV, or log-eigenvalue, plot.)

In an attempt to take some of the guesswork out of these procedures, Horn (1965) suggested generating a random data set having the same number of variables and observations as the set being analyzed. These variables should be normally distributed but uncorrelated. A SCREE plot of these roots will generally approach a straight line over the entire range. The intersection of this line and the SCREE plot for the original data should indicate the point separating the retained and deleted pc's, the reasoning being that any roots for the real data that are above the line obtained for the random data represent roots that are larger than they would be by chance alone. (Just *larger*, not *significantly*

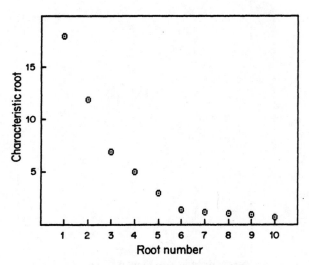

FIGURE 2.3. "Wernimont" SCREE plot.

larger.) Farmer (1971) carried out a similar study for LEV plots using both random and structured data. These procedures are sometimes called *parallel analysis*. Expressions for the expected values of these roots, using regression from these simulations, were obtained by Allen and Hubbard (1986), Lautenschlager et al. (1989), and Longman et al. (1989). Longman et al. also included the 95th percentile. Lautenschlager (1989) felt that interpolation might be superior to regression in many instances. He also included extensive tables. Lambert et al. (1990) used bootstrap techniques to study the sampling variability of the characteristic roots and their effect on the stopping rule.

2.8.7 The Broken Stick

A rather quick-and-dirty version of Horn's technique is the *broken stick* model (Jolliffe, 1986). This is based on the fact that if a line segment of unit length is randomly divided into p segments, the expected length of the kth-longest segment is

$$g_k = \frac{1}{p} \sum_{i=k}^{p} \left(\frac{1}{i} \right) \tag{2.8.4}$$

As long as the proportion explained by each l_k is larger than the corresponding g_k, retain the corresponding pc.

For example, the proportions for the case of four independent variables would be .521, .271, .146, and .062. The proportion explained by each pc for the Ballistics Missile example are .782, .112, .068, and .038. Only the first pc accounts for more than would be expected by chance alone.

2.8.8 The Average Root

An even quicker technique is to retain only those pc's whose roots exceed $\text{Tr}(S)/p$, which is the size of the average root. The rationale here is that any pc that is deleted will have a root smaller than the average. In the next chapter, one of the topics to be discussed will be PCA starting with a correlation rather than covariance matrix. In this case the average root is equal to 1 and for this reason, the average root rule is widely used in the fields of psychology and education, among others, where correlation matrices are generally employed. This method, sometimes called the *Guttman–Kaiser* criterion, has been criticized as being too inflexible. In fairness to Guttman, he derived this criterion as a lower bound for the number of common factors in factor analysis (Section 17.2.2) along with two other criteria that were better; PCA was never mentioned (Guttman, 1954).

For the Ballistics Missile example, the average root is $429.11/4 = 107.28$, which is larger than all of the roots save the first and would imply that only one pc should be retained. Jolliffe (1972) contended that this cutoff was too high because it did not allow for sampling variability and, based on some simulation studies, recommended using 70% of the average root. For the Ballistics Missile example, this would amount to 75.09, which would leave the conclusion unchanged.

2.8.9 Velicer's Partial Correlation Procedure

This procedure is based on the partial correlations among the original variables with one or more pc's removed (Velicer, 1976b). Let

$$S_k = S - \sum_{i=1}^{k} v_i v_i' \qquad k = 0, 1, \ldots, p - 1 \tag{2.8.5}$$

and

$$R_k = D_s^{-1/2} S_k D_s^{-1/2} \tag{2.8.6}$$

where D_s is a diagonal matrix made up of the diagonal elements of S_k. R_0 is the original correlation matrix, R_1 is the matrix of correlations among the residuals after one pc has been removed and so on.

Let

$$f_k = \sum_{i \neq j}\sum (r_{ij}^k)^2 / [p(p-1)] \tag{2.8.7}$$

where r_{ij}^k are the correlations in the matrix R_k. f_k is the sum of squares of the partial correlations at stage k and, as a function of k, will normally have a minimum in the range $0 < k < p - 1$ and the value of k for which this occurs will indicate the number of pc's to retain.

The logic behind Velicer's test is that as long as f_k is declining, the partial covariances are declining faster than the residual variances. This means that Velicer's procedure will terminate when, on the average, additional pc's would represent more variance than covariance. This represents a departure from the other stopping rules, which are concerned only with the characteristic roots themselves, and is somewhat of a compromise between these other procedures and the method of factor analysis that will be discussed in Chapter 17. It also precludes the inclusion of pc's that are primarily a function of one variable (Jolliffe, 1986).

To illustrate this method, consider the Ballistics Missile example with one pc retained:

$$S_k = S - v_1 v_1' = \begin{bmatrix} 29.25 & -6.77 & -5.01 & -16.25 \\ -6.77 & 18.81 & -7.01 & -11.28 \\ -5.01 & -7.01 & 13.92 & .48 \\ -16.25 & -11.28 & .48 & 31.79 \end{bmatrix}$$

whence

$$
R_1 = \begin{bmatrix}
1 & -.29 & -.25 & -.53 \\
-.29 & 1 & -.43 & -.46 \\
-.25 & -.43 & 1 & .02 \\
-.53 & -.46 & .02 & 1
\end{bmatrix}
$$

and

$$
f_1 = [(-.29)^2 + \cdots + (.02)^2]/[(4)(3)] = .14
$$

The results are: $f_0 = .50, f_1 = .14, f_2 = .38$, and $f_3 = 1.00$. The minimum occurs at $k = 1$ so one pc would be retained. This is a typical result in that fewer pc's are retained than would be under Barlett's test. In Section 2.8.2 it was suggested that Bartlett's test might retain more pc's than was realistic, particularly for large n, and it was because of this that Velicer's procedure was devised.

Reddon (1985) carried out some simulations using data generated from populations having unit variances and zero covariances. In this case, f_0 should be less than f_1. The smaller the sample size relative to the number of variables, the larger the probability that this will not happen (Type I error). His example suggests a sample size of $n = 3p$ will be required to produce $\alpha = .05$ and of $n = 5p$ to reduce this below .01. This, of course, is only part of the story and Reddon also used this technique with some real data and found the results to be consistent with those already reported in the literature.

2.8.10 Cross-validation

Yet another suggestion for determining the optimum number of pc's is the cross-validation approach advocated by Wold (1976, 1978) and Eastment and Krzanowski (1982). This approach is recommended when the initial intention of a study is to construct a PCA model with which future sets of data will be evaluated. Cross-validation will be discussed in Section 16.3 but, briefly, the technique consists of randomly dividing the sample into g groups of n/g observations each. A PCA, using only the first pc, is performed on the entire sample save the first group. Predictions of these deleted observations are then obtained using (2.7.2) and a value of Q is obtained for each of these observations. The first group is then returned to the sample, the second group is deleted, and the procedure is repeated. This continues until all g groups have had their turn. The grand average of all the Q-values, divided by p, is called the *PRESS-statistic*. The entire procedure is repeated using a two-component model, a three-component model, and so on, from which additional PRESS-statistics are formed. The stopping rule is based on some comparison schemes of PRESS-statistics. As a stopping rule, this method requires the original data for its implementation. The others require only the covariance matrix.

2.8.11 Some Other Ad Hoc Procedures

There have been a number of other stopping rules proposed from time to time. Most of these are intuitive, many of them coming from the chemometricians, and their distributional properties are generally unknown. Among them are:

1. *Indicator function* (Malinowski, 1977).

$$\text{IND} = \sqrt{\frac{\sum_{j=k+1}^{p} l_j}{n(p-k)^s}} \qquad (2.8.8)$$

   Obtain IND as a function of k and terminate when IND reaches a minimum. Droge and Van't Kloster (1987) and Droge et al. (1987) compared IND with some early cross-validation techniques. For comments on this comparison and a discussion of cross-validation, see Wold and Sjöström (1987).

2. *Imbedded error* (Malinowski, 1977).

$$\text{IE} = \sqrt{\frac{\sum_{j=k+1}^{p} l_j}{np(p-k)/k}} \qquad (2.8.9)$$

   Obtain IE as a function of k and terminate when IE is a minimum.

3. The ratio l_i/l_{i+1} (Hirsh et al., 1987).

4. Number or percentage of residuals outside of univariate limits (Howery and Soroka, 1987).

Another practice is to stop at an arbitrary k because of some prior information or hypothesis. Newman and Sheth (1985) did this in a study of primary voting behavior. They had $n = 655$ observations on $p = 88$ variables. They stopped at $k = 7$ because that is what the model called for although there were more than seven roots (from a correlation matrix) that were greater than unity. The first seven only explained 32%.

With this last example, it is perhaps appropriate to quote Morrison (1976). "It has been the author's experience that if that proportion [he had previously recommended at least 75%] of the total variability cannot be explained in the first four or five components, it is usually fruitless to persist in extracting vectors even if the later characteristic roots are sufficiently distinct ... ." If one knows in advance that the inherent variability is 60% or 70% and that there are a small number of real pc's, it is still permissible to obtain pc's up to that point but in most situations this is not known and, as Morrison suggests, obtaining a large number of pc's will invariably produce pc's that will be difficult to interpret at best and may well represent inherent variability.

2.8.12 Conclusion

In this section, a number of procedures have been presented for determining the optimum number of retained pc's. Some of these techniques have been included merely because they enjoy widespread use despite the fact that they have severe shortcomings. The Ballistic Missile example has been used as an example in this section solely for illustrative purposes because there were only four variables. Because the first pc, explaining 78% so dominates the others, some of these techniques retained only that one (Broken Stick, Average Root, and Velicer). Anderson's version of Bartlett's Test retained two as would have the SCREE plot except that one cannot really do much with only four variables. If one had used 95% of the trace of S (quite often used as a default on some computer programs), three pc's would have been retained. In the next section, a 14-variable example will be introduced that will produce a much more realistic comparison of these various methods.

In Section 2.4, it was shown that zero roots were indicative of linear constraints among the variables. For this reason, one should obtain *all* of the roots even though there is no intention of using all of the pc's in the model.

A number of studies have been carried out to compare these various procedures. Many of these were simulation studies. Among them was one by Krzanowski (1983) in which he found, for the four procedures he compared, that Bartlett's test retained the most pc's, followed by the 75% trace rule, the average root, and finally cross-validation, although the last two were similar. Krzanowski showed that cross-validation looked for gaps between the roots, thus providing an analytical version of the SCREE plot. Wold (1978) concluded that the cross-validation technique retained fewer pc's because it included a predictive procedure using the Q-statistic. Bauer (1981) concluded that the average root criterion was usually within ± 1 of the true dimensionality and never off by more than 3. The SCREE plot usually had one too many. Zwick and Velicer (1986) compared simulation results with the underlying models and concluded that, of the five procedures they compared, parallel analysis was most accurate followed by Velicer's partial correlation technique and the SCREE plot. The average root generally retained too many pc's. Bartlett's test, they felt, was erratic. The Zwick–Velicer paper refers to a number of other comparisons of these methods, many with conflicting results. This is probably the result of the different ranges of variations employed in the various experiments.

2.9 A PHOTOGRAPHIC FILM EXAMPLE

2.9.1 Introduction

The two examples used so far, the Chemical analyses and the Ballistics Missile tests, were included primarily for purpose of illustrating the mechanics of PCA since they involved two and four variables, respectively. Although a number of features associated with PCA have already proven useful, the real power of PCA is more evident when a larger number of variables is employed. The following example involves the monitoring of a photographic product manufactured by the Eastman Kodak Company and was used by Jackson and Bradley (1966) to illustrate some sequential multivariate techniques (Section 2.10.3).

Photographic products and processes are usually monitored by means of a piece of film that contains a series of graduated exposures designed to represent the entire range of exposures used in actual practice. In this present example there are 14 steps in even increments of log exposure. After the strip has been exposed and processed, optical densities are obtained for each of the steps and the resultant curve, as a function of log exposure, displays the photographic characterizations of the product or process being monitored (Carroll et al., 1980: Stroebel et al., 1986). An example of such a curve is shown in Figure 2.4,

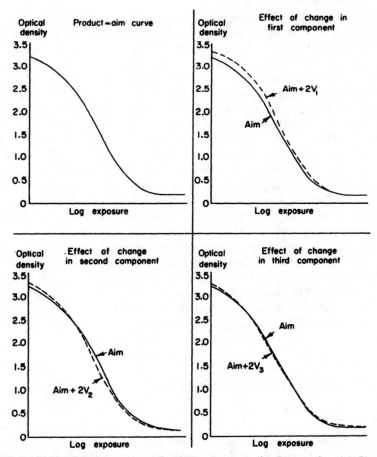

FIGURE 2.4. B&W Film example. (a) Sensitometric curve. (b) Curve $+2v_1$. (c) Curve $+2v_2$, (d) Curve $+2v_3$. Reproduced from Jackson and Bradley (1966) with permission of Academic Press.

which is the aim for the particular product under study. The high-density or "shoulder" portion of the curve represents shadow areas of a picture, the low-density or "toe" portion of the curve represents light areas and the middle densities represent the average densities in a picture. (Chapter 6 contains a similar example dealing with color film, which will have three of these curves, one for each of the primary colors. That example, throughout the book, will be referred to as the "Color Film" example while the current example under discussion will be called the "Black and White Film" or "B&W Film" example.)

There are four general sources of variability in this type of example: (1) The film itself, (2) the film developing process, (3) the device used to place the selected exposures on the film (called a sensitometer), and (4) the device used to measure the optical densities of the processed film (called a densitometer). In this example, it is the film variability that is of interest; the other three sources are considered to represent the inherent variability of the monitoring system. One thing that sets this example apart from the others is that the original variables represent a continuous function (exposure) rather than the discrete types of tests used in the other two examples. We shall see, as we proceed throughout the book, that examples of this type possess some rather unique properties with respect to PCA.

2.9.2 PCA Analysis

The covariance matrix for 232 sample film test strips representing normal manufacturing variability is given in Table 2.4. These entries have been multiplied by 10^5 for simplicity. The other displays will be in the correct units. The value for the Gleason–Staelin φ-coefficient (equation 2.8.2) is .82 so we are definitely in cobblestone country. The stopping rule adopted was the one related to individual residual variances (Section 2.8.5). It was assumed that the inherent variability associated with this monitoring process was of the order of $\sigma = .007$ at each step and three pc's were required to reduce the residual variability to this amount. The V-vectors for these three pc's, along with their characteristic roots, are shown in Table 2.5. The residual standard deviations are also included; with the maximum equal to .009 and the minimum equal to .006, it would appear that the requirements for this stopping rule had been satisfactorily met.

The reason that V-vectors are displayed is that they are in the same units as the original data, optical density. They, too, represent a continuous function that may be used as an aid to interpretation. This can be seen in the other three displays in Figure 2.4. The first of these has, in addition to the aim, the quantity (aim + $2v_1$). The coefficients for this vector are all positive except for the last two in the toe, which are negligible. Positive values of the first pc, y_1, would represent cases where all of the other steps exhibited increased density. Furthermore, if the variation in the film samples affected y_1 only, 95% of their curves should fall within the envelope (aim $\pm 2v_1$). This pc accounts for 83%

Table 2.4. B & W Film Example. Covariance Matrix (× 10^5)

x_1	x_2	x_3	x_4	x_5	x_6	x_7	x_8	x_9	x_{10}	x_{11}	x_{12}	x_{13}	x_{14}
672	667	650	633	577	487	353	228	112	50	17	-2	-6	-15
	686	686	688	649	565	424	281	144	66	22	-2	-9	-20
		726	759	748	677	530	363	200	97	32	0	-11	-26
			839	859	805	653	464	268	135	49	7	-9	-28
				924	892	745	543	325	170	65	16	-6	-28
					893	765	574	355	189	78	25	1	-20
						688	535	341	188	82	33	9	-10
							446	296	168	82	41	22	8
								211	124	65	37	25	16
									81	45	28	22	16
										32	22	20	18
											21	18	18
												21	19
													22

of the total variability and represents, essentially, increases and decreases in overall density.

The second pc has positive coefficients for the first four steps and negative coefficients for the rest. Referring to the plot of (aim + $2v_2$), it can be seen that this curve has a steeper slope than the aim. Film strips that exhibit this property will be darker than normal in the high-density region and lighter than normal in the lower densities. This will make the resultant picture appear to have more *contrast* than normal. Negative values of y_2 would represent a flatter curve than normal and would have a lower-contrast picture. This pc accounts for 14% of the total variability.

The third pc explains about 2% and appears to represent a buckle in the middle of the curve. However, a look at Table 2.5 will reveal that this pc also explains a good share of the variability in the lower-density (high-exposure) steps. (Recall that the $v_3 v_3'$ is the amount of the covariance matrix explained by y_3.) Overall, these three pc's account for 99% of the total variability but, more important, the residuals are all reduced to their prescribed amount. Some of the residuals in the toe region still amount to 25% of the original variability but this variability was relatively small to start with and what remains is considered to be inherent variability of the same magnitude as the other steps.

An important point to make is that the physical importance of a principal component is not necessarily related to the size of its characteristic root. Although y_1 explains much more variability than does y_2, variation in y_2 would be more noticeable to the human eye than would a corresponding change in y_1.

Table 2.5. B & W Film Example. Characteristic V-vectors and Residuals

Step	Characteristic Vectors			Residual Standard Deviation
	v_1	v_2	v_3	
1	.0666	.0454	.0134	.0072
2	.0728	.0383	.0071	.0059
3	.0811	.0252	−.0005	.0069
4	.0905	.0092	−.0065	.0080
5	.0946	−.0085	−.0108	.0071
6	.0908	−.0234	−.0093	.0073
7	.0756	−.0330	−.0022	.0074
8	.0556	−.0346	.0098	.0088
9	.0333	−.0281	.0130	.0066
10	.0175	−.0179	.0118	.0067
11	.0071	−.0099	.0118	.0058
12	.0021	−.0063	.0110	.0066
13	−.0000	−.0042	.0117	.0075
14	−.0021	−.0033	.0121	.0076
% Explained	83.1	13.5	2.3	

2.9.3 Comparison of Stopping Rules

In this section, the various stopping rules discussed in Section 2.8 will be compared for the film example.

a. *Bartlett's Test.* This is shown in Table 2.6. Using $\alpha = .05$, this procedure terminates after 10 pc's have been retained.

b. *Proportion of Trace Explained.* This method is not recommended but if it had been employed here, the common default of .95 would have retained two pc's.

c. *Individual Residual Variances.* This was the method used in Section 2.9.2 and led to three pc's being retained.

d. *The SCREE Test.* This is shown in Figure 2.5. Because the first few roots are so large, an LEV plot had to be used. The straight-line segment appears to begin with the fifth root so the Cattell–Jaspers rule would recommend keeping five pc's.

e. *The Broken Stick.* The ordered "broken stick" segments for $p = 14$ are: .232, .161, .125, .101, .083, .069, .057, .047, .038, .030, .023, .016, .011, and .005. The first pc explains .83, which is larger than .232, but the second pc explains only .14, which is less than .161, so the process would be terminated with only the first pc retained. This procedure would appear to break down when a system is dominated by the first pc.

f. *The Average Root.* The average root is $.062607/14 = .004472$. Only the first two roots are larger than that. If one used Jolliffe's recommendation of

Table 2.6. B & W Film Example. Characteristic Roots and Bartlett's Test

	Characteristic Root	Test for Equality of Last $(14 - i + 1)$ Roots		
i	l_i	χ^2	d.f.	Prob.
1	.052036	11 219.9	104	<.0001
2	.008434	6 667.8	90	<.0001
3	.001425	2 999.2	77	<.0001
4	.000255	909.7	65	<.0001
5	.000116	424.6	54	<.0001
6	.000075	251.4	44	<.0001
7	.000060	169.1	35	<.0001
8	.000050	109.0	27	<.0001
9	.000041	60.6	20	<.0001
10	.000031	26.0	14	.025
11	.000025	10.7	9	.30
12	.000023	6.4	5	.27
13	.000020	1.4	2	.50
14	.000017	No Test		

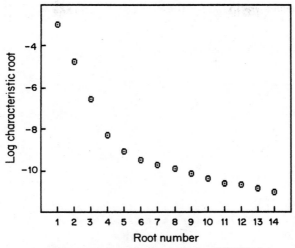

FIGURE 2.5. B&W Film example: log SCREE plot.

70% of the average trace, the cutoff value would be .003130, which would still result in two pc's being retained.

   g. *Velicer's Method.* The values of f_i are

$$f_0 = .391 \qquad f_7 = .098$$
$$f_1 = .417 \qquad f_8 = .134$$
$$f_2 = .331 \qquad f_9 = .200$$
$$f_3 = .111 \qquad f_{10} = .372$$
$$f_4 = .058 \qquad f_{11} = .362$$
$$f_5 = .066 \qquad f_{12} = .490$$
$$f_6 = .080 \qquad f_{13} = 1.0$$

A plot of these results is displayed in Figure 2.6.

   This function has its minimum at f_4, indicating that four pc's should be retained. Note that this curve has an irregularity in it at f_1 (or f_0) so that one should compute enough values of f_i to assure that the minimum has been achieved.

   Seven different stopping rules have been illustrated for this set of data and would have recommended retaining anywhere from 1 to 10 principal components— quite a spread! (The cross-validation method was not included because the original data, which it requires, are no longer available.) The method actually employed for this example—individual residual variances—retained three pc's

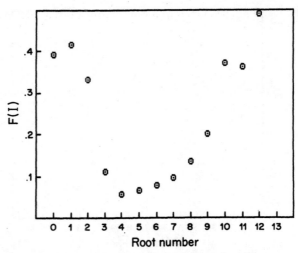

FIGURE 2.6. B&W Film example: Velicer plot.

but this must be used in conjunction with Bartlett's test to ascertain that those three roots are significantly different from the others in the first place.

2.10   USES OF PCA

2.10.1   Introduction

There have been thousands of applications of PCA over the years, hundreds of them appearing in print. It is not the intent of this section to review them all but rather to mention a few examples where PCA has been applied. Other references will be made when applicable to specific PCA techniques or enhancements throughout the book. Many of the articles referenced will have the term *factor analysis* in the title. This is a technique similar to but distinctly different from PCA in that it determines its "components" by accounting for correlations instead of variability. Much confusion has existed between the two and we shall deal with this in Sections 17.2 and 17.10. For now, suffice to say that many of these articles are, in fact, using PCA.

For reasons which will become apparent in Chapter 3, many PCA studies are not carried out on covariance matrices, the correlation matrix being the most common alternative. Many of the examples given below are in this latter mode.

2.10.2   Psychology and Education

The first and, by far, the most frequent use of PCA over the years has been related to the fields of psychology and education. A typical example will involve a battery of achievement tests given to a number of subjects such as the Seventh Grade Data to be introduced in the next chapter. Factor analysis owes its existence to this activity, in which one attempted to characterize the ability of an individual or group of individuals based on these tests. (One of the early books on factor analysis was entitled *The Vectors of the Mind*, Thurstone, 1935). With the similarity of and confusion between factor analysis and PCA, many of these studies opted for the latter, particularly before the days of mainframe computers because PCA computations were simpler. Many of these studies may be found in earlier issues of such journals as *Educational and Psychological Measurement*, *Psychological Bulletin*, and *Psychometrika*. The examples used in Harmon (1976) are typical of these.

2.10.3 Quality Control

This book makes frequent reference to quality control applications, particularly with regard to principal components control charts. There has also been some recent use of PCA with cumulative sum procedures. Early multivariate cusums were performed by carrying out a cusum for each variable and using Bonferroni bounds on the results. As these did not take into account the correlations among the variables, Woodhall and Ncube (1985) suggested transforming the data into principal components and using these for the cusums, again using Bonferroni bounds. Crosier (1988) has since proposed a procedure that applies the T^2-test to cumulative sums, thus avoiding the Bonferroni bounds. For problems with a large number of correlated variables, some combination of these techniques might be useful. Fisher et al. (1986) also used PCA cusums for diagnostic purposes during some calibration studies.

The development of methods to conduct multivariate acceptance sampling has proceeded very slowly. The principal reason for this is the fact that specifications are usually made for each variable independently of the others, which produces a p-dimensional "box" and does not take into account the fact that the variables may be correlated. This then poses the problem of how to relate the p-dimensional ellipsoid representing variability with this box. In applying some sequential multivariate schemes to some ballistic missile data, Jackson and Bradley (1961a) considered inscribing this solid within or superscribing it about the box. Neither was particularly successful because high correlations were present, which produced a highly rotated ellipsoid. For the B&W Film example of Section 2.9, Jackson and Bradley (1966) suggested establishing specifications on the principal components, that had the advantage of being uncorrelated. While this is an attractive alternative, it can be used only when the pc's are so well established in the quality engineers' minds that they can make specifications on them. Closely related to acceptance sampling are *acceptance control charts* (Freund, 1960; Schilling, 1982, p. 169), in which the control limits are a function of both the inherent variability itself and its relation to the specifications established for the product or process. The same problems exist with regard to multivariate extensions of these as with acceptance sampling.

2.10.4 Chemistry

The use of statistical methods in chemistry has always been an active field and recently acquired the name *chemometrics*. (For a review article on the subject, see Brown et al., 1988.) One widely used application of PCA is in the analysis of mass spectroscopy or gas chromatography data in which the data represent concentrations at various frequencies or times. For some representative examples, see Weiner (1973), Rozett and Petersen (1975a,b), Ritter et al. (1976), Howery (1977), Malinowski and Howery (1980), and Windig (1988).

The test battery employed in psychology and education has it counterpart in science and industry. Thomas (1961) used PCA on test data for refinery products where the variables represented such tests as flash point, specific gravity, and so on. Carey et al. (1975) used PCA in method comparison studies in place of the "referee" method. Bretaudiere et al. (1981) carried out an analysis of blood samples similar to the Ballistic Missile example in that the variables all represented similar types of measurements.

2.10.5 Photographic Science

The example in Section 2.9 involved optical densities at a number of exposure levels on black and white film. Similar studies were carried out by Simonds (1958a,b, 1963) as well as an example using color film processing that will be introduced in Chapter 6. In color densitometry, each sample is measured three times, each through a different filter to obtain density to red, green, and blue light. Densities may also be obtained as a function of wavelength similar to the mass spectroscopy analyses of chemical compounds. The use of PCA for this type of data has been discussed by Brewer and Williams (1954), Morris and Morrissey (1954), and Nayatani et al. (1983). A similar study involving reflectance of paper samples as a function of wavelength was reported by Grum and Wightman (1960). Other applications related to color perception will be discussed in Section 11.3.

2.10.6 Market Research

PCA is used a greal deal in market research as a data-reduction technique prior to performing regression or cluster analysis. A typical example due to Vavra (1972) deals with perceptions of products before and after an advertising compaign.

A tool used in long-range planning is *cross impact analysis*, which obtains estimates of the effect of the occurrence of one event on the probability of occurrence of another (Jackson and Lawton, 1976). A PCA solution for this problem was proposed by Jensen (1981).

2.10.7 Economics

Much of multivariate analysis work in economics is related to systems of equations where there are a number of correlated response variables as well as predictor variables, with some response variables also serving as predictor variables. This activity is described briefly in Sections 12.7 and 17.8.3 and requires a methodology all its own. PCA may serve in a supporting role, for instance being used to produce instrumental variables in two-stage least-squares (Kloek and Mennes, 1960). However, PCA has been used in its own right. Among the earlier applications was a study involving cotton yarn over the period 1924–1938 (Bartlett, 1948). The variables were the consumption, price, and cost of producing the yarn as well as the income of the consumers. From this, Bartlett was able to construct a demand and supply function. Another example involved the relationships among different kinds of financial transactions (Stone, 1947). Chatelin and Belaid (1987) used PCA in an analysis of 11 branches of the French government over the period 1872–1971, which includes some missing data. Econometric methods also make use of time series analysis (Chow, 1975). The application of PCA to these techniques is discussed in Section 14.6.

2.10.8 Anatomy and Biology

A situation that readily lends itself to multivariate analysis is data consisting of physical measurements on living beings. We shall see an example of this with the Physical Measurements data in Section 8.4.1, which involved eight different measurements on young girls who were then characterized by PCA. The subjects need not be alive; VanGerven and Oakland (1973) used the same technique on some 11th century human femurs. Jeffers (1967) reported on a similar study with winged aphids. He plotted the first two pc's in the form of a scatter plot and discovered a very nonrandom pattern indicating, in this case, that there were four different groups of aphids in the sample. This is a good thing to do with any PCA study because *uncorrelated* and *independent* do not mean the same thing.

PCA is also used with data of this sort to obtain measures of shape and size (see, for instance Jolicoeur and Mosimann (1960), analyzing data from painted turtles). This includes the topic of *allometry*, defined to be the study of the relative growth of a part of an organism in relation to an entire organism. Some special PCA techniques related to allometry will be discussed in Section 15.4. Growth curves will be dealt with in Section 16.7.2.

In Chapters 10 and 11, we shall examine the relationship of PCA to multidimensional scaling. Several examples of one of these techniques applied to biomedical research were given by Gabriel and Oderoff (1990).

2.10.9 Forestry and Agriculture

Some studies on fruit tree growth and cropping were given by Pearce and Holland (1960). One of their examples used pc's obtained from "within-rootstock" variations, which reduced the number of pc's required from three to two. We shall discuss subgrouping in Chapter 6. Two studies were reported by Jeffers; one dealt with trees grown from seed (Jeffers, 1962) and the other with pitprops, timbers used in mines (Jeffers, 1967).

PCA studies on soil samples have been done by Holland (1969) and Gittins (1969). Gittins also reported on a study of grasses in Wales. (Gittins' soil study actually employed a factor analysis approximation technique called the *centroid* method, which was used years ago because it was easier to employ than PCA but is now passé.) Kendall (1939) used PCA to study the geographical distribution of crop productivity in England. Holland's study also resulted in a map. In Section 13.7, we shall consider the use of PCA in univariate analysis of variance; the earliest instance of this was Fisher and MacKenzie (1923) in a study of potatoes.

Closely related to these topics is the analysis of climatic data, of which a study by Corsten and Gabriel (1976) is typical. Gabriel and his associates conducted a number of such studies to illustrate the method of *biplots*, which will be discussed in Section 10.5.

2.10.10   Other Applications

Some of the applications in the previous subsection produced results in the form of maps. A survey article on the use of PCA in "human geography" was given by Clark et al. (1974) and contained well over 100 references related to such things as business, ecology, and the social sciences. Menozzi et al. (1978) employed PCA to reduce genetic data and plotted contours of frequency on maps of Europe. They used this to trace migration of early farming in Europe. Multidimensional scaling outputs are often referred to as "maps" even though they are not usually geographical; these will be discussed in detail in Chapters 10 and 11.

A recent article by Dawkins (1989) employed PCA on some track and field data. National track records for seven distances, ranging from 100 meters to the marathon, were given for 55 nations. The first pc represented overall time differences among nations while the second pc was a contrast between the related times of short and long distances. Two data sets, one each for men and women, are included in this article.

CHAPTER 3

Scaling of Data

3.1 INTRODUCTION

In Section 1.6, the matter of scaling principal components was discussed. In that context, *scaling* referred to the scaling of the characteristic vectors used to obtain the pc's. In particular, we had:

1. U-vectors that were orthonormal and that produced pc's, $\mathbf{z}$, with variances equal to the corresponding characteristic root
2. V-vectors whose coefficients were in the same units as the data and whose pc's, $\mathbf{L}^{1/2}\mathbf{z}$, had variances equal to the square of the characteristic roots
3. W-vectors whose pc's, $\mathbf{y}$, had unit variance

It was pointed out that each of these three sets of pc's conveyed exactly the same amount of information and the choice of which one to use was strictly a matter of the use to which it was being applied.

This chapter is also concerned with scaling but in this case it will be the data which are being scaled and now the results *will* depend on the method employed. Once the method of scaling is selected, the PCA operations will proceed, for the most part, as described in Chapters 1 and 2 but there will be some modifications unique to each method. The main effect of this choice will be on the matrix from which the characteristic vectors are obtained. It will be the purpose of this chapter to describe: (1) the various methods, (2) the reasons why each might be employed, and (3) the changes in the operations described in Chapters 1 and 2 that might be required.

Specifically, the following three methods will be considered:

1. No scaling at all. The final variate vector is $\mathbf{x}$.
2. Scaling the data such that each variable has a zero mean (i.e., in terms of deviation from the mean). The final variate vector is $\mathbf{x} - \bar{\mathbf{x}}$.
3. Scaling the data such that each variable is in standard units. (i.e., has zero mean and unit standard deviation). Each variable is expressed as $(x_i - \bar{x}_i)/s_i$.

63

There will also be a discussion of weighted PCA in which each variable has a different weight, situations in which an observation is part of a two-way classification and must be corrected for means in both directions, and finally a section on complex variables.

As stated above, the choice of scale will determine the dispersion matrix used to obtain the characteristic vectors. If no scaling is employed, the resultant matrix will be a *product* or *second moment* matrix; if the mean is subtracted, it will be a *covariance* matrix; if the data are in standard units, it will be a *correlation* matrix. Although this is a logical progression for describing these methods, we shall start with the covariance matrix since that is the one described in Chapters 1 and 2.

3.2 DATA AS DEVIATIONS FROM THE MEAN: COVARIANCE MATRICES

This is the method that has been used in the preceding two chapters. It is not necessary to actually subtract the variable means from the data; the operations required to obtain the covariance matrix will take care of it. The only time the means are required are when one wishes to obtain the principal components using equation (1.6.9) or one of its equivalents or when one wishes to predict the original variables from the pc's using equation (2.7.2).

If one were to subtract the means from the data and use these deviations as a data set, say $x_d = x - \bar{x}$ with the resulting $n \times p$ data matrix X_d, the covariance matrix would be $X'_d X_d / (n - 1)$. The characteristic vectors U, V, and W would be the same as before. Equation (1.6.9) becomes $y = W' X_d$ and equation (2.7.2) becomes $\hat{x}_d = Vy$. For large problems in terms of sample size or large number of digits for the original data, this option may be preferable, numerically, to obtaining S from the raw data directly.

People generally feel most comfortable when operating with the covariance matrix. The V-vectors are in the units of the original variables and the inferential procedures in Chapter 4, sketchy as they may be in some instances, are considerably better developed for the covariance matrix than for any other situation.

3.3 DATA IN STANDARD UNITS: CORRELATION MATRICES

3.3.1 Rationale for Using Correlation Matrices

Comfortable or not, there are many occasions when one cannot use the covariance matrix. There are two possible reasons for this:

1. The original variables are in different units. In this case, the operations involving the trace of the covariance matrix have no meaning. For instance,

if a variable is expressed in inches, its variance is 1296 times what it would be if it were expressed in yards. This variable would now exert considerably more influence on the shaping of the pc's since PCA is concerned with explaining variability. When the units are different, the solution is to make the variances the same (i.e., use standard units), which makes the covariance matrix into a correlation matrix. When one departs from physical and engineering science applications, such as those displayed in Chapters 1 and 2, and begins to apply PCA to such fields as psychology, education, or business, one will find that the variables are rarely in the same units.

2. Even if the original variables are in the same units, the variances may differ widely, often because they are related to their means. If this gives undue weight to certain variables, the correlation matrix should be employed here also (unless, possibly, taking logs of the variables or the use of some other variance-stabilizing transformation will suffice). Harris (1985), in discussing some tests for the homogeneity of variance of correlated variables, recommended the use of these procedures to decide whether or not a covariance matrix may be employed.

These conditions would seem to preclude a large number of applications from utilizing a covariance matrix. The use of the correlation matrix is so widespread in some fields of application that many practitioners never use the covariance matrix at all and may not be aware that this is a viable option in some instances. In fact, some computer packages do not allow the covariance matrix option. (Others permit the input of a covariance matrix but then automatically convert it into a correlation matrix from which the PCA is obtained!) Nevertheless, when the variables *are* in the same units and do have the same amount of variability, there are some advantages in using covariance matrices, as has already been pointed out. This is particularly true in physical applications where PCA is used in building physical models. Many quality control applications will also be able to use the covariance matrix, which should help with diagnostics since, again, the V-vectors are in the original units of the variables. In addition, the covariance matrix option has more inferential procedures available, as we shall see in Chapter 4.

It is important to note that *there is no one-to-one correspondence between the pc's obtained from a correlation matrix and those obtained from a covariance matrix*. The more heterogeneous variances are, the larger the difference will be between the two sets of vectors. If the covariance matrix has $(p - k)$ zero roots, then the correlation matrix will also have $(p - k)$ zero roots. However, if the covariance matrix has $(p - k)$ equal roots, the correlation matrix will not necessarily have the same number. For example, if the correlation matrix is patterned such that all of the correlations are equal, the last $p - 1$ roots will be equal but those of the covariance matrix will be equal only if all of the variances are equal. This latter case could occur if one linear combination of the original variables accounted for everything except inherent variability.

Srivastava and Carter (1983) compared the vectors of both matrices for the example to be given in Section 6.6 and found that what was the second vector of the covariance matrix was closely approximated by the third vector of the correlation matrix but that the second vector of the correlation matrix was more like the fourth vector of the covariance matrix, this due in part to unequal variances among the variables.

3.3.2 PCA Operations with Correlation Matrices

In essence, what one does in working with correlation matrices is to first put all of the data in standard units; that is, perform the operation $(x - \bar{x})/s$ for each variable where $\bar{x}$ and s are obtained from the sample. These standardized data are then treated as observations. By doing this, all of the transformed variables will have unit variances and the resulting covariance matrix will actually be the correlation matrix of the original variables. While it is not necessary to transform the data into standard units to obtain a correlation matrix, it will be necessary to do this if one is to obtain the pc-scores or the V- and W-vectors need to be restated in terms of the original units of the variables.

Let

$$\mathbf{D} = \begin{bmatrix} s_1 & 0 & \cdots & 0 \\ 0 & s_2 & \cdots & 0 \\ \vdots & \vdots & \cdots & \vdots \\ 0 & 0 & \cdots & s_p \end{bmatrix}$$

that is, a diagonal matrix of standard deviations of the original variables so that the correlation matrix $\mathbf{R}$ becomes

$$\mathbf{R} = \mathbf{D}^{-1}\mathbf{S}\mathbf{D}^{-1} \tag{3.3.1}$$

Let $\mathbf{U}$, $\mathbf{V}$, and $\mathbf{W}$ be the various sets of characteristic vectors obtained from the correlation matrix. If the data were restated in standard units, then to obtain the y-scores, let

$$\mathbf{y} = \mathbf{W}'\mathbf{D}^{-1}(\mathbf{x} - \bar{\mathbf{x}}) = \mathbf{W}'\mathbf{x}^* \tag{3.3.2}$$

where $\mathbf{x}^* = \mathbf{D}^{-1}(\mathbf{x} - \bar{\mathbf{x}})$ is a data vector in standard units. The back substitution to the original units requires the following expression:

$$\hat{\mathbf{x}} = \bar{\mathbf{x}} + \mathbf{D}\mathbf{V}\mathbf{y} \tag{3.3.3}$$

For those who would prefer to use the data in its original form, the characteristic vectors require modification. Let

$$\begin{aligned} \mathbf{v}_i^* &= \mathbf{D}\mathbf{v}_i = \sqrt{l_i}\mathbf{D}\mathbf{u}_i \\ \mathbf{w}_i^* &= \mathbf{D}^{-1}\mathbf{w}_i = \mathbf{D}^{-1}\mathbf{u}_i/\sqrt{l_i} \end{aligned} \qquad i = 1, 2, \ldots, k \qquad (3.3.4)$$

which are the characteristic vectors restated for use with the original variables. Then,

$$\mathbf{y} = \mathbf{W}^{*\prime}(\mathbf{x} - \bar{\mathbf{x}}) \qquad (3.3.5)$$

and the back substitution becomes

$$\hat{\mathbf{x}} = \bar{\mathbf{x}} + \mathbf{V}^*\mathbf{y} \qquad (3.3.6)$$

which is the same as (3.3.3). The U-vectors are not modified since they are used only as a diagnostic tool. The sum of squares of residuals must also remain in standard units to be compatible with the test procedures described in Section 2.7. The equivalent of (2.7.4) is

$$Q = (\mathbf{x}^* - \hat{\mathbf{x}}^*)'\,(\mathbf{x}^* - \hat{\mathbf{x}}^*) = (\mathbf{x} - \bar{\mathbf{x}})'\mathbf{D}^{-1}(\mathbf{I} - \mathbf{V}'\mathbf{L}\mathbf{V})\mathbf{D}^{-1}(\mathbf{x} - \bar{\mathbf{x}})$$

$$(3.3.7)$$

which will also have the same asymptotic distribution. If one replaces $\mathbf{S}$ with $\mathbf{R}$, the relationships in Section 1.6 hold for $\mathbf{U}$, $\mathbf{V}$, and $\mathbf{W}$ but not for $\mathbf{V}^*$ or $\mathbf{W}^*$.

We return to the two-variable Chemical example of Chapter 1 whose covariance matrix was

$$\mathbf{S} = \begin{bmatrix} .7986 & .6793 \\ .6793 & .7343 \end{bmatrix}$$

which makes

$$\mathbf{D} = \begin{bmatrix} .8936 & 0 \\ 0 & .8569 \end{bmatrix}$$

and

$$\mathbf{D}^{-1} = \begin{bmatrix} 1.1191 & 0 \\ 0 & 1.1670 \end{bmatrix}$$

from whence, using (3.3.1), the correlation matrix becomes

$$R = \begin{bmatrix} 1 & .8871 \\ .8871 & 1 \end{bmatrix}$$

From this,

$$U = \begin{bmatrix} .7071 & -.7071 \\ .7071 & .7071 \end{bmatrix}$$

$$V = \begin{bmatrix} .9714 & -.2376 \\ .9714 & .2376 \end{bmatrix}$$

$$W = \begin{bmatrix} .5147 & -2.1041 \\ .5147 & 2.1041 \end{bmatrix}$$

(When operating with correlation matrices for the case $p = 2$, U will always have these values as it will for covariance matrices when the two variances are equal. V and W will depend on the correlation coefficient. See Section 15.3.) One nice property of PCA using correlation matrices is that the correlations between the pc's and the original variables is given by V directly rather then by equation (1.5.4) because the variances of the standardized variables are all equal to 1.0.

Should one wish to operate with the original data,

$$V^* = \begin{bmatrix} .8680 & -.2123 \\ .8324 & .2036 \end{bmatrix}$$

$$W^* = \begin{bmatrix} .5760 & -2.3546 \\ .6007 & 2.4555 \end{bmatrix}$$

From Table 1.2, the fifth observation

$$x - \bar{x} = \begin{bmatrix} 1.7 \\ 1.5 \end{bmatrix}$$

whence

$$x^* = D^{-1}(x - \bar{x}) = \begin{bmatrix} 1.9024 \\ 1.7505 \end{bmatrix}$$

$$y^* = W'x^* = W^{*'}(x - \bar{x}) = \begin{bmatrix} 1.880 \\ -.320 \end{bmatrix}$$

The test for equality of characteristic roots given in Section 2.8 has an asymptotic χ^2-distribution when applied to covariance matrices but loses this property for correlation matrices. This will be discussed in more detail in Section 4.7 where some alternatives are presented. Suffice to say, for now, the elements in a correlation matrix are *functions* of variances and covariances and hence do not follow the Wishart distribution. All of the other stopping rules listed in that section are applicable because they do not involve significance tests. Generally speaking, these procedures are not as sensitive for correlation matrices because, for a given number of variables, the roots are somewhat bounded a priori as the sum of them is equal to the number of variables. Keep in mind that techniques involving residual traces or residual variances require quantities that are no longer in the original units of the data. This may explain why the SCREE test is so popular in many fields of application.

3.3.3 Invariance of T^2

In the previous secton it was stated that there is no one-to-one correspondence between the pc's obtained using a covariance matrix and those obtained using a correlation matrix. However, if all of the pc's are employed, the two T^2's will be the same. Consider the Ballistic Missile example from Section 2.3, whose W-vectors are displayed in Table 2.3. The W-vectors from the correlation matrix are as follows:

$$\mathbf{W} = \begin{bmatrix} .272 & -.973 & 1.060 & .564 \\ .292 & -.225 & -1.660 & .596 \\ .299 & .164 & .103 & -1.966 \\ .270 & 1.042 & .616 & .964 \end{bmatrix}$$

Consider a data vector as difference from the mean of $(15, 10, 20, -5)$. The y-scores using the covariance matrix are $(1.09, -1.74, .56, -3.43)$ and for the correlation matrix are $(1.16, -1.80, .10, -3.43)$. However, $T^2 = 16.32$ for both of them. (Equation 1.7.2 is also invariant.) Keep in mind that this holds *only* if all of the pc's are used. Since the pc's are different, the T^2's obtained from a subset of pc's will not be equal.

3.3.4 A Historical Example

The following example is from Hotelling's original paper on principal components (Hotelling, 1933) and is based on data of Kelley (1928). The data consist of four tests out of a test battery given to 140 seventh-grade children. The four tests selected were: x_1 = reading speed, x_2 = reading power, x_3 = arithmetic speed and x_4 = arithmetic power. The correlation matrix for these data is given in Table 3.1. The U- and V-vectors along with the characteristic roots are shown in Table 3.2. The first pc (46%), not unexpectedly

Table 3.1. Seventh Grade Tests (Kelley–Hotelling). Correlation Matrix

	Reading		Arithmetic	
	Speed x_1	Power x_2	Speed x_3	Power x_4
Reading speed	1	.698	.264	.081
Reading power	.698	1	−.061	.092
Arithmetic speed	.264	−.061	1	.594
Arithmetic power	.081	.092	.594	1

Source: Hotelling (1933).

Table 3.2. Seventh-Grade Tests. Characteristic Roots, U- and V-Vectors

	u_1	u_2	u_3	u_4
x_1	.599	.366	−.402	.588
x_2	.507	.516	.398	−.564
x_3	.450	−.553	−.521	−.470
x_4	.427	−.542	.639	.340
Characteristic Root	1.847	1.464	.522	.166
	v_1	v_2	v_3	v_4
x_1	.814	.443	−.291	.240
x_2	.689	.624	.288	−.230
x_3	.611	−.670	−.376	−.191
x_4	.581	−.656	.462	.139

represents overall performance while the second pc (37%), represents the difference between reading and arithmetic tests. These two pc's account for 83% of the variability of the standardized variables. The third pc (13%) is the difference between speed and power tests and the last pc, accounting for only 4% represents the only possible contrast left—the interaction between reading–arithmetic and speed–power. Because of its size and the interpretability of the pc's, this will be added to our stable of examples.

3.3.5   Other Properties of Correlation Matrix PCA

In Section 1.5.3, the generalized variance was defined as $|S|$. A corresponding quantity exists for the correlation matrix. $|R|$ is sometimes referred to as the *scatter coefficient* (Frisch, 1929). This coefficient is bounded between 0 (all of the variables are perfectly correlated) and p (all of the variables are uncorrelated). Also available is the Gleason–Staelin descriptor (2.8.1) and the index of a matrix, $\sqrt{l_1/l_p}$.

Kullback (1959, p. 197) determined the following statistics for the correlation matrix: The *information* statistic is

$$I(1:2) = -\tfrac{1}{2} \ln (|\mathbf{R}|) = -\tfrac{1}{2} \sum_{i=1}^{p} \ln (l_i) \qquad (3.3.8)$$

which will be a minimum when the variables are all uncorrelated so that the roots will all be 1.0 and $I(1:2) = 0$. The *divergence* statistic is

$$J(1:2) = \tfrac{1}{2} \operatorname{Tr} (\mathbf{R}^{-1}) - \frac{p}{2} = \sum_{i=1}^{p} \frac{1 - l_i}{(2l_i)} \qquad (3.3.9)$$

which will also equal zero when the roots are all equal.

It is interesting to note that the U-vectors of a correlation matrix are not a function of the correlations themselves but of their ratios. If all of the off-diagonals of a correlation matrix were multiplied by the same constant, these vectors would remain the same. For instance, if the correlation coefficients in Table 3.1 are all multiplied by 1.3, viz.,

$$
\begin{bmatrix}
1 & .9074 & .3432 & .1053 \\
.9071 & 1 & -.0793 & .1196 \\
.3432 & -.0793 & 1 & .7722 \\
.1053 & .1196 & .7722 & 1
\end{bmatrix}
$$

the U-vectors will remain the same as before. Nothing else stays the same, however. The characteristic roots are now 2.101, 1.603, .379, and $-.084$ (the matrix is no longer positive definite, in this example) and so the V- and W-vectors will be affected.

3.3.6 Other Correlation Matrices

In the study of genetics, the starting matrices may be made up of genetic covariances or correlations. These matrices are not positive semidefinite so some of the characteristic roots may be negative. This requires some special handling and interpretation (Rouvier, 1966, 1969; Hashiguchi and Morishima, 1969). Correlation matrices have also been made up of tetrachoric correlations. The assumptions underlying their use, particularly with respect to factor analysis, were discussed by Muthén and Hofacker (1988). Partial correlations have also been used in place of simple correlations.

3.4   DATA ARE NOT SCALED AT ALL: PRODUCT OR SECOND MOMENT MATRICES

3.4.1   Introduction

In this case, the data are not adjusted in any way, either with regard to mean or variance. The matrix from which the characteristic vectors are obtained is merely the *product* or *second moment* matrix: $X'X/n$. (There is a somewhat parallel situation in regression where the regression line is forced through the origin rather than through the intersection of the means.) The most widely used applications of this scaling (lack of scaling, actually) is in the field of chemistry, in particular, with additive models. In some situations, the concentration of a particular chemical in a solution cannot be measured directly but instead a sample of the solution is analyzed on a spectrophotometer that measures the absorbence of the solution at various wavelengths or frequencies. The chemist must then devise some technique for estimating the concentration from these absorbences. If the solution contains only one compartment, the absorption often is directly related to the concentration (Beer's Law). This relationship could be determined by producing and analyzing several solutions having known concentrations. That done, the concentrations of unknown solutions may be estimated from the spectrophotometric curves.

3.4.2   Some Applications

Among studies using product matrices is one by Wernimont (1967) designed to study the results of spectrophotometers. Three different solutions of potassium dichromate were measured on 16 different instruments on two different days. The absorbences were obtained at $p = 20$ wavelengths. The first pc represented overall curve shape and amounts of this pc represented shifts in the absorbency curve and were used to test Beer's Law. The second pc was related primarily to instrumentation variability and was used to check for deviations in wavelength calibrations.

Sylvestre et al. (1974) carried out an experiment in which each solution was made up of two chemical components in varying amounts and used PCA to resolve the curves. To remove the mean would have removed the main shape characteristic of these curves and, although they could have been modeled in this manner, the product matrix approach seemed more simple and direct. The first pc again represented overall curve shape while the second was a contrast between the absorbences at the higher wavelengths and those at the lower wavelengths. Addition of higher concentrations of one of the chemical components added a third pc that was related to a reaction product. Spjøtvoll et al. (1982) included a restriction that the two chemical concentrations should equal a constant, which resulted in one pc instead of two. Windig et al. (1987) also required the concentrations to add to a constant but used the "correlation matrix about the origin" in which each variable was transformed by using $x_i^* = x_i/\sqrt{\Sigma x_i^2}$. They used a technique called a *Vardia* plot to identify pure components. (One must take care in performing PCA when the variables add to a constant. We shall return to this problem in Section 16.8.) Ohta (1973) gave an example with three chemical components. Cochran and Horne (1977) used PCA of product matrices to determine a lower bound for the number of related solutes and used covariance matrices to obtain estimates of linear dependence of time rates on concentration.

3.4.3 Some Properties of Product Matrices

One of the drawbacks of using product matrices is that virtually no inferential procedures associated with them exist at the present time. Two results of Corballis (1971) should be noted:

1. If the covariance matrix is of full rank, then the product matrix will also be of full rank. If the covariance matrix is of rank $r < p$, then the product matrix will generally have rank $r + 1$.
2. If $n < p$ (this does happen in some chemical applications), and the product matrix has rank n, then the covariance matrix will have rank $= n - 1$.

Another potential problem is that the first characteristic root often accounts for over 99% of the total variability so that other relevant components may account for a fraction of 1%. If the data are all positive, the first pc often represents the mean of the data and the further from the origin the mean is, the larger will be this root relative to the others.

3.4.4 A Numerical Example

To see how product matrices may be employed, consider the data set displayed in Table 3.3. These are artificial data representing the absorbance curves for 10

Table 3.3. Absorbance Data

Obs. No.	Wavelength Number						
	1	2	3	4	5	6	7
1	.5	1.0	1.5	1.0	.5	1.5	2.5
2	1.0	2.0	3.0	2.0	1.0	3.0	5.0
3	1.5	3.0	4.5	3.0	1.5	4.5	7.5
4	2.0	4.0	6.0	4.0	2.0	6.0	10.0
5	2.5	5.0	7.5	5.0	2.5	7.5	12.5
6	3.0	6.0	9.0	6.0	3.0	9.0	15.0
7	3.5	7.0	10.5	7.0	3.5	10.5	17.5
8	4.0	8.0	12.0	8.0	4.0	12.0	20.0
9	4.5	9.0	13.5	9.0	4.5	13.5	22.5
10	5.0	10.0	15.0	10.0	5.0	15.0	25.0

Table 3.4. Absorbance Data. Product Matrix

9.625	19.250	28.875	19.250	9.625	28.875	48.125
	38.500	57.750	38.500	19.250	57.750	96.250
		86.625	57.750	28.875	86.625	144.375
			38.500	19.250	57.750	96.250
				9.625	28.875	48.125
					86.625	144.375
						240.625

different samples measured at seven wavelengths. The product or second moment matrix is displayed in Table 3.4. All of the data vectors are proportional to each other and there is no error in this set of data, hence there is only one nonzero characteristic root, $l_1 = 510.125$. The various normalizations of the first characteristic vector are:

$$
\mathbf{u}_1 = \begin{bmatrix} .1374 \\ .2747 \\ .4121 \\ .2747 \\ .1374 \\ .4121 \\ .6868 \end{bmatrix} \quad
\mathbf{v}_1 = \begin{bmatrix} 3.1024 \\ 6.2048 \\ 9.3073 \\ 6.2048 \\ 3.1024 \\ 9.3073 \\ 15.5121 \end{bmatrix} \quad
\mathbf{w}_1 = \begin{bmatrix} .0061 \\ .0122 \\ .0182 \\ .0122 \\ .0061 \\ .0182 \\ .0304 \end{bmatrix}
$$

Note that these vectors represent the general shape of the data curves because the mean has not been removed.

Since these operations are about the origin rather than the mean,

$$y_1 = \mathbf{x}\mathbf{w}_1$$

and

$$\hat{\mathbf{x}} = y_1 \mathbf{v}'_1$$

For this particular set of data $\mathbf{x} - \hat{\mathbf{x}} = 0$ since there is no error.

Since the absorbences must be positive and all of the coefficients of $\mathbf{w}_1$ are positive, the y-scores must also be positive. (This would not be true of other pc's if they were to be added to the model in real-life situations.) Keep in mind that these pc's are principal components in name only; they do not have the statistical properties of pc's obtained by operating about the mean.

If Beer's Law held, the pc's should be linearly related to the concentration, c, viz., $y_1 = a + bc$. If the process were being controlled to certain limits for concentration, say c_1 and c_2, then y_1 should be held to between $a + bc_1$ and $a + bc_2$. If one wished to predict concentration, given a particular sample data vector, then

$$c = \frac{y_1 - a}{b} = \frac{\mathbf{xw}_1 - a}{b}$$

Suppose Beer's Law did not hold, but rather, say, $y_1 = ae^c$ or $y_1 = \ln(c - a)$. Then the concentrations could be predicted by $c = \ln(\mathbf{xw}_1/a)$ or $c = a + e^{\mathbf{xw}}$, respectively. In actual practice, there will be experimental error so that one pc will not explain the variability completely. There may also be more than one real component in the model, either because of instrumentation or other variability as mentioned above or because there are two or more chemical constituents in the solution. The analysis of curves such as these is often referred to as the *growth curve* problem, to which we shall return in Section 16.7.2.

3.5 DOUBLE-CENTERED MATRICES

Another scaling technique used for some specific applications is "double centering", in which not only are the variable (column) means subtracted out of the X-matrix but the row means as well. (Actually, some forms of this add the grand mean back in so that the resultant matrix resembles the residual matrix of a two-way table in the analysis of variance.) We shall see versions of this technique in multidimensional scaling (Chapters 10 and 11) as well as an adjunct to the univariate analysis of variance in Section 13.7.

3.6 WEIGHTED PCA

There are two basic ways in which weighted PCA may be carried out: (1) weight the variables and (2) weight the observations. Weighting variables has already been illustrated for correlation matrices in Section 3.3, where each variable was weighted inversely proportionally to its variance so that the weighted variables all have unit variances. Any other weighting of the variables would have the same form as (3.3.2) where $\mathbf{D}$ is now replaced by some other appropriate diagonal matrix. In the chemical example, for instance, if one wished to give x_1 a weight of 1.0 and x_2 a weight of 2.0, then

$$\mathbf{D} = \begin{bmatrix} 1 & 0 \\ 0 & 2 \end{bmatrix}$$

An example of weighted PCA in kinetics experiments may be found in Cochran and Horne (1977).

Meredith and Millsap (1985) gave a rather formal treatment of weighting and listed a number of schemes that might be considered. Let $\mathbf{D}_1$ represent a diagonal matrix made up of the diagonal elements of Σ, let $\mathbf{D}_2$ be another diagonal matrix made up of the diagonal elements of Σ^{-1} and let $\mathbf{D}_3$ be a diagonal matrix made up of error variances, possibly estimated separately or with each element taken as the average error variance. Among the schemes they considered were:

$$\mathbf{G}_1 = \alpha \mathbf{D}_1^{-1} \tag{3.6.1}$$

$$\mathbf{G}_2 = \alpha \mathbf{D}_2^{-1/2} \, \Sigma \mathbf{D}_2^{-1/2} \tag{3.6.2}$$

$$\mathbf{G}_3 = \alpha \mathbf{D}_3^{-1/2} \, \Sigma \mathbf{D}_3^{-1/2} \tag{3.6.3}$$

In each case α must be a positive quantity. The $\mathbf{G}_i$ matrices are all diagonal. If $\alpha = 1$ and Σ is replaced by S, then $\mathbf{G}_1^{1/2}$ will be the weights used in the correlation matrix procedures in Section 3.3. $\mathbf{G}_2$ will produce the image analysis procedure to be given in Section 18.2. $\mathbf{G}_3$ would be useful in weighting inversely proportionally to the expected error variability.

Gabriel and Zamir (1979) also proposed some weighting schemes, particularly with regard to singular value decomposition, which will be introduced in Chapter 10. In this context, one could weight either variables or observations.

In weighting individual observations, the operation is similar. Let $\mathbf{D}$ now be an $n \times n$ diagonal matrix relating to the weights for each observation in obtaining the sums of squares and crossproducts. To simplify things, assume that $\mathbf{X}$, $(n \times p)$ has mean zero. Then the weighted covariance matrix will be $\mathbf{S}_w = \mathbf{X}'\mathbf{D}\mathbf{X}/\text{Tr}(\mathbf{D})$. If the rank of $\mathbf{X}$ is p and the rank of $\mathbf{D}$ is $r \leqslant n$, the rank of $\mathbf{S}_w$ will be the minimum of p and r and holds for the case where $\mathbf{D}$ is not a diagonal matrix (Okamoto, 1973).

Suppose we wished to *geometrically* weight the 15 observations in the chemical example of Chapter 1. Geometric weights (also called *exponential smoothing*) are of the form $c(1 - c)^{n-m}$ where $0 < c < 1$, n is the sample size, and m is the order of the observation in the sample. In this example, $n = 15$ so for the fifteenth observation, $m = 15$ and the weight will be c. For the fourteenth observation it will be $c(1 - c)$ and the other weights back to the start will be $c(1 - c)^2$, $c(1 - c)^3, \ldots, c(1 - c)^{14}$. This scheme is used when one wishes to obtain weighted averages, or in this case a weighted sum of squares and crossproducts, where the weights decay *exponentially* back through time. c is usually fairly small, .2 or less (Brown, 1963, p. 106). For this example, using $c = .2$, the weights will be .2, .16, .128, .1024, .0819, .0655, .0524, .0419, .0336, .0268, .0215, .0172, .0137, .0110, and .0088. The sum is .9647. These become the diagonal elements of $\mathbf{D}$ in reverse order (i.e., $d(1, 1) = .0088$). In actual practice, one would probably obtain a weighted mean as well, but to keep this example simple, we shall use the deviations from the mean given in Table 1.2. From these data,

$$\mathbf{S}_w = \begin{bmatrix} .9595 & .7781 \\ .7781 & .7736 \end{bmatrix}$$

the characteristic roots are 1.6502 and .0829 and

$$\mathbf{U} = \begin{bmatrix} .7479 & -.6638 \\ .6638 & .7479 \end{bmatrix}$$

which is quite similar to the results for the unweighted data in Chapter 1, implying that the data did represent random variability.

3.7 COMPLEX VARIABLES

The occasion may arise when one may be working with complex variables. If that is the case, the covariance matrix will be a *Hermitian* matrix (i.e., a matrix equal to its conjugate transpose, see Appendix A) and instead of $\mathbf{S} = \mathbf{VV}'$, it will be $\mathbf{H} = \mathbf{VV}^{*\prime} = \mathbf{ULU}^{*\prime}$ where $\mathbf{V}^*$ and $\mathbf{U}^*$ are the complex conjugates of $\mathbf{V}$ and $\mathbf{U}$. The characteristic roots will be real.

The most likely field of application for complex PCA is in time series analysis, which will be discussed in Section 14.6. However, complex PCA can pop up in unexpected places. Gulliksen (1975) was describing a number of one-dimensional scaling techniques in a unified matrix form. Some paired comparison matrices are skew-symmetric (diagonal elements equal zero, $a_{ij} = -a_{ji}$) with rank 2. Both characteristic roots are imaginary, one positive and one negative, and the characteristic vectors are complex conjugates.

If facilities for operating with Hermitian matrices are not available, one may, instead, work with the matrix

$$\mathbf{H}^A = \begin{bmatrix} \text{Re}(\mathbf{H}) & \vdots & -\text{Im}(\mathbf{H}) \\ \hdashline \text{Im}(\mathbf{H}) & \vdots & \text{Re}(\mathbf{H}) \end{bmatrix}$$

where $\text{Re}(\mathbf{H})$ and $\text{Im}(\mathbf{H})$ are the real and imaginary parts of $\mathbf{H}$. If $\mathbf{H}$ is $p \times p$, then $\mathbf{H}^A$ will be $2p \times 2p$. If the characteristic roots of $\mathbf{H}$ are $l_1, \ldots, l_p$ with corresponding vectors $\mathbf{u}_1, \ldots, \mathbf{u}_p$, then p roots of $\mathbf{H}^A$ will also be $l_1, \ldots, l_p$. The first characteristic vector will now have $2p$ elements, the first p being equal to the real parts of $\mathbf{u}_1$ and the second p being the imaginary parts. The remaining $p - 1$ vectors follow the same pattern. There will be another p vectors having the same roots as before, but now the first p elements will be equal to $-\text{Im}(\mathbf{u}_j)$ and the last p equal to $\text{Re}(\mathbf{u}_j)$.

Consider the Hermitian matrix

$$\mathbf{H} = \begin{bmatrix} 4 & 2 + i \\ 2 - i & 3 \end{bmatrix}$$

The augmented matrix is

$$H^A = \begin{bmatrix} 4 & 2 & 0 & -1 \\ 2 & 3 & 1 & 0 \\ 0 & 1 & 4 & 2 \\ -1 & 0 & 2 & 3 \end{bmatrix}$$

The characteristic roots of H^A are $l_1 = l_2 = 5.7913$ and $l_3 = l_4 = 1.2087$.

$$U^A = \begin{bmatrix} .6790 & -.3848 & .5586 & -.2809 \\ .6244 & -.0324 & -.7895 & .0017 \\ .3848 & .6970 & .2809 & .5586 \\ .0324 & .6244 & -.0017 & -.7805 \end{bmatrix}$$

Note that $u_{11}^A = u_{32}^A$ and $u_{21}^A = u_{42}^A$—these are the real parts of $\mathbf{u}_1$ and $u_{31}^A = -u_{12}^A$ and $u_{41}^A = -u_{22}^A$—these are the imaginary parts of $\mathbf{u}_1$. The last two columns of U^A have the same properties for $\mathbf{u}_2$. This means that

$$U = \begin{bmatrix} .6790 + .3848i & .5586 + .2809i \\ .6244 + .0324i & -.7805 - .0017i \end{bmatrix}$$

and

$$V = \begin{bmatrix} 1.6341 + .9259i & .6141 + .3088i \\ 1.5026 + 0.780i & -.8580 - .0019i \end{bmatrix}$$

The complex conjugate of V, V^* is

$$= \begin{bmatrix} 1.6341 - .9259i & .6141 - .3088i \\ 1.5026 - .0780i & -.8580 + .0019i \end{bmatrix}$$

and from this, one may verify that $V^{*'}V = L$ and $VV^{*'} = H$.

Nash (1979, p. 94) recommended that for descriptive purposes, the complex vectors be normalized so that the largest element in each column is equal to 1.0. His reasoning was that a characteristic vector multiplied by a constant is still a characteristic vector and that holds for complex constants. Multiplying real vectors by real constants, as one does in going from U to V or W, leaves the general appearance of the matrix unchanged. However, multiplying complex vectors by complex constants may leave an entirely different-appearing result. Nash's recommendation would make the dominant element real and hence be less affected than the other coefficients. For the present example, U could be renormalized to be

$$
\begin{bmatrix}
1 & -.7165 - .3583i \\
.7165 - .3583i & 1
\end{bmatrix}
$$

More details may be found in Goodman (1963), Brillinger (1975), or Chow (1975). The distributions of these roots and vectors will be considered in Chapter 4.

CHAPTER 4

Inferential Procedures

4.1 INTRODUCTION

Considerable energy has been expended in the area of statistical inference related to PCA, and well it should be. It is an important topic, often ignored by practitioners, and one in which considerable work is still required. The distributions related to characteristic roots are extremely complicated and for the most part one will have to be satisfied with asymptotic results or exact distributions for a few specific cases.

Some of these inferential topics have already been introduced in Chapter 2 but this present chapter will treat them in more detail as well as introduce some other useful techniques. Topics will include point and interval estimates, tests of equality of characteristic roots, procedures involving functions of roots, and procedures related to the characteristic vectors. We shall see, in Section 4.7, that many of these results are different if the PCA is carried out on a correlation matrix. It is unfortunate that there are considerably more usable results for PCA related to covariance matrices than for correlation matrices since the majority of PCA applications are carried out on the latter.

4.2 SAMPLING PROPERTIES OF CHARACTERISTIC ROOTS AND VECTORS

4.2.1 Basic Properties

The large-sample properties associated with the characteristic roots and vectors, under the assumption that the original variables have a multivariate normal distribution, are as follows:

a. The maximum likelihood estimate of the ith population characteristic vector Υ_i is $\mathbf{u}_i$ (Girshick, 1936). (U-vectors are the same whether the divisor for the covariance matrix is n or $n - 1$.)

b. The maximum likelihood estimate of the ith characteristic root λ_i is $[(n-1)/n]l_i$ (Anderson, 1984a). This is because the maximum likelihood estimate of the population covariance matrix, Σ, has n not $n-1$ for a divisor.

c. The characteristic roots l_i are distributed independently of their associated characteristic vectors (Girshick, 1939). This is somewhat analogous to the fact that the mean and variance of a normal distribution are independent of each other.

d. The variance of l_i is asymptotically $2\lambda_i^2/n$ (Girshick, 1939). Therefore, the quantity $\sqrt{n}(l_i - \lambda_i)$ is normally distributed with mean zero and variance $2\lambda_i^2$. This means that the asymptotic 95% confidence limits on the roots would be of the form

$$\frac{l_i}{1 + 1.96\sqrt{2/n}} < \lambda_i < \frac{l_i}{1 - 1.96\sqrt{2/n}} \tag{4.2.1}$$

In the chemical example, where $n = 15$, $l_1 = 1.4465$, and $l_2 = .0864$, the asymptotic 95% confidence limits are the following:

$$.84 < \lambda_1 < 5.09$$

$$.05 < \lambda_2 < .30$$

Since the variance of l_i involves λ_i, which usually requires the sample estimate l_i to be substituted for it, Girshick suggested the use of $\ln(l_i)$, which has a variance of $2/n$. Confidence limits for λ in that case would be

$$.71 < \lambda_1 < 2.96$$

$$.04 < \lambda_2 < .18$$

which are shifted downward and considerably narrower. However, these are really limits for $\ln(\lambda_i)$ that have been converted back to λ_i.

e. The sample roots are asymptotically independent of each other subject to the constraint that the sum of the roots is equal to the sum of the variances (Girshick, 1939).

f. The elements $(\mathbf{u}_1 - \Upsilon_i)$ are normally distributed with mean $\mathbf{0}$ and covariance matrix (Girshick, 1939)

$$\frac{\lambda_i}{n} \sum_{h \neq i}^{p} \frac{\lambda_h}{(\lambda_h - \lambda_i)^2} \Upsilon_h \Upsilon_h' \tag{4.2.2}$$

To obtain this matrix in practice, the sample roots and vectors must usually be substituted for Λ and Υ.

g. The asymptotic standard error for u_{gi} (the gth element of the ith vector), with sample estimates substituted, is

$$\sqrt{\frac{l_i}{n} \sum_{h \neq i}^{p} \frac{l_h}{(l_h - l_i)^2} u_{gh}^2} \qquad (4.2.3)$$

For the Chemical example, the coefficients and their standard errors are

	Coefficient	Standard Error
u_{11}	.7236	.0463
u_{21}	.6902	.0486
u_{12}	−.6902	.0486
u_{22}	.7236	.0463

This is a quantity that needs to be exploited. We are sophisticated enough to quote the standard error of a regression coefficient along with the coefficient itself; why not do the same for characteristic vectors?

It is common practice to use only the first k vectors in equations (4.2.2) and (4.2.3). This underestimates the standard errors and will cause similar problems with the covariances, but the bias will be small if the variability explained by the first k vectors is large.

h. The covariance of the gth element of $\mathbf{u}_i$ and the hth element of $\mathbf{u}_j$, $i \neq j$, is (Girshick, 1939)

$$\frac{-\lambda_i \lambda_j \Upsilon_{gj} \Upsilon_{hi}}{n(\lambda_i - \lambda_j)^2} \qquad (4.2.4)$$

Again, sample estimates must generally be substituted. Equations (4.2.2) and (4.2.4) allow us to obtain the correlation between the coefficients both within and among vectors. (Independence of two sample pc's does not imply independence of the coefficients of their vectors. The coefficients within a vector are constrained so that $\mathbf{u}_i'\mathbf{u}_i = 1$ and between any two vectors so that $\mathbf{u}_i'\mathbf{u}_j = 0$.)

The elements of (4.2.2) and (4.2.4) form a $p^2 \times p^2$ matrix although, in practice, it is usually reported only for the k vectors retained, which would reduce it in size to $kp \times kp$. Some properties of this matrix are given in Flury (1988, p. 31). Although this matrix will be employed in this book as a descriptive tool, it is also useful in constructing tests of hypotheses about characteristic vectors.

4.2.2 Biases

Many of the results in the previous section were asymptotic. The maximum likelihood estimates for λ_i are large-sample estimates and they are biased, even beyond the usual factor $(n - 1)/n$. Lawley (1956) showed that

$$E(l_i) = \lambda_i \left[1 + \frac{1}{n} \sum_{j \neq i}^{p} \left(\frac{\lambda_j}{\lambda_i - \lambda_j} \right) \right] + O(1/n^2) \tag{4.2.5}$$

The general effect is that the larger roots will be too large and the smaller roots will be too small. Furthermore, the variance of l_i is

$$V(l_i) = \frac{2\lambda_i^2}{n} \left[1 - \frac{1}{n} \sum_{j \neq i}^{p} \left(\frac{\lambda_j}{\lambda_i - \lambda_j} \right)^2 \right] + O(1/n^3) \tag{4.2.6}$$

and the covariance between l_i and l_j is

$$\frac{2}{n^2} \left(\frac{\lambda_i \lambda_j}{\lambda_i - \lambda_j} \right)^2 + O(1/n^3) \tag{4.2.7}$$

The net effect of these is that the variances are understated and the characteristic roots are correlated. However, all of these biases and correlations decrease as the sample size increases.

To correct the roots for these biases, one may restate (4.2.5) and substitute l_i for λ_i whereever appropriate. This produces

$$\hat{l}_i = l_i \left[1 - \frac{1}{n} \sum_{j \neq i}^{p} \left(\frac{l_j}{l_i - l_j} \right) \right] \tag{4.2.8}$$

G. A. Anderson (1965) derived a similar expression using reciprocals of the roots as estimators. Dey (1988) estimated all of the roots under a sum of squared error loss and obtained an estimator that shrank or expanded the sample roots towards their geometric mean.

Table 4.1 contains, for the Ballistic Missile example, the sample characteristic roots, the corrected roots using (4.2.8) both for the sample size of $n = 40$ used in that example and, for comparison, for $n = 100$. Note that the first two

Table 4.1. Ballistic Missile Example. Biases of Estimates

| i | l_i | Equation (4.2.8) | | Standard error of l_i using | |
		$n = 40$	$n = 100$	$\sqrt{2l_i^2/n}$	Equation (4.2.6) $n = 40$
1	335.34	332.70	334.29	74.98	74.95
2	48.03	46.92	47.59	10.74	10.18
3	29.33	31.09	30.03	6.56	5.74
4	16.41	18.40	17.20	3.67	3.25

corrected roots are smaller than the originals and the last two roots are larger. For a sample size of $n = 100$, the corrections are becoming minimal. The sum of the corrected roots is still equal to the trace of the covariance matrix S. Also included in Table 4.1 are the standard errors for the l_i using both $\sqrt{2l_i^2/n}$ and the square root of (4.2.6). These discrepancies get larger as the roots get smaller but for larger problems these latter pc's would probably not be retained in the model anyhow. The sample covariance between l_1 and l_2 is 3.9283, making the correlation between them less than .01, and even the correlation between l_3 and l_4 is only .09; clearly not a problem.

Equation (4.2.8) involves *all* of the characteristic roots. If there are a large number of variables and only a small number of pc's, k, are to be retained, Lawley (1956) proposed the following approximation:

$$\hat{l}_i = l_i \left[1 - \frac{1}{n} \sum_{j \neq i}^{k} \left(\frac{l_j}{l_i - l_j} \right) - \left(\frac{p - k}{n} \right) \left(\frac{T}{l_i - T} \right) \right]$$ (4.2.9)

where T is equal to the sum of the roots of the deleted pc's divided by $p - k$ (i.e., the average of the deleted roots). The sum of the deleted roots may be obtained by subtracting the first k roots from the trace of the covariance matrix. For the Ballistic Missile example, with $k = 2$, the corrected estimates of the first two roots, using (4.2.9), are 332.69 and 47.25, which differ little from Table 4.1. The variances for these estimates do involve all of the roots and are the same as (4.2.6) except that the sign for the bias is now positive since it is being added on.

In conclusion, the message from this section is that these estimates do have biases, and for small sample sizes one may need to take them into account, particularly if they are being used for model estimation. In that case, one may also be interested in the effect on skewness and kurtosis of the distribution of small-sample estimates of λ_i. For this, one will need the third and fourth cumulants, which are (Lawley, 1956)

$$k_3(l_i) = 8\lambda_i^3/n^2 + O(1/n^3)$$
$$k_4(l_i) = 48\lambda_i^4/n^3 + O(1/n^4)$$ (4.2.10)

4.2.3 Bounds on the Roots

Much elaborate work has gone into obtaining bounds for the characteristics roots, see for example, Brauer (1953), Breckenbach and Bellman (1965, Chap. 2), and Marshall and Olkin (1979, Chap. 9). Many of these results are interesting from a mathematical point of view but will be of less use to the practitioner unless the values of the roots are an integral part of a model of some kind. Some more simple bounds are of interest as they may be used in making a cursory inspection of the sample covariance matrix.

The most simple bounds are

$$l_1 \geqslant \max(s_i^2)$$
$$l_p \leqslant \min(s_i^2)$$

(4.2.11)

for $i = 1, 2, \ldots, p$. The equalities hold when the variables are all uncorrelated. A more useful bound for both covariance and correlation matrices is

$$l_1 \leqslant \max \sum_{j=1}^{p} |s_{ij}| \quad i = 1, 2, \ldots, p$$

(4.2.12)

In other words, obtain the row sums of the absolute values of S (or R). The largest root will not exceed the greatest of these sums. This is a quick way of ascertaining whether PCA is worthwhile. This bound works best for correlation matrices and covariance matrices with equal variances (Morrison, 1976).

For the Ballistic Missile example, the row sums of S are 312.51, 398.00, 307.59, and 302.57. The maximum is 398.00, which is an upper bound for l_1. The maximum variance is 142.74, which is a lower bound for l_1. The minimum variance is 84.57, which is an upper bound on l_4. Not a whole lot of help but better than nothing.

If the true covariance matrix Σ is of full rank, then the smallest sample root, l_p will be bounded away from zero (Yin et al., 1983; Silverstein, 1984). This means that you cannot get a zero root by chance alone; if you get one, such as in Section 2.4, there is a reason for it. (However, if $\lambda_p = 0$, l_p may not be zero due to rounding.)

4.3 OPTIMALITY

Although not, strictly speaking, a topic in statistical inference, we are due some remarks that place PCA on a firm foundation. Many of these have been mentioned already but it may be convenient to formally list them all in one place. Many of these are due to Okamoto and Kanazawa (1968) and Okamoto (1969).

a. For any linear combination of X, Xa, such that $a'a = 1$, the coefficients of a that will maximise the explained variability of X is the first characteristic vector, u_1. No other linear combination will account for more.

This book is concerned with operations, not theorems and proofs but if one is worthwhile sketching out, it is this one. Let

$$z = (x - \bar{x})a$$

Then $V(z) = a'Sa$ where S is the covariance matrix of the original variables. We wish to find a such that $a'Sa$ is maximized subject to the constraint that

$a'a = 1$. To do this, use Lagrange multipliers and maximize

$$a'Sa - l(a'a - 1) \tag{4.3.1}$$

where l is a Lagrange multiplier.

Differentiating this with respect to a' yields

$$Sa - la = [S - lI]a$$

which is the same as equation (B.6) with $a = u_1$. Therefore, the explained variance is maximized by the first principal component.

As a corollary, the last principal component accounts for less than any other nonzero linear combination. Put another way,

$$l_1 = \max\left(\frac{a'Sa}{a'a}\right) \quad \text{and} \quad l_p = \min\left(\frac{a'Sa}{a'a}\right)$$

(T. W. Anderson, 1965)

b. Let A_k, $k = 1, 2, \ldots, p - 1$ be a $k \times p$ matrix subject to $A_k'A_k = I$ and $A_k'SA_k$ is diagonal. Then, the variance of XA_k will be maximized for the first k characteristics vectors of S, U_k. These two properties imply that you can do no better than use the first pc or the first k pc's. No other combination is better for any k. Explained variability holds either in terms of the trace of the explained matrix or its determinant.

c. As a corollary, for fixed k, the trace of the unexplained or residual matrix is minimized by the first k pc's (Darroch, 1965; Hudlet and Johnson, 1982). If Q as defined by (2.7.4) is computed for each row of X, this quantity will be

$$Tr[S - U_k'SU_k] = \sum_{i=1}^{n} Q_i \tag{4.3.2}$$

Rao (1964) showed that the norm of the residual matrix is also minimized.

d. For fixed k, the first k pc's will have maximum correlation with the original variables of any set of k predictors. Obenchain (1972) investigated this property for some weighting schemes.

4.4 TESTS FOR EQUALITY OF CHARACTERISTIC ROOTS

4.4.1 Tests of the Hypothesis that a Subset of the Characteristic Roots are Equal

In Section 2.6, a large-sample test was given for testing the hypothesis that the last $p - k$ roots were equal (2.6.1). This is a form of Bartlett's test due to Anderson (1963). Actually, this is a fairly general form and will handle more situations than initially described in that section. Specifically, this procedure will allow for the fact that the population roots are not distinct, (i.e., $\lambda_1 \leqslant \lambda_2 \leqslant \cdots \leqslant \lambda_p$) and can be used to test *any* subset of consecutive roots for equality, not necessarily the last $p - k$. Let us assume that we wish to test a subset of b roots, given that k pc's are already in the model. The null hypothesis is

$$H_0: \lambda_{k+1} = \lambda_{k+2} = \cdots = \lambda_{k+b}$$

For this hypothesis, (2.6.1) becomes

$$v\left[-\sum_{i=k+1}^{k+b} \ln (l_i) + b \ln\left(\sum_{i=k+1}^{k+b} \frac{l_i}{b} \right) \right] \tag{4.4.1}$$

where v denotes the degrees of freedom associated with the sample co-variance matrix. Expression (4.4.1) has an asymptotic χ^2-distribution with $(b - 1)(b + 2)/2$ d.f. If this subset happens to be the last $p - k$ roots, then $b = p - k$ and this expression reduces to (2.6.1). Fujikoshi (1977) replaced the constant v with a more complicated expression involving the population roots and showed that this had a limiting χ^2-distribution.

If one is willing to assume that the population roots *are* distinct and wishes to test the last $p - k$ roots for equality, a form due to Lawley (1956) may be used. Let

$$c = (v - k) - (1/6)\{2(p - k) + 1 + (2/[p - k])\} + \overline{l}^2 \sum_{i=1}^{k} \frac{1}{(l_i - \overline{l})^2}$$

where $\overline{l} = (l_{k+1} + \cdots + l_p)/(p - k)$.

Lawley's statistic is the same as (2.6.1) with v being replaced by c. This statistic has an asymptotic χ^2-distribution with $(p - k - 1)(p - k + 2)/2$ d.f. as before. Consider the Ballistic Missile example also given in Section 2.6. The difference between Anderson's and Lawley's procedures can be seen in Table 4.2. The Lawley statistic will generally be smaller than that of Anderson. For

Table 4.2. Ballistic Missile Example. Comparison of Anderson and Lawley Procedures for Testing Characteristic Roots

Characteristic Root	d.f.	Test that Last k Roots are Equal	
		Lawley	Anderson
$l_1 = 335.34$	9	106.2	110.7
$l_2 = 48.03$	5	10.2	10.9
$l_3 = 29.33$	2	3.0	3.2
$l_4 = 16.41$	—	No test	

the firm example in Section 2.9, Lawley's statistic is larger for only the test of the last two roots. Lawley's procedure is valid even if the first k roots are not distinct as long as they are distinct from the remaining $p - k$ roots (Radcliff, 1964).

It was mentioned earlier that the general form of the test used in this section was due to Bartlett and one will notice a similarity in form between this and his test for homogeneity of variance. If all of the roots are tested for equality (i.e., $k = 0$), this is the same as Mauchley's Test of Sphericity (1940) and sometimes goes by that name. Improvements were made to the χ^2-approximation of Bartlett's test in terms of different multiplying constants, first by Bartlett (1954) and later by the one described above due to Lawley (1956). [For an alternative derivation of Lawley's expression, see James (1969).] However, as a general all-purpose form, (4.4.1) should suffice for most applications. One should keep in mind that tests run in a sequential manner for $k = 0, k = 1, \ldots,$ are not independent and have the usual Type I error problems.

4.4.2 Test of the Hypothesis that the Last $p - k$ Roots are Equal to a Known Constant

Lawley (1956) also proposed a test for the hypothesis

$$H_0: \lambda_{k+1} = \cdots = \lambda_p = \lambda$$

where λ is known (or hypothesized in advance.) The form of this test is

$$c\left\{ -\sum_{i=k+1}^{p} \ln(l_i) + (p - k)\ln(\lambda) + \frac{1}{\lambda}\sum_{i=k+1}^{p} l_i - (p - k) \right\}$$

$$(4.4.2)$$

where

$$c = (v - k) - (1/6)\{2(p - k) + 1 - [2/(p - k + 1)]\}$$

$$- \frac{1}{(p - k + 1)}\left\{ \sum_{i=1}^{k} \frac{l_i}{(l_i - \lambda)} \right\}^2 + \lambda^2 \sum_{i=1}^{k} \frac{1}{(l_i - \lambda)^2}$$

Expression (4.2.2) has an asymptotic χ^2-distribution with $(p - k)(p - k + 1)/2$ d.f.

Returning to the Ballistic Missile example, let us assume that the variance associated with uncontrollable measurement error is $\sigma^2 = 10$ for each of the four variables. The sum of these, $4\sigma^2 = 40$ is assumed to be the residual trace. Under the hypothesis

$$H_0: \lambda_3 = \lambda_4 = \lambda$$

λ would have a value of 20. Recalling, for this example, that $n = 40$, $l_1 = 335.34$, $l_2 = 48.03$, $l_3 = 29.33$, and $l_4 = 16.41$, the constant $c = 34.22$, and the statistic $\chi^2 = 3.49$. $\chi^2_{3,.05} = 7.81$ so the hypothesis of known residual variability would be accepted. A similar procedure for any subset of consecutive roots was given by Fujikoshi (1977).

4.5   DISTRIBUTION OF CHARACTERISTIC ROOTS

4.5.1   The Joint Distribution of Sample Characteristic Roots

To construct inferential procedures about characteristic roots such as the tests in Section 4.4, one generally needs to know something about the joint distribution of these roots. These distributions are extremely complicated and obtaining usable expressions for them has probably attracted more attention on the part of mathematical statisticians than any other topic in PCA.

The joint distribution of the characteristic roots of a covariance matrix is based on the null hypothesis that the roots are equal, that is, $\Sigma = I$ or $\Sigma = \sigma^2 I$. Actually, this problem is a subset of a larger class of eigenvalue problems, namely those associated with the multivariate analysis of variance, canonical correlation, and tests for covariance matrices. These generally have determinantal equations of the form $|A - lB| = 0$ where, for instance, A might represent explained variability in a MANOVA model and B would be the residual variability. Another possibility would be for A to represent the sample covariance matrix, S, and B the hypothetical covariance matrix Σ_0. If we let $\Sigma_0 = I$, we have the PCA solution. This gives us a different perspective on PCA, that is, a characterization of how S differs from a unit matrix. There have been a number of specifications of a general model in which PCA, MANOVA, and canonical correlations are special cases; see for instance Robert and Escoufier (1976) and Nomakuchi and Sakata (1984).

The year 1939 was a great one for multivariate analysis. Not only was the paper of Girshick, referenced in Section 4.2, published in that year but also three papers produced independently by R.A. Fisher, P.L. Hsu and S.N. Roy dealing with the distribution of the roots of $|A - lB| = 0$, Hsu pointing out that PCA was a special case. The form of the distribution of the sample roots

$$l_1 \leqslant l_2 \leqslant \cdots \leqslant l_p \leqslant 0$$

when the population roots are equal is

$$f(l) = c \left\{ \exp\left(-\tfrac{1}{2} \sum_{j=1}^{p} l_j \right) \right\} \prod_{j=1}^{p} l_j \prod_{i<j}^{p} (l_i - l_j) \tag{4.5.1}$$

where c is a constant. The expression for unequal population roots is considerably more involved. (A.M. Mood had produced an alternative derivation of equation (4.5.1) about the same time but it was not published since all of the others had already appeared. However, Hotelling thought that Mood's derivation of the normalizing constant c was of interest so the paper was ultimately published; Mood, 1951, 1990.)

To obtain the marginal distributions of one or more roots would require obtaining one or more integrals of (4.5.1) and this is a formidable task for all save a few special cases. It is this task that has attracted the mathematical statisticians over the years but, even to the present time, the majority of the results are asymptotic and generally quite complicated. Early workers employed *pseudo determinants* to obtain more workable forms but these were still useful only in special cases. James (1960, 1964, 1966) restated (4.5.1) in terms of hypergeometric functions of matrix argument making use of zonal polynomials. Although this allowed researchers a fresh start on this problem the expressions are still quite unwieldy for most practical applications. Moreover, even though the use of hypergeometric functions, in theory, made it possible to obtain expressions that could be computed, many of these expressions converged quite slowly and this produced new efforts to obtain asymptotic expressions designed to speed up this process. Muirhead (1978) made the distinction between *asymptotic distributions* of the roots and *asymptotic expressions* of the distributions themselves. The latter retain the linkage factors related to the influence of the population roots that the asymptotic distributions do not.

Rather than review all of this work in detail, it will suffice to refer the reader to review articles by Pillai (1976, 1977), Krishnaiah (1978, 1980) and Muirhead (1978, 1982, Chap. 9), which are quite complete and contain a great many references. For those who find that they do have to deal with some of these distributions, some publications dealing with hypergeometric functions and integrals by Exton (1976, 1978) and Mathai and Saxena (1978) may be of use in addition to the work of James, referenced above, on zonal polynomials. Some worthwhile results have been obtained for some special applications, namely, the distribution of individual roots and the ratios of roots or partial sums of roots to the trace of the covariance matrix. These will be described in the remainder of this section.

4.5.2 The Distribution of the Individual Roots

One of the major accomplishments in characteristic root distributions has been with regard to the distribution of the largest root l_1 or the smallest one, l_p. Percentiles of this distribution have been tabulated for both l_1 and l_p for $p = 2(1)10$ for $p < v < 100$ by Hanumara and Thompson (1968) where v represents the degrees of freedom associated with the covariance matrix. These tables have been reproduced in Appendix G.6. Similar tables may be found in Pearson and Hartley (1972, Table 12) and, for $p = 2(1)15$, in Krishnaiah (1980, Tables 13 and 14). Additional tables for l_1, only, have been obtained for $p = 11(1)20$ for $2 < v < 200$ by Pillai and Chang (1970). Krishnaiah and Waiker (1971) obtained an expression for the distribution of any l_i for $2 \leqslant i \leqslant p - 1$

that may also be found in Krishnaiah (1980, Table 15) for $p = 2(1)7$. For the complex case, tables for the largest root were prepared by Pillai and Young (1971) for $p = 2(1)11$ and for *all* of the roots by Schuurmann and Waikar (1974) for $p = 3, 4$ and 5. All of these tables are based on the null hypothesis: $\Sigma_0 = I$.

For an example of the use of these tables, suppose one had a five-variable problem with a sample size of 51 (i.e., $p = 5$, $v = 50$) and wished to test the hypothesis

$$H_0: \lambda_1 = 1$$

against the hypothesis

$$H_1: \lambda_1 > 1$$

at the level $\alpha = .05$. Compute the ratio $\lambda_1 R_\alpha / v$ where R_α is the tabular value from Appendix G.6. For this particular example, $R_{.05} = 91.08$ and the ratio is $(1)(91.08)/50 = 1.82$. If the largest sample root $l_1 > 1.82$, then one would accept the alternative hypothesis.

In most actual situations, the assumption $\Sigma_0 = I$ is probably not realistic but might conceivably be replaced by $\Sigma_0 = \sigma^2 I$, that is, the population variances of all the variables are equal but not constrained to be equal to 1.0. As an example, again consider the Ballistic Missile example. The sample variances range from 69.42 to 142.74. Let us assume, for the moment, that the population variances are all equal to 100 so $\mathrm{Tr}(\Sigma_0) = 400$. The null hypothesis could be restated as

$$H_0: \lambda_1 = \mathrm{Tr}(\Sigma_0)/p = 100$$

and the alternative as

$$H_1: \lambda_1 > 100$$

Using $\alpha = .05$ and $v = 39$, the ratio for this example is $(100)(71.82)/39 = 184.15$. The first sample characteristic root $l_1 = 335.34$, which is significantly larger than 184.15 so H_1 is accepted.

For tests involving the smallest root, the inequality in the alternative hypothesis is *reversed* and that hypothesis will be accepted if the smallest root is *less* than the adjusted tabular value. This situation may occur when it is suspected that the smallest root would be zero except for rounding error because of a linear constraint among the variables.

Even though these distributions involve only a single root, they are still quite complicated. For example, Muirhead (1975) derived an expression for the exact distribution of the largest root for the case $p = 2$. This still involved an infinite series, each term of which contained *two* hypergeometric series.

Muirhead (1974) also gave some bounds for the distribution of l_1 and l_p of the form

$$P\{l_1 \leqslant x\} \leqslant \prod_{i=1}^{p} P\{\chi_v^2 \leqslant (x/\lambda_i)\} \qquad (4.5.2)$$

and

$$P\{l_p \leqslant x\} \geqslant 1 - \prod_{i=1}^{p} P\{\chi_v^2 \geqslant x/\lambda_i\} \qquad (4.5.3)$$

However, these bounds are fairly wide. For the Ballistic Missile example just given, using $x = 184.10$ and $\lambda_1 = \cdots = \lambda_4 = 100$ in the right-hand side of (4.5.2):

$$\sum_{i=1}^{4} P\{\chi_{39}^2 \leqslant (39)(184.15)/(100)\}$$

$$= [P\{\chi_{39}^2 \leqslant 71.80\}]^4$$

$$\approx (.999)^4 = .996$$

whereas $P\{l_1 \leqslant 184.15\} = .95$.

4.5.3 The Extreme Roots

Another topic of interest is the joint distribution of the largest root l_1 and the smallest root l_p. Hanumara and Thompson (1968) constructed simultaneous tests for the extreme roots of the form

$$P\{L < l_p < l_1 < U\} = 1 - \alpha \qquad (4.5.4)$$

under the null hypothesis $H_0: \Sigma = \sigma^2 I$. If L was fixed such that $P\{l_p > L\} = 1 - \alpha/2$, they showed for $\alpha \leqslant .05$ that their tabular values for the maximum root were sufficiently close to U that they could be used in (4.5.4). That being the case, Appendix G.6 will suffice for both the tests for l_1 and l_p separately and the joint test of all the roots. For the Ballistic Missile example, again using the interpolated tabular values for $v = 39$, 14.94 and 75.25, dividing by 39 and multiplying by 100 results in

$$P\{38.31 < l_4 < l_1 < 192.95\} = .95$$

under the null hypothesis. $l_4 = 16.41$ is smaller than 38.31 and $l_1 = 335.34$ is larger than 192.95 so the hypothesis is rejected. Since the roots must sum to $\text{Tr}(S)$, one would expect l_p to be significantly small if l_1 was significantly large.

One may wish to obtain joint confidence limits for λ_1 and λ_p, which in effect is also a bound for all of the roots. Large-sample limits of the form

$$\frac{l_p}{a_1} \leqslant \lambda_p \leqslant \lambda_1 \leqslant \frac{l_1}{a_2} \tag{4.5.5}$$

were first obtained by Roy (1954, 1957, p. 104). Somewhat sharper bounds of the same form were obtained by T.W. Anderson (1965, 1984a, p. 472). For Anderson's limits, a_1 and a_2 are such that

$$P\{va_2 \leqslant \chi_v^2\} P\{\chi_{v-p+1}^2 \leqslant va_1\} = 1 - \alpha \tag{4.5.6}$$

for $(1 - \alpha)\%$ confidence limits. There are an infinite number of combinations of a_1 and a_2 such that (4.5.5) holds. Presumably one could design an algorithm to obtain the shortest limits. At present the most common methods employed are to use equal probabilities for both parts of (4.5.5) or let $a_1 = 1/a_2$, a strategy proposed by Clemm et al. (1973) for which tables by Krishnaiah (1980, Table 12) are available.

For the Ballistic Missile example, $l_1 = 335.34$, $l_4 = 16.41$, $p = 4$, and $v = 39$. To obtain 95% confidence limits, we shall determine a_1 and a_2 such that each probability is equal to .975. For this example, $\chi_{39,.975}^2 \simeq 23.66$ and $\chi_{36,.025}^2 \simeq 54.43$, from whence $a_1 = 1.40$ and $a_2 = .61$ and the resulting limits will be

$$11.7 \leqslant \lambda_4 \leqslant \lambda_1 \leqslant 549.7$$

4.5.4 Ratios Involving Single Roots

The most common example of ratios involving single roots are the ratio of the largest root or the smallest root to the sum of all the roots. In particular, the ratio $l_1/\text{Tr}(S)$ is of interest as a test for nonadditivity in the univariate analysis of variance and will be discussed more fully in Section 13.7. A wide range of percentiles for this ratio have been obtained for $p = 3,4$ by Krzanowski (1979a) drawing on earlier work by Mandel (1971) and Johnson and Graybill (1972). Schuurmann et al. (1973) tabled the .05 and .01 values for this ratio for $p = 3(1)6$ for $0 \leqslant (v - p - 1)/2 \leqslant 25$. For values outside these tables, equation (4.5.7), in the next section, may be used with $k = 1$. In Section 13.7 we shall see that these tables may also be employed for the distribution of $l_k/(l_k + \cdots + l_p)$. Krishnaiah and Schuurmann (1974) tabulated (for $\alpha = .05$) the critical values of the ratio to $\text{Tr}(S)$ of l_1, l_2, l_{p-1}, and l_p for $p = 3(1)6$ and $0 \leqslant (v - p - 1)/2 \leqslant 25$ for both the real and complex cases.

In Section 2.8.2, the index of a matrix was defined to be $\sqrt{l_1/l_p}$. The distribution of l_1/l_p, itself, was obtained by Sugiyama (1970) and Waikar

(1973). Some rather limited tables have been prepared by Waikar and Schuurmann (1973), Mandel and Lashof (1974), and Krishnaiah and Schuurmann (1974), the latter for the complex case as well. Pillai et al. (1969) gave the distribution of the ratio of *any* two roots. An extension of this procedure is the use of the ratio of the maximum root of one covariance matrix to the minimum root of another and is used as a test that one matrix is larger than the other (Krishnaiah and Pathak, 1968).

4.5.5 Ratios Involving Sums of Roots

This section deals with the quantity

$$\frac{\sum_{i=1}^{k} l_i}{\text{Tr}(S)}$$

the ratio of the sum of the roots related to the pc's in the model to the sum of all the sample roots. This is an estimate of the proportion of the total variability of the original variables accounted for by the first k pc's. We have already commented in Section 2.8.4 that this quantity is not recommended as a stopping rule. Nevertheless, it may be of interest purely as a measure of information. Sugiyama and Tong (1976) obtained an asymptotic distribution related to the statistic

$$\frac{\sum_{i=1}^{k} l_i}{\text{Tr}(S)} - \frac{\sum_{i=1}^{k} \lambda_i}{\text{Tr}(\Lambda)} \tag{4.5.7}$$

that includes the first six derivatives of the normal distribution. The formulas for the coefficients of this expression would easily fill this page but they are quite straightforward, involving only functions of the populations roots, and would be fairly easy to program. This quantity could be used to test the hypothesis that the first k pc's explain a certain proportion of the sum of the variances. Sugiyama and Tong also gave a procedure for obtaining confidence limits for this ratio. Some large-sample results for these situations were given by T. W. Anderson (1965) and Flury (1988, p. 28).

Krishnaiah and Lee (1977) extended this to the general case:

$$\frac{\sum_{i=1}^{k} a_i l_i}{\sum_{i=1}^{p} b_i l_i} - \frac{\sum_{i=1}^{k} a_i \lambda_i}{\sum_{i=1}^{p} b_i \lambda_i} \tag{4.5.8}$$

where a_i and b_i are arbitrary coefficients. This amounts to weighting the pc's. If $a_1 = \cdots = a_k = 1$, $a_{k+1} = \cdots = a_p = 0$, and $b_1 = \cdots = b_p = 1$, (4.5.8) reduces to (4.5.7) as a special case.

Both sets of these results hold for both the real and complex cases.

4.6 SIGNIFICANCE TESTS FOR CHARACTERISTIC VECTORS: CONFIRMATORY PCA

4.6.1 Introduction

Although most of the inferential procedures related to characteristic vectors are in the nature of standard errors and confidence limits discussed in Section 4.2, there may be occasions when one might want to test whether one or more sample vectors are significantly different from some hypothetical vectors. These occasions might include hypothetical vectors that may be proposed in relation to various physical or chemical models or may be approximations to sample vectors already obtained. These techniques may be referred to as *confirmatory* PCA, corresponding to a similar but more involved set of procedures used in factor analysis that will be given in Section 17.8.

In general, these test procedures employ vectors of unit length.

4.6.2 Testing a Single Vector

Let a_i be a hypothesized vector for which a test of the hypothesis H_0: $\Upsilon_i = a_i$ is desired. A procedure, due to Anderson (1963), is the statistic

$$A_1 = v\{l_i a_i' S^{-1} a_i + (1/l_i) a_i' S a_i - 2\} \tag{4.6.1}$$

which has an asymptotic χ^2-distribution with $p - 1$ degrees of freedom. The population roots must be distinct.

For an example, let us return to the Chemical example and test the hypothesis that

$$\Upsilon_1 = a_1 = \begin{bmatrix} .707 \\ .707 \end{bmatrix}$$

Using $l_1 = 1.4465$, the covariance matrix introduced in Section 1.2 and $v = 14$ d.f., $A_1 = .107$, which is not significant when compared with $\chi^2_{1,.05} = 3.84$. This means that a_1 is not significantly different from the first population vector. A test of the hypothesis

$$\Upsilon_2 = a_2 = \begin{bmatrix} .707 \\ -.707 \end{bmatrix}$$

using the second root, $l_2 = .0864$, also produces $A_1 = .107$—again, not significant.

Schott (1987a) showed that the Type I error associated with this test becomes large for small sample sizes. He suggested, in the spirit of the Bartlett and Lawley corrections found elsewhere in this chapter, replacing the multiplier v in (4.6.1) with an expression involving functions of all of the characteristic roots.

This does a better job of stabilizing the Type I error except for small sample sizes, sizes on which the practitioner should not rely in the first place. For vectors associated with the larger roots, Schott concluded that the use of $v - p - 1$ as a multiplier produced similar results. Using this multiplier and rearranging some terms for computational ease, (4.6.1) may be restated as

$$A_2 = (v - p - 1)\mathbf{a}_i'[l_i\mathbf{S}^{-1} + (1/l_i)\mathbf{S} - 2\mathbf{I}]\mathbf{a}_i \tag{4.6.2}$$

For the Chemical example above, this would have the effect of decreasing both test results to .091.

Another test, which does *not* depend on the characteristic roots, has been given in various forms by Mallows (1961), James (1977), and Jolicoeur (1984) and may be expressed as

$$A_3 = \frac{(v - p)}{(p - 1)}[\mathbf{a}_i'\mathbf{S}\mathbf{a}_i\mathbf{a}_i'\mathbf{S}^{-1}\mathbf{a}_i - 1] \tag{4.6.3}$$

which has an F-distribution with $v_1 = p - 1$ and $v_2 = v - p$ degrees of freedom. In the Chemical example, for a test of $\mathbf{a}_1$, $A_3 = .092$—again, not significant.

A situation may exist where there is only one distinct characteristic root, with the remaining $p - 1$ population roots being equal. In this case, the first pc is called *nonisotropic*. Tests for this condition have been derived by Kshirsagar (1961, 1966) and Schott (1987a). Fang and Krishnaiah (1986) showed that the small-sample test for the mean of the nonisotropic pc has an asymptotic t-distribution.

Two-sample procedures will be discussed in Section 16.6.3. There is no two-sample procedure corresponding to (4.6.1) but there will be some useful techniques for comparing the sample vectors from two or more matrices.

4.6.3 Confidence Limits

Jolliffe (1986, p. 43) showed that (4.6.1) or (4.6.2) may be inverted to obtain $(1 - \alpha)\%$ confidence limits for Υ_i, the population vectors normalized to 1.0. Using (4.6.2), these limits are of the form

$$(v - p - 1)\Upsilon_i'[l_i\mathbf{S}^{-1} + (1/l_i)\mathbf{S} - 2\mathbf{I}]\Upsilon_i \leqslant \chi^2_{p-1,x} \tag{4.6.4}$$

For the chemical example, the 95 % confidence limits for Υ_1 would be of the form

$$(11)\Upsilon_1'[1.4465\mathbf{S}^{-1} + .6913\mathbf{S} - 2\mathbf{I}]\Upsilon_1 < \chi^2_{1,.05}$$

or

$$[\Upsilon_{11}\ \Upsilon_{12}]\begin{bmatrix} 7.0519 & -7.3936 \\ -7.3936 & 7.7517 \end{bmatrix}\begin{bmatrix} \Upsilon_{11} \\ \Upsilon_{12} \end{bmatrix} \leqslant .3491$$

Since $\Upsilon_{11}^2 + \Upsilon_{12}^2 = 1$, the solution may be obtained in terms of Υ_{11}, which becomes

$$.609 < \Upsilon_{11} < .821$$

In other words, the limits are represented by a curve from $\Upsilon_{11} = .609$, $\Upsilon_{12} = .793$ to $\Upsilon_{11} = .821$, $\Upsilon_{12} = .571$. If $p = 3$, the limits would be in the form of a curved strip in three-space and for even moderate p would become rather unwieldy.

4.6.4 Tests of k Vectors

There have been some extensions to the procedures discussed in Section 4.6.2 for testing the first k vectors. Anderson's procedure has been generalized by Tyler (1981, 1983b) and Flury (1988, p. 33). James (1977) gave a large-sample extension of (4.6.3) to become

$$\mathbf{J} = \mathbf{A}_1' \mathbf{S} \mathbf{A}_1 \mathbf{A}_1' \mathbf{S}^{-1} \mathbf{A}_1 - \mathbf{I}_k \qquad (4.6.5)$$

where $\mathbf{J}$ is a $k \times k$ asymmetric matrix, $\mathbf{A}_1$ is now $p \times k$ and $\mathbf{I}_k$ is a $k \times k$ identity matrix. As with (4.6.3), this procedure does not require the characteristic roots of $\mathbf{S}$. James' procedure obtains the characteristic roots of $\mathbf{J}$, $l_1, \ldots, l_k$ and from these, $r_i^2 = l_i/(1 + l_i)$, which are the squares of *canonical* correlation coefficients. (Canonical correlation will be discussed in detail in Section 12.6.2.) If the r_i^2 are not significantly different from zero, the corresponding pc's are said to span the covariance matrix.

In the Ballistic Missile example of Section 2.6, we had concluded that only the first two characteristic roots were distinct and that the first pc represented overall shifts in level and the second pc represented gauge differences. Let the null hypothesis be

$$H_0: \mathbf{A}_1 \text{ spans the covariance matrix}$$

where

$$\mathbf{A}_1 = \begin{bmatrix} .5 & -.5 \\ .5 & -.5 \\ .5 & .5 \\ .5 & .5 \end{bmatrix}$$

Substitution of this and the covariance matrix in (4.6.4) produces

$$\mathbf{J} = \begin{bmatrix} .0954 & .0432 \\ .0126 & .1876 \end{bmatrix}$$

with characteristic roots .1932 and .0898. $r_1^2 = .16$ and $r_2^2 = .08$. r_1^2 is significant so A_1 does not span the covariance matrix. The problem is probably with the second characteristic vector of S,

$$\mathbf{u}_2 = \begin{bmatrix} -.62 \\ -.18 \\ .14 \\ .58 \end{bmatrix}$$

whose coefficients differ in size to some extent.

Extended tests for nonisotropic pc's have been developed by Schott (1986) for the case $k = 2$ and by Kshirsagar and Gupta (1965) and Gupta (1973) for the general case. Schott (1987b) developed a conservative test for the hypothesis that k pc's are the *first* k pc's and incorporated (4.6.3) as a first step in this procedure. Tables are included for the case $k = 1$. Schott (1989) also produced an adjustment for this test similar in form to the multiplying constant in (4.4.2). Gupta (1967) gave a large sample procedure for jointly testing the first k vectors *and* roots.

4.7 INFERENCE WITH REGARD TO CORRELATION MATRICES

4.7.1 The Rules Change

Unfortunately, a lot of the material presented in Sections 4.2 through 4.6 hold only for characteristic roots and vectors obtained from covariance matrices, not from correlation matrices. One problem is that the correlations are themselves functions of the elements that make up the covariance matrix. Another is that the sum of the characteristic roots of a correlation matrix is equal to the number of variables, p. These combine to complicate things.

To illustrate how different things may be, consider the covariance matrix of the characteristic roots. The variance of the ith characteristic root of a covariance matrix as we have seen in Section 4.2.1 is approximately $2\lambda_i^2/n$, but when the same root is obtained from a correlation matrix it has an asymptotic variance of

$$V(l_i) = (2/n)(\lambda_i^2 + \mathbf{a}_i'\mathbf{Ba}_i - 2\lambda_i\mathbf{a}_i'\mathbf{a}_i) \tag{4.7.1}$$

where the matrix A is constructed by replacing each element of the population V-matrix, $\boldsymbol{\Phi}$, by its square; that is $a_{ij} = (\varphi_{ij})^2$. B bears the same relationship to the population correlation matrix P: $b_{ij} = (\rho_{ij})^2$. Worse yet, the covariance between the ith and jth roots, which is asymptotically zero for covariance matrices is, for correlation matrices, asymptotically equal to

$$\text{Cov.}(l_i, l_j) = (2/n)[\mathbf{a}_i'\mathbf{Ba}_j - (\lambda_i + \lambda_j)\mathbf{a}_i'\mathbf{a}_j] \tag{4.7.2}$$

[Both results are due to Girshick (1939).] The characteristic roots are, themselves, correlated because the sum of the roots is equal to p. In practice, one would be normally be working with sample estimates so that $\mathbf{P}$, Λ, and Φ would be replaced by $\mathbf{R}$, $\mathbf{L}$, and $\mathbf{V}$, respectively. Note that apart from the correlation matrix and the sample size, (4.7.1) and (4.7.2) are functions of only the relevant characteristic roots and vectors. This means that if one has obtained only the first k characteristic roots and vectors, the covariance matrix for the first k roots may be obtained without having to obtain the rest of them.

Konishi (1978) obtained asymptotic expansions of the distributions of the characteristic roots and of the ratio of a partial sum of these roots to the total when these roots are obtained from a correlation matrix. These expressions are something like Cornish–Fisher expansions with coefficients that are *extremely* complicated although they have been worked out for the case $p = 2$.

4.7.2 Procedures for Testing the Last $p—k$ Roots for Equality

Another problem with correlation matrices is that (2.6.1) cannot be used in testing the last $p-k$ roots for equality. The test procedure is essentially the same but its distribution is no longer asymptotically chi-square. The best that can be done is present some approximate procedures along with some caveats about using them.

There is one test that does have a χ^2-distribution and that is the one for testing that *all* of the characteristic roots are equal. For a correlation matrix, this corresponds to the case where there is no correlation among any of the variables. That is,

$$H_0: \lambda_1 = \cdots = \lambda_p$$

is equivalent to the hypothesis

$$H_0: \mathbf{P} = \mathbf{I}$$

where $\mathbf{P}$ is the population correlation matrix. The test for this is

$$\chi^2 = [n - (1/6)(2p + 5)] \ln |\mathbf{R}|$$
$$= [n - (1/6)(2p + 5)] \sum_{i=1}^{p} \ln (l_i) \qquad (4.7.3)$$

with $p(p - 1)/2$ degrees of freedom (Box, 1949; Bartlett, 1950). The nonnull distribution for this statistic was given by Konishi (1979).

Other than testing all of the roots for equality, there are many problems. As noted before, the comparable test to (2.6.1) does not have a χ^2-distribution but can approximate it under some rather stringent conditions. The test statistic is

$$c\left\{-\sum_{i=k+1}^{p} \ln{(l_i)} + (p-k)\ln\left[\sum_{i=k+1}^{p} \frac{l_i}{(p-k)}\right]\right\} \tag{4.7.4}$$

where $c = n - (1/6)(2p + 5) - (2/3)k$ (Bartlett, 1951a, 1954). This will have a χ^2-distribution *only* if $\lambda_1, \ldots, \lambda_k$ are large relative to the remaining roots. Even if that is so, the degrees of freedom are a function of those roots. Bartlett (1951b) stated that the upper limit for the degrees of freedom was $(1/2)(p - k - 1)(p - k + 2)$ but would have to be corrected for the values of the roots themselves.

Let the matrix of characteristic vectors be partitioned as $U = [U_1 | U_2]$ where U_1 (the retained vectors) is $(p \times k)$ and U_2 is $(p \times (p - k))$ and let $C = U_2 U_2'$. Further, let

$$F = 2(p - k - 1)\bar{l}\sum_{i=1}^{p} c_{ii}^2 \quad \text{where } \bar{l} = (l_{k+1} + \cdots + l_p)/(p - k)$$

$$G = (p - k) \sum_{i=1}^{p} \sum_{j=1}^{p} c_{ij}^2 r_{ij}^2$$

and

$$H = \sum_{i=1}^{p} \sum_{j=1}^{p} c_{ii} c_{jj} r_{ij}$$

Lawley (1956) determined that the degrees of freedom associated with (4.7.4) is equal to

$$\tfrac{1}{2}(p - k - 1)(p - k + 2) - \frac{1}{(p - k)}\{F - G + H\} \tag{4.7.5}$$

The first term is the maximum degrees of freedom and the second is the correction due to the characteristic roots. Another approximate χ^2-test was developed by Schott (1988).

Two special cases for nonisotropic pc's have been worked out. The first of these is the case where only the first root λ_1 is distinct. In this situation, all of the population correlation coefficients, ρ_{ij} are equal. That is to say, the null hypotheses

$$H_0: \lambda_2 = \cdots = \lambda_p$$

$$H_0: \rho_{ij} = \rho \quad \text{for all } i \neq j$$

are equivalent. If that is the case, then $\lambda_1 = 1 + (p - 1)\rho$ and each of the others is equal to $1 - \rho$. Bartlett (1951a,b) gave the degrees of freedom as

$$[(p - 2)(p + 1)/2] - [(p - 1)(p - 2)(1 - \rho^2)/p]$$

which can be verified from (4.7.6). Lawley (1963) gave the form of the test as $n \to \infty$ and showed that it has a χ^2-distribution with $(p + 1)(p - 2)/2$ d.f.

The second case is the rather unusual situation where the first k population roots are equal and the last $p - k$ roots are equal. Let

$$\bar{l}_1 = \frac{1}{k} \sum_{i=1}^{k} l_i$$

and

$$\bar{l}_2 = \frac{1}{p - k} \sum_{i=k+1}^{p} l_i = \frac{p - k\bar{l}_1}{p - k}$$

λ_1 and λ_2 are similarly defined. Anderson (1963) showed that the quantity

$$\left[\frac{\bar{l}_2 - \lambda_2}{\lambda_2 [p - (p - k)\lambda_2]} \right] \sqrt{\frac{npk(p - k)}{2}}$$

is normally distributed $(0,1)$ as $n \to \infty$. Confidence limits for λ_1 and λ_2 were given by Morrison (1976, p. 298). Anderson worked out the distributions for these cases for $p = 3,4$.

4.7.3 Numerical Example

Some of the techniques described in this section will now be illustrated with the Kelley–Hotelling Seventh-Grade example that was introduced in Section 3.3.4. Recall from Section 4.7.1 that while the sample characteristic roots from a covariance matrix are uncorrelated, this is not the case with correlation matrices. Table 4.3 shows the covariance matrix for these characteristic roots

Table 4.3. Seventh-Grade Tests. Covariance Matrix of Characteristic Roots (Correlations in Parentheses)

	l_1	l_2	l_3	l_4
		$n = 140$		
l_1	.0197	−.0162	−.0028	−.0006
l_2	(−.86)	.0180	−.0017	−.0001
l_3	(−.31)	(−.19)	.0043	.0002
l_4	(−.19)	(−.04)	(.14)	.0005

Table 4.4. Seventh-Grade Tests. Bartlett's Test for Equality of Roots

		Test Using (4.7.4) (Correct)		Test Using (2.6.1) (Incorrect)	
i	Root		d.f.		d.f.
1	1.847	199.67	5.17	201.36	9
2	1.464	146.30	3.86	148.26	5
3	.522	42.41	2.15	43.19	2
4	.166		No test		

using equations (4.7.1) and (4.7.2) with the sample estimates in Tables 3.1 and 3.2 being substituted for P, Λ, and Φ. (The correlations among the roots are displayed in the off-diagonal elements below the diagonal.) If one were to assume independence and use $2l_i^2/n$ as the variance, the results would be .0487, .0307, .0039, and .0004. Note that the first two diagonal elements in Table 4.3 are markedly smaller than these and furthermore, the correlation between l_1 and l_2 is $-.86$. From this, it would appear that reduced variances and high correlations are the norm for the larger roots but as the roots become smaller and more alike, they become less correlated and the variances begin to approach those one would obtain assuming independence.

Table 4.4 gives the test results for equality of roots using (4.7.4) with this same set of data. Table 4.4 also gives the results if (2.6.1) had been used. Notice that the χ^2-values are slightly smaller for the correct test but that there is a marked difference in the degrees of freedom, particularly for the case $k = 0$. For a larger number of variables, the degrees of freedom for (4.7.4) will approach $(1/2)(p - k - 1)(p - k + 2)$ fairly rapidly. In the next chapter, an eight-variable example will be examined and by the time the test procedure reached $k = 3$, this had been accomplished.

4.8   THE EFFECT OF NONNORMALITY

The majority of PCA applications are descriptive in nature. In these instances, distributional assumptions are of secondary importance. However, if one wishes to make probability statements, then knowledge of the underlying distributions becomes important. Much of the results in this chapter rest on the assumption that the underlying distribution is multivariate normal. What if it is not? Waternaux (1976) and Davis (1977) showed that the quantity $\sqrt{n}(l_i - \lambda_i)$ is still normally distributed so long as Σ and the fourth cumulants of the underlying univariate distributions are finite. Later, Waternaux (1984) showed that for a class of elliptical distributions, Bartlett's test was quite sensitive to positive kurtosis—not surprising since Bartlett's test for homogeneity of variance has the same problem—and she proposed a multiplying constant, a function of kurtosis, that did continue to have an asymptotic χ^2-distribution. Tyler (1983a,b) obtained asymptotic distributions of the roots and also obtained a test for the characteristic vectors when sampling from elliptic distributions. Fujikoshi (1980) obtained an asymptotic expression for the joint distribution of the characteristic roots under nonnormality as well as the distribution of $(l_1 + \cdots + l_k)/\text{Tr}(S)$. Fujikoshi (1978) also obtained some results for symmetric functions of roots that required only that the fourth moments be finite. Fang and Krishnaiah (1982) obtained an expression for the joint distribution of the roots for both the covariance and correlation matrices and investigated the behavior, for the case $p = 2$, of l_1, $l_1 - l_2$, l_1/l_2, and $(l_1 + l_2)^2/(l_1 l_2)$ for some nonnormal distributions. Davis (1977) gave a test for a hypothetical vector in the presence of nonnormality.

Jensen (1986, 1988) investigated cases in which the underlying distributions (in particular, heavy-tailed distributions) are such that they do not possess finite second moments. By using other measures of scatter, he derived a system he called *principal variables* (not the same meaning as the topic of Section 14.4.3) of which PCA becomes a special case. A number of the standard significance tests for roots and vectors hold as well.

Confidence limits for characteristic roots and vectors as well as tests for vectors using bootstrap procedures (Efron and Tibshirani, 1986) were given by Beran and Srivastava (1985). Daudin et al. (1988) evaluated four data sets for stability using different bootstrap procedures. The data for two of the examples were included; one dealing with the composition of milk is a rather large data set.

In a simulation study, Dudzinski et al. (1975) compared results for normal and nonnormal data and concluded that, if the PCA model is strong enough, it will prevail through nonnormality. They included examples of what is "strong enough." Dudzinski (1975) also investigated some animal data that was nonnormal, using PCA. Some jackknife estimates of vector coefficients and the proportion accounted for by the first k pc's of a correlation matrix were obtained by Nagao (1988).

For the case where the underlying distribution is uniform, Malinowski (1987) proposed a stopping rule that is a variant of Jolliffe's "broken stick" (Section 2.8.7). This consists of comparing the proportion explained by the ith pc with the quantity

$$\frac{(n - i + 1)(p - i + 1)}{\sum_{j=1}^{p} (n - j + 1)(p - j + 1)}$$

Bru (1989) investigated the effect of Brownian perturbations in the data on PCA estimates.

One of the topics in Chapter 16 deals with robust PCA. Robust estimates may be obtained either by obtaining robust estimates of Σ or by obtaining robust estimates of the vectors directly.

4.9   THE COMPLEX DOMAIN

There have been references throughout this chapter to operations in the complex domain. Operations for and uses of complex PCA were discussed in Section 3.7. The distributional aspects of complex PCA are, for the most part, parallel to those of real PCA and these are summarized in another review paper by Krishnaiah (1976). Gupta (1972) derived a goodness-of-fit procedure for one or two theoretical pc's in the complex domain.

CHAPTER 5

Putting It All Together—
Hearing Loss I

5.1 INTRODUCTION

The purpose of this chapter is to bring together, in a single example, many of the techniques that have been discussed in the preceding chapters. The example used in this chapter deals with *audiometry*, specifically, a study carried on within the Eastman Kodak Company involving the measurement of hearing loss. This same topic will be discussed in considerable detail in the case history presented in Chapter 9. In this chapter, we shall begin with a small subsample of original data and carry out all of the steps in detail so that the reader may be able to check them out, although, for simplicity in display, fewer digits will be presented than would normally be required. The data and analysis from this chapter will also allow the reader the opportunity to evaluate the software at his or her disposal. Do not be alarmed if the signs of the coefficients you obtain for one or more characteristic vectors are reversed from those given here. This is merely the results of the algorithm employed. The interpretation will be the same except that the signs of the corresponding pc's will be changed.

In addition to air and water pollution, concern is building over *noise* pollution. Within the industrial community, this refers to the workplace environment. An individual who has worked in a sheet metal shop for 40 years may well have suffered some loss in hearing as a result of it. On the other hand, an individual who has worked in an office environment for 40 years may also have experienced some hearing loss, not due to the environment of the workplace but merely because of normal aging. The first type of hearing loss, if severe enough, would have the potential for a compensation award; the second would not. The problem, then, is to be able to distinguish between normal and induced hearing loss. This will be discussed in more detail in Chapter 9 but, briefly, this was accomplished by building a database of individuals who represent normal hearing experience. These are people who have no history of noise exposure or disease that might affect their hearing. These people are subdivided by sex and by age. Then an individual or group of individuals may be compared with a peer group composed of people with normal hearing of the same age and sex. ("Normal" in this context refers to the usual effect of aging rather than to the distribution of test results.)

5.2   THE DATA

The example in this chapter is strictly for illustrative purposes and consists of 100 males, age 39, who presumably had no indication of noise exposure or hearing disorders. The common method of measuring hearing is with the use of an instrument called an *audiometer* in which an individual is exposed to a signal of a given frequency with an increasing intensity until the signal is perceived. These threshold measurements are calibrated in units referred to as *decibel loss* in comparison to a reference standard for that particular instrument. Observations are obtained, one ear at a time, for a number of frequencies. In the present example, the frequencies used were 500 Hz, 1000 Hz, 2000 Hz, and 4000 Hz, which for two ears results in an eight-variable problem. The data for these 100 individuals are shown in Table 5.1. (The limits of the instrument are -10 to 99 decibels. A negative value does not imply better than average hearing; the audiometer had a calibration "zero" and these observations are in relation to that.)

The covariance matrix for these 100 observations is given in Table 5.2. The corresponding correlation matrix is given in Table 5.3, which also includes the mean and standard deviation of this sample for each variable. Note that the means increase with frequency. This is the prevailing situation regardless of age; increasing age merely changes the relative shape of these "curves."

Although the units of these eight variables are all in Hertz (Hz), it will be seen in Table 5.2 that the variances of the highest frequencies are roughly nine times those of the lower frequencies and hence a PCA on the covariance would result in pc's heavily dominated by 4000 Hz. While this *is* an important frequency, in that normal hearing loss usually occurs earlier at higher frequencies, this frequency is not that important, clinically, relative to the others. Since the use of the correlation matrix will weight all variables equally, this was the approach used instead. Note that the highest correlations occur between adjacent frequencies on the same ear, that is, x_1 vs. x_2, x_5 vs. x_6, and so on, and between corresponding frequencies on both ears such as x_1 vs. x_5.

Another approach might have been to operate with a covariance matrix of the logarithms of the original observations. However, the use of a reference standard does produce some negative values and it would be an arbitrary decision as to how to transform these data to produce all positive values so that logarithms could be obtained. One approach was to fit these data for each variable with a three-parameter log–normal distribution, using the translation parameter estimate to recode the data, but this effort was not very successful.

Table 5.1. Audiometric Data.

ID	Left Ear				Right Ear			
	500	1000	2000	4000	500	1000	2000	4000
1	0	5	10	15	0	5	5	15
2	−5	0	−10	0	0	5	5	15
3	−5	0	15	15	0	0	10	25
4	−5	0	−10	−10	−10	−5	−10	10
5	−5	−5	−10	10	0	−10	−10	50
6	5	5	5	−10	0	5	0	20
7	0	0	0	20	5	5	5	10
8	−10	−10	−10	−5	−10	−5	0	5
9	0	0	0	40	0	0	−10	10
10	−5	−5	−10	20	−10	−5	−10	15
11	−10	−5	−5	5	5	0	−10	5
12	5	5	10	25	−5	−5	5	15
13	0	0	−10	15	−10	−10	−10	10
14	5	15	5	60	5	5	0	50
15	5	0	5	15	5	−5	0	25
16	−5	−5	5	30	5	5	5	25
17	0	−10	0	20	0	−10	−10	25
18	5	0	0	50	10	10	5	65
19	−10	0	0	15	−10	−5	5	15
20	−10	−10	−5	0	−10	−5	−5	5
21	−5	−5	−5	35	−5	−5	−10	20
22	5	15	5	20	5	5	5	25
23	−10	−10	−10	25	−5	−10	−10	25
24	−10	0	5	15	−10	−5	5	20
25	0	0	0	20	−5	−5	10	30
26	−10	−5	0	15	0	0	0	10
27	0	0	5	50	5	0	5	40
28	−5	−5	−5	55	−5	5	10	70
29	0	15	0	20	10	−5	0	10
30	−10	−5	0	15	−5	0	10	20
31	−10	−10	5	10	0	0	20	10
32	−5	5	10	25	−5	0	5	10
33	0	5	0	10	−10	0	0	0
34	−10	−10	−10	45	−10	−10	5	45
35	−5	10	20	45	−5	10	35	60
36	−5	−5	−5	30	−5	0	10	40
37	−10	−5	−5	45	−10	−5	−5	50
38	5	5	5	25	−5	−5	5	40
39	−10	−10	−10	0	−10	−10	−10	5
40	10	20	15	10	25	20	10	20
41	−10	−10	−10	20	−10	−10	0	5
42	5	5	−5	40	5	10	0	45
43	−10	0	10	20	−10	0	15	10

continued

Table 5.1. *Continued.*

	Left Ear				Right Ear			
ID	500	1000	2000	4000	500	1000	2000	4000
44	−10	−10	10	10	−10	−10	5	0
45	−5	−5	−10	35	−5	0	−10	55
46	5	5	10	25	10	5	5	20
47	5	0	10	70	−5	5	15	40
48	5	10	0	15	5	10	0	30
49	−5	−5	5	−10	−10	−5	0	20
50	−5	0	10	55	−10	0	5	50
51	−10	−10	−10	5	−10	−10	−5	0
52	5	10	20	25	0	5	15	30
53	−10	−10	5	25	−10	−10	−10	40
54	5	10	0	−10	0	5	−5	15
55	15	20	10	60	20	20	0	25
56	−10	−10	−10	5	−10	−10	−5	−10
57	−5	−5	−10	30	0	−5	−10	15
58	−5	−5	0	5	−5	−5	0	10
59	−5	5	5	40	0	0	0	60
60	5	10	30	20	5	5	20	10
61	5	5	0	10	−5	5	0	10
62	0	5	10	35	0	0	5	20
63	−10	−10	−10	0	−5	0	−5	0
64	−10	−5	−5	20	−10	−10	−5	5
65	5	10	0	25	5	5	0	15
66	−10	0	5	60	−10	−5	0	65
67	5	10	40	55	0	5	30	40
68	−5	−10	−10	20	−5	−10	−10	15
69	−5	−5	−5	20	−5	0	0	0
70	−5	−5	−5	5	−5	0	0	5
71	0	10	40	60	−5	0	25	50
72	−5	−5	−5	−5	−5	−5	−5	5
73	0	5	45	50	0	10	15	50
74	−5	−5	10	25	−10	−5	25	60
75	0	−10	0	60	15	0	5	50
76	−5	0	10	35	−10	0	0	15
77	5	0	0	15	0	5	5	25
78	15	15	5	35	10	15	−5	0
79	−10	−10	−10	5	−5	−5	−5	5
80	−10	−10	−5	15	−10	−10	−5	5
81	0	−5	5	35	−5	−5	5	15
82	−5	−5	−5	10	−5	−5	−5	5
83	−5	−5	−10	−10	0	−5	−10	0
84	5	10	10	20	−5	0	0	10
85	−10	−10	−10	5	−10	−5	−10	20
86	5	5	10	0	0	5	5	5
87	−10	0	−5	−10	−10	0	0	−10
88	−10	−10	10	15	0	0	5	15

Table 5.1. *Continued.*

	Left Ear				Right Ear			
ID	500	1000	2000	4000	500	1000	2000	4000
89	−5	0	10	25	−5	0	5	10
90	5	0	−10	−10	10	0	0	0
91	0	0	5	15	5	0	0	5
92	−5	0	−5	0	−5	−5	−10	0
93	−5	5	−10	45	−5	0	−5	25
94	−10	−5	0	10	−10	5	−10	10
95	−10	−5	0	5	−10	−5	−5	5
96	5	0	5	0	5	0	5	15
97	−10	−10	5	40	−10	−5	−10	5
98	10	10	15	55	0	0	5	75
99	−5	5	5	20	−5	5	5	40
100	−5	−5	−10	−10	−5	0	15	10

Table 5.2. Audiometric Example. Covariance Matrix

Left Ear				Right Ear			
500	1000	2000	4000	500	1000	2000	4000
41.07	37.73	28.13	32.10	31.79	26.30	14.12	25.28
	57.32	44.44	40.83	29.75	34.24	25.30	31.74
		119.70	91.21	18.64	31.21	71.26	68.99
			384.78	25.01	33.03	57.67	269.12
				50.75	30.23	10.52	18.19
					40.92	24.62	27.22
						86.30	67.26
							373.66

Table 5.3. Audiometric Example. Correlation Matrix

	Left Ear				Right Ear			
	500	1000	2000	4000	500	1000	2000	4000
	1	.78	.40	.26	.70	.64	.24	.20
		1	.54	.27	.55	.71	.36	.22
			1	.42	.24	.45	.70	.33
				1	.18	.26	.32	.71
					1	.66	.16	.13
						1	.41	.22
							1	.37
								1
Average	−2.8	−0.5	2.0	21.4	−2.6	−0.7	1.6	21.4
Standard deviation	6.4	7.6	10.9	19.6	7.1	6.4	9.3	19.3

This, plus the fact that the original data are already in the form of logarithmic measurements, prompted the use of the correlation matrix of the original data.

5.3 PRINCIPAL COMPONENT ANALYSIS

The stopping rule employed was based on the inherent variability expected for replicate examinations. Using this rule, four pc's were deemed sufficient to adequately describe the variability of these 100 individuals. The other four roots in Table 5.4 will be required to obtain the limits for the sums of squares of the residuals, Q, unless equations (2.7.8)–(2.7.10) are employed. The sample covariance matrix of the first four characteristic roots and their correlations are given in Table 5.5 using equations (4.7.11) and (4.7.12). (Remember that the roots of a correlation matrix are themselves correlated.) The first four pc's explained 87% of the trace of the correlation matrix.

It is interesting to contemplate what would have been the effect of using some of the other stopping rules given in Chapter 2. The large-sample significance test for the equality of characteristic roots, using (4.7.4) and (4.7.5) are given in Table 5.4 and indicate that the first six roots are distinct. Stopping with 95% of the trace would also have retained six pc's. Using only the pc's whose roots were greater than 1 would have resulted in using only two pc's, which, as we shall see, would have deleted two rather important ones. The SCREE plot for these roots is given in Figure 5.1 and would indicate that four or five pc's be retained. The cross-validation method will be illustrated for these data in Section 16.3.2 and implies that four pc's should be retained. Considering that the retention of at least four pc's appears pretty solid, it is interesting to note that the Gleason–Staelin statistic is $\varphi = .45$. The index of the correlation matrix is 5.04.

The U-vectors along with their standard errors are displayed in Table 5.6. Note that in nearly all cases, the elements of these vectors are considerably larger than their standard errors, which will provide comfort when attempts are made to interpret the pc's.

The characteristic vector corresponding to the first pc is similar to many of those already encountered in this book. The coefficients are all positive and about the same size. (Anytime that all of the covariances or correlations in a matrix are positive, the first characteristic vector will have all positive coefficients. The converse is not true.) The corresponding pc represents, essentially, the overall hearing level of a respondent and implies that individuals suffering hearing loss at certain frequencies will more than likely suffer loss at the other frequencies as well. This first pc explains 49% of the total variability. The second characteristic vector, representing 20% of the total variability, has coefficients that are more or less the same for each ear but show a "contrast" between the high frequencies (2000 Hz and 4000 Hz) and the low frequencies (500 Hz and 1000 Hz). It is well known in the case of normal hearing that hearing loss as a function of age is first noticeable in the higher frequencies. The third pc

Table 5.4. Audiometric Example. Large-sample Test for Characteristic Roots

Characteristic Root	Residual Trace	% Explained by l_i	% Cumulative Explained	χ^2	d.f.	Significance Level
$l_1 = 3.9290$	4.0710	49.11	49.11	448.21	25.14	.01
$l_2 = 1.6183$	2.4527	20.23	69.34	212.64	23.89	.01
$l_3 = .9753$	1.4773	12.19	81.53	107.24	19.04	.01
$l_4 = .4668$	1.0106	5.84	87.37	35.29	13.97	.01
$l_5 = .3401$	.6705	4.25	91.62	19.18	9.07	.05
$l_6 = .3159$	.3546	3.95	95.57	12.47	4.94	.05
$l_7 = .2001$	.1545	2.50	98.07	1.55	1.94	N. Sig.
$l_8 = .1545$	—	1.93	100.00		No test	

Table 5.5. Audiometric Example. Covariance Matrix of First Four Characteristic Roots (Correlations in Parentheses)

l_1	l_2	l_3	l_4
.0839	−.0396	−.0185	−.0091
(−.68)	.0402	−.0029	.0007
(−.47)	(−.11)	.0181	.0008
(−.43)	(.05)	(.08)	.0054

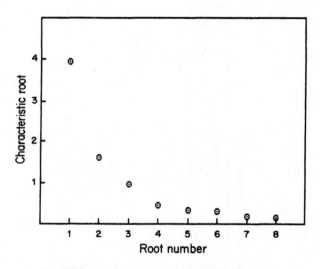

FIGURE 5.1. Audiometric example: SCREE plot.

Table 5.6. Audiometric Example. U-Vectors (Standard Error in Parentheses)

Frequency	u_1	u_2	u_3	u_4
500L	.40(.04)	−.32(.06)	.16(.08)	−.33(.04)
1000L	.42(.03)	−.23(.06)	−.05(.08)	−.48(.03)
2000L	.37(.04)	.24(.10)	−.47(.06)	−.28(.07)
4000L	.28(.06)	.47(.09)	.43(.10)	−.16(.07)
500R	.34(.05)	−.39(.07)	.26(.10)	.49(.05)
1000R	.41(.03)	−.23(.05)	−.03(.07)	.37(.02)
2000R	.31(.05)	.32(.11)	−.56(.08)	.39(.08)
4000R	.25(.06)	.51(.09)	.43(.10)	.16(.07)

(12%) represents another contrast, this one primarily between the two higher frequencies, and the fourth pc (6%) represents the difference between left and right ears. This interpretation is partially borne out by the results of Table 5.7, which displays the amount of each variable accounted for by each principal component. Table 5.7 also shows that the inherent variability, in standard units, is rather evenly distributed over all eight variables. The residual correlation matrix is given in Table 5.8. In the full case study described in Chapter 9, there was much better agreement after screening between the coefficients for the two ears for a given frequency, than that found in Table 5.5, as one would expect with a larger sample size with screened data.

The correlation matrix of the vector coefficients is given in Table 5.9. To conserve space, the correlations are rounded off to the nearest tenth (rounding any correlation higher than .95 up to 1.0), correlations smaller than $\pm.1$ are recorded as zero while correlations where $.1 < |r| < .6$ show only the sign. (This is not meant to allude to any sort of significance test—it is merely a simplification in display.) Note that the coefficients of adjacent vectors, say $\mathbf{u}_1$

Table 5.7. Audiometric Example. Proportion of Each Variable Explained by Each Principal Component

	Principal Component				Unexplained
Frequency	1	2	3	4	Variation
500L	.63	.16	.02	.05	.13
1000L	.70	.08	.00	.11	.11
2000L	.53	.09	.22	.04	.13
4000L	.31	.36	.18	.01	.13
500R	.46	.24	.07	.11	.12
1000R	.67	.09	.00	.06	.18
2000R	.38	.16	.31	.07	.08
4000R	.25	.42	.18	.01	.13

Table 5.8. Audiometric Example. Residual Correlation Matrix After Four Principal Components

Left Ear				Right Ear			
500	1000	2000	4000	500	1000	2000	4000
.13	−.07	−.02	−.03	−.01	−.06	.06	.03
	.11	−.07	−.03	−.03	.02	.02	.04
		.13	.01	.08	−.02	−.08	−.02
			.13	.02	.03	−.01	−.13
				.12	−.11	−.01	−.03
					.18	−.05	−.01
						.08	.00
							.13

Table 5.9. Audiometric Example. Correlation of Characteristic Vector Elements

Source: Jackson (1981a) with permission of the American Society for Quality Control.

Table 5.10. Audiometric Example. V- and W-Vectors

Frequency	v_1	v_2	v_3	v_4	w_1	w_2	w_3	w_4
500L	.795	−.403	.156	−.224	.202	−.249	.160	−.480
1000L	.834	−.287	−.051	−.329	.212	−.177	−.053	−.705
2000L	.726	.304	−.465	−.193	.185	.188	−.476	−.413
4000L	.557	.603	.424	−.110	.142	.373	.435	−.236
500R	.680	−.491	.256	.333	.173	−.303	.263	.714
1000R	.816	−.295	−.028	.254	.208	−.182	−.029	.545
2000R	.618	.403	−.556	.267	.157	.249	−.570	.573
4000R	.504	.654	.421	.109	.128	.404	.432	.233

Table 5.11. Audiometric Example. V*- and W*-Vectors

Frequency	v_1^*	v_2^*	v_3^*	v_4^*	w_1^*	w_2^*	w_3^*	w_4^*
500L	5.10	−2.58	1.00	−1.44	.032	−.039	.025	−.075
1000L	6.32	−2.17	−.39	−2.49	.028	−.023	−.007	−.093
2000L	7.95	3.32	−5.08	−2.11	.017	.017	−.044	−.038
4000L	10.92	11.83	8.32	−2.16	.007	.019	.022	−.012
500R	4.85	−3.50	1.82	2.37	.024	−.043	.037	.100
1000R	5.22	−1.89	−.18	1.63	.032	−.028	−.005	.085
2000R	5.74	3.75	−5.16	2.48	.017	.027	−.061	.062
4000R	9.74	12.63	8.14	2.10	.007	.021	.022	.012

and u_2 are more highly correlated than those farther apart such as u_1 and u_4. The patterns of the correlations are affected by the signs of the coefficients themselves. Of particular interest are the high correlations among the coefficients within vectors, indicating that although a similar set of data should produce the same general set of vectors, the coefficients within a vector may vary quite a bit from one data set to another. More will be said about this type of matrix in Section 7.5.

In order to obtain the y-scores for the individuals in the sample, we shall need to obtain the V- and W-vectors and, since they were obtained from a correlation matrix, the corresponding V*- and W*-vectors as well. These vectors are displayed in Tables 5.10 and 5.11.

5.4 DATA ANALYSIS

The next step in this process is to screen the 100 observations for outliers and abnormal behavior. All tests were carried out at the $\alpha = .05$ level. While this may be considered, by some, a rather large Type I error to be using in screening data, particularly when multiple tests are carried out, it will serve to furnish some examples. The following quantities are required:

1. The y-scores, $y_1, \ldots, y_4$ for the sample. For this example, with $n = 100$, the limits for these components are ± 1.99.
2. T^2. The upper limit for $T^2 = y'y$, using equation (1.7.4), with $k = 4$ substituted for p, is 10.23.
3. Reconstructed original data in original units for display and in standard units for the Q-statistic.
4. Q, the sums of squares of the residuals in standard units. $Q = (x - \hat{x})'$ $(x - \hat{x})$. The upper limit for Q, equation (2.7.6), with $\theta_1 = 1.01$, $\theta_2 = .2794$, $\theta_3 = .082556$, $h_0 = .287$, and $c_{.05} = 1.645$ is 2.47.

Table 5.12 contains the original data for individual #1 along with all of the quantities listed above. Nothing is significant; one would conclude that this individual is a rightful member of his peer group.

The values of Q, T^2, and the four pc's for all 100 men are shown in Table 5.13. The reason they are listed in this order is because this is the order in which the decisions about the data should be made. The first thing to consider is Q, because if this is significant it means that the four-component model does not fit that particular individual for whatever reason and this should be investigated first before considering T^2 or any of the pc's. There are 9 cases out of 100 with a significant value of Q—not alarming considering that the test was carried out at the $\alpha = .05$ level. Many times, the cause of a significant value of Q may be ascertained merely by referring to the data. However, when a correlation matrix is employed, using the data may be misleading because of the heterogeneity of the variances. Table 5.14 contains the individual residuals *in standard units* for these nine individuals. Some of the results indicate left–right

Table 5.12. Audiometric Example. Results for Individual #1

Frequency	Original Values	Predicted Values	Residual in Standard Units
500L	0	.98	−.15
1000L	5	4.88	.02
2000L	10	8.85	.11
4000L	15	15.52	−.03
500R	0	.65	−.09
1000R	5	3.44	.24
2000R	5	6.65	−.18
4000R	15	14.24	.04

$$y_1 = .60 \qquad T^2 = 1.19$$
$$y_2 = -.54$$
$$y_3 = -.74 \qquad Q = .136$$
$$y_4 = -.07$$

Table 5.13. Audiometric Data. Principal Component Analysis (* = Significant at the $\alpha = .05$ level)

ID	Q	T^2	Principal Components			
			y_1	y_2	y_3	y_4
1	.14	1.19	.60	−.54	−.74	−.07
2	.97	3.75	−.15	−.85	−.29	1.71
3	.55	1.82	.37	.35	−1.11	.58
4	1.15	2.88	−1.08	−.84	−.03	−1.01
5	2.90*	3.66	−.72	.21	1.74	.27
6	1.73	2.26	.44	−1.32	−.53	−.20
7	.29	2.17	.41	−.81	−.08	1.16
8	.25	2.93	−1.34	−.15	−.70	.79
9	1.18	2.36	.02	−.48	1.12	−.94
10	.27	2.27	−.97	−.05	.79	−.85
11	1.54	3.49	−.69	−1.04	.42	1.33
12	.83	4.10	.38	−.04	−.53	−1.92
13	.61	5.90	−.90	−.42	.68	−2.11*
14	.61	6.91	1.54	.19	1.80	−1.11
15	2.25	.67	.31	−.55	.40	−.33
16	.64	3.66	.37	.09	.16	1.87
17	2.75*	1.86	−.63	−.03	1.12	−.44
18	.36	10.61*	1.43	.36	2.10*	2.01*
19	.62	1.43	−.60	.51	−.84	−.33
20	.20	2.02	−1.31	−.10	−.50	.23
21	.32	2.29	−.62	.22	1.20	−.66
22	.77	2.69	1.17	−.96	.05	−.62
23	.32	3.24	−1.20	.50	1.24	.13
24	.45	1.83	−.48	.70	−.95	−.46
25	1.39	.30	.06	.45	−.27	−.15
26	.69	1.73	−.45	−.18	−.26	1.20
27	.58	2.34	.74	.61	1.06	.54
28	1.38	9.26	.52	1.89	1.48	1.79
29	4.28*	2.73	.54	−1.23	.35	−.90
30	.20	3.12	−.34	.51	−.84	1.43
31	.47	11.09*	−.20	.46	−1.79	2.77*
32	.54	1.53	.19	.10	−.91	−.80
33	.80	3.52	−.20	−.68	−.79	−1.55
34	.77	5.63	−.79	1.91	1.02	.55
35	2.24	12.62*	1.81	2.10*	−1.71	1.42
36	.67	2.66	−.02	.93	.28	1.31
37	.41	4.99	−.54	1.57	1.47	−.23
38	1.66	2.46	.46	.39	.25	−1.43
39	.06	2.81	−1.64	−.18	.05	−.31
40	.94	16.51*	2.59*	−2.54*	−.26	1.81
41	.73	1.99	−1.32	.47	−.12	.06
42	.76	4.95	1.08	−.38	1.73	.80

continued

Table 5.13. *Continued.*

ID	Q	T^2	Principal Components			
			y_1	y_2	y_3	y_4
43	.67	4.35	−.09	.80	−1.91	.21
44	.95	4.19	−1.01	.65	−1.63	−.32
45	.99	5.48	−.31	.72	2.17*	.38
46	.54	2.21	1.10	−.86	.09	.50
47	2.69*	3.77	1.22	1.44	.40	−.22
48	.91	2.90	1.02	−1.20	.59	.27
49	1.48	1.74	−.73	.01	−1.03	−.38
50	.42	4.29	.41	1.84	.49	−.72
51	.14	2.49	−1.55	−.06	−.26	−.12
52	.16	3.65	1.40	.10	−1.14	−.61
53	1.47	3.70	−.97	1.28	.73	−.76
54	1.60	3.98	.38	−1.76	−.15	−.85
55	1.65	13.97*	2.77*	−1.82	1.73	−.04
56	.44	2.97	−1.62	−.27	−.48	−.24
57	.48	2.40	−.65	−.28	1.38	.04
58	.15	.74	−.65	−.21	−.52	.01
59	1.34	2.88	.59	1.00	1.24	.00
60	.79	7.91	1.61	−.32	−2.26*	−.36
61	1.16	1.98	.31	−1.02	−.28	−.87
62	.30	.86	.61	.09	−.16	−.68
63	.46	3.43	−1.14	−.65	−.23	1.29
64	.41	2.30	−1.18	.30	−.06	−.90
65	.23	2.54	.83	−1.18	.50	−.46
66	.77	8.02	.06	2.36*	1.36	−.77
67	.69	12.44*	2.28*	1.62	−2.04*	−.68
68	.55	2.46	−1.15	−.00	1.03	−.31
69	.86	.56	−.53	−.36	−.22	.33
70	.20	1.17	−.60	−.54	−.44	.57
71	.56	13.95*	1.85	2.34*	−1.69	−1.48
72	.17	1.49	−.92	−.72	−.33	−.05
73	3.86*	8.45	2.00*	1.59	−1.34	−.35
74	2.50*	7.57	.30	2.27*	−1.12	1.04
75	3.11*	12.87*	.76	.73	2.16*	2.66*
76	.94	1.98	−.05	.59	−.42	−1.20
77	1.03	.87	.51	−.58	.08	.52
78	1.67	10.38*	1.71	−2.35*	.83	−1.12
79	.05	2.36	−1.23	−.31	.02	.86
80	.25	2.11	−1.36	.32	−.14	−.37
81	1.34	.56	−.07	.49	−.15	−.54
82	.06	.91	−.81	−.44	.00	−.23
83	.30	3.05	−1.04	−1.36	.15	.34
84	.06	6.06	.53	−.64	−.50	−2.26*
85	.53	2.08	−1.34	.08	.47	.23
86	.36	3.35	.58	−1.22	−1.17	−.38
87	1.21	3.73	−.95	−.85	−1.45	−.02

Table 5.13. *Continued.*

ID	Q	T^2	Principal Components			
			y_1	y_2	y_3	y_4
88	1.31	3.68	−.30	.35	−.86	1.65
89	.40	.94	.05	.22	−.88	−.34
90	1.43	6.04	−.01	−2.16*	.10	1.17
91	.72	.97	.18	−.92	−.20	.23
92	.15	2.60	−.86	−.98	−.06	−.94
93	1.88	2.38	−.07	.18	1.35	−.72
94	3.50*	.64	−.74	−.26	−.15	.06
95	.30	1.73	−1.04	−.04	−.64	−.48
96	1.44	1.71	.38	−1.05	−.49	.47
97	2.95*	2.29	−.93	.69	.26	−.93
98	1.58	9.21	1.66	1.20	1.51	−1.66
99	1.65	.43	.43	.40	−.16	.24
100	1.95	5.98	−.51	−.41	−1.36	1.92

Table 5.14. Audiometric Example. Residuals in Standard Units for Respondents with Significant Q

Respondent No.	Left Ear				Right Ear			
	500	1000	2000	4000	500	1000	2000	4000
5	.1	.2	.2	−1.0	.4	−.8	.0	1.0
17	.7	−.8	.7	−.2	.6	−.8	−.1	.1
29	−.8	1.0	−.2	.1	1.0	−1.2	.4	−.1
47	.7	−.6	−.4	.7	−.5	.4	.4	−.7
73	−.4	−.7	1.3	−.1	.2	.6	−1.1	.0
74	.7	.1	−.5	−.8	−.2	−.5	.5	.7
75	.4	−.7	.6	.5	.9	−.9	.1	−.6
94	−.6	−.0	.4	.1	−.6	1.4	−.8	.0
97	−.4	−.6	.7	.8	.2	.5	−.6	−.8

differences for one or two frequencies, 4000 Hz for #5 and primarily 2000 Hz for #73. Respondents #17 and #75 suggest model failure for the lower frequencies. It would be difficult to diagnose #29, #47, #74, and #97—all of whom appear to have a frequency–ear interaction at all frequencies, as anything more than excessive variability—and that may well be true of the others as well. Nevertheless, one should always be on the lookout for clerical, recording, or data-entry errors. High values of Q could also be caused by a respondent being "asleep at the switch," particularly for high frequencies, and then suddenly realizing that the signal is on.

Another possibility for residual analysis is to plot the residuals as a function of their predicted values. Such a plot for x_8, 4000 Hz–right ear, is given in Figure 5.2. The residuals are plotted in standard units and the abscissa represents the predicted values, $\hat{x}_8$, also in standard units, using four pc's. The use of

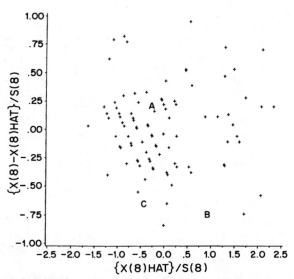

FIGURE 5.2. Audiometric example: $(x_8 - \hat{x}_8)/s_8$ vs. $\hat{x}_8/s_8$.

$\hat{x}_8$ rather than x_8 follows from accepted procedures in regression analysis (Jackson and Lawton, 1967). There will be eight of these plots, one for each variable, and graphics techniques such as *brushing* (Becker and Cleveland, 1987), which track aberrant deviations from one plot to another, may be of use. (Brushing will be illustrated in connection with the y-scores themselves.)

Note a pattern in Figure 5.2 of slanting "lines" of these residuals. The vertical distance between these lines corresponds, in original units, to 5 decibels, the discreteness of the original data. The two points farthest to the left represent the two individuals whose original data were − 10 decibels. The next line of points represents individuals with values of 0 decibels (there were no occurrences of − 5 decibels). The next line represents individuals with values of 5 decibels, and so on. Other than the discreteness effect, the residuals appear to be fairly normally distributed.

Nine individuals had significant values of T^2 that were fairly evenly distributed over the four pc's. One of them, #75, also had a significant residual sum of squares. There were eight others with significant y-scores although their T^2-values were not significant. None of these are very large and can probably be considered to be the random deviations one expects by chance, with the possible exceptions of #73 and #74 which also had significant values of Q.

In the study from which these data were taken, cases such as these were followed up and in many cases it was found that the medical records were incomplete, the individual having forgotten, or not having been asked for, some relevant piece of information or not realized that it was relevant and hence should not have been in the control group in the first place. Others were simply

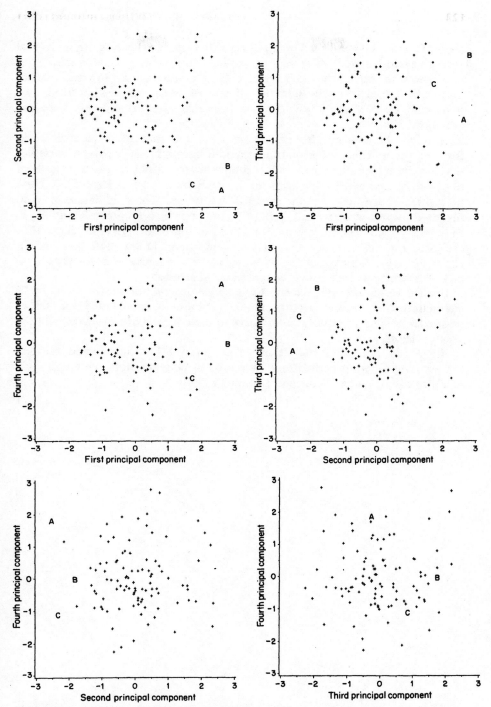

FIGURE 5.3. Audiometric example: pc scatterplots. (a) y_1 vs. y_2. (b) y_1 vs. y_3. (c) y_1 vs. y_4. (d) y_2 vs. y_3. (e) y_2 vs. y_4. (f) y_3 vs. y_4.

cases of people who had aged prematurely with respect to hearing. An individual with an excessive value of Q might be asked to repeat the examination.

Those who feel that $\alpha = .05$ was too large a value to use for this analysis may take solace from the fact that if $\alpha = .01$ had been used instead, two individuals (#29 and #73) would have been rejected by Q and only one more (#40) by T^2.

Although the four pc's from this sample are uncorrelated, as we shall see in Sections 14.3 and 16.2, some nonrandom patterns can produce pairs of variables that are linearly uncorrelated. It is a simple matter to obtain pairwise plots of the y-scores and these are shown for all six pairs of pc's in Figure 5.3. Note that three individuals are represented by letters. These individuals, #40 (denoted by A), #55 (B) and #78 (C), may appear to be possible outliers in the y_1 vs. y_2 plot and have been "brushed" in the other five plots. Since they are suspect in this first plot, they may also be suspect in any plots that contain either y_1 or y_2. A glance at the y_3 vs. y_4 plot will indicate that for those two pc's, at least, these individuals exhibited no aberrations.

Section 16.4.3 discusses the effect of a single observation on the estimates of characteristic roots and vectors. Hocking (1984) recommended the use of pc-scatterplots as a graphical alternative to discover the observations with the greatest effect.

Recall, in Section 2.10.8, that Jeffers (1967) had discovered clusters representing different populations by means of pc-scaterplots. This technique was also discussed by Weihs and Schmidli (1990).

CHAPTER 6

Operations With Group Data

6.1 INTRODUCTION

Most of the examples used to this point have been related to quality control applications. Quality control practitioners will realize that one basic element has been missing from all of these examples; this is the ability to test hypotheses about the *variability* of a process. Standard univariate quality control methodology requires the simultaneous maintenance of control charts for means and some measure of dispersion, either the range or standard deviation. So far, we have had neither because no provision has been made for dividing the data into rational subgroups. This chapter will be concerned with subgrouping multivariate data. Hotelling's generalized T-statistics will be introduced and it will be shown how PCA can be applied to this methodology. While the applications in this chapter will again feature quality control, these techniques may be applied to any multivariate data analysis where the data have been subgrouped. Similarities of these techniques and the multivariate analysis of variance will be discussed. This is not to be confused with the situation where two or more *samples* representing different populations are being analyzed. Methods of comparing PCA estimates among different samples will be presented in Section 16.6 and the possible use of PCA in discriminant and cluster analysis will be discussed briefly in Section 14.5.

6.2 RATIONAL SUBGROUPS AND GENERALIZED T-STATISTICS

The term *rational subgroup* dates back to the early days of statistical quality control (Shewhart, 1931). A subgroup usually represents some homogeneous unit of time or similar division of production effort. This information enables one to work not only with averages, which are more precise estimates and tend to normality regardless of the underlying distribution, but also some measure of dispersion such as the standard deviation or range. (For a more thorough discussion of univariate control charts, see, for example, Grant and Leavenworth,

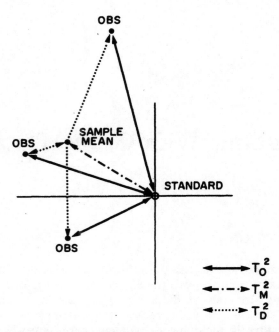

FIGURE 6.1. Generalized T^2-statistics. Reproduced from Jackson (1981a) with permission of The American Society for Quality Control and Jackson (1985) with permission of Marcel Dekker.

1988). It is conceivable that a process may exhibit excess variability but in a random pattern such that the mean may still be close to standard even though an unacceptable proportion of the produced goods are out of specification. A test for dispersion would resolve this situation.

A multivariate generalization of this subgroup approach was developed by Hotelling (1947) in which, for each subgroup of size n observations, he obtained three measures. The overall variability, T_0^2, of the subgroup about an aim or grand mean is broken into the sum of T_M^2, a measure of the distance of the subgroup mean from this target and T_D^2, a measure of the variability of the subgroup about its own mean. This is the multivariate analog of control charts for means and ranges. A geometric interpretation of these statistics for a two-dimensional example may be seen in Figure 6.1. T_M^2 tests the hypothesis that the process mean of the subgroup is equal to its aim; T_D^2 tests that the process variability of the subgroup is equal to a measure of dispersion, **S**; and T_0^2 is a combination of these two tests.

The formulas for these T-statistics based on subgroups of size n are

$$T_0^2 = \sum_{i=1}^{n} T_i^2 \qquad (6.2.1)$$

where

$$T_i^2 = (\mathbf{x} - \mathbf{x}_a)'\mathbf{S}^{-1}(\mathbf{x} - \mathbf{x}_a) \qquad i = 1, 2, \ldots, n$$

is the T^2-statistic computed for each observation and $\mathbf{x}_a$ is the aim or grand mean. T_0^2 can also be expressed as

$$T_0^2 = \text{Tr}[\mathbf{S}_n \mathbf{S}^{-1}]/n \qquad (6.2.2)$$

where $\mathbf{S}_n$ is the covariance matrix of the subgroup sample (using n, not $n - 1$, as a divisor since the deviations are in terms of the hypothesized mean rather than the sample mean). In this form, it is known as the *Lawley–Hotelling trace statistic* (Lawley, 1938; Hotelling, 1951). Several tabulations of this statistic have been produced, in particular by Davis (1970a,b, 1980) and is found in Appendix G.5 for $p = 2, \ldots, 10$ and $\alpha = .01$ and $.05$. This statistic is also related, asymptotically, to the F-distribution (Seber, 1984) and, ultimately, has an asymptotic χ^2-distribution with pn degrees of freedom. The χ^2-approximation is the one generally used in quality control, partially because of its simplicity and the ready availability of tables. (The F-approximation appears a bit messy, although really no problem in this day and age.) A normal approximation was given by Mijares (1990).

$$T_M^2 = n(\bar{\mathbf{x}} - \mathbf{x}_a)'\mathbf{S}^{-1}(\bar{\mathbf{x}} - \mathbf{x}_a) \qquad (6.2.3)$$

where $\bar{\mathbf{x}}$ is the mean of the subgroup. If $\mathbf{S}$ is estimated from g groups of n observations each (i.e., $g(n - 1)$ degrees of freedom), then under the assumption that the process mean equals its aim, T_M^2 will have the same distribution as the individual T_i^2 that make up T_0^2:

$$T_{p,g(n-1),\alpha}^2 = \frac{pg(n - 1)}{[g(n - 1) - p]} F_{p,[g(n-1)-p],\alpha}$$

which is the same form as (1.7.5) except for different degrees of freedom. T_M^2 also has an asymptotic χ^2-distribution with p degrees of freedom.

$$T_D^2 = \sum_{i=1}^{n} (\mathbf{x}_i - \bar{\mathbf{x}})'\mathbf{S}^{-1}(\mathbf{x}_i - \bar{\mathbf{x}})$$

$$= T_0^2 - T_M^2 \qquad (6.2.4)$$

This is also a Lawley–Hotelling statistic (but in terms of the sample mean, not the aim, and with divisor $n - 1$) and will have an asymptotic χ^2-distribution with $p(n - 1)$ degrees of freedom. The likelihood ratio test of Anderson (1984a) has been proposed as an alternative to T_D^2 (Alt, 1985) but it does not have the additive property of (6.2.4). Although it may be more powerful than T_D^2, just how much more powerful and against what alternatives has not been established at this time.

Hotelling illustrated these generalized procedures in his 1947 paper with some bombsight tests carried out during World War II, and a sequential procedure developed by Jackson and Bradley (1961a,b) involved ballistic missile testing. Hotelling (1954) cited J. S. Hunter as an early user of these techniques when he was at Aberdeen Proving Ground. A photographic example using PCA will be forthcoming in Section 6.5. In an example dealing with the growth of fruit trees, Pearce and Holland (1960) reduced the number of pc's from three to two by obtaining the variability about rootstock means rather than the grand mean of all rootstocks. This also says something about the effect of rootstocks. Their example points out that the distinction between subgroups of observations from a single population and samples representing different populations may be fine indeed.

6.3 GENERALIZED T-STATISTICS USING PCA

It is a simple matter to adapt the formulas in the previous section for use with principal components. Equation (6.2.1) for T_0^2 remains the same except that $T_i^2 = y'y$ as initially given in (1.7.1). To test the subgroup mean,

$$T_M^2 = n\bar{y}'\bar{y} \tag{6.3.1}$$

where $\bar{y}$ is the average of each of the k y-scores over the n observations in the subgroup. Given these, $T_D^2 = T_0^2 - T_M^2$ as in (6.2.3). The degrees of freedom associated with the χ^2- and F-distributions as well as the Lawley–Hotelling statistic remain the same except that p is replaced by k, the number of retained pc's.

As an illustration, let us return to the Ballistic Missile example, first presented in Section 2.3. Table 6.1 combines the same sample observation used in that chapter to illustrate the Q-statistic along with two more observations to form a subgroup of size 3. Also included are the first two y-scores for each observation, the associated T^2's, and some other computations that will be required for the discussion of residual procedures in Section 6.4. For this example, then, $k = 2$ and $n = 3$. From these data, we obtain the following generalized statistics.

a. *Overall*

$$T_0^2 = 4.24 + 1.90 + 1.13 = 7.27$$

To use the tables of the Lawley–Hotelling statistic in Appendix G.5, one needs to work with T_0^2/n. The degrees of freedom associated with S are $m_E = 39$, $m_H =$ our $(n = 3)$ and $d =$ our $(k = 2)$. For $\alpha = .05$, the critical value for T_0^2 is approximately 14.28 so T_0^2 is not significant. (The corresponding χ^2-approximation for $nk = 6$ degrees of freedom is 12.592.)

Table 6.1. Ballistic Missile Example. Observations and Averages

	Observations			Averages

$$\mathbf{x} = \begin{bmatrix} 15 & 10 & 10 \\ 10 & 15 & 5 \\ 20 & 15 & 10 \\ -5 & 10 & 0 \end{bmatrix} \qquad \begin{bmatrix} 11.67 \\ 10.00 \\ 15.00 \\ 1.67 \end{bmatrix} = \bar{\mathbf{x}}$$

$$\mathbf{y} = \begin{bmatrix} 1.094 & 1.374 & .672 \\ -1.744 & .099 & -.825 \end{bmatrix} \qquad \begin{bmatrix} 1.047 \\ -.824 \end{bmatrix} = \bar{\mathbf{y}}$$

$$T^2 = \quad 4.238 \qquad 1.898 \qquad 1.132$$

$$\hat{\mathbf{x}} = \begin{bmatrix} 16.9 & 11.4 & 9.3 \\ 14.3 & 15.2 & 8.5 \\ 7.5 & 11.6 & 4.9 \\ -.1 & 11.8 & 1.2 \end{bmatrix} \qquad \begin{bmatrix} 12.52 \\ 12.67 \\ 8.01 \\ 4.31 \end{bmatrix} = \hat{\bar{\mathbf{x}}}$$

$$\mathbf{x} - \hat{\mathbf{x}} = \begin{bmatrix} -1.9 & -1.4 & .7 \\ -4.3 & -.2 & -3.5 \\ 12.5 & 3.4 & 5.1 \\ -4.9 & -1.8 & -1.2 \end{bmatrix} \qquad \begin{bmatrix} -.85 \\ -2.67 \\ 6.99 \\ -2.64 \end{bmatrix} = \bar{\mathbf{x}} - \hat{\bar{\mathbf{x}}}$$

$$Q \qquad 202.36 \qquad 16.80 \qquad 40.19$$

$$p_i \qquad\quad .015 \qquad\quad .680 \qquad\quad .403$$

b. *Subgroup Mean*

$$T_M^2 = n\bar{\mathbf{y}}'\bar{\mathbf{y}} = 3[(1.047)^2 + (-.823)^2] = 5.32$$

The limit for T_M^2 (1.7.5) is

$$T_{2,40,.05}^2 = [(2)(39)/38]F_{2,38,.05} = 6.66$$

so this is not signficant.

c. *Subgroup Dispersion*

$$T_D^2 = T_0^2 - T_M^2 = 1.95$$

The Lawley–Hotelling critical value, using $m_H = (n - 1) = 2$, is approximately 10.68 so T_D^2 is not significant for $\alpha = .05$. (The critical value of χ^2 for $(n - 1)k = 4$ degrees of freedom is 9.488.)

It has been my experience as well as that of my colleagues that for multivariate quality control T_0^2 is of very little use other than to compute T_D^2. If it is significant, T_M^2 and/or T_D^2 will be significant anyhow and this is the information on which action may be taken.

6.4   GENERALIZED RESIDUAL ANALYSIS

6.4.1   Generalized Q-Statistics

We now need to add some generalized statistics that take into account the residuals arising from the inability of the retained pc's to exactly reproduce the original data. The Q-statistic was introduced in Section 2.7.2 to test the size of the residuals and the purpose of this section will be to generalize Q for subgroup data. Two statistics will be developed: one, Q_0, is for the *overall* residual variability and the other, Q_M, is for the *average* residual variability (Jackson and Mudholkar, 1979).

Unlike the T^2-statistics, which may be added to obtain the overall T_0^2, the Q-statistics are not additive. However, the desired effect may be obtained using Fisher's *combination of independent tests of significance* (Fisher, 1932). Recall that expression (2.7.6) for the critical value of Q, given α, can be inverted to (2.7.5) to obtain the normal deviate, c_i, corresponding to the probability associated with the ith sample value of Q. Let the one-tail probability associated with c_i be denoted by p_i. Then

$$Q_0 = -2 \sum_{i=1}^{n} \ln (p_i) \tag{6.4.1}$$

which, under the null hypothesis, is distributed like χ^2 with $2n$ degrees of freedom. Table 6.1 displays both the individual Q_i's and the p_i's associated with them. From these,

$$Q_0 = -2[\ln (.015) + \ln (.680) + \ln (.403)] = 10.99$$

On comparing this with $\chi^2_{(2)(3)} = 12.592$, it is seen that Q_0, representing the overall residual variability, is not significant. Other properties related to Fisher's combination of tests as well as some of its competitors are discussed by Littell and Folks (1971, 1973), Brown (1975), and Mudholkar and George (1979). For situations where the tests are independent, as they are in this example, Fisher's procedure is asymptotically optimal.

The other statistic, Q_M, tests the hypothesis that the principal component model holds for an average of the n observations. Fortunately, Q_M has the same distribution and the same degrees of freedom as Q itself. The individual observations in Q are now replaced with averages, viz.,

$$Q_M = n(\bar{\mathbf{x}} - \hat{\bar{\mathbf{x}}})'(\bar{\mathbf{x}} - \hat{\bar{\mathbf{x}}}) \tag{6.4.2}$$

where $\hat{\bar{\mathbf{x}}}$ is the predicted average vector using the expression

$$\hat{\bar{\mathbf{x}}} = \mathbf{V}\bar{\mathbf{y}} \tag{6.4.3}$$

From the data in Table 6.1, where both $\bar{\mathbf{x}}$ and $\hat{\bar{\mathbf{x}}}$ are displayed,

$$Q_M = 3[(-.85)^2 + \cdots + (-2.64)^2] = 191.04$$

For this example, with $\alpha = .05$, the critical region for Q_M is $Q \geq 140.45$ as it was in Section 2.7.2 for individual observations. This result is significant and the principal contributor appears to be x_3, mostly coming from the first observation.

As was the case in Section 2.7.2, there is also an order in which the tests should be conducted. A flow chart for this procedure is shown in Figure 6.2. The first test should be for Q_0, which, if significant, should make the subgroup suspect, and the probable cause can be ascertained by looking at the individual Q-statistics. If one or more of them is significant, the data should be reexamined and/or deleted. Next in line is T_D^2, which is a measure of the within-subgroup variability of the pc's. If the process makes it that far, Q_M is next, and if that is significant, it indicates that although the individual residuals were not important, by the time one had averaged n of them the pc model now did show some inadequacy. If Q_M is not significant, it means that no evidence of excess testing or measurement variability has been found and one can proceed with the main business at hand, T_M^2, which tells us if the process level conforms to the aim. If it does not, then the y-scores for the individual pc's should be examined for the cause.

6.4.2 Generalized Hawkins Statistics

The Hawkins test for residuals, given in Section 2.7.3, can also be generalized for subgroup data. Recalling that this test was the sum of squares of the last $p - k$ y-scores, that is,

$$T_H^2 = \sum_{i=k+1}^{p} y_i^2$$

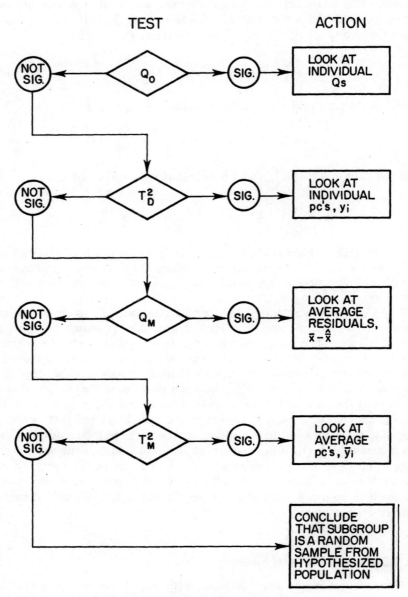

FIGURE 6.2. PCA procedure for subgroup data.

the generalized statistic corresponding to Q_0 *is* additive and will be

$$T^2_{0(H)} = \sum_{i=1}^{n} T^2_{Hi} \qquad (6.4.4)$$

This is also a Lawley–Hotelling trace statistic having the same distribution as (6.2.1) with p being replaced by $p - k$ and with an asymptotic χ^2-distribution with $n(p - k)$ d.f. For the data in Table 6.1, the y-scores for the deleted pc's are

$$y_3 \quad .559 \quad -.100 \quad .664$$

$$y_4 \quad 3.431 \quad .988 \quad 1.301$$

yielding values of T^2_H of 12.084, .986, and 2.134; from these, $T^2_{0(H)} = 15.204$, which is significant when compared with a critical value of 14.28 for $\alpha = .05$ using Appendix G.5. (Recall, from the previous section, that Q_0 was *not* significant. Actually, Q_0 would have been significant for $\alpha = .09$ while $T^2_{0(H)}$ has a probability of about .02, so the discrepance is not great but as they are on the opposite sides of .05 the conclusions would be different for this particular set of data.)

Corresponding to Q_M, we have

$$T^2_{M(H)} = n \sum_{i=k+1}^{p} \bar{y}_i^2 \qquad (6.4.5)$$

with the same distribution as T^2_H. For these data, $\bar{y}_3 = .374$ and $\bar{y}_4 = 1.907$ so $T^2_{M(H)} = 11.330$, which is significant when compared with its limit for $\alpha = .05$, 6.66. (This conclusion *does* agree with Q_M of the previous section.)

The other Hawkins statistic,

$$\underset{i > k}{\text{Max}} |y_i|$$

could be generalized to $\text{Max}|\bar{y}_i|$ to correspond to Q_M or $T^2_{M(H)}$. A test corresponding to Q_0 or $T^2_{0(H)}$ would require another combination-of-tests procedure similar to that employed for Q_0.

6.5 USE OF HYPOTHETICAL OR SAMPLE MEANS AND COVARIANCE MATRICES

Although the subject of targets has been discussed before, it is worthwhile repeating in the context of these generalized operations. The generalized statistic T^2_M is a measure of the distance between a subgroup mean and some reference point or target. This target is not restricted. It may be the mean of a set of base period data, in which case the null hypothesis is that the true mean of a particular subgroup is the same as the sample mean of the base period based on the within-subgroup variability of the base period. However, the target may, instead, be an arbitrary aim related to the specifications of a particular product or process. In this case, the null hypothesis relates the subgroup mean to this aim and a significant value of T_M^2 indicates that the mean of this subgroup is significantly different from this aim based, again, on the within-subgroup variability of the base period. The same reasoning holds for T_0^2.

T_D^2, meanwhile, is a measure of the variability of a subgroup around its own mean and hence is independent of the choice of target. However, one might wish to replace a sample covariance matrix with some hypothesized matrix Σ_0, possibly related to tolerances. Then the variability with a particular subgroup can be compared to Σ_0 and can also be used in T_0^2 and T_M^2.

6.6   NUMERICAL EXAMPLE: A COLOR FILM PROCESS

The numerical example used to illustrate these generalized statistics is due to Jackson and Morris (1957) and is similar to the black-and-white film example introduced in Section 2.9. As in that example, this one monitors film by means of a developed strip of film that has previously received a series of exposures of varying intensity. In the black-and-white example, the density measurements of these exposures are plotted as a function of exposure, an example being displayed in Figure 2.4. The present example will deal with a *color* film product and although the test exposures are still made with white light, each developed exposure will now be measured with respect to the density to red, green, and blue light. Rather than a single response curve as in Figure 2.4, there will now be three of them, one for each color density. This example dates back to an earlier, precomputer, era in which only a few exposure levels were generally used for control. For this example, three levels were used. One of these was taken from the high-density or "shoulder" portion of these curves, another from the low-density or "toe" portion, and the third, called "middle-tone" from the middle section of the curves, which represented the normal picture-taking range. Three colors and three exposure levels result in a nine-variable problem.

This particular example deals with the control of a film process used to develop customer film—a typical photofinishing operation. However, photofinishing 40 years ago was somewhat different from the methods employed today, which involve equipment that is much less complex with regard to the chemical solutions involved. Formerly, the nature of the equipment required that it generally be run continuously rather than on a batch basis as is commonplace today. Monitoring of a photofinisher process was, and still is, carried out by means of preexposed film strips that are processed along with the customer film. The frequency of these tests depends on the nature of the process and the quality control philosophy in force. The data for this example came from a quality assurance operation in which each processing machine within the Eastman Kodak Company throughout the world was monitored once a day using film strips exposed on a single sensitometer on a homogeneous film coating and measured on a single densitometer. The data for this example represent a single machine.

This example used rotational subgroups, in this case within-weeks. The machines were started up on Monday morning and were run until they were shut down at the end of the week. This Monday morning start-up could be quite complicated as there were over a 100 chemical variables in the various solutions. It was very difficult to exactly match one week's level of operation the following week, but once the process had attained equilibrium for the week, so that processing of customer film could commence, it was fairly stable about that level. There were, then, two components of variability—the inherent variability within weeks and the variability between weeks; the latter had an assignable cause—the Monday morning start-up. In following the philosophy of Shewhart control charts, the goal was to control the process within the limits with which it was expected to vary *within* a week.

The base period for this analysis consisted of 31 weeks. The within-week covariance matrix based on 108 degrees of freedom is displayed in Table 6.2. (One would expect $31 \times 4 = 124$ degrees of freedom based on a five-day week, but this particular machine did not always have enough customer orders to keep it going all week. Film usage was much more seasonal then.) It can be seen that most of the covariances are positive and that the variances change quite a bit from one part of the curve to another. The corresponding correlation matrix is given in Table 6.3. Note that high correlations exist among many of the variables, particular for the shoulder and middle-tone densities, but that the toe densities are somewhat independent of the others.

The stopping rule employed here was that one would retain pc's until the residual variability was equal to that attributable to the sensitometer and densitometer and that resulted in five pc's. The W-vectors corresponding to

Table 6.2. Color Film Example. Within-week Covariance Matrix ($\times 10^{-5}$)

Shoulder			Middle-Tone			Toe		
Red	Green	Blue	Red	Green	Blue	Red	Green	Blue
177	179	95	96	53	32	−7	−4	−3
	419	245	131	181	127	−2	1	4
		302	60	109	142	4	4	11
			158	102	42	4	3	2
				137	96	4	5	6
					128	2	2	8
						34	31	33
							39	39
								48

Table 6.3. Color Film Example. Within-week Correlation Matrix

Shoulder			Middle-Tone			Toe		
Red	Green	Blue	Red	Green	Blue	Red	Green	Blue
1	.66	.41	.57	.34	.21	−.09	−.05	−.03
	1	.69	.51	.76	.55	−.02	.01	.03
		1	.27	.54	.72	.04	.04	.09
			1	.69	.30	.05	.04	.02
				1	.72	.06	.07	.07
					1	.03	.03	.10
						1	.85	.82
							1	.90
								1

these pc's, and their associated roots, are given in Table 6.4. The interpretations of the pc's are as follows:

y_1 is essentially a density shift in the shoulder and middle-tone portions of the curves; the coefficients are all positive. This is much the same as the first pc for the black-and-white example.

y_2 is, in photographic parlance, a warm–cold color-balance shift. Excessive blue density will result in the picture being too yellow, which is considered a "warm" color. Lack of blue density will produce a blue or "cold" picture.

Table 6.4. Color Film Example. W-Vectors

	w_1	w_2	w_3	w_4	w_5
Shoulder					
Red	3.25	−10.96	11.47	8.92	12.55
Green	6.97	−3.41	5.09	1.29	−22.75
Blue	5.16	13.27	6.57	3.58	14.89
Middle-tone					
Red	2.79	−11.09	−12.74	−3.37	16.18
Green	3.45	−.85	−13.80	−8.70	−5.59
Blue	2.89	8.43	−7.49	−7.27	5.41
Toe					
Red	.03	1.29	−7.13	13.62	−1.44
Green	.06	1.21	−7.41	15.69	−2.41
Blue	.15	1.99	−7.87	17.36	−1.93
Characteristic root					
l_i	.00879	.00196	.00129	.00103	.00081

Excessive red density will produce a blue–green or cyan picture, which is considered "cold," and lack of red density will produce a red or "warm" picture. This pc represents, essentially, the relationship between the red and blue densities, independent of the green for the shoulder and middle-tone.

y_3 has positive coefficients in the shoulder and negative for the middle-tone and toe. This describes photographic contrast. High contrast results in darker shadow or lighter highlights, so that much detail is lost in the picture. This is comparable to the second pc in the black-and-white example.

y_4 has positive coefficients for both the toe and shoulder and negative for the middle-tone. This is similar to the third pc in the black-and-white example and, like it, pertains primarily to the toe densities. This can be verified by looking at the diagonal elements of $v_4 v_4'$, the amount explained by the fourth pc.

y_5 represents a green–magenta color-balance shift relative to the other colors. This is a second color-balance parameter and is orthogonal to the warm–cold one. (Customarily, the specification of color-balance requires two dimensions.)

Observe that y_1, y_3, and y_4 were similar to the three components obtained in the black-and-white example and the other two are related to color-balance. It was noted in Section 2.9.2 that the physical importance of a pc is not necessarily related to the relative size of its characteristic root. The point is worth repeating in connection with this example. Shifts in the two color-balance pc's can be detected much more readily by the human eye than can corresponding shifts in density or contrast. Because this example was among the first to be published, it has been used in a number of multivariate textbooks. Some of these authors have arbitrarily set the number of retained pc's at two or three on the basis of the proportion explained of the trace, and in doing so would have eliminated several important pc's. [Srivastava and Carter (1983, p. 291) analyzed these same data using the correlation matrix and found, in that case, that the first pc was the same but that what was the fourth pc in the covariance matrix was the *second* pc in the correlation matrix owing to the fact that the toe densities now had equal weight, which they had not had using the covariance matrix. The warm–cold pc came in third.]

Figure 6.3 shows the original data for an 8-week period of operation subsequent to the base period. The central lines for these control charts are the standards for the process. The limits for this example were 99.7%, the so-called "three-sigma" limits. The red and green shoulders lost density until the sixth week, when they went out of control and after which they behaved quite normally. The blue shoulder was more variable. The red and green middle-tone appear to be satisfactory except for the sixth week, but again the blue was more variable. The toe densities all appeared to be on the high side for the first 6 weeks and then appeared normal thereafter.

The y-scores for these data are shown in Figure 6.4. They are no longer in order by size of the root but are grouped so that the color-balance pc's are together. As has already been noted, the process got lighter until the sixth week,

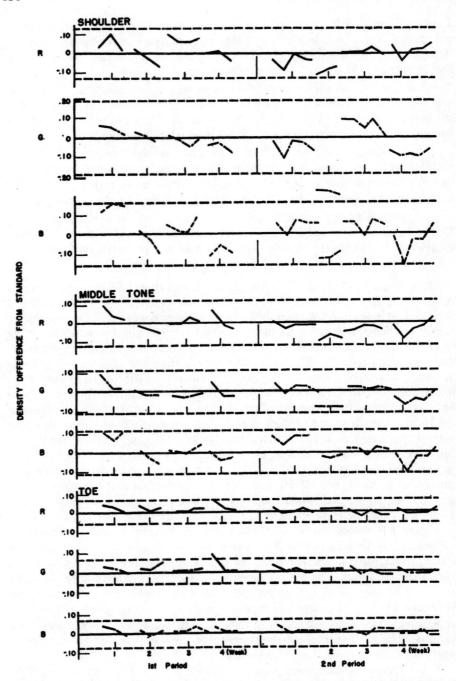

FIGURE 6.3. Color Film example: control charts for original variables; 99.7% limits. Reproduced from Jackson and Morris (1957) with permission of the American Statistical Association.

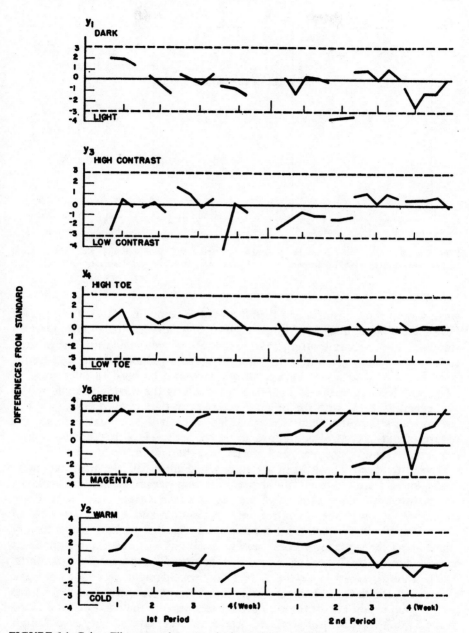

FIGURE 6.4. Color Film example: control charts for principal components; 99.7% limits. Reproduced from Jackson and Morris (1957) with permission of the American Statistical Association.

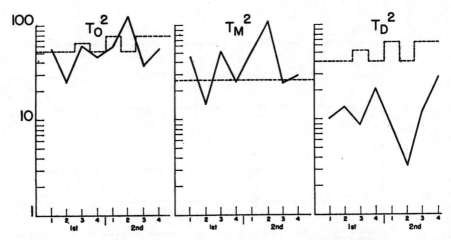

FIGURE 6.5. Color Film example: generalized T^2-statistics; 99.7% limits. Reproduced from Jackson and Morris (1957) with permission of the American Statistical Association.

when it went out of control and then returned to being in control. The contrast shows some level shifts. The toe pc appears high for the first month, then on standard. The more interesting charts are those for color-balance. The process runs intermittently green and magenta, with green predominating, and trends from warm to cold about once a month. It would be hard to infer some of these from looking at the original data, particularly the warm–cold effect. The pc's generally agree with the original variables in that when one set is out of control, so is the other, and vice versa. However, it is much easier to see the nature of out-of-control conditions using the pc's. These parameters reflect, much more readily, chemical and mechanical changes in the process.

The generalized T^2-statistics are shown in Figure 6.5. The irregular limits for T_O^2 and T_D^2 are the result of unequal subgroup sizes from week to week. T_O^2 was flirting with the limit most of the first 6 weeks, going out of control the sixth week. T_M^2 was out of control most of the time and T_D^2 was never out of control, verifying the statement made earlier that the process had difficulty maintaining the same level from week to week but was quite stable about its level for the week. This example predates the use of acceptance control charts, which were discussed in Section 2.10.3 and which might or might not have regarded these shifts with more amiability. Clearly, this process would have been a good candidate for these control charts because the condition indicated by the T_M^2 in Figure 6.5 still produced product of acceptable quality even though the mean level shifted significantly from week to week.

Some illustrations of the residual statistics can be seen in Table 6.5. Three sets of three observations each, as difference from standard, are shown. Because of the availability of tables, we shall use 95% limits this time. The limits for all of the statistics given here are as follows (because of the large sample size

Table 6.5. Color Film Example. Data in Terms of Deviation from Standard

	Set #1				Set #2				Set #3	
	x_1	x_2	x_3	$\bar{x}$	x_1	x_2	x_3	$\bar{x}$	x_3	$\bar{x}$
Shoulder										
Red	.02	−.03	−.08	−.030	−.11	−.09	−.08	−.093	.22	.007
Green	.02	.01	−.01	.007	−.28	−.29	−.31	−.293	.50	−.023
Blue	.01	−.03	−.10	−.040	−.12	−.12	−.09	−.110	.20	−.013
Middle-tone										
Red	−.01	−.03	−.05	−.030	−.11	−.06	−.09	−.087	.15	−.007
Green	.00	−.01	−.01	−.007	−.08	−.08	−.08	−.080	.15	−.003
Blue	.01	−.03	−.06	−.027	−.02	−.03	−.01	−.020	.05	.000
Toe										
Red	.03	.00	.02	.017	.02	.02	.02	.020	.02	.020
Green	.02	.01	.05	.027	.02	.02	.02	.020	.02	.020
Blue	.02	−.01	.01	.007	.02	.01	.03	.020	.01	.013
Principal components										
y_1	.26	−.39	−1.19	−.438	−3.56*	−3.46*	−3.43*	−3.484*	6.32*	−.231
y_2	.14	−.02	−.25	−.046	1.78	.93	1.83	1.513*	−2.76*	−.015
y_3	−.07	.26	−.99	−.267	−1.27	−1.57	−1.29	−1.368*	1.66	−.385
y_4	1.27	.03	.84	.733	.37	.27	.74	.467	1.91	.848
y_5	−.29	−1.65	−3.51*	−1.827*	1.65	2.90*	3.51*	2.676*	−3.87*	.219
Summary statistics for individuals										
T^2	1.79	2.93	15.50*		20.32*	23.79*	29.68*		68.91*	
Q	.00041	.00038	.00116		.00280*	.00145	.00397*		.00194	
P_i	.432	.459	.109		.012	.070	.003		.035	
Summary statistics for subgroups										
T^2				20.22				73.79*		113.02*
T_0^2				12.42*				71.02*		2.90
T_M^2				7.80				2.77		110.12
T_D^2				7.67				25.78*		20.86*
Q_0										
Q_M				.00074				.00785*		.00100

* Starred quantities are significant at 5% level.

in the base period, these are all large-sample approximations):

y-scores: ± 1.96
y-score averages: ± 1.132
T^2: 11.06
T_0^2: 24.99
T_M^2: 11.06
T_D^2: 18.31
Q and Q_M: .0017
Q_0: 12.59

Set #1 is a case where only the means have shifted and the only significant results are T_M^2 and two of the pc averages. For set #2, T_M^2 is again significant and T_0^2 is not, which might lead one to believe that the process had again shifted level but was still stable. However, both Q_M and Q_0 are significant, which indicates that there are some consistent problems with the residuals in this sample and that the PCA model did not adequately characterize the process at the time this sample was obtained. Note, in particular, that this is not a general deterioration of the process, because T_D^2 is still small, indicating that whatever has happened to the process, it is a real shift, not the result of a single outlying observation or excessive variability within the set.

By contrast, consider set #3. The first two observations are the same as set #2 but x_3 has been contrived to bring the average of the sample close to the standard. In so doing, both T^2 and Q for this third observation are very large. For this set, T_M^2 is not significant, as was intended, and neither is Q_M, which implies that the PCA model holds for the averages of this set. However, both

Table 6.6. Color Film Example. Residuals for Data in Table 6.5

	Set #1			Set #2			Set #3
	x_1	x_2	x_3	x_1	x_2	x_3	x_3
Shoulder							
Red	.008	−.007	−.009	.028	.020	.034	−.022
Green	−.002	.001	.002	−.012	−.008	−.015	.010
Blue	−.006	.006	.008	−.015	−.010	−.016	.012
Middle-tone							
Red	−.006	.005	.006	−.025	−.018	−.031	.020
Green	.001	−.001	.000	.019	.014	.027	−.017
Blue	.011	−.010	−.013	.025	.017	.025	−.017
Toe							
Red	.011	.000	−.004	.001	.004	−.002	.009
Green	−.002	.009	.021	.003	.006	.000	.000
Blue	−.005	−.010	−.018	.001	−.005	.007	−.011

T_D^2 and Q_0 are significant indicating that the process is unstable, the PCA model not holding for the individual observations and the within-set variability being excessive besides. The residuals for these examples are shown in Table 6.6.

6.7 GENERALIZED *T*-STATISTICS AND THE MULTIVARIATE ANALYSIS OF VARIANCE

The relationship of PCA to the multivariate analysis of variance (MANOVA) will be discussed in Chapter 13. While not a substitute for MANOVA, some of the techniques discussed in this chapter can be used either as a cursory examination of the data or to enhance the other type of analysis. This can, of course, be accomplished without the use of PCA, but PCA—particularly with the use of residual analysis—might enhance the results.

Hotelling's 1947 bombsight paper had two response variables: horizontal and vertical deviations of a bomb from its target. There were a myriad of predictor variables, many of which were accounted for by multiple regression. The resultant residual matrix was used as a basis for the various T^2 tests presented above and the other predictors were studied using these same statistics in control chart form. PCA was not used in this study but might be useful in a similar type of study with more variables.

CHAPTER 7

Vector Interpretation I: Simplifications and Inferential Techniques

7.1 INTRODUCTION

Most of the examples presented so far in this book have produced characteristic vectors whose coefficients were readily interpretable and whose pc's were fairly easy to identify. The Chemical example in Chapter 1 had pc's that were identified as "process" and "testing and measurement." The interpretation of the first two pc's of the Ballistic Missile example in Chapter 2 were straightforward; the last two were not, but that was expected because their characteristic roots were not significantly different. The first four pc's of the Audiometric example in Chapter 5 were easily identifiable as were the first three pc's of the Black-and-White Film example in Chapter 2 and the first five pc's of the Color Film example in Chapter 6. However, in general, life is not this simple. All five of these examples had the advantage of coming from the physical and engineering sciences and used sets of variables that all had the same units. Moreover, in the case of the audiometric and film examples, these variables represented values of continuous functions: frequency in the first case and exposure in the other two. These examples were deliberately chosen because they did have "happy endings" and hence were useful in illustrating the various techniques presented. The majority of examples from these fields of application do have readily interpretable vectors. In such fields as psychology, education, the social sciences, and market research, the variables are seldom in the same units and easy interpretation is by no means a sure thing.

Many applications of PCA are associated with "test batteries," a collection of tests such as school examinations and aptitude tests, of which the Seventh Grade example in Chapter 3 is typical. The units may or may not be the same but the variables certainly do not represent sampling of a continuous function.

This situation is also applicable to science and industry where the variables are now physical tests but, again, quite possibly in different units.

The next two chapters will be devoted to techniques useful when the characteristic vectors appear to be either difficult to interpret or difficult to work with as well as some inferential techniques used to better characterize a data set. This chapter will concentrate on methods of simplifying vectors, not only as an aid to interpretation but as an expedient to operations. Also considered in this chapter are some properties related to vector coefficients, including some tests of hypotheses. Chapter 8 will consider an alternative technique called *rotation* that can be used to investigate the structure of a set of multivariate variables when the PCA model adequately characterizes the data but is not readily interpretable.

Some special techniques that have been suggested for continuous functions will be discussed in Section 16.7.

7.2 INTERPRETATION. SOME GENERAL RULES

Before discussing these techniques, a few general comments about vector interpretation may be in order.

a. If the covariance or correlation matrix from which the vectors are obtained contains all positive elements, the first characteristic vector will also have all positive elements.

b. If the number of variables is small, there are only a few patterns of coefficients possible, since the vectors are constrained to be orthogonal. The Ballistics Missile and Seventh Grade examples are cases in point. Each of them had four variables and in both cases the first vector had all positive coefficients and the second was a contrast between two pairs of variables, gauges in one case and reading vs. arithmetic in the other. With only two vectors left, one of the interpretations may be that those are the only orthogonal contrasts left. For the case $p = 2$, there are only two possibilities for the vectors, a weighted sum and a weighted difference of the original variables.

c. Quite often, vectors corresponding to the smaller roots may have a single large coefficient. This means that there is still some variability left to be accounted for but that the off-diagonals of the residual matrix are quite small, relative to the diagonals. Practically all of the structure has been accounted for and there is nothing left but some individual residual variances. As we shall see in Chapter 17, factor analysis allows for this situation as part of its model.

d. The interpretation of characteristic vectors associated with zero roots was discussed in Section 2.4 and were shown to represent linear constraints among the variables. Box et al. (1973) pointed out that rounding error can cause these roots to be positive and gave some rules for identifying such situations. They suggested that any roots that were the same order of magnitude as the rounding error should be suspect and that the corresponding vectors may be a clue to possible dependencies among the variables.

e. As will be demonstrated in Section 7.5, the coefficients within a vector and to some extent between vectors are correlated and this should be kept in mind when attempting to interpret them.

f. Flury (1988, p. 35) gave a procedure for what he termed the *redundancy* of variables on components. A variable is said to be redundant for a pc if its coefficient for that variable is zero. This procedure tests whether or not certain coefficients, or blocks of coefficients over several vectors, are different from zero. (Redundancy, in this context, does not have the same definition as it will in Section 12.6.3 in connection with the relationship between two sets of variables.)

7.3 SIMPLIFICATION

In this section, we shall obtain some simple approximations to characteristic vectors and see how adequately they represent the data. The Chemical example in Chapter 1 had two variables: x_1 = analysis method 1 and x_2 = analysis method 2. From a set of 15 pairs of observations, the principal components were found to be

$$\mathbf{y} = \begin{bmatrix} y_1 \\ y_2 \end{bmatrix} = \mathbf{W}'[\mathbf{x} - \bar{\mathbf{x}}] = \begin{bmatrix} .602 & .574 \\ -2.348 & 2.462 \end{bmatrix} \begin{bmatrix} x_1 - \bar{x}_1 \\ x_2 - \bar{x}_2 \end{bmatrix}$$

If a routine quality control operation was to be established in which the means or standard values could be stored in a microcomputer along with $\mathbf{W}$, $\mathbf{V}$ and the limits for T^2 and the y-scores, then the data as collected could also be fed into the computer and all of the relevant statistics easily computed and tested. If one did not have a computer and there were several variables, not just two, much more effort would be involved, even with the advent of pocket calculators that can handle matrix algebra. Recall that the use of $\mathbf{W}$-vectors produces pc's that are uncorrelated and have unit variances, the latter property being the main advantage of $\mathbf{W}$-vectors over $\mathbf{U}$ or $\mathbf{V}$. Someone having to do these by hand might prefer a more simple procedure even at the expense of weakening some of the properties of the pc's.

The first row of $\mathbf{W}'$ has coefficients that are nearly equal. What would happen if one substituted, for this pc, the average of $x_1 - \bar{x}_1$ and $x_2 - \bar{x}_2$ or, even more simply, their sum? To study this effect, a matrix $\mathbf{T}$ is made up of the first characteristic vector $\mathbf{w}_1$ and the approximation to it that one might wish to use, $\mathbf{a}_1$. In this case we shall use, as our approximation, the sum, so

$$\mathbf{T} = [\mathbf{w}_1 \mid \mathbf{a}_1] = \begin{bmatrix} .602 & 1 \\ .574 & 1 \end{bmatrix}$$

The product $T'ST$ will produce the covariance matrix of these quantities. Using S from Section 1.2,

$$T'ST = \begin{bmatrix} 1.00 & 1.70 \\ 1.70 & 2.89 \end{bmatrix}$$

The variance of y_1 is 1.00 as expected, the variance of the approximation is 2.89 and the covariance between them is 1.70. From this, we can determine that the correlation between the first pc and this approximation, based on the original data set, is essentially 1.0. By the same procedure, it can be determined that the correlation between the second pc and the *difference* between the original variables is .995. If these appear to be satisfactory, we can then form the product $A'SA$ where

$$A = \begin{bmatrix} 1 & -1 \\ 1 & 1 \end{bmatrix}$$

$$A'SA = \begin{bmatrix} 2.8915 & -.0643 \\ -.0643 & .1743 \end{bmatrix}$$

This tells us that the correlation between these approximations is only

$$r = -.0643/\sqrt{(2.8915)(.1743)} = -.09$$

There are three criteria for an acceptable approximation:

1. It should have a high correlation with the principal component that it is approximating.
2. It should have low correlations with the approximations to the other principal components.
3. It should be simple to use.

This particular set appears to fulfill all three requirements. The use of these approximations for this situation could, because of its simplicity, result in a more effective operation than the original principal components. However, as the new "pc's" are no longer uncorrelated, the formula for T^2 must be changed to

$$T^2 = y_s'[A'SA]^{-1}y_s \tag{7.2.1}$$

where y_s are the scores of the "pc's" using the simplified vectors. If desired, they could be scaled to have unit variance.

Before examining an example with a larger number of variables, recall from the previous section that for small p there are usually only a few possibilities for the characteristic vectors because of the constraints of orthogonality. This

Table 7.1. Color Film Example. U-Vectors and Approximations

	u_1	a_1	u_2	a_2	u_3	a_3	u_4	a_4	u_5	a_5
Shoulder										
Red	.30	1	−.49	−1	.41	1	.29	0	.36	1
Green	.65	1	−.15	0	.18	1	.04	0	−.65	−2
Blue	.48	1	.59	1	.24	1	.12	0	.42	1
Middle-tone										
Red	.26	1	−.49	−1	−.46	−1	−.11	0	.46	1
Green	.32	1	−.04	0	−.50	−1	−.28	0	−.16	−2
Blue	.27	1	.37	1	−.27	−1	−.23	0	.15	1
Toe										
Red	.00	0	.06	0	−.26	0	.44	1	−.04	0
Green	.01	0	.05	0	−.27	0	.50	1	−.07	0
Blue	.01	0	.09	0	−.28	0	.56	1	−.06	0
Characteristic root	.00879		.00196		.00129		.00103		.00081	

implies that success with approximations for small p is quite likely. In addition to the example below, a detailed example of approximations for the Ballistic Missile example may be found in Jackson (1959) and another example was given by Bibby (1980).

We shall return to the Color Film Process example introduced in Section 6.6. This example proved to be a way of installing multivariate quality control in a production environment, that, in that day and age, would not have been possible otherwise. The initial PCA produced a set of five pc's out of a set of nine variables. Table 7.1 shows both the U-vectors and the simple approximations proposed for use as a quality control procedure. The approximations were certainly simple, exceedingly so, because they were designed for the "pencil and paper" operations of the early 1950s. How well they did in fulfilling the other conditions can be seen in Table 7.2, which shows the correlations of the simplified "pc's" both with the actual pc's and with each other. Note that for even these simple approximations, the correlations with the pc's are fairly good, the lowest being for the third component, $r = .74$. There were a couple of disturbing correlations among the approximations, somewhat alleviated later by the replacement of the third approximation with a new form designed to make it less correlated with the first. Jackson and Morris (1957) included control charts for these simplified variables for the same data as displayed in Figure 6.3. To the casual user, these would appear to be little different from Figure 6.4. Weihs and Schmidli (1990) advocated making scatter plots of each pc with its corresponding approximation. When the correlations between approximations are disturbingly high, scatter plots of these quantities may also be of use.

The use of these simplified vectors to reduce the computational effort may not seem as impressive as it did 40 years ago, but there are still cases where it is worthwhile. Such an example will be presented in Chapter 9. Moreover, simplification along the lines presented above may still be useful in interpreting vectors, particularly for larger problems and for confirmatory PCA models, the subject of the next section. They may also be of some use to a statistician in

Table 7.2. Color Film Example. Correlation of Pc's and Simplified Pc's

Pc's produced by	a_1	a_2	a_3	a_4	a_5
u_1	.99	.20	−.56	−.04	−.54
u_2	−.04	.97	−.05	−.15	.15
u_3	.06	−.01	.74	−.50	−.18
u_4	.03	.11	.35	.84	−.16
u_5	.08	−.08	.09	.08	.78
a_1	1	.16	.48	.04	−.49
a_2		1	.13	.07	−.04
a_3			1.	−.05	−.19
a_4				1.	−.02
a_5					1

explaining the concept of PCA to someone who is being exposed to it for the first time, even though the simplified vectors are not used after that.

Before leaving this section, a word should be said for the most basic simplification of all—rounding. Some of the examples in this book show the vector coefficients displayed with only two digits. This may not be enough if one wishes to check some computations but it is certainly enough for interpretation and will probably enhance it. (Occam's razor: Less is more.) Even to use severely rounded coefficients operationally may not cause much of a problem. The pc's will no longer be exactly uncorrelated and the variances may change a bit, but this should not be of any consequence. The quantitative effects of rounding will be discussed in Section 16.4.4.

7.4 USE OF CONFIRMATORY PCA

It may seem reasonable, particularly in quality control applications, to test the hypothesis that a simplified vector is not significantly different from its companion characteristic vector, and if that hypothesis is accepted to use the simplified pc's. After all, in modeling a system,

$$y = x_1 - x_2$$

might *seem* more appropriate for use than

$$y = .6016x_1 - .5739x_2$$

Such hypotheses may be tested using the procedures presented in Section 4.6 and, in fact, this was one of the examples used there. Both $x_1 - x_2$ and $x_1 + x_2$ were found to be adequate representations of the characteristics vectors.

Let us revisit the Color Film example of the previous section and submit these approximations to a statistical test. To do so, the simplifications in Table 7.1 must be normalized to unit length. These become:

a_1	a_2	a_3	a_4	a_5
.408	−.5	.408	0	.289
.408	0	.408	0	−.577
.408	.5	.408	0	.289
.408	−.5	−.408	0	.289
.408	0	−.408	0	−.577
.408	.5	−.408	0	.289
0	0	0	.577	0
0	0	0	.577	0
0	0	0	.577	0

Using Anderson's test with Schott's more simple multiplier (4.6.2), the first approximation, a_1, produces a value of $A_2 = 149.46$, which is very large when compared with $\chi^2_{8,.05} = 15.51$. Similarly, the alternative test (4.6.3) produced a value of $A_3 = 16.65$, compared with $F_{8,99,.05} = 2.0$. Recall that this alternative does not require the characteristic roots.

The other approximations produced the following results:

$$a_2: A_2 = 31.59$$
$$a_3: A_2 = 40.92$$
$$a_4: A_2 = 12.95$$
$$a_5: A_2 = 200.16$$

The first two quadratic forms in (4.6.2) appeared to have contributed about equally in this example except for a_1 and a_5 where the second term is nearly triple the first and these were the two that had extremely high values of A_2.

With the exception of a_4, all of these results are significant, implying that only the fourth approximation could be considered to be a pc. This is due to the fact that the approximations, as was pointed out before, are quite simple, all using one-digit coefficients. The procedure appears to be rather sensitive, particularly for larger sample sizes as is the case here. This does not detract from the usefulness of the approximations in this example but does reinforce some of the evidence supplied in Table 7.2. In the actual example, a different approximation was used for a_3 after the article was written that was not quite as simple but increased its correlation with u_3 and decreased it with a_1.

These tests were carried out on a one-at-a-time basis with the usual effect on the Type I error, but it was done in this manner to ascertain the relationships for each vector separately. If one wished to test the set of approximations as a whole, the techniques referenced in Section 4.6.4 could be employed.

Another measure which may be of use is the *congruence* coefficient whose distribution was investigated by Korth and Tucker (1975). If both the hypothetical and actual vectors are scaled to unit length, this coefficient is the sum of their crossproducts. For the Color Film example, the values of this coefficient for the first five vectors are .93, .97, .84, .87, and .87. Note that this measure does not involve the data itself, other than that the data produced the sample vectors, and should be used with care (Davenport, 1990). Nevertheless, it may be useful in developing approximations in the first place.

7.5 CORRELATION OF VECTOR COEFFICIENTS

Jackson and Hearne (1973) suggested that a study of the correlations among characteristic vector coefficients might be useful in interpreting the vectors. The variances and covariances of the coefficients of a single vector were given in (4.2.2) and the covariances among the coefficients of different vectors in (4.2.4). The sample estimates of these quantities may be obtained by substituting the estimates $\mathbf{L}$ and $\mathbf{U}$ in these expressions. From these, a composite covariance matrix of all the coefficients may be obtained as well as the corresponding correlation matrix.

If k characteristic vectors have been retained in the PCA model, these correlation matrices are of size $(p \times k) \times (p \times k)$ so they can be large even for a modestly sized problem. For example, the Audiometric example displayed in Table 5.9 was 32×32; the Color Film example of Chapter 6 would be 45×45; and the Black and White Film example from Chapter 2, which will be shown in Table 7.4, is 42×42. For a more simple illustration, the Ballistic Missile example of Chapter 2 is displayed in Table 7.3. Even using all four vectors, this is only 16×16. The same ruthless rounding employed in Table 5.9 will be used here: correlations are rounded to the nearest tenth; correlations greater than $\pm.95$ will be rounded to ± 1.0; those smaller than .1 in absolute value will be recorded as zero; and those for which $.1 < |r| < .6$ will show only the sign.

Note, first, that the correlations among the coefficients *within* a vector are generally fairly sizable. This is to be expected since the coefficients are constrained to have their sum of squares be of unit length. This condition closely parallels that of multiple regression with correlated predictors. In the case of multiple regression, the estimated regression equation is the best least-squares fit but fits nearly this good may be obtained from equations with quite different coefficients. A change in one coefficient because of sampling fluctuations could produce compensating changes in the others that are highly correlated with it. Second, the correlation between the coefficients of any two *adjacent* vectors, say $\mathbf{u}_1$ and $\mathbf{u}_2$ are higher than those further apart, say $\mathbf{u}_1$ and $\mathbf{u}_4$, implying that these correlations are inversely related to the separation of the corresponding characteristic roots. (It may come as somewhat of a surprise that these coefficients are correlated at all, since the corresponding pc's are uncorrelated, but when one reflects that vectors with nearly equal roots are not well defined one should expect them to be *highly* correlated.) Finally, it can be seen that patterns of the correlations are affected by the signs of the coefficients themselves.

The correlation matrix for the three vectors used in the Black-and-White Film example are displayed in Table 7.4. Again, there are high correlations for the coefficients within a vector. Note, for this larger example, the patterns of the correlations between $\mathbf{u}_1$ and $\mathbf{u}_2$ and between $\mathbf{u}_2$ and $\mathbf{u}_3$. The first pair have the higher correlations for roughly the first half of the variables, which represent the higher densities, and the second pair shows more of its strength with the lower-density variables. A glance at the correlation matrix for the original variables in Table 7.5 (obtained from the covariance matrix in Table 2.4) will indicate that each variable is most highly correlated with the variable on either side of it, a phenomenon fairly prevalent when the variables are locations on a continuous curve.

These two examples, as well as the one in Chapter 5, are typical of results that one would obtain using this technique. Once a few of them have been constructed and studied carefully, a better appreciation of these relationships should have been obtained that will be useful in the interpretation of future

Table 7.3. Ballistic Missile Example. Correlations of Characteristic Vector Elements

	u₁				u₂				u₃				u₄			
	u_{11}	u_{21}	u_{31}	u_{41}	u_{12}	u_{22}	u_{32}	u_{42}	u_{13}	u_{23}	u_{33}	u_{43}	u_{14}	u_{24}	u_{34}	u_{44}
u₁ u_{11}	1.0															
u_{21}	–	1.0														
u_{31}	–	–	1.0													
u_{41}	–.6	–	0	1.0												
u₂ u_{12}	0	0	0	–	1.0											
u_{22}	0	0	0	–	–.9	1.0										
u_{32}	+	0	0	–	+	–	1.0									
u_{42}	+	0	0	–	.9	–.7	0	1.0								
u₃ u_{13}	0	0	0	0	–.9	.9	+	.9	1.0							
u_{23}	0	0	0	0	–.7	.7	–	–.9	.9	1.0						
u_{33}	0	0	0	0	+	–	0	+	–	–.6	1.0					
u_{43}	0	0	0	0	–.7	–.6	0	.7	–.7	–	–	1.0				
u₄ u_{14}	0	0	0	0	0	0	+	–	–	+	–	–	1.0			
u_{24}	0	0	0	0	0	0	0	0	+	+	–	–	–.8	1.0		
u_{34}	0	0	–	0	+	0	–	.8	.8	–1.0	.8	+	.6	–.9	1.0	
u_{44}	0	0	0	0	–	0	+	–	–	+	–	–	0	–.6	.8	1.0

151

Table 7.4. B & W Film Example. Correlation of Characteristic Vector Elements

u_1 u_2 u_3

u_1 u_2 u_3

Source: Jackson and Hearne (1973) with permission of the American Statistical Association.

Table 7.5. B & W Film Example. Correlation Matrix

	x_1	x_2	x_3	x_4	x_5	x_6	x_7	x_8	x_9	x_{10}	x_{11}	x_{12}	x_{13}	x_{14}
x_1	1	.98	.93	.84	.73	.63	.52	.42	.30	.21	.11	−.01	−.05	−.13
x_2		1	.97	.91	.82	.72	.62	.51	.38	.28	.14	−.02	−.08	−.16
x_3			1	.97	.91	.84	.75	.64	.51	.40	.21	.00	−.09	−.20
x_4				1	.98	.93	.86	.76	.64	.52	.30	.06	−.07	−.21
x_5					1	.98	.93	.85	.74	.62	.38	.11	−.04	−.19
x_6						1	.97	.91	.82	.70	.46	.19	.01	−.15
x_7							1	.97	.90	.79	.56	.28	.08	−.08
x_8								1	.96	.88	.69	.42	.23	.08
x_9									1	.95	.80	.56	.38	.23
x_{10}										1	.88	.68	.52	.37
x_{11}											1	.86	.76	.66
x_{12}												1	.85	.82
x_{13}													1	.89
x_{14}														1

examples. Most practitioners do not, as a rule, compute these correlation matrices for each problem but they should do one every now and then—they are useful. In particular, it will reinforce the notion that coefficients within a vector are highly correlated and that the coefficients may also be highly correlated if their roots are close together even though the corresponding pc's are uncorrelated.

CHAPTER 8

Vector Interpretation II: Rotation

8.1 INTRODUCTION

Principal components may be used for such applications as data analysis or quality control whether or not the vectors can be interpreted since the procedures for T^2, the residuals, and so on, still hold. The problem is that if T^2 is significant, reference is often made back to the principal components and these may not be too helpful if they are difficult to interpret. Similarly, as suggested in Chapter 1, PCA may also be used as a diagnostic device in determining generalized components of variance. In this case, if the pc's are difficult to interpret, about all the procedure will do is estimate the number of pc's to be retained and provide a check on the residuals. One possible solution to either of these situations is to try some of the approximation techniques suggested in Chapter 7. However, when the vectors are difficult to interpret, it will probably also be difficult to fulfill most of the requirements of approximate vectors. By the time simple vectors are obtained, they will probably be well correlated with each other and not well correlated with the original vectors. What has been proposed to alleviate this situation is to perform a rotation on the characteristic vectors, in effect, a *second* rotation, producing some new "components" that may be useful although they are obtained by a different criterion from PCA. It will be the purpose of this chapter to discuss what these criteria might be and describe some methods of obtaining rotated vectors.

155

8.2   SIMPLE STRUCTURE

The rotation methods usually employed are designed to attain the properties
of what Thurstone (1947) called *simple structure*. If one is using U-vectors, the
purpose of these methods is to produce rotated vectors whose coefficients are
as close to 0 or 1 as possible. V- and W-vectors may also be rotated under the
same philosophy. Suppose the rotated matrix is denoted by

$$\mathbf{B} = [\mathbf{b}_1 \mid \mathbf{b}_2 \mid \ldots \mid \mathbf{b}_k]$$

The requirements for simple structure are as follows (Harmon, 1976):

1. Each row of **B** should contain at least one zero. This means that each of
   the original variables should be uncorrelated with at least one of the
   rotated components.
2. Each column of **B** should have at least k zeros. For interpretation,
   obviously, the more zeros in a vector, the better; this specifies a goal.
3. For each pair of columns of **B**, there should be several variables that have
   zeros in one column but not the other, and, if $k \geqslant 4$, a large number of
   variables with zeros in both columns and a small number of variables
   with nonzero coefficients in both columns. This is an attempt to obtain
   some independence among the variables produced by the rotated vectors.

The object of simple structure is to produce a new set of vectors, each one
involving primarily a subset of the original variables with as little overlap as
possible so that the original variables are divided into groups somewhat
independent of each other. This is, in essence, a method of clustering variables
and some computer packages do cluster variables in this manner. Ideal simple
structure is difficult to obtain and the rotated variables are no longer
uncorrelated. Nevertheless, these procedures may be quite useful in trying to
resolve a complex group of related variables and that is the main purpose of
simple structure rather than to obtain new "components" with which to carry
on subsequent operations.

This is not to imply that rotation should be carried out for every PCA.
Section 8.5.1 will contain an example where the unrotated pc's were much more
useful than those obtained by rotation, a situation more typical in the scientific
domain. Rotation is much more likely to be employed in such fields as
psychology and educational measurement, where the original variables represent
a battery of tests and simple structure attempts to cluster this battery into
groups of tests having the most in common.

If the rotation of the characteristic vectors is done by means of an orthogonal
matrix, the new correlated variables will explain exactly the same amount of
variability of the original variables as did the pc's from which they were derived.

8.3 SIMPLE ROTATION

As an illustration of this technique, consider the following artificial seven-variable problem from which two characteristic vectors have been obtained:

$$\mathbf{U} = \begin{bmatrix} .15 & .42 \\ .30 & .36 \\ .38 & .66 \\ .15 & -.07 \\ .38 & -.24 \\ .45 & -.12 \\ .61 & -.44 \end{bmatrix}$$

The first pc evidently represents variation common to all seven variables since the coefficients are all positive. The second pc represents the difference between the first three variables and the last four. These coefficients are plotted as a scatter diagram in Figure 8.1.

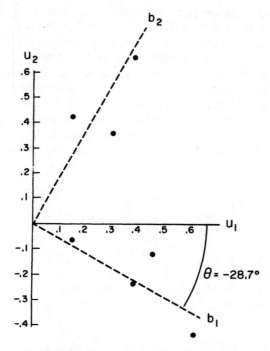

FIGURE 8.1. Artificial example: $\mathbf{u}_1$ vs. $\mathbf{u}_2$. Reproduced from Jackson (1981a) with permission of the American Society for Quality Control.

Now let us rotate both axes (representing the characteristic vectors) through an angle of $\theta = -28.7°$. Let

$$\Theta = \begin{bmatrix} \cos(\theta) & -\sin(\theta) \\ \sin(\theta) & \cos(\theta) \end{bmatrix} = \begin{bmatrix} .88 & .48 \\ -.48 & .88 \end{bmatrix}$$

The rotated vectors then become

$$\mathbf{B} = \mathbf{U}\Theta = \begin{bmatrix} -.07 & .44 \\ .09 & .46 \\ .02 & .76 \\ .16 & .01 \\ .45 & -.03 \\ .45 & .11 \\ .75 & -.09 \end{bmatrix}$$

($\mathbf{B}'\mathbf{B} = \mathbf{I}$, that is the rotated vectors are still orthonormal.) These rotated vectors are represented by dashed lines in Figure 8.1.

Note that these new vectors have not been driven to zeros and ones but they at least show some tendency towards simple structure. (In fact, there cannot be very many large coefficients at all because the vectors are of unit length.) There appear to be three groups of variables: the first three, the last three, and the fourth variable, which has little effect on either component. Admittedly, this example is not very impressive because these conclusions could have been deduced from the original vectors without going through the rotation. Nevertheless, Figure 8.1 demonstrates what one wishes to achieve in simple structure. The points representing the variables should closely adhere to one vector or the other or else be in close to the vertex. To be avoided are points that have large coefficients on both vectors and there are none such here.

If the vectors of our two-variable Chemical example from Section 1.3 are rotated through $\theta = 1.35°$ then,

$$\mathbf{B} = \mathbf{U}\theta = \begin{bmatrix} .7236 & -.6902 \\ .6902 & .7236 \end{bmatrix} \begin{bmatrix} .9997 & -.0236 \\ .0236 & .9997 \end{bmatrix}$$

$$= \begin{bmatrix} .7071 & -.7071 \\ .7071 & .7071 \end{bmatrix}$$

$\mathbf{B}$ is the same as the simplification matrix $\mathbf{A}$ in Section 7.2 except that the column vectors have been normalized to unit length. The covariance matrix of the new variables is

$$\mathbf{S}_B = \mathbf{B}'\mathbf{S}\mathbf{B} = \begin{bmatrix} 1.4457 & -.0321 \\ -.0321 & .0872 \end{bmatrix}$$

where S is the covariance matrix displayed in Section 1.2. Note that

$$|S_B| = .1250 = |S| = l_1 l_2$$

and

$$Tr(S_B) = 1.5329 = Tr(S) = l_1 + l_2$$

demonstrating that an orthonormal transformation of the characteristic vectors leaves the total variability unchanged. However, S_B is not a diagonal matrix, which indicates that the new variables are no longer uncorrelated. The correlation between them is

$$r = -.0321/\sqrt{(1.4457)(.0872)} = -.09$$

as was the case in Section 7.2. The only linear transformation of the original variables that produces uncorrelated variables is the set of characteristic vectors in their original form before rotation.

For two-pc problems, graphical rotation may be fairly easy and effective. For three or more vectors, one has to do this by pairs of vectors, and, for many more than three, this can be rather difficult to do.

The seven-variable example in Figure 8.1 conveniently had the two groups of coefficients at right angles to each other. Suppose the best angle between the rotated axes required to obtain simple structure was either less than or greater than 90°. Rotation to a nonorthogonal solution is permissible. This is called *oblique* rotation. The result is a tradeoff between more simple structure and the loss of orthogonality. As we shall see in Chapter 17, factor analysis is concerned with the structure of a model rather than explaining variability, and, for this reason, correlated factors are acceptable if they help explain the structure. Oblique methods will be described in Section 8.4.3.

Rotated components are not invariant with change in the number of components. If one adds a $(k + 1)$st pc to the model, the first k pc's remain the same but the first k rotated components will not, because the rotation is now in $(k + 1)$-space rather than k-space.

8.4 ROTATION METHODS

8.4.1 Introduction

With the exception of two-component examples, such as the one in Figure 8.1, rotation methods can become quite complicated and the earlier graphical procedures can break down rather rapidly. With the advent of modern computers, a number of algorithms have been proposed for both orthogonal and oblique rotation. Most computer packages will have several options. This section will describe a few of the more popular methods. While it would be desirable to have just one procedure, each of these methods has a particular property that might make it preferable under certain circumstances. The point is that the only judgment now required is in the selection of the procedure.

To illustrate these procedures, a new example will be introduced because it has only two components and hence can be easily displayed. This is a classic set of data taken from Harmon (1976) and has been widely used in the literature. Eight physical measurements were obtained on 305 girls as follows: x_1-height, x_2-arm span, x_3-length of forearm, x_4-length of lower leg, x_5-weight, x_6-bitrochanteric diameter (related to upper thigh), x_7-chest girth, and x_8-chest width. The first four variables are measures of "lankiness" and the last four of "stockiness." Since the units of these measurements differ, the correlation matrix will be employed and is shown in Table 8.1. Note that the two types of measurements are more highly correlated within types than between them. The first two V-vectors, accounting for 80%, their associated characteristic roots, and residual variances are given in Table 8.2. (The remaining six roots are .48, .42, .23, .19, .14 and .10.) V-vectors will be employed in this section as this is the form commonly used for simple structure rotation.

Table 8.1. Physical Measurements. Correlation Matrix

Height	1	.85	.80	.86	.47	.40	.30	.38
Arm span	.85	1	.88	.83	.38	.33	.28	.42
Length of forearm	.80	.88	1	.80	.38	.32	.24	.34
Length of lower leg	.86	.83	.80	1	.44	.33	.33	.36
Weight	.47	.38	.38	.44	1	.76	.73	.63
Bitrochanteric diameter	.40	.33	.32	.33	.76	1	.58	.58
Chest girth	.30	.28	.24	.33	.73	.58	1	.54
Chest width	.38	.42	.34	.36	.63	.58	.54	1

Source: Harmon (1977) with permission of the University of Chicago. All rights reversed.

Table 8.2. Physical Measurements. V-Vectors

	v_1	v_2	Residual Variance
Height	.86	−.37	.12
Arm span	.84	−.44	.10
Length of forearm	.81	−.46	.13
Length of lower leg	.84	−.40	.14
Weight	.76	.52	.15
Bitrochanteric diameter	.67	.53	.26
Chest girth	.62	.58	.28
Chest width	.67	.42	.38
Characteristic root	4.67	1.77	

Many of these procedures are either of the form (8.4.1) or (8.4.2), which involve quartic terms of one form or another. Clarkson and Jennrich (1988) have unified this approach, which may be useful in writing general-purpose rotation programs.

8.4.2 Orthogonal Methods

A great many of the orthogonal rotation schemes start with the general expression

$$Q_1 = \sum_{j=1}^{k} \left\{ \sum_{i=1}^{p} b_{ij}^4 - \frac{c}{p} \left(\sum_{i=1}^{p} b_{ij}^2 \right)^2 \right\} \tag{8.4.1}$$

where c is an arbitrary constant. These methods obtain a rotation such that Q_1 is maximized and this general procedure is called *orthomax* rotation.

By far the most popular orthogonal rotation method is the *varimax* procedure of Kaiser (1958, 1959), which utilizes (8.4.1) with $c = 1$. The sums of squares of **B** are maximized columnwise. This form is actually called *raw* varimax. If the individual b_{ij}^2 are divided by the corresponding diagonal element of **VV'**, the variability explained by the PCA model, the resultant form is called *normal* varimax, which Kaiser recommended. This weighting is sometimes used for other methods as well and is used as a default in SAS PROC FACTOR, the computer package used for the examples in this section. A different weighting was suggested by Cureton and Mulaik (1975) to take care of certain pathologic situations such as an overabundance of very small coefficients in one or more of the original vectors, particularly the first.

Using SAS PROC FACTOR, the varimax rotation for the vectors in Table 8.2 is

$$\mathbf{\Theta} = \begin{bmatrix} .771 & .636 \\ -.636 & .771 \end{bmatrix}$$

shown in Figure 8.2, which also includes the plot of the coefficients of **V**. The resultant rotated vectors are

$$\mathbf{B} = \begin{bmatrix} .90 & .26 \\ .93 & .20 \\ .92 & .16 \\ .90 & .23 \\ .25 & .89 \\ .18 & .84 \\ .11 & .84 \\ .25 & .75 \end{bmatrix}$$

There are four large coefficients in each vector; the first vector representing the "lanky" variables and the second vector representing the "stocky" ones. These results are expected, considering the structure of the correlation matrix in Table 8.1.

A number of things are of interest. First, the rotation in Figure 8.2 does not go through the two configurations of points but it is the best orthogonal fit, subject to maximizing Q_1. Second, all of the coefficients are positive because the rotation vectors are both outside the configuration of points. Finally, the amount explained by these rotated components (which are no longer uncorrelated) is exactly the same as it was before rotation. The sums of squares of the coefficients in **B** are equal to $l_1 + l_2$. This is because the rotation is carried out in a two-dimensional plane defined by **V**, imbedded in the eight-dimensional configuration defined by the correlation matrix.

Because of its wide use, varimax has received more attention in the nature of refinements and investigations. Gebhardt (1968) concluded, by simulation, that Kaiser's algorithm would not ordinarily have trouble with local maxima. TenBerge (1984) showed that varimax rotation was among a class of solutions for the simultaneous diagonalization of a set of symmetric matrices such as one would employ in individual multidimensional scaling (Section 11.6) and concluded that varimax and an equivalent procedure by DeLeeuw and Pruzansky (1978) were the best available. Varimax was used in the original

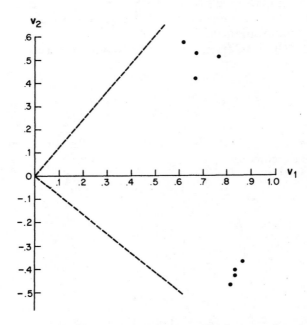

FIGURE 8.2. Physical measurements: varimax rotation.

Little Jiffy, described in Section 17.9 and is also used as a default in some computer packages.

Varimax was preceded by *quartimax* (Neuhaus and Wrigley, 1954), which is obtained by maximizing Q_1 with $c = 0$. This has the effect of maximizing the sums of squares across the *rows* of **B**. For the present example, the results are nearly identical to those of varimax. (This is not always the case—see, for instance, Section 8.5.1. Quartimax, generally, is more likely to produce a "general" component than varimax.) If $c = k/2$, (8.4.1) becomes the *equimax* rotation (Saunders, 1961). In the present example, for which $k = 2$, the result is the same as from varimax. A method for obtaining a single rotation to simultaneously obtain the simple structure for two sets of vectors was obtained by Hakstian (1976).

8.4.3 Oblique Methods

For oblique rotation, (8.4.1) may be replaced by

$$Q_2 = \sum_{g < j = 1}^{k(k-1)/2} \left\{ p \sum_{i=1}^{p} b_{ij}^2 b_{ig}^2 - c \left(\sum_{i=1}^{p} b_{ij}^2 \right) \left(\sum_{i=1}^{p} b_{ig}^2 \right) \right\} \qquad (8.4.2)$$

again, c being an arbitrary constant. This is the form for *oblimin* rotation and to obtain oblique rotation in this form an appropriate value of c is chosen and Q_2 is *minimized*. The *quartimin* method (Carroll, 1953; Jennrich and Sampson, 1966) results for $c = 0$. If $c = 1$, it is the *covarimin* method (Carroll, 1953; Kaiser, 1958). For those who feel that quartimin solutions are too oblique and covarimin solutions not oblique enough, there is *biquartimin* with $c = .5$ (Carroll, 1957). Jennrich (1979) showed that negative values of c were not only admissible but, in his judgment, desirable. He concluded that convergence problems were most likely to occur when $c > (2k - 1)/(k - 1)^2$.

Today, however, these have been replaced in popularity by some nonquartic methods. Two of these are the *orthoblique* or *HK*-method (Harris and Kaiser, 1964), which is used in the second-generation *Little Jiffy*, and *promax* (Hendrickson and White, 1964). Both of these are two-stage procedures that start with an orthogonal rotation, such as varimax, and extend it to an oblique solution. For the present example, these methods produce nearly identical results. Cureton and Mulaik (1975) also developed their weighting scheme for promax. The *HK*-rotation of the Physical Measurement vectors, using SAS PROC FACTOR, is

$$\boldsymbol{\Theta} = \begin{bmatrix} .676 & .496 \\ -.886 & .999 \end{bmatrix}$$

and the rotated vectors are

$$\mathbf{B} = \begin{bmatrix} .91 & .06 \\ .96 & -.02 \\ .96 & -.06 \\ .92 & .02 \\ .05 & .90 \\ -.02 & .87 \\ -.10 & .88 \\ .08 & .75 \end{bmatrix}$$

While the general pattern is the same as for the orthogonal varimax solution, the differences between the large and small coefficients are more pronounced. The amount explained by these rotated components is still the same as for the original pc's even though it is an oblique rotation; we have not left our temporary home in two-space.

This rotation is shown in Figure 8.3. Note that there are *two* sets of rotated vectors. Those labeled P_1 and P_2 are called *primary* vectors, which do go through the configuration of points. R_1 and R_2 are *reference* vectors, R_i and P_i being

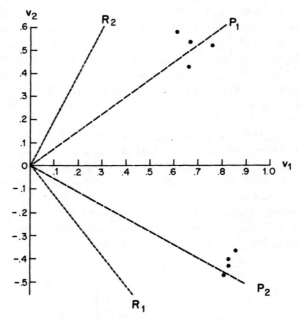

FIGURE 8.3. Physical measurements: HK rotation.

orthogonal to each other. It is the projections of the points on the reference vectors that produce the rotated vectors **B**. In the orthogonal rotations, this problem does not exist.

Cattell has criticized most oblique methods for the fact that they will not allow for two simple structure components with similar patterns, the so-called *cooperative* components. A procedure called *maxplane* was designed to allow for this (Cattell and Muerle, 1960; Cattell and Khanna, 1977). A similar procedure called *functionplane* was derived by Katz and Rohlf (1974).

Hakstian (1971) and Hakstian and Abell (1974) compared HK, maxplane, oblimin, and promax with both artificial examples and real data sets. Maxplane did not fare very well in these comparisons. The other three methods all have an arbitrary constant of some sort and it was their conclusion that the choice of this constant outweighed the other differences among the methods, although the HK method appeared to do a little better on real examples. They recommended trying several values of the constant for whichever method was employed. For HK, these suggestions were: $c = 0$ (tends to be biased towards oblique), .35, .5, .75 (tends to be biased towards orthogonality). For oblimin, $c = -2, 0$ (orthogonal bias), .1, .2, .3, .4, .5, and .6 (oblique bias). For promax, $c = 2$ (orthogonal bias), 3, 4, 5, 6 (oblique bias), although, as was observed earlier, Jennrich had the most success with negative values for promax.

8.5 SOME COMMENTS ABOUT ROTATION

8.5.1 Is Rotation for Everyone?

The answer is "no." Rotation was designed to obtain simple structure by clustering variables into groups that might aid in the examination of the structure of a multivariate data set. This has proven quite useful in such fields as market research, psychology, and education. In physical applications, rotation is usually of secondary interest if it is appropriate at all. As an example, we shall return to the Audiometric example of Chapter 5. Four principal components were retained from an eight-variable situation. The V-vectors for this example are repeated in Table 8.3. The rotation matrices for both varimax and quartimax,

Table 8.3. Audiometric Example. V-Vectors

Frequency	v_1	v_2	v_3	v_4
500L	.80	−.40	.16	−.22
1000L	.83	−.29	−.05	−.33
2000L	.73	.30	−.46	−.19
4000L	.56	.60	.42	−.11
500R	.68	−.49	.26	.33
1000R	.82	−.29	−.03	.25
2000R	.62	.40	−.56	.27
4000R	.50	.65	.42	.11

Table 8.4. Audiometric Example. Rotation Matrices

Varimax				Quartimax			
.58	.40	.47	.53	.82	.38	.43	.04
−.52	.70	.39	−.29	−.56	.72	.42	.04
.23	.59	−.77	−.01	.15	.58	−.79	−.10
.57	−.00	.18	−.80	−.00	−.02	.11	−.99

Table 8.5. Audiometric Example. Rotated Vectors

Frequency	Varimax				Quartimax			
500L	.58	.13	.06	.71	.90	.11	.03	.22
1000L	.44	.10	.27	.79	.83	.09	.25	.36
2000L	.05	.22	.78	.45	.35	.22	.79	.28
4000L	.04	.89	.15	.20	.18	.89	.15	.11
500R	.91	.08	−.00	.23	.87	.05	−.07	−.35
1000R	.77	.10	.34	.31	.82	.08	.28	−.23
2000R	.17	.19	.93	−.00	.19	.19	.91	−.17
4000R	.11	.91	.19	−.02	.11	.90	.17	−.10

using SAS PROC FACTOR, are shown in Table 8.4, and the rotated vectors in Table 8.5.

Both rotation methods identify 500 Hz and 1000 Hz (both ears) as one "cluster" of variables, both 4000 Hz measurements as another, and both 2000 Hz measurements as a third. Varimax, operating on the columns of **V**, has the information about 500 Hz and 1000 Hz included in both the first and fourth rotated vectors. Quartimax, operating on the rows, manages to accomplish all of this in the first three vectors and restates the left/right ear difference in the fourth. This is suggested by the fourth varimax vector but it is only a suggestion. Neither of these results are very useful in dealing with the problem posed in Chapter 5, nor in the upcoming extension of it in Chapter 9. However, if there were 30 or 40 variables instead of only 8, these methods might be useful in reducing the number of variables employed. Jolliffe (1989) suggested that rotation might be useful for a group of pc's (not necessarily the first few) whose characteristic roots are nearly equal.

8.5.2 Rotation and Factor Analysis

Chapter 17 will consider *factor analysis*, which is a technique quite similar in concept to PCA. The principles of rotation and simple structure had their genesis with factor analysis and most of the material on this subject will be found in factor analysis texts such as Harmon (1976) and Cureton and D'Agostino (1983). The majority of factor analysis solutions are in terms of rotated factors. For some of the rotation methods, slight modifications may be required when using factor analysis but the use of normal rather than raw rotations will eliminate most of this. Most computer packages make no distinction whatsoever.

8.5.3 Standard Errors

The formulas for the standard errors of the coefficients of the rotated vectors are more complicated than (4.2.3) because the rotation criteria are a function of the coefficients themselves. Expressions for the standard errors when orthomax rotation is employed (including varimax and quartimax as special cases) were given by Archer and Jennrich (1973) and for the oblique oblimin series by Jennrich (1973a) and Clarkson (1979). Jennrich (1973b) also investigated the stability of rotated coefficients and empirically verified the so-called "Wexler phenomenon": If there is good simple structure, rotated coefficients may be surprisingly stable and vice versa. The example in his paper produced rotated coefficients whose standard errors were nearly equal, whereas the standard errors associated with the original vectors varied considerably. Archer and Jennrich (1976) considered the asymptotic distribution of coefficients obtained by quartimax rotations.

8.5.4 Ad Hoc Rotation

One can, of course, create rotation schemes based on whatever criteria seem appropriate, as was the practice in the 1930s and 1940s. The simplification schemes used in the previous chapter are, in fact, rotations as are any other linear combinations of either the vectors or the original variables. The point to be made about ad hoc procedures is that some sort of check should be made regarding the statistical properties associated with them. Some special procedures still exist, particularly in the field of chemometrics.

As an example, Lawton and Sylvestre (1971), working with product matrices in a curve-resolution problem, obtained two pc's and then rotated them in such a way that the coefficients of both vectors would all be positive. Since they were working with the raw data, rather than deviations from the mean, the pc-scores would also be positive. This made for a more intuitive model for adding together the effect of the two constituents.

8.6 PROCRUSTES ROTATION

8.6.1 Introduction

The rotational procedures discussed so far in this chapter have been designed to transform the matrix of characteristic vectors into a new set of vectors that possesses some desired property. This section deals with the comparison of two matrices and may be stated as follows:

Given two $n \times m$ matrices, **A** and **B**, what $(m \times m)$ transformation matrix **T** will best transform **A** into **B**? In other words, what matrix **T** will make **AT** most like **B**?

Within this context, **B** is usually referred to as the *target* matrix. Although this procedure is often carried out for matrices of characteristic vectors, it is not restricted to these alone. This solution holds for matrices in general.

The problem is formulated as follows. Let

$$E = AT - B \tag{8.6.1}$$

where **E** is the difference between **B** and the approximation for it (i.e., $AT = B + E$). Then the object will be to obtain **T** such that $\text{Tr}(E'E)$ will be a minimum. If more than one matrix is being compared to a target matrix, the quantity $\text{Tr}(E'E)$ may be used as a comparison measure of goodness of fit.

The name "procrustes" is due to Cattell, the same person who brought us SCREE plots. In Greek mythology, when Thesus was cleaning up Greece's highways of highjackers and other undesirables, one of those he terminated was Procrustes, the Stretcher. Procrustes had an iron bed on which he tied any traveller who fell into his hands. If the victim was shorter than the bed, he or she was stretched out to fit; if too long, Procrustes lopped off whatever was necessary. Cattell felt that this rotation procedure used about the same philosophy (Hurley and Cattell, 1962).

TenBerge and Knol (1984) established a set of criteria for classifying these procedures. They are:

1. Optimality criterion
2. Orthogonal or oblique
3. Simultaneous or successive solution
4. Two matrices or more than two
5. Whether or not the matrices are all the same size

Not all combinations of these criteria have solutions at this time.

8.6.2   Orthogonal Rotation

If **T** in (8.6.1) is orthonormal such that $TT' = T'T = I$, then this is called the *orthogonal procrustes problem* and will have a least-squares solution (Schönemann, 1966). To obtain **T**, form the $(m \times m)$ product $S = A'B$. Let the U-vectors of $S'S$ be denoted by **U** and of SS' by U^*. Then

$$T = U^*U' \tag{8.6.2}$$

Since U and U* are both orthonormal, T will also be orthonormal. (In Sections 10.2 and 10.3, we shall see that U and U* can be obtained from S directly in a single operation using *singular value decomposition*.) Schönemann's technique is a generalization of a more restrictive solution by Green (1952), although the notion goes back to Mosier (1939).

In Chapter 9, the audiometric problem will be revisited and some simple approximations for the U-vectors will be made. These approximations, which are found in Table 9.6, are previewed along with the U-vectors in Table 8.6. The transformation matrix is found to be

$$
T = \begin{bmatrix}
.996 & -.062 & -.041 & -.057 \\
.058 & .993 & -.099 & .009 \\
.048 & .096 & .994 & .021 \\
.056 & -.015 & -.022 & .998
\end{bmatrix}
$$

The product UT is given in Table 8.7 along with the sum of squares of deviations from the target matrix for each column. $Tr(E'E) = .150$. Note that the sums of squares of deviations increase with each successive vector.

8.6.3 Extensions of the Procrustes Model

The procrustes model has been extended by Schönemann (1968) to *two-sided orthogonal* procrustes procedures, which are used to test whether or not a $(p \times p)$ matrix A is a permutation of another $(p \times p)$ matrix B. This is particularly useful in looking for particular patterns in matrices (see for example, Section 11.5), which may obscured by a permutation of the rows and/or columns. If A and B are both symmetric, the solution is the same as above. There are, however, some uniqueness problems. To obtain the minimum $Tr(E'E)$, the columns of U and U* must be in order by the size of the root. A reflection of either U or U* will affect T but not the fit. Finally, although some of the roots may be negative, they must be distinct.

Table 8.6. Audiometric Case Study. U-Vectors and Target Matrix B

Frequency	u_1	b_1	u_2	b_2	u_3	b_3	u_4	b_4
500L	.40	.36	-.32	-.33	.16	.19	-.33	-.36
1000L	.42	.36	-.23	-.22	-.05	0	-.48	-.36
2000L	.37	.36	.24	.22	-.47	-.57	-.28	-.36
4000L	.28	.36	.47	.55	.43	.38	-.16	-.36
500R	.34	.36	-.39	-.33	.26	.19	.49	.36
1000R	.41	.36	-.23	-.22	-.03	0	.37	.36
2000R	.31	.36	.32	.22	-.56	-.57	.39	.36
4000R	.25	.36	.51	.55	.43	.38	.16	.36

Table 8.7. Audiometric Case Study. Predicted Target Matrix

Frequency	Orthogonal				Oblique			
500L	.37	−.32	.18	−.35	.34	−.29	.17	−.33
1000L	.38	−.25	−.04	−.51	.35	−.22	−.04	−.48
2000L	.34	.17	−.50	−.31	.35	.19	−.49	−.29
4000L	.32	.50	.37	−.16	.36	.53	.38	−.16
500R	.36	−.39	.27	.47	.33	−.35	.26	.44
1000R	.42	−.26	−.03	.35	.39	−.23	−.04	.33
2000R	.32	.24	−.61	.36	.34	.25	−.61	.35
4000R	.31	.53	.36	.16	.35	.56	.37	.15
Residual SS	.009	.012	.016	.114	.003	.005	.015	.110
Tr(E′E)			.151				.133	

If **A** and **B** are asymmetric, then the following procedure may be employed.

1. Obtain the following sets of characteristic vectors:

$$U_a \text{ from } A'A,$$

$$U_a^* \text{ from } AA',$$

$$U_b \text{ from } B'B,$$

$$U_b^* \text{ from } BB'$$

(Again, singular value decomposition will shorten this procedure.)

2. Form the products

$$T_1 = U_a^* U_b^{*\prime}$$
$$T_2 = U_a U_b' \tag{8.6.3}$$

3. The least-squares estimate of **B** is $T_1' A T_2$ or

$$B = T_1' A T_2 + E \tag{8.6.4}$$

Schönemann implied that this method may be generalized to rectangular **A** and **B**.

There have been a number of other modifications and extensions of the orthogonal procrustes procedure. Expression (8.6.1) has been expanded to allow for translation, reflection, and contraction or dilation by Schönemann and Carroll (1970), Lingoes and Schönemann (1974), and Gower (1975). Other criteria have been proposed in place of min $Tr(E'E)$. One of these is the *inner product* criterion (Cliff, 1966), which maximizes $Tr(T'A'B)$. Another is the *maximum congruence* $Tr\{[Diag(T'A'AT)]^{-1/2}T'A'B[Diag(B'B)]^{-1/2}$. This is like the inner product criterion except that it has a denominator that it attempts to minimize at the same time. Maximum congruence is felt to be useful when A and B are themselves related to characteristic vectors (Brokken, 1983), although the examples used in Brokken's paper did not appear to have any clear advantage over the classical procrustes fit. Maximum congruence does require more computational effort, the iterative solution of $m(m + 1)/2 + m^2$ equations. A discussion of some problems experienced with procrustes analysis may be found in Huitson (1989).

Another extension is the simultaneous comparison of more than two matrices (Evans, 1971; TenBerge, 1977; Van de Geer, 1984; Brokken, 1985). Brokken found that the maximum congruence procedure, which showed little promise for the two-matrix case, did, in fact, produce some better results for a large number of matrices. Finally, Borg (1978), TenBerge and Knol (1984), Peay (1988), and Ramsay (1990) gave some procedures for comparing two or more matrices having the same number of rows but differing numbers of columns or vice versa. Some studies of robustness of the procrustes procedure were carried out by Sibson (1978, 1979).

In the field of chemometrics, Rozett and Petersen (1975b) defined four different types of components: (They referred to them as "factors" but, in fact, used PCA.)

1. *Principal.* Directly related to experimental measurements. Used to define dimensionality.
2. *Typical.* Same but direct, whereas principal components may have them lumped together. They would pick out spectra representative of the "factor" properties and obtain them by target rotation.
3. *Basic.* Molecular property that accounts for the observed variability in the set of mass spectra. Also obtained by target rotation.
4. *Partial.* These provide analytical insight into a set of mass spectra and are obtained by varimax rotation.

Examples of target rotation may be found in Rozett and Petersen (1975a,b) and Malinowski and McCue (1977), the latter using it to detect whether or not a suspected substance was present in a mixture.

8.6.4 Oblique Procrustes

As one would expect, oblique solutions to (8.6.1) may produce better fits but are harder to do. A general solution by TenBerge and Nevels (1977) will also handle the case where the matrix A is not of full column rank. Some procedures using weighted oblique procrustes rotation have been developed by Meredith

(1977) and Nevels (1979). Browne (1972) produced a technique where only some of the elements of **B** are specified. There had been some earlier versions of this but the specified elements were restricted to be zero; Browne's procedure is general. A summary of oblique congruence methods may be found in TenBerge (1979).

The oblique procrustes solution for the Audiometric example given in Section 8.6.2 was obtained using SAS PROC FACTOR. This solution is also shown in Table 8.7. Note a reduction in the sum of squares for every column. $\mathrm{Tr}(\mathbf{E}'\mathbf{E}) = .133$.

CHAPTER 9

A Case History—Hearing Loss II

9.1 INTRODUCTION

In Chapter 5 ("Putting it Altogether"), a numerical example was presented to illustrate many of the techniques that had been introduced in the book up to that point. The example dealt with a sample of 100 observations from a much larger study of audiometric examinations. In this chapter, the larger study will be discussed in detail. The purpose of this chapter is to illustrate a case history in which the method of principal components played a prominent part and of the various modifications and additions to the PCA methodology that were required by the specific nature of this application.

As the reader will recall from Chapter 5, the purpose of this study was to be able to characterize normal hearing loss as a function of age and sex so as to be able to distinguish between normal hearing loss due to aging and abnormal hearing loss due to noise exposure, illness, and so on. Comparisons of this sort might be carried out either on individuals or groups of individuals.

This study is presented here solely for the purpose of illustrating the uses and extensions of PCA and *not* for the purposes of making specific comments about hearing experience or comparing this with other studies. This particular study was carried on in the 1960s. The norms presented here represent that period of time for a single group of people—employees of the Eastman Kodak Company—and should in no way be construed to represent any other point in time or any other group of people. Changes in instrumentation alone could render these data useless for comparison if one did not have detailed information about these changes. There has been a change in the instrumentation standard since that time (ANSI, 1969). In addition, it is well known that hearing norms do vary in different parts of the world [see, for instance, Taylor et al. (1965), Kell et al. (1970), and Kryter (1973)]. It is also apparent that a portion of the younger population of the United States, and possibly other parts of the world as well, has shown a significant change in average hearing threshold since the 1960s due to increased exposure to loud noise (Mills, 1978).

A number of studies had been carried out previous to the one being described. Riley et al. (1961) discussed a study carried out in the 1950s relating normal hearing loss to age. The variability associated with audiometric examinations including instrument and operator variability as well as the inherent subject variability was discussed in Jackson et al. (1962). This last component of variability was obtained by subjecting a number of persons taking preemployment physicals to two examinations, one at the beginning and one at the end of the physical. Some preliminary investigation of the use of multivariate methods indicated that PCA would be a useful technique in the analysis and reporting of audiometric data and that four pc's would adequately represent the variability of normal individuals from the same age group.

9.2   THE DATA

I expect that most of our readers, at one time or another, have taken an audiometric examination. This generally consists of the subject sitting in a soundproof booth and wearing earphones through which are sent a series of signals of varying frequencies, one ear at a time. The intensity of these signals is increased until the subject indicates that it has been perceived. The data, then, consist of the intensity required for each ear at each frequency. These are known as *thresholds* and are reported in units of "decibel loss." These frequencies are usually transmitted in a random order rather than going from the lowest frequency to the highest. At the time this study was conducted, nine frequencies were employed for routine examinations, 250, 500, 1000, 2000, 3000, 4000, 6000, and 8000 Hz, but this study concentrated on only four of them. The reason for this was primarily one of speed and capacity of the first-generation mainframe computers in use at that time. It had been demonstrated that given three frequencies (1000 Hz, 2000 Hz, and 4000 Hz) the other six could be predicted with reasonable precision. However, 500 Hz was also included in the study because this frequency, along with 1000 Hz and 2000 Hz, was used by the New York State Worker's Compensation Board in determining hearing impairment caused by exposure to occupational noise.

To put things in perspective, 500 Hz corresponds roughly to "middle C" on a piano and the other three frequencies are successive octaves of 500 Hz (i.e., 500 Hz to 1000 Hz is one octave, 1000 Hz to 2000 Hz is another, and so on).

Earlier studies had usually broken the population into five age groups each representing 10 years covering the range of age 16–65. One problem with this stemmed from the monitoring of individuals over time. For example, a person aged 25 would be among the oldest in the first age group and being compared to the average for that group, would generally appear to be worse off. The next year, this individual would be 26 and among the youngest in age group 26–35 and hence would appear to be better than average. The resolution of this problem, which will be discussed in Section 9.4, resulted in the age groups being restructured into 3-year increments, these being the smallest age brackets that would still produce adequate sample sizes, particularly for the older age groups.

The data were obtained from the Company medical records and included everyone who would be considered "normal" from the standpoint of hearing. Excluded from this normal group were:

1. Those who had any significant noise exposure in the present job, in a prior job, or while pursuing noisy avocations
2. Those who were subjected to significant noise while in military service
3. Anyone indicating any history of ear trouble
4. Anyone who exhibited a significantly high hearing threshold or any other evidence of an abnormal audiogram

The screening of individuals for steps 1, 2, and 3 was done from medical records. Step 4 was done by examining the Q- and T^2-statistics for each individual during the PCA phase of the analysis in the manner illustrated in Chapter 5. After this screening, 10 358 audiograms remained in the male normal group and 7672 in the female normal group. The sample sizes by age groups are shown in Table 9.1.

Because this is a description of statistical techniques rather than a clinical report, the procedures in the following sections will be illustrated for the male data only, but exactly the same steps were carried out for the female data. The results differed somewhat because, *for these particular populations*, the males

Table 9.1. Audiometric Case Study. Sample Sizes for Three-Year Age Groups of Normals

Age Group	Sample Size	
	Males	Females
17–19	2215	3173
20–22	2243	1708
23–25	1424	619
26–28	796	388
29–31	515	273
32–34	405	233
35–37	337	294
38–40	300	243
41–43	312	224
44–46	344	180
47–49	323	128
50–52	363	85
53–55	276	45
56–58	232	40
59–61	156	28
62–64	117	11

Table 9.2. Audiometric Case Study. Sample Means—Males

Age Group	Left Ear				Right Ear			
	500	1000	2000	4000	500	1000	2000	4000
17–19	−3.6	−4.0	−3.9	.5	−4.0	−4.0	−3.9	.0
20–22	−3.2	−3.3	−3.8	1.3	−3.6	−3.3	−3.5	1.1
23–25	−3.4	−3.3	−3.1	4.5	−3.8	−3.4	−3.4	3.4
26–28	−3.4	−2.7	−2.1	8.7	−3.9	−2.8	−2.5	7.0
29–31	−3.0	−2.3	−1.3	11.7	−3.7	−2.4	−1.9	10.9
32–34	−2.9	−2.4	−1.7	15.3	−3.5	−2.0	−1.7	14.3
35–37	−2.7	−1.8	1.3	17.9	−3.0	−2.0	.1	16.2
38–40	−2.2	−.5	1.4	21.4	−2.7	−.6	.9	19.3
41–43	−2.4	−.3	3.4	24.8	−2.4	−.2	2.0	23.5
44–46	−2.0	.0	3.3	25.2	−2.3	−.4	2.6	23.7
47–49	−1.2	.6	4.6	27.0	−1.6	.9	3.4	25.4
50–52	−.9	2.1	6.9	32.2	−.6	1.9	5.0	27.7
53–55	−.1	2.1	8.8	32.8	−.9	1.6	7.1	29.3
56–58	1.3	4.3	13.3	38.4	.7	4.7	10.4	36.6
59–61	1.8	5.8	14.1	40.1	2.2	5.2	11.0	36.4
62–64	3.9	7.4	17.1	43.6	2.9	7.1	16.8	40.9

Table 9.3. Audiometric Case Study. Sample Standard Deviations—Males

Age Group	Left Ear				Right Ear			
	500	1000	2000	4000	500	1000	2000	4000
17–19	5.45	5.46	6.79	10.93	5.27	5.20	5.95	8.89
20–22	5.69	5.77	6.95	11.62	5.44	5.35	6.12	9.68
23–25	5.58	5.66	7.43	12.94	5.37	5.44	6.23	11.18
26–28	5.63	5.81	8,15	15.75	5.70	5.71	6.77	13.65
29–31	5.78	6.12	8.27	17.81	5.57	5.94	6.93	15.86
32–34	5.86	6.21	8.44	17.72	5.56	5.95	7.34	16.60
35–37	6.31	6.72	9.54	17.87	6.18	6.32	8.99	16.91
38–40	6.38	7.19	10.13	19.12	6.60	6.79	8.93	18.14
41–43	6.51	6.96	10.49	18.78	6.43	6.56	9.08	18.94
44–46	6.77	7.49	11.41	18.88	6.37	6.91	10.65	18.01
47–49	6.78	7.07	11.36	19.63	6.99	7.23	10.17	18.89
50–52	7.27	7.67	11.82	18.64	7.56	7.85	10.57	18.78
53–55	7.36	8.43	14.65	19.83	7.29	7.63	11.56	18.75
56–58	7.58	9.11	15.75	20.04	7.25	8.70	13.76	19.73
59–61	8.80	9.93	15.99	18.39	8.71	10.23	13.69	18.90
62–64	9.68	11.40	18.62	19.46	9.20	11.11	16.30	19.88

had higher average hearing loss than the females for all age levels as well as greater variability at all age levels. The age functions described in Section 9.4 were different in some cases, due primarily to the fact that, owing to insufficient data, the female data for ages 47–64 were omitted and therefore the functions, which were generally nonlinear, were fitted over a considerably shorter range. Both groups used four principal components and these were quite similar.

The means and standard deviations for each male age group are displayed in Tables 9.2 and 9.3. Note that both quantities increase as a function of age.

9.3 PRINCIPAL COMPONENT ANALYSIS

The principal component analysis was carried out for each age group separately. The correlation matrix for each age group was used rather than the covariance matrix because of the large differences in variability between the lower and higher frequencies. Previous studies had indicated that four principal components would leave, as a residual, that variability expected from repeat examinations on individuals. The U-vectors were quite similar to those obtained from the small data set in Chapter 5. The first pc represented overall hearing loss; the second pc represented the difference between the high frequencies (2000 Hz and 4000 Hz) and the low frequencies (500 Hz and 1000 Hz); the third pc was related primarily to the difference between 2000 Hz and 4000 Hz, while the fourth pc represented the left–right ear differences.

In fact, the U-vectors differed very little between age groups. What did change was the proportion of the total variability explained by the associated pc's. Section 16.6 contains a test of the hypothesis that certain unit vectors are not significantly different among groups—only their roots. The present situation, although relevant, is different to the extent that these groups are a function of a continuous variable, age. (Also, the techniques listed in Section 16.6 were developed nearly 20 years after this study was carried out.) The characteristic roots, l_i, are given in Table 9.4 along with the amount unexplained, which is equal to the four remaining roots.

Since the correlation matrices all have the same trace, $\text{Tr}(\mathbf{R}) = p = 8$, they may be compared directly. While it appears that the older the age group, the better the fit, this is due to the fact that the older the group, the more variability there is. This additional increment in variability is, by and large, being accounted for by the first four pc's and although the inherent variability is also growing, it is doing so more slowly, which accounts for the appearance of Table 9.4. Note that although l_1, the root associated with the first pc, overall hearing loss, increases steadily with age, l_2 (high–low) is constant while l_3 (2000–4000 Hz) and l_4 (left–right) decrease. This does not mean, for instance, that left–right differences decrease with age but merely that they do not increase as rapidly as the overall hearing loss. This illustrates one of the difficulties of using correlation matrices but the alternative of using covariance matrices would have yielded vectors heavily dominated by the higher frequencies.

Table 9.4. Audiometric Case Study. Sample Characteristic Roots—Males

Age Group	l_1	l_2	l_3	l_4	Residual Trace
17–19	3.175	1.523	.898	.665	1.739
20–22	3.262	1.498	.929	.618	1.693
23–25	3.234	1.471	.984	.639	1.672
26–28	3.333	1.643	.898	.609	1.517
29–31	3.567	1.432	.923	.613	1.465
32–34	3.499	1.524	.991	.596	1.390
35–37	3.642	1.655	.900	.576	1.227
38–40	3.689	1.533	.935	.543	1.300
41–43	3.693	1.778	.769	.575	1.185
44–46	3.798	1.663	.877	.494	1.168
47–49	3.901	1.539	.798	.553	1.209
50–52	3.940	1.567	.841	.532	1.120
53–55	4.107	1.670	.671	.487	1.065
56–58	4.189	1.545	.827	.430	1.009
59–61	4.707	1.359	.627	.494	.813
62–64	4.820	1.510	.555	.424	.691

9.4 ALLOWANCE FOR AGE

As stated in the introduction to this chapter, the purpose of this study was to be able to characterize normal hearing as a function of age and sex. One would then be able to compare an individual with his or her peers to determine whether there was any evidence of abnormal hearing. One would also wish to be able to compare groups of individuals, say those representing a certain occupation or environment, to determine whether that group as a whole is different from the normal population. In this latter case, the group might cover a large gamut of ages. Since a PCA has been carried out on each of the 3-year age groups, y-scores may be obtained for each individual, and in the case of groups these scores may be combined to obtain information for the group as a whole.

However, when one is working with large databases such as these with over 18 000 in the control groups alone, some simplification of the analytical procedures is desirable. For this application, the simplification was done in two parts: (1) simplification of the characteristic vectors and (2) expression of these new vectors directly as a function of age. It was stated in the previous section that, for a given sex, the U-vectors differed little among age groups. Further, quantities such as the mean and standard deviations of the original variables as well as the characteristic roots appeared to be monotonically related to age. The intent, then, is to combine these two results into an efficient procedure for carrying out the desired comparisons.

To do this, three assumptions were made:

Assumption 1. The coefficients for the left and right ears were the same (except for sign on u_4) and could be represented by the average of the two ears for each frequency. It was assumed that this would also hold for the means and standard deviations of each frequency. There did appear to be a small but consistent bias between ears for 2000 Hz and 4000 Hz for the older age groups but, clinically, there was no reason why this should be so.

Assumption 2. The means, standard deviations and characteristic roots could be related to age by fairly simple functions. Clinically, there is no reason why this should not be so.

Figures 9.1, 9.2 and 9.3 show how the original quantities appeared as a function of age for the males and the functions used to relate them. Each age group was weighted equally in fitting these functions rather than taking the sample sizes into account. The means required exponential functions except for 4000 Hz, where a linear relationship was adequate. Exponential functions were also used for the standard deviations although, in the case of 4000 Hz, the sign of the exponential coefficient was negative, reflecting an upper asymptote. Linear functions were used for all four characteristic roots. (The values of l_1 for the two highest age groups were not used in fitting this line; it was assumed that their departure from a linear model was due to small sample sizes rather than any fundamental cause.) By using these functions to relate these various parameter estimates to age, it was felt that the precision of the system was probably enhanced over what it would have been using the original estimates for each age group. It was also necessary to obtain the limit for Q as a function of age and this is shown in Figure 9.4.

Assumption 3. The characteristic vectors could be approximated by a set of "streamlined" vectors, in the manner of the examples in Chapter 6. Recalling from that chapter, the criteria for such streamlining were that the approximations be simple, that the new "pc's" correlate well with the old, and that they be relatively uncorrelated with each other. It can be seen from Table 9.5 that the approximations *are* simple. Subsequently, it will be seen that the other two criteria are fulfilled as well.

For the applications in Chapter 6, approximations of this sort were sufficient. However, the complexity of this operation requires that these be restated into something equivalent to U-, V-, and W-vectors. These will be designated by U^a, V^a, and W^a, respectively.

To obtain u_i^a, simply divide the elements of each column in Table 9.5 by the square root of the sum of squares of the coefficients. For instance, for u_1^a, the coefficients are all 1's; hence, the sum of squares is 8 so each coefficient is divided by $\sqrt{8}$ and becomes .3536. The sums of squares for the other three vectors are 84, 28, and 8 respectively and U^a is displayed in Table 9.6.

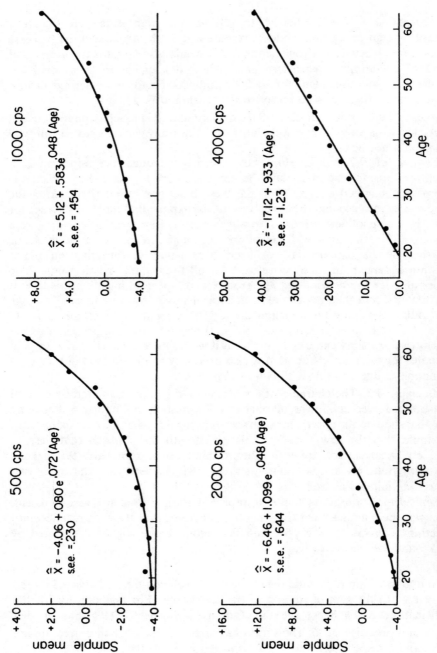

FIGURE 9.1. Audiometric Case Study: male sample means vs. age. Reproduced from Jackson and Hearne (1978) with permission from Biometrie-Praximetrie.

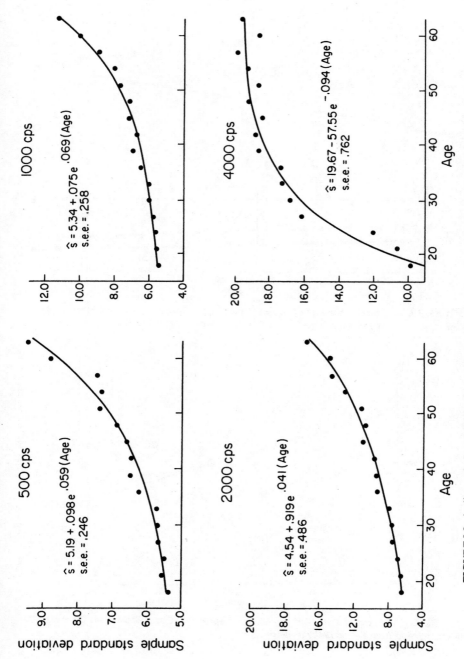

FIGURE 9.2. Audiometric Case Study: male sample standard deviations vs. age. Reproduced from Jackson and Hearne (1978) with permission from Biometrie-Praximetrie.

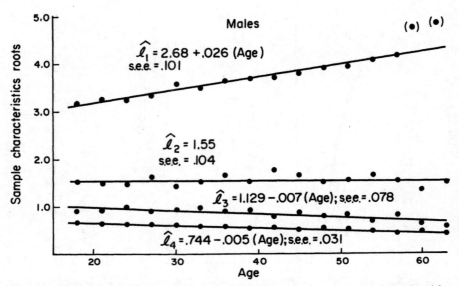

FIGURE 9.3. Audiometric Case Study: male sample characteristic roots vs. age. Reproduced from Jackson and Hearne (1978) with permission from Biometrie-Praximetrie.

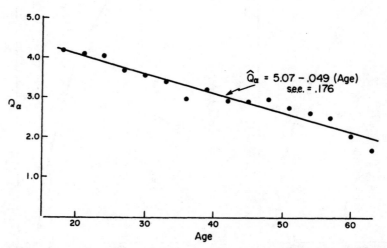

FIGURE 9.4. Audiometric Case Study: Q vs. age. Reproduced from Jackson and Hearne (1978) with permission from Biometrie-Praximetrie.

Table 9.5. Audiometric Case Study. Simplified Vectors, Basic Form—Males

Frequency	Vectors			
	1	2	3	4
500L	1	−3	1	−1
1000L	1	−2	0	−1
2000L	1	2	−3	−1
4000L	1	5	2	−1
500R	1	−3	1	1
1000R	1	−2	0	1
2000R	1	2	−3	1
4000R	1	5	2	1

Table 9.6. Audiometric Case Study. Simplified U-Vectors—Males

Frequency	u_1^a	u_2^a	u_3^a	u_4^a
500L	.3536	−.3273	.1890	−.3536
1000L	.3536	−.2182	0	−.3536
2000L	.3536	.2182	−.5669	−.3536
4000L	.3536	.5455	.3780	−.3536
500R	.3536	−.3273	.1890	.3536
1000R	.3536	−.2182	0	.3536
2000R	.3536	.2182	−.5669	.3536
4000R	.3536	.5455	.3780	.3536

Let D^a be a diagonal matrix made up of the *predicted* standard deviations, obtained as a function of age, from the relationships displayed in Figure 9.2. Let L^a be the diagonal matrix of *predicted* characteristic roots as obtained from Figure 9.3. Then following (1.6.1) and (1.6.2):

$$V^a = [L^a]^{1/2} U^a D^a \qquad (9.4.1)$$

and

$$W^a = [L^a]^{-1/2} U^a [D^a]^{-1} \qquad (9.4.2)$$

This means that there are different sets of V^a- and W^a-vectors for each age from 17 through 64, 48 in all, but since the quantities D^a and L^a are functionally related to age, these vectors need not be stored but could be computed as required by storing the functions instead. V^a and W^a correspond to V^* and W^* (Section 3.3) in that they may be used with the data in original rather than standard units.

9.5 PUTTING IT ALL TOGETHER

To see how this all fits together, let us consider the example of person aged 39. The V^a- and W^a-vectors for age 39 are given in Table 9.7.

Considering the entries for 500 Hz for the first characteristic vector, U^a_{11}, from Table 9.6, is .3536. From Figure 9.2, the standard deviation for 500 Hz for a 39-year old is $5.19 + .098 \text{EXP}\{(.059)(39)\} = 6.168$. The characteristic root associated with the first vector is, from Figure 9.3, $2.68 + (.026)(39) = 3.694$, whose square root is 1.922. Then $v^a_{11} = (.3536)(1.922)(6.168) = 4.19$. $w^a_{11} = (.3536)/\{(1.922)(6.168)\} = .0298$. Later we shall need the $\alpha = .05$ critical value for Q, which from Figure 9.4 becomes $5.07 - (.049)(39) = 3.16$. With the large combined sample employed in this study, one would feel quite comfortable in using large-sample limits for the other quantities as well, resulting in limits of ± 1.96 for the pc's and 9.49 for T^2.

Table 9.8 contains the data for one 39-year old male. The values for $\bar{x}_{39}$ are obtained from the expressions in Figure 9.1. Also included are the predicted $\hat{x}$ based on the y-scores for this individual ($y_1 = -.52$, $y_2 = .55$, $y_3 = -.92$, and

Table 9.7. Audiometric Case Study. Simplified V- and W-Vectors—Males Age 39

Frequency	v^a_1	v^a_2	v^a_3	v^a_4	w^a_1	w^a_2	w^a_3	w^a_4
500L	4.19	−2.51	1.08	−1.62	.0298	−.0426	.0330	−.0773
1000L	4.38	−1.75	0	−1.69	.0285	−.0272	0	−.0739
2000L	6.17	2.47	−4.78	−2.38	.0203	.0193	−.0673	−.0525
4000L	12.36	12.36	6.38	−4.77	.0101	.0241	.0224	−.0262
500R	4.19	−2.51	1.08	1.62	.0298	−.0426	.0330	.0773
1000R	4.38	−1.75	0	1.69	.0285	−.0272	0	.0739
2000R	6.17	2.47	−4.78	2.38	.0203	.0193	−.0673	.0525
4000R	12.36	12.36	6.38	4.77	.0101	.0241	.0224	.0262

Table 9.8. Audiometric Case Study. Sample Data for 39-year-old Male

Frequency	x	$\bar{x}_{39}$	$x - \bar{x}_{39}$	$\hat{x} = \bar{x}_{39} + Vy$	$\text{Res.} = \dfrac{x - \hat{x}}{s_{39}}$	$(\text{Res.})^2$
500L	−10	−2.7	−7.3	−7.07	−.47	.22
1000L	0	−1.3	1.3	−4.35	.67	.45
2000L	0	.7	−.7	3.51	−.39	.15
4000L	15	19.3	−4.3	14.33	.04	.00
500R	−10	−2.7	−7.3	−7.43	−.42	.18
1000R	−5	−1.3	−3.7	−4.73	−.04	.00
2000R	5	.7	4.3	2.99	.22	.05
4000R	15	19.3	−4.3	13.28	.09	.01
						1.06 $= Q$

$y_4 = .11$), and the components that made up the residual sum of squares, Q. Table 9.9 compares these statistics with those for the same individual using characteristic vectors based only on the age group 38–40. These results are fairly typical. Table 9.10 gives the correlations between the principal components using the simplified vectors and those using the original vectors as obtained from each age group. With a few exceptions related to the fourth component, these correlations are quite high. The few low correlations occur quite randomly throughout the age span and it may be reasonable to assume that the approximations might, in fact, be superior to the original vectors in those instances.

Table 9.9. Audiometric Case Study. Comparison of New Method Adjusted for Age with Vectors Obtained for 38–40 Age Group for Data Shown in Table 9.8

	New Method	Age Group 38–40
y_1	−.52	−.60
y_2	.55	.51
y_3	−.92	−.84
y_4	−.11	−.33
T^2	1.42	1.43
Q	1.06	.62

Table 9.10. Audiometric Case Study. Males: Correlations Between pc's Using Original and Simplified Vectors (Correlations > .995 = .99+)

Age Group	First pc	Second pc	Third pc	Fourth pc
17–19	.99	.99+	.99	.98
20–22	.99	.99+	.99	.98
23–25	.99	.99+	.99	.89
26–28	.99	.99+	.94	.91
29–31	.99+	.99+	.97	.95
32–34	.99	.97	.92	.60
35–37	.99	.99+	.97	.96
38–40	.99	.99+	.97	.95
41–43	.99+	.99+	.97	.90
44–46	.99+	.99+	.96	.54
47–49	.99+	.99+	.97	.99+
50–52	.99+	.99	.98	.96
53–55	.99+	.99	.96	.97
56–58	.99+	.99	.96	.59
59–61	.99+	.99	.88	.86
62–64	.99+	.99	.96	.96

Table 9.11. Audiometric Case Study. Males: Correlations Among pc's Using Simplified Vectors

Age Group	Components					
	1–2	1–3	1–4	2–3	2–4	3–4
17–19	.07	−.08	−.02	.02	.01	.02
20–22	.07	−.09	.00	.00	.01	−.02
23–25	.09	−.09	−.02	.05	.00	−.03
26–28	.09	−.16	−.02	.11	.00	−.01
29–31	.14	−.07	−.02	.09	.01	−.06
32–34	.07	−.11	.02	.16	.04	.03
35–37	.10	−.09	−.04	.09	.01	−.06
38–40	.10	−.12	−.04	.10	−.03	.02
41–43	.15	−.14	−.03	.04	−.01	−.03
44–46	.15	−.13	−.03	.03	−.02	.06
47–49	.13	−.14	.00	.06	−.02	−.02
50–52	.14	−.10	−.02	.08	.06	.02
53–55	.19	−.16	.00	−.03	.04	−.04
56–58	.20	−.14	−.04	.00	.12	.07
59–61	.15	−.20	.01	.19	.12	−.08
62–64	.18	−.16	.01	.01	−.07	−.04

How well the final criterion was handled can be seen in Table 9.11, which gives the correlations among all of the approximate pc's. These are quite low. It was concluded that this system of approximations had produced a viable procedure for the routine analysis of audiometric data that adjusted for the aging process and, as will be seen in the next section, may also be used in studying groups of individuals.

9.6 ANALYSIS OF GROUPS

Even though a group of individuals may represent several different age groups, the use of PCA and the generalized statistics described in Chapter 6 allows these data to be combined for purposes of significance tests. Some of the generalized statistics can be used directly, in particular T_0^2 (equation 6.2.1) and T_D^2 (equation 6.2.4). Q_0 (equation 6.4.1) can be employed but requires some additional work. Computation of Q_0 requires the use of equation 2.7.5 and to take age into account, $\theta_1, \theta_2, \ldots$ would have to be expressed as a function of age. (These quantities involve the last four characteristic roots and, for this particular example, that information was no longer available when the statistic was developed.) Q_M (equation 6.4.2) can also be employed but since $Q_{.05}$ used for this test is a function of age, one would have to assume that use of the average age of the group is adequate. T_M^2 (equation 6.3.1) is required to obtain T_D^2 but must be modified somewhat to report on its own. The reason for this is due to the fourth pc, which represents left–right ear difference. For a group of individuals, one would expect, on the average, as many people with trouble in only the left ear as those with only trouble in the right ear and that these differences would cancel each other out. For a summary statistic, one would want an indication of the overall left–right difficulties. Accordingly, T_M^2 must be modified as follows: Equation (6.3.1) is

$$T_M^2 = n\bar{y}'\bar{y}$$

$$= n\bar{y}_1^2 + n\bar{y}_2^2 + n\bar{y}_3^2 + n\bar{y}_4^2$$

Each of these terms is distributed in the limit as χ^2 with one degree of freedom, so T_M^2, as used here, would be distributed as χ_4^2. What is required is the substitution of some function of $|y_4|$ for $n\bar{y}_4^2$. Jackson and Hearne (1979a) showed that the quantity

$$g = n\left[1.6589\left(\sum_{i=1}^{n}|y_{4i}|/n\right) - 1.3236\right]^2 \tag{9.6.1}$$

is approximately distributed as χ^2 with one degree of freedom. Therefore, T_M^2 may be replaced with

$$T_{M'}^2 = n\sum_{i=1}^{3}\bar{y}_i^2 + g \tag{9.6.2}$$

which is asymptotically distributed as χ_4^2 and life goes on as before.

Table 9.12 shows the data for five individuals, one being the same 39-year old that was used in the previous section. Also included are the y-scores, T^2, Q, and the residuals expressed in standard units. Nothing is significant with the exception of y_2 for respondent #1.

T_0^2 is equal to the sum of the five values of T^2, which for this example is 23.05. The limit, $\chi^2_{(4)(5),.05}$, is equal to 31.4 so T_0^2 is not significant. From Table 9.12, T_M^2 is seen to be 14.42. T_D^2, the difference, is equal to 8.63 and is also not significant when compared with its limit, $\chi^2_{(4)(4),.05} = 26.3$.

To obtain $T_{M'}^2$, first obtain g from (9.6.1), which is equal to

$$g = 5[(1.6589)(.236) - 1.3236]^2 = 4.3440$$

and, as seen in Table 9.12, produces a $T_{M'}^2$-value of 18.58. This is significant when compared to $\chi^2_{4,.05} = 9.49$. The sums of squares of the residual averages is .3551, which when multiplied by 5, is equal to $Q_M = 1.78$. The average age of these five individuals is 37.8, which would produce a value of $Q_{.05} = 3.22$, so that is not significant. Q_0 as stated above, cannot be computed for this particular data set.

Table 9.12. Audiometric Case Study. Statistics for Group of Five Males of Differing Ages

	Age of Individual					
Frequency	18	25	39	47	60	Ave.
Original Data						
500L	−10	0	−10	10	10	
1000L	0	5	0	15	15	
2000L	10	10	0	20	35	
4000L	15	15	15	30	55	
500R	−5	0	−10	5	5	
1000R	−5	5	−5	15	15	
2000R	5	10	5	20	40	
4000R	10	15	15	25	60	
T^2	7.44	5.00	1.42	5.63	3.56	
Q	1.43	.49	1.06	.89	.45	
Principal Components						
y_1	1.08	1.74	−.52	1.75	1.43	1.096
y_2	2.14	.34	.55	−.87	.72	.576
y_3	−1.18	−1.37	−.92	−1.26	−1.00	−1.146
y_4	−.57	0	−.11	−.50	0	−.192

$$(\sum |y_4|)/5 = .236$$

Residuals in Standard Units						
500L	−.89	−.09	−.47	.24	.34	−.174
1000L	.41	.39	.68	.42	.12	.404
2000L	.05	−.17	−.38	−.27	−.28	−.210
4000L	−.16	−.23	.04	−.16	−.38	−.178
500R	.35	−.09	−.42	−.24	−.24	−.128
1000R	−.16	.39	−.04	.67	.12	.196
2000R	−.39	−.17	.22	−.02	.04	−.064
4000R	−.38	−.23	.09	−.17	−.13	−.164

$$T_M^2 = 5[(1.096)^2 + \cdots + (−.192)^2] = 14.42$$
$$T_{M'}^2 = 5[(1.096)^2 + \cdots + (−1.146)^2] + 4.3440 = 18.58$$
$$Q_M = 5[(−.174)^2 + \cdots + (−.164)^2] = 1.78$$

Overall, nothing is significant except $T_{M'}^2$ and from that one would conclude that the overall average of these individuals is significantly different from the normal group against which they are being compared. A glance at the y-score averages, whose limits are $\pm 1.96/\sqrt{5} = \pm.88$, show that the first and third pc's are significant; the overall hearing loss is greater and the differences between 2000 Hz and 4000 Hz are less than would be expected from the normal group.

CHAPTER 10

Singular Value Decomposition; Multidimensional Scaling I

10.1 INTRODUCTION

The purpose of this chapter is threefold:

1. To present an alternative form of obtaining characteristic roots and vectors called *singular value decomposition.*
2. To use this technique to analyze two-dimensional data arrays.
3. To introduce multidimensional scaling as a special case.

We shall see that singular value decomposition (SVD) is a more general form of PCA than what has been so far described. SVD may not only be employed to accomplish many of the tasks described in the earlier chapters but may also be used to handle other problems more simply than the standard PCA procedures presented so far. In the designation of the variants of PCA to be discussed in this chapter, the nomenclature of Okamoto (1972) will be employed.

10.2 *R*- AND *Q*-ANALYSIS

Most of the PCAs presented so far are examples of *R-analysis. n* observations are observed on p variables. A $p \times p$ covariance or correlation matrix is computed from these data and from this the characteristic roots and vectors are obtained. These, in turn, are used to represent each observation vector in terms of principal components. Mathematically, it is possible to do it the other way around, that is, to obtain the $n \times n$ covariance or correlation matrix of the observations, compute the characteristic roots and vectors of this matrix, and use these to obtain the principal components for each of the variables. This is sometimes referred to as *Q-analysis.*

189

We shall see later, in Chapter 11, why someone might want to employ Q-analysis but for now let us establish the relationships between the two methods. If $p < n$, there will be at most p nonzero characteristic roots from the dispersion matrix of the variables and there will be the same number of nonzero roots for the dispersion matrix of the observations. In other words, the upper bound on the rank of either matrix will be p. Conversely, if $n < p$, the upper bound of either matrix will be n so, overall, the upper bound will be min (p, n).

Consider the following 4×3 data matrix where $n = 4$ observations on $p = 3$ variables:

$$
\begin{array}{c}
\text{(Variables)} \\
\mathbf{X} = \begin{bmatrix} -2 & -2 & -1 \\ 0 & 1 & 0 \\ 2 & 2 & 2 \\ 0 & -1 & -1 \end{bmatrix} \text{(Observations)}
\end{array}
$$

Variable means   0   0   0

Note that an individual observation is represented by a row, not a column as has been our custom in previous chapters. This is the normal convention for data matrices so we shall conform. In subsequent chapters, the data may be represented in either manner; it is a manner of watching ones p's and n's! No matter which orientation $\mathbf{X}$ may have, the characteristic vectors $\mathbf{U}$, $\mathbf{V}$, and $\mathbf{W}$ will always be expressed in column form. To keep things simple, we have made the means of each variable equal to zero. We shall also work with the sums-of-squares matrix rather than the covariance or correlation matrix because of the fact that the covariance matrix for the variables will be $\mathbf{X'X}/(n-1)$ while that for the observations will be $\mathbf{XX'}/(p-1)$. Dividing by the degrees of freedom is a convenience but not a necessity for PCA so to keep the units the same for both matrices, we shall omit this operation. Therefore, the dispersion matrix for the variables will simply be $\mathbf{X'X}$ and for the observations, $\mathbf{XX'}$.

For this set of data:

$$
\begin{array}{cc}
\text{(Variables)} & \text{(Observations)} \\
\mathbf{X'X} = \begin{bmatrix} 8 & 8 & 6 \\ 8 & 10 & 7 \\ 6 & 7 & 6 \end{bmatrix} & \mathbf{XX'} = \begin{bmatrix} 9 & -2 & -10 & 3 \\ -2 & 1 & 2 & -1 \\ -10 & 2 & 12 & -4 \\ 3 & -1 & -4 & 2 \end{bmatrix}
\end{array}
$$

$$
\text{Trace}(\mathbf{X'X}) = 24 \qquad\qquad \text{Trace}(\mathbf{XX'}) = 24
$$

Characteristic roots of $\mathbf{X'X}$	Characteristic roots of $\mathbf{XX'}$
$l_1 = 22.2819$	$l_1 = 22.2819$
$l_2 = 1.0000$	$l_2 = 1.0000$
$l_3 = .7181$	$l_3 = .7181$
	$l_4 = 0$
$\mathbf{X'X}$ is positive definite with rank $= 3$	$\mathbf{XX'}$ is positive semidefinite with rank $= 3$

Note that the first three roots are identical for both matrices and this will extend to any size of matrix. Both matrices will have the same rank, which will be less than or equal to the minimum of p or n. The remaining roots will be zero. Since the dispersion matrix made up from the observations is usually much larger than that for the variables, it will normally have a large number of zero roots.

The characteristic vectors of $\mathbf{X'X}$ will be denoted by $\mathbf{U}$, $\mathbf{V}$, and $\mathbf{W}$, while those for $\mathbf{XX'}$ will be denoted by $\mathbf{U^*}$, $\mathbf{V^*}$, and $\mathbf{W^*}$. For this data set:

$$\mathbf{U} = \begin{bmatrix} -.574 & .816 & -.066 \\ -.654 & -.408 & .636 \\ -.493 & -.408 & -.768 \end{bmatrix} \quad \mathbf{U^*} = \begin{bmatrix} .625 & -.408 & -.439 \\ -.139 & -.408 & .751 \\ -.729 & 0 & -.467 \\ .243 & .816 & .156 \end{bmatrix}$$

(Only three characteristic vectors are obtained for $\mathbf{XX'}$ because the fourth characteristic root is zero and, although a fourth vector can be obtained orthogonal to the first three, it has no length and would be of no use in the present discussion. For the record, all of the coefficients of $\mathbf{u_4^*}$ are equal to .5.) $\mathbf{U}$ and $\mathbf{U^*}$ are both orthonormal, that is, $\mathbf{U'U} = \mathbf{I}$ and $\mathbf{U^{*'}U^*} = \mathbf{I}$.

Let the nonzero characteristic roots be

$$\mathbf{L} = \begin{bmatrix} 22.2819 & 0 & 0 \\ 0 & 1.0000 & 0 \\ 0 & 0 & .7181 \end{bmatrix}$$

and their square roots

$$\mathbf{L}^{1/2} = \begin{bmatrix} 4.7204 & 0 & 0 \\ 0 & 1.0000 & 0 \\ 0 & 0 & .8474 \end{bmatrix}$$

From this,

$$\mathbf{V} = \mathbf{U}\mathbf{L}^{1/2} = \begin{bmatrix} -2.707 & .816 & .056 \\ -3.089 & -.408 & -.539 \\ -2.326 & -.408 & .651 \end{bmatrix}$$

and

$$\mathbf{V}^* = \mathbf{U}^*\mathbf{L}^{1/2} = \begin{bmatrix} 2.9549 & -.408 & -.372 \\ -.656 & -.408 & .636 \\ -3.441 & 0 & -.396 \\ 1.147 & .816 & .132 \end{bmatrix}$$

One can then also verify that $\mathbf{V}\mathbf{V}' = \mathbf{X}'\mathbf{X}$ and $\mathbf{V}^*\mathbf{V}^{*'} = \mathbf{X}\mathbf{X}'$.
    Lastly,

$$\mathbf{W} = \mathbf{U}\mathbf{L}^{-1/2} = \begin{bmatrix} -.122 & .816 & -.078 \\ -.139 & -.408 & .751 \\ -.104 & -.408 & -.906 \end{bmatrix}$$

and

$$\mathbf{W}^* = \mathbf{U}^*\mathbf{L}^{-1/2} = \begin{bmatrix} .132 & -.408 & -.518 \\ -.029 & -.408 & .886 \\ -.154 & 0 & -.552 \\ .051 & .816 & .184 \end{bmatrix}$$

Finally, one can verify the principal component scores for the observations,

$$\mathbf{Y} = \mathbf{X}\mathbf{W} = \mathbf{U}^* \tag{10.2.1}$$

and the principal component scores of the variables,

$$\mathbf{Y}^* = \mathbf{X}'\mathbf{W}^* = \mathbf{U} \tag{10.2.2}$$

This says that the principal component scores for the n observations are equal to the characteristic vectors obtained from the observation sums-of-squares matrix and the principal component scores for the p variables are equal to the characteristic vectors obtained from the variables sums-of-squares matrix. Further, given one set of characteristic vectors, one may obtain the other without having to obtain the other sums-of-squares matrix.

10.3   SINGULAR VALUE DECOMPOSITION

We may carry the relationships in Section 10.2 one step farther. Recall from (2.7.1) that if all of the principal components are used, the original observations may be obtained directly from the pc's, viz.,

$$\mathbf{X} = \mathbf{YV}' \tag{10.3.1}$$

which restates that relationship for a data matrix rather than a data vector. Recall, also, from (1.6.1) that $\mathbf{V} = \mathbf{UL}^{1/2}$ and from (10.2.1) that $\mathbf{Y} = \mathbf{U}^*$; then $\mathbf{X}$ can be rewritten as

$$\mathbf{X} = \mathbf{U}^* \mathbf{L}^{1/2} \mathbf{U}' \tag{10.3.2}$$

where $\mathbf{X}$ is $(n \times p)$, $\mathbf{U}$ and $\mathbf{L}^{1/2}$ are $(p \times p)$ and $\mathbf{U}^*$ is $(n \times p)$. Note that one obtains $\sqrt{l_i}$, not l_i, because this operation involves the data itself rather than some function of their sums of squares as in the case of covariance or correlation matrices.

Equation (10.3.2) is the fundamental identity defining *singular value decomposition* (SVD). (Okamoto called this *N-analysis* where N refers to *natural* or *naive*.) In SVD, a data matrix is decomposed into a product of the characteristic vectors of $\mathbf{X}'\mathbf{X}$, the characteristic vectors of $\mathbf{XX}'$ and a function of their characteristic roots. Equally important, by (10.2.1) and (10.2.2),

$$\mathbf{X} = \mathbf{YL}^{1/2} \mathbf{U}' = \mathbf{U}^* \mathbf{L}^{1/2} \mathbf{Y}^* \tag{10.3.3}$$

so that one may get either set of vectors *and* the corresponding principal component scores in one operation. From these quantities, everything else can be obtained such as other scalings of the vectors and the principal components. This being so, one may obtain $\mathbf{U}$ and $\mathbf{Y}$ by using:

1. The standard PCA procedure given in Chapter 1, a two-stage procedure, or
2. Equation (10.3.2) which is a single-stage procedure but which requires a different algorithm.

Do *not* obtain $\mathbf{Y}$ as the vectors $\mathbf{U}^*$ of $\mathbf{XX}'$. The signs of $\mathbf{U}^*$ may be the opposite of what they would have been for the other two methods since $\mathbf{U}$ and $\mathbf{U}^*$ are obtained independently of each other. This would be disastrous for the scaling techniques to be described in Sections 10.5 and 10.6 and might cause problems in verifying (10.3.2).

The main virtue of SVD is that all three matrices are obtained in one operation *without* having to obtain a covariance matrix. If the data are such that a correlation matrix should be employed, SVD may still be used but on data transformed to standard units.

For the data set introduced in the previous section, (10.3.2) is

$$U*L^{1/2}U' =$$

$$
\begin{bmatrix}
.625 & -.408 & -.439 \\
-.139 & -.408 & .751 \\
-.729 & 0 & -.467 \\
.243 & .816 & .156
\end{bmatrix}
\begin{bmatrix}
4.72 & 0 & 0 \\
0 & 1.00 & 0 \\
0 & 0 & .85
\end{bmatrix}
\begin{bmatrix}
-.574 & -.654 & -.493 \\
.816 & -.408 & -.408 \\
-.066 & .636 & -.768
\end{bmatrix}
$$

$$
=
\begin{bmatrix}
-2 & -2 & -1 \\
0 & 1 & 0 \\
2 & 2 & 2 \\
0 & -1 & -1
\end{bmatrix}
= X
$$

Reinterpreted as (10.3.3), the Chemical data of Chapter 1 become:

$$YL^{1/2}U' =$$

$$
\begin{bmatrix}
.40 & 1.72 \\
.13 & -1.43 \\
-.18 & .70 \\
-.12 & .95 \\
1.88 & -.30 \\
1.06 & -.38 \\
-1.47 & .10 \\
-.70 & -.55 \\
-.28 & -1.71 \\
-.47 & -.05 \\
.53 & -.19 \\
-1.06 & -.58 \\
1.70 & .89 \\
-.05 & -.73 \\
-1.36 & 1.55
\end{bmatrix}
\begin{bmatrix}
1.203 & 0 \\
0 & .294
\end{bmatrix}
\begin{bmatrix}
.724 & .690 \\
-.690 & .724
\end{bmatrix}
=
\begin{bmatrix}
.0 & .7 \\
.4 & -.2 \\
-.3 & .0 \\
-.3 & .1 \\
1.7 & 1.5 \\
1.0 & .8 \\
-1.3 & -1.2 \\
-.5 & -.7 \\
.1 & -.6 \\
-.4 & -.4 \\
.5 & .4 \\
-.8 & -1.0 \\
1.3 & 1.6 \\
.1 & -.2 \\
-1.5 & -.8
\end{bmatrix}
= X
$$

where Y are the y-scores displayed in Table 1.2, which also includes X, the original data in terms of deviations from the column means $\bar{x}_1 = 10.0$ and $\bar{x}_2 = 10.0$.

If fewer than p pc's are retained, (10.3.2) still holds but U is now $(p \times k)$, $L^{1/2}$ is $(k \times k)$, and U^* is $(n \times k)$. There are a number of variants on (10.3.2), some of which will be discussed later in this chapter, but the common form is $X = AB$, where the characteristic roots have been absorbed into A, B, or both. For $k < \min(p, n)$, SVD amounts to minimizing

$$\sum_{i=1}^{n} \sum_{j=1}^{p} (x_{ij} - a_i' b_j)^2 \tag{10.3.4}$$

One can also perform a *weighted* SVD by minimizing

$$\sum_{i=1}^{n} \sum_{j=1}^{p} c_{ij}(x_{ij} - a_i' b_j)^2 \tag{10.3.5}$$

(Gabriel and Zamir, 1979)

Singular value decomposition was first proposed by Eckart and Young (1936) and Householder and Young (1938), although the notion of SVD appears to go back as far as J.W. Gibbs in 1884 and J.J. Sylvester in 1890 (Johnson, 1963; Good, 1970). Among earlier developers and users of SVD were Whittle (1952) and Wold (1966a,b), the latter developing an algorithm called NILES (later called NIPALS for *nonlinear iterative partial least squares*). PLS, standing for *partial least squares* has also been called *projected latent structures*. Unlike the principal component regression to be discussed in Chapter 12, *least squares* in NIPALS refers to the regression of the pc's on the original data from which they were obtained. For a discussion of partial least squares, see Geladi (1988) or Lohmöller (1989).

The data set with which we have worked in this chapter is *singly-centered*, meaning that each observation has been corrected for its own *variable* mean although it could, conceivably, be corrected for the observation mean. Another variant is to correct for both row and column means (double-centering) and a SVD of this type of matrix, called *M-analysis* (for *mean-adjusted*) is due to Gollob (1968a). An individual item in this two-way array is

$$\begin{aligned}
x_{ij}^M &= [x_{ij} - (\bar{x}_{i.} - \bar{x}_{..}) - (\bar{x}_{.j} - \bar{x}_{..}) - \bar{x}_{..})] \\
&= (x_{ij} - \bar{x}_{i.} - \bar{x}_{.j} + \bar{x}_{..})
\end{aligned} \tag{10.3.6}$$

where $\bar{x}_{..}$ is the grand mean of the data, $\bar{x}_{i.}$ is the mean of the ith row and $\bar{x}_{.j}$ is the mean of the jth column. x^M is exactly what one obtains for the residuals in a two-way analysis of variance, so M-analysis may be used for the analysis of interactions, which will be taken up in Section 13.7. A very similar form of (10.3.6) dealing with similarities in multidimensional scaling will be seen in Chapter 11.

Recall from equation (1.6.11) that $WW' = S^{-1}$ so that PCA may be used to obtain inverses. In the same manner, SVD may be used to obtain *generalized* inverses of rectangular matrices, in particular, data matrices. The Moore–Penrose generalized inverse of **X** is (Good, 1969, 1970; Eubank and Webster, 1985)

$$\mathbf{X}^+ = \mathbf{U}\mathbf{L}^{-1/2}\mathbf{U}^{*\prime} \tag{10.3.7}$$

For the 4 × 3 matrix used in this chapter, **X**⁺ is

$$
\begin{bmatrix}
-.574 & .816 & -.066 \\
-.654 & -.408 & .636 \\
-.493 & -.408 & -.768
\end{bmatrix}
\begin{bmatrix}
.212 & 0 & 0 \\
0 & 1.000 & 0 \\
0 & 0 & 1.180
\end{bmatrix}
\begin{bmatrix}
.625 & -.139 & -.729 & .243 \\
-.408 & -.408 & 0 & .816 \\
-.439 & .751 & -.467 & .156
\end{bmatrix}
$$

$$
=
\begin{bmatrix}
-.375 & -.375 & .125 & .624 \\
-.250 & .749 & -.249 & -.250 \\
.499 & -.500 & .499 & -.500
\end{bmatrix}
$$

X⁺**X** = **I** as one would expect. **XX**⁺ is a matrix made up of .75's on the diagonal and −.25 everywhere else. Both of these matrix products are symmetric as required by the Moore–Penrose generalized inverse. This inverse also requires that **XX**⁺**X** = **X** and **X**⁺**XX**⁺ = **X**⁺; both are satisfied in this example.

10.4   INTRODUCTION TO MULTIDIMENSIONAL SCALING

It has been shown in equation (10.3.2) that a data matrix **X** may be decomposed into a product of its characteristic vectors, characteristic roots, and principal component scores. This relationship may be used to represent both the vectors and the pc's on the same plot, a multidimensional representation of the data. These plots are generally considered to be a form of *multidimensional scaling* (MDS). The subject of scaling can consume a full text on its own but our purpose in this chapter and the next is to provide a brief description of MDS and show where PCA fits in.

First, a word about unidimensional scaling. A *scale* is nothing more than a representation of a number of objects or other phenomena on a line such that the positioning of these points is related in some mathematical sense to one or more criteria associated with these objects. The real numbers that are employed in these scales possess one or more of the following properties:

1. *Order.* The numbers are ordered. An object with a scale value of 7 is higher in some criterion than an object which has a value of 6.
2. *Distance.* Differences between numbers are ordered. Two objects with scale values of 8 and 12 are farther apart than two others with scale values of 5 and 6.

3. *Origin.* The series of real numbers has a unique origin, the number *zero.* A scale value of zero would indicate the complete absence of the criterion in question.

Any scaling techniques with which we will be concerned, will, at the very least, possess the order property and may or may not possess one or both of the others. A scale possessing the order property is called an *ordinal* scale. One that possesses both order and interval properties is called an *interval* scale, and one that possesses all three is called a *ratio* scale. In a word, ordinal scales will indicate that one object has more of something than another but not by how much. Interval scales do possess this property. If a scale contains a real zero it will generally be defined only for positive numbers, hence, a ratio of any two points on the scale would be meaningful. Temperature scales such as Fahrenheit or Celsius are not ratio scales because zero is an arbitrary definition in both cases.

Some scales are nothing more than the report of some physical measurements. Others are attempts to quantify judgment data—an area in the field of psychometrics. We will be concerned not so much in this chapter or the next with how the data will be collected as with how these data are used to produce a scale. In particular, we will be concerned with either data that are already in the form of a multiple response or data in which a single response represents a multivariate phenomenon. These comprise *multidimensional* scaling. Section 11.3 will include some data-collection techniques for similarities scaling. Most dominance scaling customarily uses data consisting of an $n \times p$ matrix of ratings, ranks, and the like, for each stimulus by each respondent, although the method of paired comparisons is also widely employed.

One other classification of scaling methods is that of similarity or dominance. In *similarity* scaling, the closer together two stimuli appear, the most similar they are judged to be by whatever criterion is chosen, either by the investigator or the respondent. Tasks involve similarity or dissimilarity judgments; the degree of the desirability of a stimulus is not taken into account. In *dominance* scaling, each stimulus or object is *compared* with others in terms of one or more specific characteristics. The most common is *preference*—so common, that these are often referred to in the literature as *preference scales.* However, any characteristic that may be used to describe the dominance of one stimulus or object over another is appropriate. Most physical scales as well as value and choice models in market research use dominance scales. Sometimes, only a thin line separates similarity and dominance and, in fact, dominance scales may be considered *similarity scales of dominance.* This chapter will be concerned with dominance scaling applications that involve PCA in one form or another. Chapter 11 will discuss the relationship of PCA to multidimensional similarity scaling. It should be remarked that the material in Sections 10.5–10.8 as well as Chapter 11 is used primarily for exploratory data analysis. There are very few inferential procedures to accompany these techniques except of a very rudimentary nature.

There are a number of books now available on MDS. Green et al. (1989) focus, primarily, on a number of algorithms that were developed within the Bell

Laboratories and include many of them on two floppy disks for use with PC-compatible computers. A similar book by Green and Rao (1972) contains a data set that is employed throughout the text. Coxon (1982) is similar in nature, being organized around the MDS(X) library of programs assembled at the University of Edinburgh. A companion book (Davies and Coxon, 1982) contains 17 classic papers on MDS. Another huge collection of papers (Lingoes et al. 1979) represents work related primarily to the Guttman–Lingoes series of MDS programs (Lingoes, 1972, 1973, 1979; Roskam, 1979b), although most of the 33 papers would be of use whether or not one used those particular algorithms. This book goes into considerably more detail than the others, particularly with regard to theoretical considerations. Young (1987) also contains some theory in addition to a historical review. This book features the ALSCAL series of algorithms (Young and Lewyckyj, 1979; SAS Institute, Inc., 1986) and includes six applications papers. Another computer-oriented book by Schiffman et al. (1982) compares algorithms of the Bell Labs series, the Guttman–Lingoes series, ALSCAL, and MULTISCALE—a maximum likelihood procedure which will be discussed in Section 11.4. Some data sets are included. As an introduction, a Sage publication by Kruskal and Wish (1978) may be of use although it covers only similarity scaling. Carroll (1980) prepared a survey article on dominance scaling including a catalog of general procedures available at that time.

There are a number of techniques for obtaining multidimensional dominance plots. Most of these are related to one of two models, the *linear compensatory* model, which uses SVD, and the *unfolding* model, which will be discussed in Section 10.7. The most widely used linear compensatory models are the method of *biplots* due to Gabriel (1971) and an algorithm called MDPREF developed by Chang and Carroll (1969). There is very little difference between these techniques but they were developed within entirely different environments. Biplots have been used primarily within the scientific community while MDPREF has been the workhorse of market research. Section 10.5 will deal with biplots and Section 10.6 will then point out what makes MDPREF different. Both of these methods are generally employed to produce interval scales. However, the starting data may be comprised of ranks and, despite the claims of some of their proponents, the results of these methods, when using ranks, are not interval scales except in the special case that the stimuli being represented by these ranks are evenly spaced. Among other alternatives is a technique by MacKay and Zinnes (1986) that estimates both location and scale parameters for preference data, PROSCAL, a multivariate generalization of the Thurstone paired comparison model (Zinnes and MacKay, 1983, Büyükkurt and Büyükkurt, 1990) and LINMAP (Srinivasan and Shocker, 1973b), which uses linear programming with pair comparison data.

Many marketing studies obtain judgments by means of a technique called *conjoint* analysis (Green and Wind, 1973). This is nothing more than a method of obtaining judgments on multiattribute objects except that they have been arranged in the nature of a factorial, or fractional factorial, experiment so that one may study the *trade-offs* between pairs of attributes. It may be fair to assume that a major employment of fractional factorial designs, particularly those employing mixed levels, may be found in the area of market research.

10.5 BIPLOTS

Biplots may be used for any number of dimensions but, as originally conceived by Gabriel, biplots meant two-dimensional plots only, the assumption being that the data had rank 2 or nearly so. Anything for $k > 2$ was referred to as *bimodal*. While the examples that Gabriel used were essentially rank 2, the same does not apply to a lot of other examples that have since appeared in the literature. Two-dimensional plots are very popular because they are easy to work with but should always include some statement with regard to the proportion of the total variability explained by the first two characteristic roots. Unless this quantity is sufficiently large, the interpretation of the plot is suspect.

Now that the preliminaries are out of the way, how does one use equation (10.3.2) to obtain MDS plots? Not only can $\mathbf{X}$ be written as $\mathbf{U}^*\mathbf{L}^{1/2}\mathbf{U}'$ but it can also be expressed as a simple product of two matrices (i.e., $\mathbf{X} = \mathbf{AB}'$), both of which are used for the plot. This means that $\mathbf{L}^{1/2}$ is somehow included with $\mathbf{U}^*$, $\mathbf{U}$ or both. It will be convenient to rewrite (10.3.2) as

$$\mathbf{X} = \mathbf{AB}' = \mathbf{U}^*\mathbf{L}^{c/2}\mathbf{L}^{(1-c)/2}\mathbf{U}' \tag{10.5.1}$$

where $0 \leqslant c \leqslant 1$, $\mathbf{A} = \mathbf{U}^*\mathbf{L}^{c/2}$ and $\mathbf{B}' = \mathbf{L}^{(1-c)/2}\mathbf{U}'$. (If less than a full set of components are used, as is normally the case with biplots, equation (10.5.1) will produce only an estimate of $\mathbf{X}$, $\hat{\mathbf{X}}$.) Using the original concept of biplots, $\mathbf{A}$ will be $n \times 2$ and $\mathbf{B}$ will be $p \times 2$. Gabriel and Odoroff (1986c) proposed three choices of c. Their paper contains several examples including automobiles, alcohol density, and Fisher's iris data as well as a list of biplot programs. Many of Gabriel's other publications contain examples dealing with meteorology.

One specification of $\mathbf{A}$ and $\mathbf{B}$, called the SQ biplot, sets $c = .5$ making $\mathbf{A} = \mathbf{U}^*\mathbf{L}^{1/4}$ and $\mathbf{B} = \mathbf{UL}^{1/4}$. This case is advocated for diagnostic purposes because $\mathbf{A}$ and $\mathbf{B}$ have the same weights. For the 4×3 data set used throughout this chapter,

$$\mathbf{A} = \begin{bmatrix} 1.36 & -.41 \\ -.30 & -.41 \\ -1.58 & 0 \\ .53 & .82 \end{bmatrix} \qquad \mathbf{B} = \begin{bmatrix} -1.25 & .82 \\ -1.42 & -.41 \\ -1.07 & -.41 \end{bmatrix}$$

These are plotted in Figure 10.1(a). The same biplot is shown in Figure 10.1(b). The only difference is that $\mathbf{B}$ is now represented by vectors. Notice the projections of all the points of $\mathbf{A}$ on the vector representing the first row of $\mathbf{B}$, $\mathbf{b}_1$. These projections are perpendicular to that vector. The intersections of those projections

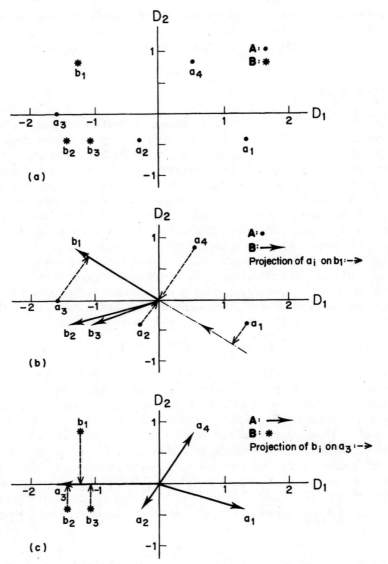

FIGURE 10.1. Example of SQ biplot. (a) SQ plot. (b) A projected on B. (c) B projected on A.

on b_1 correspond very closely to the first column of **X**. They would correspond exactly except that there is a small amount of variability accounted for by the third dimension, which is not represented in this biplot. The projections on the other two vectors will correspond similarly to the other two columns of **X**, b_3 fitting a little better than b_2. Figure 10.1(c) shows the same information except now **A** is represented by vectors. Note that the projections of **B** on a_3 are nearly all the same, corresponding to the third row of **X**, again with a slight variation due to the representation of a rank-3 matrix in two dimensions. Figures 10.1(b) and 10.1(c) are called *point–vector* plots and it is in this manner that biplots should be interpreted. It is arbitrary which matrix is represented by vectors and which one by points. The main thing is that one set is projected on the other so that once the decision has been made, the vectors represent direction and the projection of the points represents distance.

In the point–vector model, the objects represented by the vectors each collapse the space by projecting the objects represented by the points onto a single dimension oriented towards the dominant region of the space (Coxon, 1974). The angle between a vector and each of the axes defined by the SVD measures the importance of the contribution that particular dimension makes on the data from which the vector is derived.

Within each set of points, those from **A** or those from **B**, the distances between pairs of points are related to the similarity associated with each pair. For instance, in Figure 10.1(a), points b_2 and b_3 are quite close together, which implies that the second and third columns of **X** are similar. The fact that a_1 and a_3 are quite a distance apart implies that the first and third rows of **X** are quite dissimilar. Since these are point–vector plots, **NEVER** attempt to interpret the distances between the points of **A** and the points of **B** as shown in Figure 10.1(a). (In Section 10.7, we shall introduce *point–point* plots, where such a comparison *is* valid, but those plots require a different algorithm to produce them.)

If $c = 0$, then $\mathbf{A} = \mathbf{U}^*$ and $\mathbf{B} = \mathbf{U}\mathbf{L}^{1/2}$. Gabriel and Odoroff (1986c) called this a GH biplot and referred to it as the *column metric preservation* biplot. This is because **B** is also equal to **V** and $\mathbf{V}\mathbf{V}' = \mathbf{X}\mathbf{X}'$. Remember that **B** is also equal to $\mathbf{Z}^*$ and $\mathbf{A} = \mathbf{Y}$. If this is plotted with **A** being represented by vectors and **B** by points, the interpoint distances between the b_j are *euclidean* since $\mathbf{Z}^*$ are in the same units as the original data and have been called *distance plots* (TerBraak, 1983). If **B** are the vectors and **A** the points, the distances between the a_i are *Mahalinobis distances*. (Recall that euclidean distances are unweighted and Mahalinobis distances are weighted, hence downweighting variables that have larger variances.)

The cosine of the angle between vectors b_j and $b_{j'}$ will approximate the correlation between column j and column j'. The length of vector b_j is related approximately to the standard deviation of the jth column of **X**. These approximations become better the larger the proportion of the total variability that is explained by the first two characteristic roots.

For the same 4×3 example,

$$\mathbf{A} = \begin{bmatrix} .62 & -.41 \\ -.14 & -.41 \\ -.73 & 0 \\ .24 & .82 \end{bmatrix} \qquad \mathbf{B} = \begin{bmatrix} -2.71 & .82 \\ -3.09 & -.41 \\ -2.33 & -.41 \end{bmatrix}$$

(The second column of both matrices is the same as for the case $c = .5$ because $l_2 = 1.0$. Several examples of GH biplots may be found in Gabriel, 1981).

If $c = 1$, we have the JK or *row metric preserving* biplot. This is just the opposite of the GH biplot. $A = V^* = Z$ and $B = U' = Y^*$. For this same example, A and B become:

$$A = \begin{bmatrix} 2.95 & -.41 \\ -.66 & -.41 \\ -3.44 & 0 \\ 1.15 & .82 \end{bmatrix} \qquad B = \begin{bmatrix} -.57 & .82 \\ -.65 & -.41 \\ -.49 & -.41 \end{bmatrix}.$$

It is now the a_i that have euclidean distances and b_j the Mahalinobis distances. The lengths of the vectors a_i are related approximately to the standard deviation of the ith row. The a_i may also be interpreted as the projections of the points in higher dimensions on the least-square plane. The original axes projected on that same plane become the b_j.

It should be emphasized that these three biplots, SQ, GH, and JK, differ only in geometric representation. All three will produce the identical estimate of the original data. For this example,

$$\hat{X} = \begin{bmatrix} -2.0 & -1.8 & -1.3 \\ .0 & .6 & .5 \\ 2.0 & 2.2 & 1.7 \\ .0 & -1.1 & -.9 \end{bmatrix}$$

as compared with the original data

$$X = \begin{bmatrix} -2 & -2 & -1 \\ 0 & 1 & 0 \\ 2 & 2 & 2 \\ 0 & -1 & -1 \end{bmatrix}$$

Remember, there is nothing magical about two-dimensional representations except that they are convenient for display. If more than two dimensions are required to obtain an adequate representation, this is information the user of the analysis should have.

For a larger example of a JK biplot, we return to the Chemical example displayed earlier in this section. $A = Z$, which is displayed in Table 1.2, and $B = U'$, first displayed in Section 1.3. The biplot is shown in Figure 10.2. Note that this plot is identical to Figure 1.6 except for orientation, the coordinate

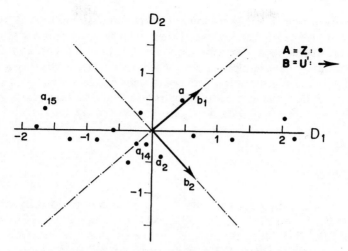

FIGURE 10.2. Chemical Example: JK biplot.

axes in Figure 10.2 being the rotated axes in Figure 1.6 and vice versa, a property of JK biplots if $p = 2$.

Although the vectors in any of these cases were obtained from the singular value decomposition, they are, in fact, least-squares estimates (Gabriel, 1978) as are the SVD solutions of all the other cases defined by Okamoto. Because of this, it may be useful to identify with each vector the correlation coefficient associated with it. This will serve to point out any vectors that are fitted with less precision than the others and will also serve to remind the reader that these plots are, in general, not perfect fits. This is seldom done in practice and this is unfortunate. The casual reader may well accept a particular representation as gospel when, in fact, the fit is quite poor. Since the computations are relatively simple, such as correlation coefficients, it is hoped that this situation will be corrected in the days to come.

The situation may occur when one wishes a biplot in which one or more columns (or rows) of **X** are to be fitted exactly, with the others doing the best they can given this restriction. A solution for this case is given by Gabriel (1978). Gabriel and Haber (1973) suggested using a biplot on the Moore–Penrose generalized inverse of the data matrix **X** as a possible method of detecting outliers. Essentially, this involves working with the two smallest components of **X**. Gheva (1988) combined several biplots representing different timepoints into an "aggregate" plot and analyzed the deviations of the individual biplots from this. Corsten and Gabriel (1976) used biplots as an adjunct to tests for comparing two or more covariance matrices.

As stated at the beginning of this section, the term "biplots" originally meant two-dimensional plots only. Obviously, they can be carried out in any dimensionality. The operations are the same except that for (10.5.1) **A** is now $(n \times k)$ and **B** is $(p \times k)$. The main problem for $k > 2$ is how to represent these pictorially. Some work has been done on three-dimensional biplots by Gabriel and Odoroff (1986a,b). The second of these papers has a number of examples using polarized color plots. Winsberg and Ramsay (1983) proposed using spline transformations of each column in cases where nonlinearity was suspected. They suggested that this might result in reduced dimensionality of the final representation and also that the nature of the transformations themselves might be informative. Nonlinear biplots were also considered by Gower and Harding (1988).

10.6 MDPREF

Very similar to the biplot is MDPREF, a technique developed by Chang and Carroll (1969). (See also Carroll, 1972, and Green and Rao, 1972.) As stated in Section 10.5, MDPREF has been used primarily for market research. Columns of the data matrix, **X**, generally refer to some stimuli such as variations of a product, different brands, and so on, and are represented by points whose distances are the Mahalinobis distances. The rows refer to the responses of a sample of n respondents to these p stimuli and are represented by vectors. These responses may be either scores or ranks. The MDPREF algorithm can also handle paired comparison data. A nonmetric procedure, which produces point–vector estimates but only requires an ordinal scale, was given by Kruskal and Shepard (1974).

MDPREF differs from the biplot in two ways, one operational and the other cosmetic. Whereas biplots work with a data matrix composed of deviations from the *column* means, MDPREF employs deviations from the *row* means. (These are the respondents.) Given that transformation, the form of the SVD is the same as a JK-biplot; the product will be **V*U′** (or **ZU′**).

The second difference relates to the vectors representing the respondents. These vectors are normalized to have unit length and are customarily represented only by arrows on the unit circle. This produces "cleaner" plots, which may be required by the larger number of rows in the data matrix of a typical market research study than is usually found in scientific examples. (Quite often, these large samples are clustered first and it is the cluster means that become the "respondents" for the MDPREF operation.) The rationale for doing this is that the vectors represent, primarily, direction. Their length, which is related to variability, is considered of secondary importance as are the euclidean distances between these endpoints. Making the vectors of unit length stresses the directional aspect and the fact that the stimuli are then projected on these vectors and (not incidentally) that this is *not* to be interpreted as a point–point plot. Another advantage of unit vectors will be seen in a three-dimensional example displayed later in this section. Again, correlations indicating the fit of these vectors should be included.

The data in Table 10.1 come from a classroom exercise in a Psychometrics class given within the Eastman Kodak Company in 1976. These data consist

Table 10.1. Presidential Hopefuls. Original Data. Graphic Rating Scales 0–100

Respondent #	Carter	Ford	Humphrey	Jackson	Reagan	Wallace	Respondent Mean
1	60	40	70	50	30	10	43.3
2	60	45	40	35	25	5	35.0
3	50	70	30	50	50	60	51.7
4	25	70	60	55	62	30	50.3
5	55	45	5	65	90	95	59.2
6	75	70	30	35	40	5	42.5
7	25	80	80	40	20	15	43.3
8	35	85	50	45	55	25	49.2
9	40	80	25	50	70	30	49.2
10	30	65	55	40	55	35	46.7
11	15	65	60	50	70	35	49.2
12	70	50	45	60	55	40	53.3
13	70	60	20	50	50	10	43.3
14	7	50	23	24	96	26	37.7
15	50	40	30	80	70	60	55.0
16	60	15	55	30	4	3	27.8
17	50	70	60	20	50	15	44.2
18	10	100	85	50	50	30	54.2
Candidate average	43.7	61.1	45.7	46.0	52.3	29.4	46.4
S.D.	21.3	20.2	21.9	14.6	23.2	23.6	

of ratings given by $n = 18$ students for $p = 6$ presidential hopefuls. The students were asked to rate each of these candidates on a scale of 0–100. This is called the method of *graphic rating scales*. (The data could have been in terms of rank order, where each respondent ranks the candidates from 1 to 6, or paired comparisons, where the choice is made pairwise using all possible pairs.) There were no other restrictions on the ratings, that is, they were not required to give one candidate a score of 100, ties were permitted, and so on. (Every time we performed this exercise during an election year, the data were usually collected in February and by the time the results were discussed in class a month or so later, most of those whom the class considered to be major candidates had invariably been eliminated in the early primaries!) The candidates for this exercise were Gov. Jimmy Carter of Georgia, the ultimate victor; incumbent President Gerald Ford; Sen. Hubert Humphrey of Minnesota; Sen. Henry Jackson of Washington; Gov. Ronald Reagan of California and Gov. George Wallace of Alabama. Ford and Reagan were Republicans; the rest were Democrats. The means and standard deviations for each candidate are included in Table 10.1. Note that the mean for Jackson was about equal to the average of the whole set of data and that his standard deviation was considerably smaller than the rest.

An analysis of variance of the data in Table 10.1 is given in Chapter 13 (Table 13.6) and indicates that no significant differences existed among the means of the respondents but that a significant difference did exist among the candidates. There is no test for the respondent × candidate interaction although one would surmise that there probably was an interaction, human nature being what it is. There is a test for nonadditivity, which is given in Section 13.7.6. An MDPREF plot may prove useful in describing this situation.

In accordance with MDPREF methodology, the data have the respondent means subtracted out and the resultant deviations are submitted to a singular value decomposition, producing $\mathbf{V}^*$ and $\mathbf{U}$. The characteristic roots are $l_1 = 20\,973$, $l_2 = 14\,849$, $l_3 = 8877$, $l_4 = 2452$, $l_5 = 1665$, and the sixth root is essentially zero. The first three pc's account for nearly 92%. The first three columns of $\mathbf{U}$ and $\mathbf{V}^* = \mathbf{Z}$ (before normalizing to unit length) are given in Tables 10.2 and 10.3.

Table 10.2. Presidential Hopefuls. First Three Characteristic Vectors

	$\mathbf{u}_1$	$\mathbf{u}_2$	$\mathbf{u}_3$
Carter	.145	.644	−.577
Ford	−.557	−.345	−.210
Humphrey	−.486	.320	.589
Jackson	.126	.058	−.011
Reagan	.144	−.595	−.251
Wallace	.629	−.083	.461
l_i	20973	14849	8877
% Explained	43.0	30.4	18.2

Table 10.3. Presidential Hopefuls. First Three z-scores

Respondent #	z_1	z_2	z_3
1	−30.7	31.5	−5.3
2	−24.7	22.7	−24.9
3	4.9	−14.1	−11.4
4	−29.9	−25.0	3.8
5	61.3	−36.1	−17.8
6	−29.4	11.6	−48.5
7	−62.5	3.3	17.3
8	−37.3	−22.9	−11.4
9	−15.7	−35.0	−29.5
10	−23.7	−18.8	3.3
11	−24.9	−35.2	11.0
12	1.0	9.7	−20.5
13	−13.2	3.1	−49.8
14	−4.9	−63.2	−13.4
15	28.2	−13.9	−10.4
16	−20.2	50.2	−5.4
17	−41.8	−2.6	−14.1
18	−63.3	−30.1	23.9

To normalize the first respondent vector to unit length, first compute

$$\sqrt{z_1^2 + z_2^2} = \sqrt{(-30.7)^2 + (31.5)^2} = 43.99$$

and divide z_1 and z_2 by this quantity, obtaining −.698 and .716, whose squares sum to 1.0. These are plotted as a vector on the MDPREF two-dimensional display in Figure 10.3. These first two dimensions explain 73.4%.

Although the respondents are represented only by arrow-heads, remember that they *are* vectors and extend on both sides of the origin. These vectors do represent direction and the tail end of the vector is just as important as the head end because it indicates the unfavorable as well as the favorable. The projections of candidates on the vectors for the ith respondent may be obtained by the relation

$$\hat{\mathbf{x}}_i = \bar{\mathbf{x}}_i + z_i \mathbf{U}' \qquad (10.6.1)$$

using, in this case, the first two columns of $\mathbf{Z}$ and $\mathbf{U}$. For respondent #1, whose average was 43.3, these are: Carter = 59.1, Ford = 49.6, Humphrey = 68.4, Jackson = 41.3, Reagan = 20.2, and Wallace = 21.4. These are in the same order as the original ratings with the exception that Ford and Jackson are reversed and also Reagan and Wallace, though just barely. The correlation between these projections and the original ratings is $r = .91$. The correlations for all the respondents are given in Table 10.4. As an example of the effect of low ratings,

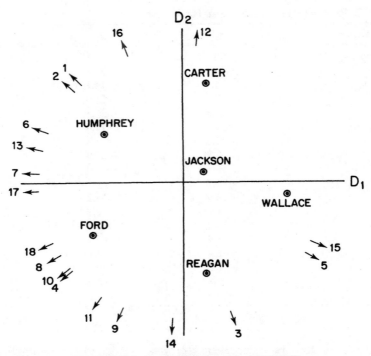

FIGURE 10.3. Presidential Hopefuls: MDPREF.

one would expect the projection of Humphrey on respondent #18 to be closer to Ford than it is since the ratings were similar. However, the low rating for Carter had the effect of pushing the vector closer to Ford, so one should pay attention to both ends of these vectors.

The interpretation of the first dimension may be a bit puzzling as it puts Ford and Humphrey on the same side of the center. The second dimension seems to be primarily a Democrat–Republican contrast. Whatever properties these axes represent, Jackson does not have them because he is almost in the center—recall that he had a very small standard deviation. Note that the ratio $l_1/l_2 = 1.41$, which is not very large, and with a sample size of only 18 the confidence limits for these roots overlap considerably. This means that given another sample of the same size, a similar configuration might result but with a different rotation. A 45° rotation of these axes would produce a new set of axes, the first being a liberal–conservative contrast and the second still Democrats vs. Republicans. However, there will be many cases on preference plots where the configuration of the stimuli may not be readily interpretable until one realizes that the distances among the stimuli represents similarity of preference, not overall similarity. In the present example, two candidates can appear close together on a preference plot because people will vote either for

Table 10.4. Presidential Hopefuls. Correlation of Predicted and Actual Ratings

Respondent #	Number of Components	
	Two	Three
1	.91	.92
2	.80	1.00
3	.50	.63
4	.95	.95
5	.97	1.00
6	.54	.99
7	.95	.98
8	.95	.98
9	.78	.98
10	.99	.99
11	.92	.95
12	.40	.94
13	.26	.99
14	.89	.91
15	.75	.79
16	.97	.97
17	.85	.90
18	.93	.99

them or against them but for different reasons. Although interpretation is desirable, it is often secondary to the main purpose of such a plot, which is to segment the sample into groups of similar preference behavior. In this particular example, the majority of respondent's vectors cluster around Humphrey and Ford and if the contest really had been between these two, it would behoove the campaign managers to find out what kind of voters respondents #7 and #17 represent and what sort of promotion would push their vectors one way or the other. Although this example was, for the purpose both of the class and this book, merely to demonstrate the technique, parallel situations in marketing are commonplace.

In *market segmentation*, one attempts to break the population of potential consumers into clusters of similar buying behavior so that a particular product might be optimally positioned. Unfortunately, most marketing studies are proprietary and very few of them ever get published. A fairly detailed example concerning choices of various bakery items at breakfast or for snacks, albeit in an experimental environment, may be found in Green and Rao (1972).

In the presidential hopefuls example, the first two components explained less than 75% of the total variability. What would the inclusion of the third component do? The third component, accounting for 18.2% was primarily a contrast between Carter on one hand and Humphrey and Wallace on the other

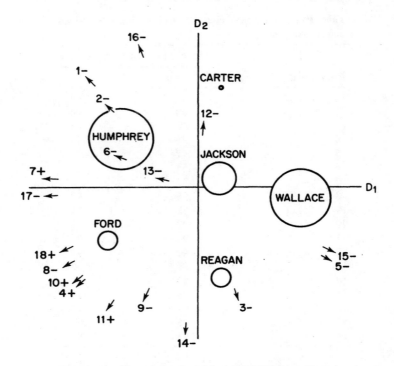

FIGURE 10.4. Presidential Hopefuls: MDPREF-three dimensions.

and might represent the feeling of Carter being a "middle-of-the-roader." A three-dimensional MDPREF plot is shown in Figure 10.4. Trying to represent three dimensions on a flat piece of paper always calls for resourcefulness, particularly if this is a point–vector plot. This demonstrates the advantage of making the respondent vectors of unit length because now the shorter these vectors appear to be, the more they are projected in the third dimension. The sign indicates in which direction from center it is heading. The stimuli are represented by spheres of different sizes, the size being proportional to the position in the third dimension. The correlations for the three-component case are also included in Table 10.4. Respondents #6, #12, and #13 had marked increases in correlation by adding the third component. The only respondents with positive direction in the third dimension, numbers 4, 7, 10, 11, and 18, all gave Carter low scores, so they are, in effect, pointing away from him. We shall return to this example in Section 13.7.6 with an MDPREF plot of residuals.

As with biplots, there are few examples of MDPREF in the literature with more than two components and virtually none of them include correlations for the individuals or much of an indication of how well two dimensions really do. These are useful techniques but *caveat emptor*.

10.7 POINT–POINT PLOTS

Passing reference has been made to *point–point* plots in the last two sections and, although they do not use SVD as such, they are a viable alternative to point–vector plots and for that reason deserve a mention at this time.

In point–point models, both the stimuli and respondents (or whatever the rows and columns of X represent) appear as points. For these models, an examination of the interpoint distances between elements of the two sets *is* valid and quite desirable. If this is a stimulus space, then the respondent points are called *ideal* points. This implies that if a stimulus is located exactly at that position, that is the one the respondent would most prefer. If not, it means that if a stimulus could be produced that would fall on that point, the respondent would choose it over all other alternatives; that is, there is room for a new product or product modification.

The most common methods of obtaining point–point representation employ a technique called *unfolding* (Coombs, 1950, 1964). To illustrate this technique, let us consider a one-dimensional hypothetical Presidential Hopefuls example. This time, we have only four candidates, A, B, C, and D, and only three respondents, X, Y, and Z. Figure 10.5 contains a one-dimensional representation of this joint space. Positions on the line represent whatever quality it is that distinguishes the candidates and we shall assume for the moment that it represents political leaning, with the left end of the line being liberal and the right end conservative. The candidates and the respondents are both located on this line. The respondent's points represent their ideal points. X would prefer someone between A and B, closer to A but not quite as liberal. Y prefers a middle-of-the-roader while Z is more conservative and would like a compromise between C and D.

Assume that this scale is on a piece of transparent material so that if it were folded back on itself the points on the back would show through. If it were folded on X's ideal point, this operation would produce the scale labeled *X*. (The asterisk on A indicates that A has been folded over and is on the back.)

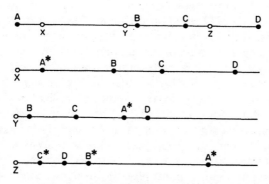

FIGURE 10.5. Unfolding demonstration.

This would show not only that X preferred candidates A, B, C, and D in that order but by how much. Scales Y and Z are obtained by folding the original scale on their ideal points. For Z, we note that not only is candidate A in last place but that the other three are quite closely grouped.

In practice, however, we have the reverse situation. We have scales X, Y, and Z and would like to *unfold* them and recreate the original scale. This procedure, essentially, fits the joint space by minimizing the sums of squares of deviations between the actual and estimated interpoint distances between each respondent and each stimulus.

Multidimensional extension of unfolding was first developed by Bennett and Hays (Bennett, 1956; Bennett and Hays, 1960; Hays and Bennett, 1961). In terms of deviations from the mean, the prediction of the data matrix using the point–vector or linear compensatory model, from equation (10.6.1) is

$$\mathbf{D} = \mathbf{ZU}' \tag{10.7.1}$$

where **D** is $n \times p$, **Z** is $n \times k$ and **U** is $p \times k$. For an individual term of that matrix, d_{ij}, representing the ith respondent and the jth stimuli, this becomes

$$d_{ij} = \sum_{h=1}^{k} z_{ih} u_{jh} \tag{10.7.2}$$

where z_{ih} represents the hth component for the ith respondent and u_{jh} is the jth coefficient of the hth characteristic vector. For the point–point or unfolding model, this becomes

$$d_{ij} = - \sum_{h=1}^{k} z_{ih} (u_{jh} - t_{ih})^2 \tag{10.7.3}$$

where z_{ih} and u_{jh} have the same interpretation as before, even though they no longer are determined by SVD, and t_{ih} is the hth coordinate of the ideal point for the ith respondent. (Again, if less than a full set of components is employed, **D** will be an estimate, $\hat{\mathbf{D}}$.) A geometric interpretation of unfolding models may be found in Davidson (1972, 1973) and Coxon (1982).

The Presidential Hopefuls data is represented by a point–point plot in Figure 10.6. Note that the configuration of the candidates is little changed from Figure 10.4. The only respondent that is really outside this configuration is #14, whose score for Reagan was nearly double the next highest contender (Ford) who is on the same side of the space. Some marketing applications of point–point plots may be found in Moore et al. (1979). Their study consisted of data from 450 housewives made up of paired comparison responses for various brands of cake mixes, liquid cleaners, and toothpaste. Cluster analysis was performed on these data and the ideal points were determined for the cluster means. A procedure for obtaining point–point plots from binary choice data was given by DeSarbo and Hoffman (1987). Schönemann and Wang (1972) gave a

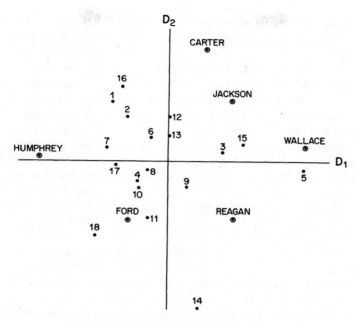

FIGURE 10.6. Presidential Hopefuls: unfolding.

procedure for use with pair comparison data using the Bradley–Terry–Luce model.

If the unfolding model adequately represents the data in k dimensions, the point–vector model will require $k + 1$ dimensions to do as well (Coombs, 1975; Roskam, 1979a). Coombs and Kao (1960) called this extra dimension a "social utility" dimension in the sense that the closer a point is to all the other points in point–point space, the higher its projection on the extra dimension in the point–vector space. In the Presidential Hopefuls example, the MDPREF correlations of respondents 6, 12, and 13 all increased dramatically with the addition of the third dimension and one notices that these three individuals are all near the center of the point–point plot in Figure 10.6.

For other examples of point–point and point–vector plots on the same data, see Green and Rao (1972), Coxon (1974), and Boeckenholt and Gaul (1986). Ross and Cliff (1964) state that double-centering the X-matrix (*M*-analysis) will remove the extra dimension but, in this case, it would be a matter of throwing out the baby with the bathwater because all information on overall candidate differences would be low. *M*-analysis *is* useful in investigating interactions, a matter to be dealt with in Section 13.7.

Unfortunately, the results do not always match up to the expectations. Point–point models are harder to fit than the corresponding point–vector model. Part of the problem is analytical; degeneracy and local minima are fairly common—so much so that Schiffman et al. (1981) purposely gave unfolding models only a brief mention in their book. In addition to that, point–point procedures appear to work better when each stimulus has at least one first-place vote, a situation that may well not happen (Roskam, 1979a). Another problem that may occur is the situation where all of the ideal points are outside the configuration of the stimuli. In this case, the point–vector model (Schiffman et al. called this "the more the better" model) will probably be more useful. Bechtel et al. (1971) used unfolding for some presidential hopefuls data they obtained in 1968 and concluded that while the point–point model did produce an adequate fit for those data, a point–vector model would probably be more appropriate for something like a Miss America contest, where a judge would have difficulty in defining an ideal point for beauty or talent. It is unfortunate that the unfolding model has all of these operational problems, because it is a very appealing model, particularly in the field of market research where it forms the basis of a multiattribute attitude model for predicting consumer preference (Wilkie and Pessemier, 1973; Miller, 1976).

There are a number of algorithms available, among them, ALSCAL (Young and Lewyckyj, 1979; Greenacre and Browne, 1986), SSAR-II (Lingoes, 1979), MINI-RSA and MINI-CPA (Roskam, 1979b), and ORDMET (McClelland and Coombs, 1975). An algorithm for metric unfolding using SVD was given by Schönemann (1970) in enough detail that it may easily be programmed; some restraints were added by Gold (1973) to obtain a unique solution. LINMAP, mentioned in Section 10.4 to produce point–vector estimates using linear programming, can also be used to produce point–point estimates (Srinivasan and Shocker, 1973a). The solution for Figure 10.6 was obtained using ALSCAL, option CMDU.

10.8 CORRESPONDENCE ANALYSIS

10.8.1 Introduction

Correspondence analysis is very similar to the techniques discussed in Sections 10.5 and 10.6 except that it is used primarily for the analysis of frequency data, in particular data in the form of contingency tables. Typical contingency tables are displayed, using fictitious data, in Tables 10.5 and 10.6. Of primary interest is whether or not the frequencies in the rows are independent of the columns

Table 10.5. Contingency Table. Brand vs. Sex

	Brands			Sex
Sex	A	B	C	Total
Male	50	54	46	150
Female	30	36	34	100
Brand total	80	90	80	250

Table 10.6. Contingency Table. Brand vs. Age

Age	Brands				Age Total
	A	B	C	D	
Under 25	10	15	5	70	100
25–55	90	70	60	80	300
Over 55	60	20	10	10	500
Age total	160	105	75	160	100

in which they are located or whether they are *contingent* upon the column identification. A number of tests are available for testing the hypothesis of independence, the two most widely used being the Pearsonian chi-square test and log–linear modeling (Bishop et al., 1975). If the null hypothesis of independence is upheld, it means that all of the relevant information may be found in marginal totals of the table and no further use is required of the table itself. (Contingency tables, often called "cross-tabs," are widely used in such fields as market research and the social sciences but all too often these tests are not employed and the client is left buried in printout that could well be reduced to a smaller and more informative report.) On the other hand, the rejection of this null hypothesis implies an interaction between the rows and columns that must be further clarified.

For the data in Table 10.5, there are $r = 2$ rows and $c = 3$ columns. Let the matrix of frequencies be denoted by $\mathbf{F}$, the ith row total be denoted by r_i and the jth column total by c_j. The expected frequency for the ijth cell under the null hypothesis of independence is $\hat{f}_{ij} = r_i c_j / n$. For the first cell, $f_{11} = (150)(80)/250 = 48$. The Pearson chi-square test of independence is

$$\chi^2_{(r-1)(c-1)} = \sum_{i=1}^{r} \sum_{j=1}^{c} [(f_{ij} - \hat{f}_{ij})^2 / \hat{f}_{ij}] \tag{10.8.1}$$

For this example, $\chi^2 = .417$ which, with $(r - 1)(c - 1) = 2$ degrees of freedom is not significant. This says that the market shares among the brands are not a function of sex and that the individual cell entries in Table 10.5 provide no new information. On the other hand, the data in Table 10.6 produces a χ^2-value of 118.8 with 6 d.f., which is highly significant. In this case the brand market shares are a function of the age group and the individual cell entries are relevant. For a contingency table of this size, it is a simple matter to determine the nature of this interaction. Brand D appeals to the youth, Brand A to the senior citizens, and the people between the ages of 25 and 55 are apportioned about equally among all four brands. While probably nothing new can be gained here by multidimensional scaling, this example is small enough to illustrate how it may be employed.

As mentioned above, two fields of application that employ vast amounts of count data are market research and the social sciences and hence these would also be potential applications for correspondence analysis. Hayashi (1980) used this technique in evaluating some questionnaire data related to a comparison of American and Japanese customs as perceived by respondents from both countries.

10.8.2 Mechanics of Correspondence Analysis

From the previous sections, one might surmise that an MDS plot could be obtained from a SVD of the contingency table itself. What differentiates correspondence analysis from the other methods is that rather than subtracting out row and/or column means, the table is multiplied by functions of the row and column totals, obtaining what is sometimes referred to as a *chi-square* metric. There are number of methods proposed to do this but among the most widely used is one or another form of the procedure described in Greenacre (1984) and Carroll et al. (1986, 1987) among others. Correspondence analysis is useful when the contingency tables are large. Tables the size of the ones in Tables 10.5 and 10.6 can be diagnosed by inspection if the rows and columns are not independent of each other. For the larger tables, some help is required and correspondence analysis may prove quite useful in these instances.

Let D_R and D_c be diagonal matrices whose diagonal elements are made up of the row totals r_i and the column totals c_j. Then, the matrix that will be subjected to SVD is

$$H = D_R^{-1/2} F C_C^{-1/2} \qquad (10.8.2)$$

from which the usual singular value decomposition

$$H = U*L^{1/2}U' \qquad (10.8.3)$$

may be obtained. For the example in Table 10.6, (10.8.1) is

$$
\begin{bmatrix} .1 & 0 & 0 \\ 0 & .0577 & 0 \\ 0 & 0 & .1 \end{bmatrix}
\begin{bmatrix} 10 & 15 & 5 & 70 \\ 90 & 70 & 60 & 80 \\ 60 & 20 & 10 & 10 \end{bmatrix}
\begin{bmatrix} .0791 & 0 & 0 & 0 \\ 0 & .0976 & 0 & 0 \\ 0 & 0 & .1155 & 0 \\ 0 & 0 & 0 & .0791 \end{bmatrix}
$$

$$
= \begin{bmatrix} .0791 & .1464 & .0577 & .5534 \\ .4108 & .3944 & .4000 & .3651 \\ .4743 & .1952 & .1155 & .0791 \end{bmatrix} = H
$$

The SVD procedure (10.8.3) produces

$$\mathbf{U}^* = \begin{bmatrix} .4472 & -.8097 & .3799 \\ .7746 & .1383 & -.6172 \\ .4472 & .5703 & .6891 \end{bmatrix} \qquad \mathbf{U} = \begin{bmatrix} .5657 & .5867 & .5432 \\ .4583 & .1054 & -.2802 \\ .3873 & .1658 & -.7640 \\ .5657 & -.7856 & .2068 \end{bmatrix}$$

and

$$\mathbf{L}^{1/2} = \begin{bmatrix} 1.0000 & 0 & 0 \\ 0 & .4487 & 0 \\ 0 & 0 & .1903 \end{bmatrix}$$

If one wishes to use a subset of k characteristic vectors, then the quantity

$$\mathbf{F}_k = \mathbf{D}_R^{1/2}\mathbf{U}_k^*\mathbf{L}_k^{-1/2}\mathbf{U}_k'\mathbf{D}_C^{1/2} \tag{10.8.4}$$

where $\mathbf{U}_k^*$ is $r \times k$, $\mathbf{U}_k$ is $c \times k$ and $\mathbf{L}_k$ is $k \times k$, should be computed and compared with the original frequencies.

Those wishing to plot these results are faced with the same problem of what to do with $\mathbf{L}^{1/2}$ as in Section 10.5. In addition, one must include the effect of the rows and columns. One other peculiarity with (10.8.2) is that the first columns of $\mathbf{U}$ and $\mathbf{U}^*$ are not used, as they become vectors of constants (the so-called "trivial" vector). In fact, the sum of all of the characteristic roots save this first one is equal to the value of the Pearson chi-square test divided by the sample size. For this example, $l_2 + l_3 = .2014 + .0362 = .2376 = 118.8/500$. This quantity is called *mean-square contingency*.

One can avoid the trivial vector altogether by replacing $\mathbf{F}$ in (10.8.2) with $[\mathbf{F} - \hat{\mathbf{F}}]$, where $\hat{\mathbf{F}}$ is the matrix of expected frequencies. If this is done, the rank of $\mathbf{H}$ is reduced by 1. The first root obtained using $\mathbf{F}$ is replaced by a zero root and the first columns of $\mathbf{U}$ and $\mathbf{U}^*$ are now associated with the zero root and all of the other roots and vectors remain unchanged but move up one position in order. This variation is also useful in that $[\mathbf{F} - \hat{\mathbf{F}}]$ is available for inspection, although, if the significance test had been run, that information would have been available anyhow. The use of $[\mathbf{F} - \hat{\mathbf{F}}]$ is quite prevalent, as is the practice of dividing $\mathbf{F}$ by the sample size before starting the operation. This does not alter the results either, as both $\mathbf{D}_R$ and $\mathbf{D}_C$ will be similarly affected.

Let the coordinates corresponding to the row attributes (age groups, for this example) be denoted by $\mathbf{X}$ and the column attributes (brand) by $\mathbf{Y}$. Then the choices for scaling these plots include:

1.
$$X = D_R^{-1/2} U * L^{1/4}$$
$$Y = D_C^{-1/2} U L^{1/4}$$
(10.8.5)

This is philosophically equivalent to the SQ biplot where both sets of attributes are weighted equally but neither will produce Mahalanobis or euclidean distances. Although both sets of attributes would be plotted as points, it is not a point–point plot. Distances may be compared only within attributes. It is not strictly a point–vector plot in terms of projections either, because of the reweighting by the rows and columns, but, conceptually, it is somewhat like a point–vector plot.

2.
$$X = D_R^{-1/2} U * L^{1/2}$$
$$Y = D_C^{-1/2} U$$
(10.8.6)

This is equivalent to the JK biplot. The distances between the age groups are euclidean and between the brands are Mahalinobis.

3.
$$X = D_R^{-1/2} U *$$
$$Y = D_C^{-1/2} U L^{1/2}$$
(10.8.7)

This is equivalent to the GH biplot. The distances between the age groups are Mahalinobis and between the brands are euclidean.

4.
$$X = D_R^{-1/2} U * L^{1/2}$$
$$Y = D_C^{-1/2} U L^{1/2}$$
(10.8.8)

In this one, both sets of distances are euclidean. Unlike the first three options, (10.8.8) does not partition $L^{1/2}$ but uses it for both parts so the results are essentially a "vector–vector" plot and the only way to compare members of the two sets of vectors is to examine the angle between them. This option is quite popular among French practitioners and is likely to be the one employed in computer programs emanating from there.

5.
$$X = D_R^{-1/2} U * [L^{1/2} + I]^{1/2}$$
$$Y = D_C^{-1/2} U [L^{1/2} + I]^{1/2}$$
(10.8.9)

Although it will not be apparent until Section 10.8.5, (10.8.9) produces a point–point plot. This scaling was proposed by Carroll et al. (1986) and is sometime referred to as CGS scaling. There has been some controversy over this model (Greenacre, 1989; Carroll et al., 1989), which extends over into multiple correspondence analysis, the topic of Section 10.8.5.

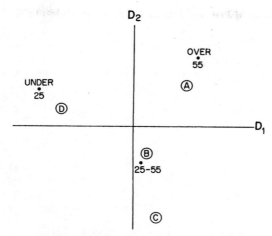

FIGURE 10.7. Brands vs. age using (10.8.9).

Eliminating the trivial vector, using (10.8.9) produces, for the Brand vs. Age data,

$$
\mathbf{X} = \begin{bmatrix} -.0975 & .0414 \\ .0096 & -.0389 \\ .0686 & .0752 \end{bmatrix} \qquad
\mathbf{Y} = \begin{bmatrix} .0558 & .0469 \\ .0124 & -.0298 \\ .0230 & -.0962 \\ -.0748 & .0178 \end{bmatrix}
$$

and is displayed in Figure 10.7.

For this example, the plot for (10.8.8) would look very much like Figure 10.7 and that for (10.8.5) would differ only in the relative spacing for **X** and **Y**. Equations (10.8.6) and (10.8.7) would appear different because of the unequal weighting given the sets of attributes. (Greenacre, 1984, called these *asymmetric* displays.) Again, a caveat about two-dimensional plots: More than two dimensions may be required to adequately represent a contingency table, although you will find precious few examples of this in the literature.

10.8.3   A Larger Example

The data for this example are found in Table 10.7 and consist of reports of personal assaults in England and Wales for the years 1878–1887 broken down by the quarters of the year. These data are taken from an essay by Leffingwell (1892) in which he was studying the effect of season on the incidence of crime. (This table was labeled "Crimes Against Chastity, Assaults, etc.") It is evident that the incidence of these crimes increased steadily throughout this 10-year period and that there is a higher incidence in the warmer months. The value

Table 10.7. Occurrence of Personal Assault. England and Wales; 1878–1887 by Quarters. F-Matrix

Year	Jan.– Mar.	Apr.– Jun.	Jul.– Sep.	Oct.– Dec.	Year Total
1878	127	168	167	105	567
1879	99	143	174	126	542
1880	140	181	182	123	626
1881	106	168	210	144	628
1882	139	206	254	101	700
1883	141	179	212	133	665
1884	143	248	256	156	803
1885	133	219	267	234	853
1886	234	368	442	271	1315
1887	215	345	379	259	1198
Quarter total	1477	2225	2543	1652	7897

Source: Leffingwell (1892).

for the chi-square test is $\chi^2 = 70.65$, which for $(9)(3) = 27$ d.f. is highly significant. The contribution of each cell towards this total is shown in Table 10.8, except that the value displayed is the square root of this quantity along with the appropriate sign indicating the difference between the actual and expected frequency. From (10.8.2), $\mathbf{H}$ is obtained and is displayed in Table 10.9. The characteristic roots associated with this matrix are $l_1 = 1$, $l_2 = .006486$, $l_3 = .001879$, and $l_4 = .000581$. Recall that the first characteristic vector is a constant and is not used. The sum of the last three roots is .008946, which is equal to the mean square contingency, 70.65/7897. The first two of these roots account for 94% of this value so we may feel justified in preparing a two-dimensional plot.

Table 10.8. Occurrence of Personal Assault. England and Wales; 1878–1887. $(f_{ij} - \hat{f}_{ij})/\sqrt{\hat{f}_{ij}})$

Year	Jan.– Mar.	Apr.– Jun.	Jul.– Sep.	Oct.– Dec.
1878	2.0	.7	−1.2	−1.2
1879	−.2	−.8	−.0	1.2
1880	2.1	.3	−1.4	−.7
1881	−1.1	−.7	.6	1.1
1882	.7	.6	1.9	−3.8
1883	1.5	−.6	−.1	−.5
1884	−.6	1.4	−.2	−.9
1885	−2.1	−1.4	−.5	4.2
1886	−.8	−.1	.9	−.2
1887	−.6	.4	−.3	.5

Table 10.9. Occurrence of Personal Assault. England and Wales; 1878–1887. H = $D_R^{-1/2}FD_C^{-1/2}$

Year	Jan.– Mar.	Apr.– Jun.	Jul.– Sep.	Oct.– Dec.
1878	.1388	.1496	.1391	.1085
1879	.1106	.1302	.1482	.1332
1880	.1456	.1534	.1442	.1210
1881	.1101	.1421	.1662	.1414
1882	.1367	.1651	.1904	.0939
1883	.1423	.1472	.1630	.1269
1884	.1313	.1855	.1791	.1354
1885	.1185	.1590	.1813	.1971
1886	.1679	.2151	.2417	.1839
1887	.1616	.2113	.2171	.1841

The two relevant vectors from SVD of **H** are

$$
U^* = \begin{bmatrix}
-.290 & -.456 \\
.184 & -.063 \\
-.217 & -.556 \\
.217 & .191 \\
-.531 & .508 \\
-.134 & -.249 \\
-.122 & .155 \\
.681 & -.048 \\
.015 & .310 \\
.090 & .006
\end{bmatrix}
$$

and

$$
U = \begin{bmatrix}
-.464 & -.658 \\
-.263 & .039 \\
-.078 & .695 \\
.841 & -.286
\end{bmatrix}
$$

Substituting these values of **U*** and **U** in (10.8.9) produces the point–point correspondence analysis plot shown in Figure 10.8. One can interpret this chart by referring to Table 10.8. For instance, the two years closest to the point for

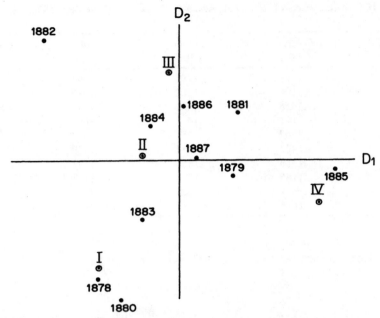

FIGURE 10.8. Personal Assault: correspondence analysis. *Code*: I = Jan.–Mar., II = Apr.–June, III = July–Sept., IV = Oct.–Dec.

Jan.–Mar., 1878 and 1880 do not represent years that had the highest incidence of these crimes in the first quarter, but rather years in which the incidence for the first quarter was higher than their expected values, which takes both row and column into account. 1885 was a higher than expected year for the fourth quarter and, coincidentally, a low year for the first quarter. The location of 1882 is partially because of a relatively high occurrence in the third quarter but more because of large drop in the fourth quarter. Keep in mind that this plot, like the chi-square test with which it is associated, represents the interaction between quarters and years. The fact that the overall incidence of these crimes was climbing over this period or that there is a definite seasonal pattern is not reflected in this plot.

10.8.4   Historical Background

Much of the early development as well as the application of correspondence analysis was carried out in France, motivated primarily by the work of Benzecri (1973, 1977). The first text devoted primarily to this subject was by Lebart et al. (1977) and an English translation (Lebart et al., 1984) which is essentially a second edition, differing primarily in the examples employed. These texts contain a great many FORTRAN programs with which to carry out these procedures. Another text is by Greenacre (1984). A geometric interpretation of correspondence analysis was given by Greenacre and Hastie (1987). A more analytical discussion may be found in Heiser and Meulman (1983). A general tutorial paper on the subject by Van der Heijden et al. (1989) may be of interest because of the remarks of a number of discussants. In a history of correspondence analysis, DeLeeuw (1983), showed that Pearson (1904, 1906) was on the verge of developing the technique and suggested that he might have done so had the SVD procedure been available in his day.

These techniques have also existed under different names or as special cases of other techniques. One of these is *dual* or *optimal scaling* (Nishisato, 1980), in which both sets of principal components are obtained in such a way that their sum of squares is a maximum. Because both sets of characteristic vectors are obtained simultaneously, there is a translation relationship between the plotted elements, X and Y of the form

$$X = D_R^{-1} FYL^{-1/2} \tag{10.8.10}$$

and

$$Y = D_C^{-1} F'XL^{-1/2} \tag{10.8.11}$$

A similar reasoning produced the method of *reciprocal averages* (Hill, 1974). Hoffman and Franke (1986) suggested using these equations to include additional data where only the row or column attributes are known. Another approach related to simultaneous linear regression is due to Lingoes (1968) and we shall see in Section 10.8.5, that correspondence analysis is also a special case of canonical correlation. Unifying treatments of these various concepts may be found in Greenacre (1984) and Tenenhaus and Young (1985). Another term sometimes used for correspondence analysis is *homogeneity* analysis.

Because of the weighting structure, the so-called chi-square metric, correspondence analysis has been used for applications other than contingency tables. For example, Bretaudiere et al. (1981) used it, rather than PCA, in the analysis of some blood samples analyzed by several different test procedures because the first vector using correspondence analysis represented differences among tests while the first pc represented the variability of the samples. This was due less, probably, to the chi-square metric than to the fact that correspondence analysis double-centers the data, which classical PCA does not. However, the techniques discussed in Section 13.7 do double-center the data and those might have sufficed for their problem. Both Lebart et al. (1984) and Greenacre (1984) gave other examples of its use.

Although correspondence analysis is generally associated with the Pearson chi-square test, Van der Heijden and DeLeeuw (1985), Van der Heijden and Worsley (1988), Goodman (1986), and Takane (1987) showed its relationship to loglinear models. Prior to that, Gabriel and Zamir (1979) had produced biplots from weighted logs of the frequencies. Although more of an analytical tool than a graphical procedure, the use of *partial least-squares regression*, which will be discussed in Chapter 12, was advocated for contingency tables by Bertholet and Wold (1985). Fitting correspondence analysis models by least absolute deviations (the "city-block" method, which will be discussed in Section 11.4.1) was proposed by Heiser (1987). DeLeeuw and Van der Heijden (1988) gave a method of handling missing cells and structural zeros—cells that, for some reason, represent nonexistent conditions.

A *sociomatrix* is a matrix of individual relationships, $a_{ij} = 1$ if i is operationally linked to j. Otherwise it is equal to zero. Noma and Smith (1985) used correspondence analysis to produce *sociograms* where the individuals are represented by points and the links as vectors.

Closely related to correspondence analysis is a technique called *spectral map analysis* (Lewi, 1982). In this approach, the original entries are converted into logarithms and the resultant matrix is double-centered before SVD. The results are then in terms of log ratios. This technique was used by Thielmans et al. (1988) on some Belgian cancer data. They also analyzed the data using correspondence analysis for comparison. The original counts for 12 types of cancer at 43 locations are included in the article.

10.8.5 Multiple Correspondence Analysis

Correspondence analysis may be extended to the case of a general p-way contingency table. To show how this may be done, it will first be necessary to examine an alternative method of carrying out correspondence analysis using *indicator matrices*. This will be illustrated with the Brand data displayed in Table 10.6 and analyzed in Section 10.8.2. Suppose that the original 500 observations were available. We could represent them in the 500 × 7 indicator matrix sketched in Table 10.10. The first three columns in this matrix represent the three age groups and the last four columns represent the four brands. Each row represents one respondent and will consist of two "1's", one corresponding to his or her age group and the other to the preferred brand. For instance,

Table 10.10. Brand vs. Age. Indicator Matrix

Respondent #	Under 25	25–55	Over 55	Brands A	B	C	D
1	1	0	0	0	0	0	1
2	0	1	0	0	1	0	0
.	.	.	.	.	.	.	.
.	.	.	.	.	.	.	.
500	0	1	0	0	0	1	0
Column totals	100	300	100	160	105	75	160

respondent #1 is under 25 and prefers Brand D. The other elements will all be zero so each row will sum to 2. The columns will sum to the entries given in the body of Table 10.6. If this matrix is denoted by Z, then a SVD of Z will result in

$$Z = U^*L^{1/2}U'$$

and, in this case, the first three elements in each column of U will be made up of the elements of U* displayed in Section 10.8.2 while the last four elements will come from U in that section. The U* derived from Z represents the principal components for the 500 respondents. The characteristic roots will be related to, but not equal to, the roots in Section 10.8.2 because each observation is counted twice, once for age and once for brand.

Although working with Z may have some advantages, in many cases it will be easier to work with Z'Z, which is called the *Burt* matrix (Burt, 1950). The Burt matrix for these data is given in Table 10.11 and is obviously a structured matrix. If we let $B = Z'Z$, then B may be partitioned as

$$B = \begin{bmatrix} D_R & F \\ \hline F' & D_C \end{bmatrix} \qquad (10.8.12)$$

All of these submatrices can be determined from Table 10.6 so that the Burt matrix may be obtained whether or not the original data are available. To obtain the characteristic vectors, it will first be necessary to obtain

$$B_s = (1/p)D_B^{-1/2}BD_B^{-1/2} \qquad (10.8.13)$$

where D_B is a diagonal matrix formed from the diagonal elements of B and p is the number of category variables, which in this example is two—age and brand. This factor of $1/2$ is because of the double counting of the observations. The matrix B_s is displayed in Table 10.12. Since the B_s is symmetric, only a classical PCA solution, rather than a SVD, is required. As this is a 7×7 matrix,

Table 10.11. Brand vs. Age. Burt Matrix

Under 25	25–55	Over 55	Brand A	B	C	D
100	0	0	10	15	5	70
0	300	0	90	70	60	80
0	0	100	60	20	10	10
10	90	60	160	0	0	0
15	70	20	0	105	0	0
5	60	10	0	0	75	0
70	80	10	0	0	0	160

Table 10.12. Brand vs. Age. B_S-Matrix (Equation 10.8.13)

Under 25	25–55	Over 55	A	B	C	D
			for Brand			
.5	0	0	.040	.073	.029	.277
0	.5	0	.205	.254	.200	.183
0	0	.5	.237	.098	.058	.040
.040	.205	.237	.5	0	0	0
.073	.254	.098	0	.5	0	0
.029	.200	.058	0	0	.5	0
.277	.183	.040	0	0	0	.5

there will be seven characteristic roots. These are: $l_1 = 1$, $l_2 = .7244$, $l_3 = .5951$, $l_4 = .5$, $l_5 = .4049$, $l_6 = .2756$, and $l_7 = 0$. Note that $l_5 = 1 - l_3$ and $l_6 = 1 - l_2$, and for that matter, $l_7 = 1 - l_1$. If we denote these roots by $l_i^{(B)}$, then the relation between these and the roots obtained in Section 10.8.2 is

$$l_i = (2l_i^{(B)} - 1)^2 \tag{10.8.14}$$

and these pairwise roots (e.g., l_2 and l_6) will have the same value.

The U-vectors are shown in Table 10.13. Since only two components were required in Section 10.8.2, it is obvious that most of this matrix is useless ("trivial" is the preferred term). u_1 is the "trivial" vector. Note that the coefficients for u_7, u_6, and u_5 are the same as u_1, u_2, and u_3 except that the signs have changed for the age coefficients. A formal account of the expected makeup of these matrices may be found in Greenacre (1984) but the rule of thumb is that the nontrivial vectors are those whose roots are less than 1 and greater than $1/p$. As $p = 2$ in this example, that leaves us with u_2 and u_3. Since these are scaled to unit length, they must be multiplied by $\sqrt{2}$ to agree with those in Section 10.8.2.

Table 10.13. Brand vs. Age. Burt Matrix—U-Vectors

	u_1	u_2	u_3	u_4	u_5	u_6	u_7
Under 25	.32	−.57	.27	.00	−.27	.57	−.32
25–55	.55	.10	−.44	.00	.44	−.10	−.55
Over 55	.32	.40	.49	.00	−.49	−.40	−.32
Brand A	.40	.42	.38	.20	.38	.41	.40
Brand B	.32	.07	−.20	−.84	−.20	.07	.32
Brand C	.27	.12	−.54	.49	−.54	.12	.27
Brand D	.40	−.56	.15	.14	.15	−.56	.40
l_i	1	.72	.60	.50	.40	.28	0

Finally, to obtain the correspondence analysis plot, let U_2 be a matrix made up of u_2 and u_3 and L_2, their corresponding roots. Then

$$p D_B^{-1/2} U_2 L_2^{1/2} = \begin{bmatrix} -.0975 & .0414 \\ .0096 & -.0389 \\ .0686 & .0752 \\ .0558 & .0469 \\ .0124 & -.0298 \\ .0230 & -.0962 \\ -.0748 & .0178 \end{bmatrix} \qquad (10.8.15)$$

which are the same coefficients one would have obtained using (10.8.9), the first three rows corresponding to X and the last four to Y. Recall stating that (10.8.9) yielded a point–point plot. This now makes sense since all seven variables were treated as equals even though they represented a two-way classification. These point–point plots are more in the philosophy of the similarity plots to be discussed in Chapter 11 rather than the model given by (10.7.3). The analogy between this technique and canonical correlation along with some others is given by Tenenhaus and Young (1985). A unified treatment of multiple correspondence analysis, PCA, and some forms of nonmetric multidimensional scaling may be found in Bekker and DeLeeuw (1988).

An exciting consequence of the use of indicator matrices and Burt matrices is that they may be extended to three- or higher-order classifications. In Table 10.14, the Brands data has been further subdivided by sex. The corresponding Burt matrix is given in Table 10.15 and from this, B_s is obtained using (10.8.13)

Table 10.14. Brand vs. Age and Sex. Contingency Table

		Brand				Age–Sex Totals
		A	B	C	D	
Under 25	Male	7	5	2	36	50
	Female	3	10	3	34	50
Age–brand totals		10	15	5	70	100
25–55	Male	58	20	20	42	140
	Female	32	50	40	38	160
Age–brand totals		90	70	60	80	300
Over 55	Male	26	5	4	5	40
	Female	34	15	6	5	60
Age–brand totals		60	20	10	10	100

Table 10.15. Brand vs. Age and Sex. Burt Matrix

Under 25	25–55	Over 55	Brands				Sex	
			A	B	C	D	M	F
100	0	0	10	15	5	70	50	50
0	300	0	90	70	60	80	140	160
0	0	100	60	20	10	10	40	60
10	90	60	160	0	0	0	91	69
15	70	20	0	105	0	0	30	75
5	60	10	0	0	75	0	26	49
70	80	.10	0	0	0	160	83	77
50	140	40	91	30	26	83	230	0
50	160	60	69	75	49	77	0	270

with $p = 3$. Note that the upper left-hand 7×7 submatrix is the same as Table 10.11. Note, also, that the Burt matrix technique considers only two-way interactions, so if there were a three-way interaction between age groups and sex with brand it would not be taken into account. The characteristic roots and vectors of $\mathbf{B}_s$ are given in Table 10.16. There are still a number of trivial vectors but the relationships among them are no longer as clear cut. Since this is a three-way contingency table, the rule of thumb is to consider all vectors nontrivial whose roots are less than 1 and greater than $1/3$. This gives us three candidates and it appears that we may need them all, as the third root accounts for nearly 30%. To obtain the final coordinates for the plot, again use (10.8.15). (The multiplier is now 3 because of the three-way classification, not because three vectors were retained.) These coordinates are given in Table 10.17 and the plot is shown in Figure 10.9.

An alternate method using Burt matrices entails the use of the CANDECOMP option of a procedure called INDSCAL, which will be described in Section 11.6 (Carroll and Green, 1988). This method provides the weight that each category of each variable associates with each dimension. Van der Burg et al. (1988) gave a generalized form of MCA for *sets* of variables, each variable being categorical in nature. A log–linear MCA model that will handle missing cells was developed by Choulakian (1988). Greenacre (1988) suggested that the use of Burt or indicator matrices may be inadequate. He proposed, instead, working with all of the two-way tables simultaneously, which ignores the diagonal elements of the Burt matrix.

The levels for a particular variable may be ordered and, in fact, they may be grouped frequencies such as the age variable in Tables 10.10–10.17. In these situations, the additional information that they are ordered, as well as the nature of the grouping, if any, allows one to treat them as continuous functions and fit them with splines or other interpolating functions (Van Rijckevorsel, 1988; DeLeeuw and Van Rijckevorsel, 1988).

Table 10.16. Brand vs. Age and Sex. U-Vectors

	u_1	u_2	u_3	u_4	u_5	u_6	u_7	u_8	u_9
Under 25	.26	−.57	.09	−.25	−.19	.23	−.56	.08	.36
25–55	.45	.11	−.25	.37	.31	−.29	.08	.13	.62
Over 55	.26	.39	.34	−.40	−.34	.28	.42	.08	.36
Brand A	.33	.36	.49	.03	.02	−.33	−.45	.34	−.31
Brand B	.26	.13	−.38	−.43	.54	.38	−.03	.28	−.25
Brand C	.22	.15	−.38	.48	−.53	.40	−.08	.23	−.21
Brand D	.33	−.56	.08	−.02	−.09	−.25	.53	.34	−.31
Male	.39	−.11	.39	.35	.30	.40	.08	−.53	−.17
Female	.42	.10	−.36	−.32	−.27	−.37	−.07	−.57	−.18
$I_i^{(B)}$	1	.485	.423	.366	.310	.233	.183	0	0

229

Table 10.17. Brand vs. Age and Sex. Coefficients for Correspondence Analysis Plot using (10.8.15)

	DIM 1	DIM 2	DIM 3
Under 25	−.120	.018	−.045
25–55	.013	−.028	.039
Over 55	.081	.066	−.072
Brand A	.060	.075	.005
Brand B	.026	−.073	−.076
Brand C	.036	−.085	.101
Brand D	−.093	.012	−.002
Males	−.015	.050	.041
Females	.013	−.043	−.035

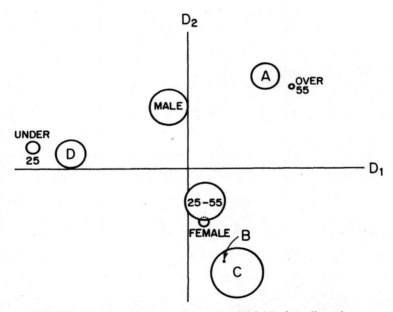

FIGURE 10.9. Brand vs. age and sex using (10.8.14), three dimensions.

10.9   THREE-WAY PCA

With the exception of multiple correspondence analysis, all of the material presented so far in this chapter has dealt with two-dimensional data arrays. Very often the columns of these arrays represent p different stimuli and the rows represent n respondents, or the rows may represent a number of characteristics for each stimulus on which it is rated or scored. Often, there will be a combination of all three; that is, n respondents scoring or rating c

characteristics on each of p stimuli. Another possibility would be for the respondents to rate each stimulus on more than one occasion, making time the third classification. In either case, the data array is now three-dimensional. Historically, PCA has been carried out for data of this sort by combining two of the classifications, such as stimulus and characteristic, leaving respondent as the other dimension and reducing the problem to two dimensions. More recently, however, some techniques have been developed to obtain direct solutions for three-way data sets. These dimensions are often referred to as *modes* and the technique is generally referred to as *three-mode PCA*. Procedures of this sort were first proposed by Tucker (1966, 1972). As one would expect, these operations are more involved than SVD.

Recall from the previous sections that the singular value decomposition is written as

$$\hat{\mathbf{X}} = \mathbf{U}^*\mathbf{L}^{1/2}\mathbf{U}'$$

where $\hat{\mathbf{X}}$ is $(n \times p)$, $\mathbf{U}^*$ is $(n \times k)$, $\mathbf{L}^{1/2}$ is $(k \times k)$, and $\mathbf{U}$ is $(p \times k)$ with $\hat{\mathbf{X}} = \mathbf{X}$ if $k = \text{rank}(\mathbf{X})$. To simplify what is to follow, let us use new symbols, $\mathbf{E} = \mathbf{U}^*$, $\mathbf{C} = \mathbf{L}^{1/2}$, and $\mathbf{F} = \mathbf{U}'$. Then

$$\hat{\mathbf{X}} = \mathbf{ECF} \tag{10.9.1}$$

This may be rewritten as

$$\hat{x}_{ij} = \sum_{l=1}^{k} \sum_{m=1}^{k} e_{il} c_l{}^m f_{mj} \tag{10.9.2}$$

If the data array is augmented to include a third dimension, (10.9.2) will also be augmented by the inclusion of an extra term, to become

$$\hat{x}_{hij} = \sum_{l=1}^{s} \sum_{m=1}^{t} \sum_{q=1}^{u} e_{hl} f_{im} g_{jq} c_{lmq} \tag{10.9.3}$$

Note a number of changes from (10.9.2). First, there is another term and another index of summation. Second, c_{lmq} is now a three-dimensional array as is $\hat{x}_{hij}$. The other arrays are all two-dimensional. $\mathbf{E}$ is $(n \times s)$, $\mathbf{F}$ is $(p \times t)$, and $\mathbf{G}$ is $(r \times u)$. Third, the sums no longer run from 1 to k but to three different values: s, t, and u.

Equation (10.9.3) is a general form with $\hat{\mathbf{X}} = \mathbf{X}$ if s, t, and u are all equal to rank$(\mathbf{X})$. This expression may be rewritten in matrix form as

$$\hat{\mathbf{X}} = \mathbf{EC}(\mathbf{F}' \otimes \mathbf{G}') \tag{10.9.4}$$

where $\otimes$ denotes a *direct product* or *Kronecker* matrix (see Appendix A). Since this is a matrix equation, both $\hat{\mathbf{X}}$ and $\mathbf{C}$ have to be restated as two-dimensional arrays, $\hat{X}$ being ($n \times pr$) and C being ($s \times tu$). E, F, and G are orthonormal matrices as before and have the same interpretation as the characteristic vectors did for the SVD. However, C, which is now called the *core matrix*, is no longer a diagonal matrix of characteristic roots. If the three modes of X were, for example, respondents, stimuli, and characteristics, the elements of C would be thought of as s pc-scores for respondents on t components for stimuli and u components for characteristics. Algorithms for obtaining solutions to (10.9.4) using alternating least-squares (ALS) may be found in Kroonenberg and DeLeeuw (1980), Sands and Young (1980), van der Kloot and Kroonenberg (1985), TenBerge et al. (1987), Wold et al. (1987), and Lohmöller (1989, p. 225). Kroonenberg (1983) has also written a book on three-mode PCA. There are other forms of three-mode PCA, the most widely used being INDSCAL (see Section 11.6), in which the numbers of retained components s, t, and u are equal (Carroll and Chang, 1970a, b).

If the third mode represents time, there may be some other considerations in the formulation of the problem. A desirable property of time series models is *stationarity*, meaning that the model should be independent of the time interval sampled. The three-mode model presented above would produce stationary pc's, that is, the correlations of each specific pc between sampling periods should be high. Meredith and Tisak (1982) opted instead for stationarity of the characteristic vectors and used a technique related to canonical correlation for three groups of variables. Carroll (1980) gave generalizations of both the point–vector and point–point models. The same n respondents judge the same p stimuli as before but now the process is repeated on r occasions, each representing a different scenario or time. See also DeSarbo and Carroll (1985).

10.10   N-MODE PCA

There is no theoretical limit in the number of dimensions a data array may have, although there are some limits of practicality. Lastovicka (1981) extended the three-mode model to four modes when, as one would expect, equation (10.9.3) is expanded to include five terms with four summations, C becomes four-dimensional, and the equivalent of equation (10.9.4) has a Kronecker product imbedded within another Kronecker product. Lastovicka included a four-mode example where the modes represented two different types of advertisements (transportation and insurance), different groups of respondents, and whether this was the first, second, … time they had seen the advertisement. Extention, in a general form, to the n-mode case was done by Kapteyn et al. (1986). In theory, the CANDECOMP option of INDSCAL will handle up to seven-way arrays, but to quote Coxon (1982): "Users are advised to proceed beyond three-way data with considerable caution. They are largely in uncharted territory."

CHAPTER 11

Distance Models;
Multidimensional Scaling II

11.1 SIMILARITY MODELS

The topic of multidimensional scaling (MDS) was introduced in Chapter 10 in conjunction with singular value decomposition. The MDS models described in that chapter were *dominance* or *preference* models. The starting data generally consisted of *n* scores, ranks, or measurements on *p* stimuli, similar to a two-way classification. The resultant MDS representation displayed both the stimuli (usually as points) and the observation vectors (usually as vectors). The closer together any two stimuli appeared, the more alike they were with respect to preference or whatever dominance characteristic was employed. It was pointed out that two stimuli might both be quite unpopular (and hence appear close together) but for different reasons.

In this chapter, we shall consider a second MDS model: *similarity* scaling. In these models, the dominance of one stimulus over another is not taken into account. Of interest in these models is the relative similarity of two or more stimuli in appearance or concept. The focal point of this MDS model is the *similarity* or *dissimilarity* matrix. Generally, these are symmetric matrices and, in the case of a similarity matrix the larger the *ij* term the more similar are stimuli *i* and *j*. (A correlation matrix is an example of a similarity matrix.) A dissimilarity matrix is just the opposite; the larger the *ij* term, the more dissimilar are stimuli *i* and *j*. This is the same interpretation as the interpoint distances between the stimuli on the resultant MDS plot. Regardless of the starting data, one ultimately ends up with a dissimilarity matrix because it is more appealing to produce plots that have the most dissimilar stimuli the farthest apart. While identification of the axes themselves, for dominance scaling, may be of secondary interest, in this chapter, it will assume more importance.

11.2   AN EXAMPLE

The same data set used in Chapter 10 to introduce singular value decomposition will now be used to illustrate distance models. The data set was

$$
X = \begin{bmatrix} -2 & -2 & -1 \\ 0 & 1 & 0 \\ 2 & 2 & 2 \\ 0 & -1 & -1 \end{bmatrix}
$$

Originally, this represented four observations on three variables. It will now be considered to be four observers, respondents, or case scores on three stimuli but, in actuality, nothing has changed. The data still are corrected for their variable (stimulus) means. We shall also make use of the characteristic vectors related to the cases, V^*, these being vectors in the units of the original data rather than those of unit length:

$$
V^* = \begin{bmatrix} 2.949 & -.408 & -.372 \\ -.656 & -.408 & .636 \\ -3.441 & 0 & -.396 \\ 1.147 & .816 & .132 \end{bmatrix}
$$

Let δ_{ij}^2 be the square of the distance, and hence a measure of dissimilarity, between the ith and jth cases. This can be defined either in terms of the original data, X, or the characteristic vectors, V^*, viz.,

$$
\delta_{ij}^2 = \sum_{k=1}^{3} (x_{ih} - x_{jh})^2 = \sum_{h=1}^{3} (v_{ih}^* - v_{jh}^*)^2 \tag{11.2.1}
$$

(The symbol δ_{ij} is used here with great reluctance since it represents a sample estimate, not a population parameter. However, the notation that will be used in this chapter is fairly standard in the relevant literature and there is an old expression about fighting City Hall!) Consider the squared distance between the first two cases:

$$
\begin{bmatrix} -2 & -2 & -1 \\ 0 & 1 & 0 \end{bmatrix}
$$

Difference -2 -3 -1

(Difference)^2 4   9   1   Sum of squares = 14 and similarly for the first two rows of $\mathbf{V}^*$:

$$\begin{bmatrix} 2.949 & -.408 & -.372 \\ -.654 & -.408 & .636 \end{bmatrix}$$

Difference   3.603   0.000 -1.008

(Difference)^2 12.98   0.00   1.02   Sum of squares $= 14.00$

These two quantities will be equal only if all three vectors are used. In practice, only enough vectors (i.e., dimensions) are used to adequately characterize the data. If only two vectors were used, the sum of squares would be 13, not 14, and would be considered an estimate of δ_{ij}^2.

The matrix of all these quantities is

$$[\delta_{ij}^2] = \begin{bmatrix} 0 & 14 & 41 & 5 \\ 14 & 0 & 9 & 5 \\ 41 & 9 & 0 & 22 \\ 5 & 5 & 22 & 0 \end{bmatrix}$$

Note that the diagonals are all zero, since this represents the distance between a case and itself.

Using "dot" notation, the row means of $[\delta_{ij}^2]$, $\bar{\delta}_{i.}^2$ are 15, 7, 18, and 8. Since this is a symmetric matrix, these are equal to the column means, $\bar{\delta}_{.j}^2$. The grand mean $\bar{\delta}_{..}^2 = 12$.

Now, define another symmetric matrix $\mathbf{B}$:

$$b_{ij} = \tfrac{1}{2}[\bar{\delta}_{i.}^2 + \bar{\delta}_{.j}^2 - \bar{\delta}_{..}^2 - \delta_{ij}^2] \tag{11.2.2}$$

This is called the matrix of *scalar products* and is associated with the M-analysis of Section 10.3. Note that an individual term in $\mathbf{B}$ is obtained in the same manner as an individual residual in a two-way analysis of variance.

$$\mathbf{B} = \begin{bmatrix} 9 & -2 & -10 & 3 \\ -2 & 1 & 2 & -1 \\ -10 & 2 & 12 & -4 \\ 3 & -1 & -4 & 2 \end{bmatrix}$$

and, it turns out, the matrix $\mathbf{B} = \mathbf{XX}'$ and its characteristic vectors are equal to $\mathbf{V}^*$.

Although similarities MDS goes back to Young and Householder (1938), the techniques, as generally applied today, had their start with Torgerson

(1952, 1958). The process of obtaining **B** and its characteristic roots is often referred to as the method of *principal coordinates* (Gower, 1966a), who also gave a method for adding a new point to an already existing similarity representation (Gower, 1968).

The various operations discussed in this section provide the following optional paths shown in Figure 11.1. Given the original data, one can obtain either characteristic vectors, distances, or a sum of squares matrix, **B**. Given the vectors, one can obtain distances or **B** and, given the distances, one can obtain (through **B**) the vectors but one cannot obtain the original data **X**. In multidimensional scaling, some methods start with **X** and some start with δ_{ij}, depending on the data-collection method. The main point is that, ultimately, the process ends up with characteristic vectors being obtained from **B** and that is the analogy to PCA. Parsimony is again the goal so one will wish to represent this space in as few dimensions as possible.

Another way of looking at this is to consider the points i and j in Figure 11.2, which shows them both in terms of the original data where the origin is known and in terms of dissimilarities where it is not. The distance model does adequately represent the relation among a set of stimuli but cannot directly provide an estimate of the origin. For some models, this may be accomplished by means of a separate step in the MDS procedure, usually of an iterative nature.

In summary, to obtain an MDS representation of similarities data, one must (by one means or another) obtain estimates of the dissimilarities, δ_{ij}, and from them obtain **B** and its characteristic vectors. The coefficients of the vectors will be the MDS plot. (If one had wished to scale the variables instead of the cases, the operations would have been on the columns rather than the rows and would have involved **V** instead of **V***.) Although this sounds quite straightforward, we shall see in the next two sections that it is not that simple in practice.

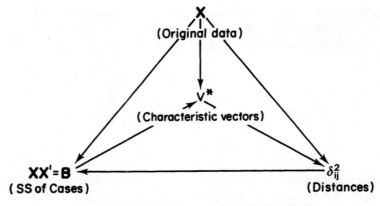

FIGURE 11.1. Possible paths in similarities MDS.

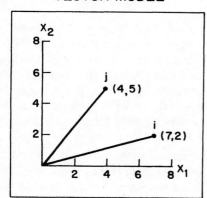
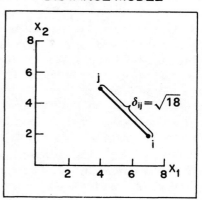

FIGURE 11.2. Vector and distance models.

11.3 DATA COLLECTION TECHNIQUES

Most of the techniques discussed in this book assume that the data are used either in their original form or are transformed by some of the methods discussed in Chapter 3. The keystone of similarities MDS is the matrix of interpoint distances, δ_{ij}^2, and a digression into various methods of obtaining these estimates may be of interest. For a more complete description of these and other techniques, the reader is advised to consult some general references on psychometric methods and/or scaling, among which some of the classics are Torgerson (1958), Coombs (1964), and Green and Rao (1972). Among the possibilities are:

1. *Direct Magnitude Estimation.* Here the respondent is asked to make a direct judgment about the distance between each pair of objects or stimuli. Unless it is obvious to the respondent what the unit of measurement is (pitch or actual distance in feet, for instance), a common technique is to use a scale of 0–100. If there are p stimuli, there will be $p(p-1)/2$ distances to be estimated. The square of each of these estimates is one entry in this matrix. Conceptually the easiest method, operationally it is usually the most difficult. Indow and Uchizono (1960) and Indow and Kanazawa (1960) used magnitude judgments versus a standard in studies of the Munsell color system. In another color experiment, reported by Gnanadesikan and Wilk (1969), the *differences* of pairs of stimuli were rated in this manner.

2. *Conditional Ranks.* In this method the respondent is told to consider a particular stimulus to be the "standard" and to rank the other $p-1$ stimuli in order from most similar to the standard to least similar. This procedure is repeated, in turn, for the other $p-1$ stimuli. A program called TRICON (Carmone et al., 1968), based on an algorithm of Coombs (1964) that converts these judgments to distances, is described and illustrated in Green and Rao (1972).

3. *Ranked Pairs.* The respondent is presented with all $p(p - 1)/2$ pairs of stimuli and is asked to rank them in order from smallest difference to largest difference. Since the average rank for each pair is an ordinal measure of distance, no special algorithm is required. This is one place where a statistical test can be employed. The Friedman test for ranks (Lehmann, 1975, p. 262) may be applied to the original data before doing any MDS at all. If the average ranks are not significantly different, there is no reason to go any further. An example using ranked pairs will be given in Section 11.4.2.

4. *Triads.* This takes various forms but the most common is the method of *triadic combinations*. In this, the respondent is presented three stimuli at a time and asked, of these three, which two are the most alike and which two are the least alike. There will be $p(p - 1)(p - 2)/6$ such combinations on which judgments must be made. This method was employed by Torgerson (1958) in a study of the Munsell color system. Another color example using triads was conducted by Helm (1964).

5. *Tetrads.* The respondent is presented with *pairs of pairs* and asked which pair has the greater difference. Although easy to operate in concept, it requires a huge number of responses for any save the smallest problems. (For $p = 10$ stimuli, the respondent would have to make 990 judgments.)

6. *Bipolar.* The respondent is asked to rate each stimulus with regard to a number of attributes, commonly on scales such as -10 to $+10$. This would result in a three-mode data array involving stimuli, attributes, and respondents and has the property that the attributes may be included on the plot of the stimuli. Green and Rao (1972) included algorithms both for obtaining the distances (DISTAN) and fitting the attributes to the similarity plot (PROFIT). The distances obtained by methods 1–5 are all referred to as *direct* similarities since the respondent is making distance judgments with regard to the stimuli. The bipolar technique produces *derived* similarities because the distances are obtained indirectly from the attribute ratings. The difference between the results produced will depend in great part on how well the attribute set chosen describes the stimuli. The bipolar technique is widely used in market research and was illustrated by Green and Rao (1972). Some other examples in connection with social psychology were given by Forgas (1979).

There are many other techniques, a number involving sorting the stimuli in one manner or another [see, for example, Rao and Katz (1971) and Frankel et al. (1984)]. In another color example, carried out by Boynton and Gordon (1965) and reanalyzed by Shepard and Carroll (1966), the respondents were asked to indicate which of four standard colors each of the experimental colors most closely resembled. The choice of data-collection technique will depend on a number of things, including the difficulty of the task, the nature of the stimuli, and problems encountered in presenting the stimuli. Rao and Katz compared several methods in their paper.

All of the techniques described so far assume that the matrix of interpoint distances is symmetric, and in the vast majority of applications this is the case. However, the occasion may arise when the distances are asymmetric. One such situation is in the study of brand identification where the number of people misidentifying Brand A as Brand B will not be the same as the number of people misidentifying B as A. Another example is a Morse code experiment reported by Rothkopf (1957) and reanalyzed by Kruskal and Wish (1978), in which the data included not only the probability of each letter being correctly identified but the probabilities associated with each of the alternatives. The probability of interpreting a "G" as a "W" is not necessarily the same as the other way around. Some other examples and their analyses may be found in Gower (1977), Constantine and Gower (1978), Escoufier and Grorud (1980), Harshman et al. (1982), Weeks and Bentler (1982), Hutchinson (1989), and Kiers (1989).

11.4 ENHANCED MDS SCALING OF SIMILARITIES

11.4.1 Introduction

The basic elements in the MDS of similarities data, as outlined in Section 11.2, consist of obtaining an interpoint distance matrix δ_{ij}, the matrix **B** using (11.2.2), and the characteristic vectors of **B** which form the coordinates for the MDS plot. However, most of the programs used to obtain these plots now do much more than that and involve nonlinear estimation of one sort or another. These enhancements will be discussed in this section and some of them will be illustrated with an example.

There are a number of competing algorithms for performing similarities MDS. One of the more prominent is KYST (Kruskal, 1977), which was produced by the Bell Laboratories and is a combination and updating of two earlier procedures, TORSCA and M-D-SCAL. The background for these procedures may be found in Kruskal (1964a,b). (KYST stands for Kruskal, Young, Shepard, and Torgerson, the principal contributors for these various procedures.) ALSCAL (Young and Lewyckyj, 1979; SAS Institute Inc., 1986) includes some options for dealing with similarities data. There are also the SSA options in the Guttman–Lingoes series (Guttman, 1968; Lingoes, 1972) or the MINISSA series (Roskam, 1979b). All of these will generally produce similar results but there are some differences in the algorithms employed and if one is going to use any of them, it would be worthwhile discovering what these differences may be. The book by Schiffman et al. (1981) may prove useful in this respect. A Monte Carlo evaluation of three of these methods was carried out by Spence (1972), who concluded that the differences were too slight to be of practical importance with a possible exception of some one-dimensional solutions that are not likely to be of practical consideration anyhow.

The principal enhancement in all of these procedures is the inclusion of one or more steps after the characteristic vectors of **B** have been obtained that will result in better agreement, in some form, between the final interpoint distances as represented on the MDS plot and the original dissimilarities δ_{ij}. The $p(p-1)/2$ interpoint distances are more than sufficient to obtain an MDS solution, only $(2p - k - 2)(k + 1)/2$ distances being required (Ross and Cliff, 1964) because of the constraints on the distances. For example, once the distances between x_1 and x_2 and between x_1 and x_3 are given, the distance between x_2 and x_3 is constrained. The more points, the more constraints. Shepard (1962) showed that preserving only the rank order of these distances was sufficient to obtain a suitable MDS representaton and, as a result, all of these schemes now have a nonmetric option that is widely exercised.

Most of the procedures include another option, the *Minkowski metric*. When estimating distances from the characteristic vectors, (11.2.1) is replaced by the expression

$$d_{ij} = \left[\sum_{h=1}^{k} (v_{ih} - v_{jh})^q \right]^{1/q} \tag{11.4.1}$$

where q is arbitrary. If $q = 2$, this is the euclidean distance, which is most commonly used. Setting $q = 1$ produces the "city-block" metric; d_{ij} is the sum of the absolute deviations. The distinction between these is shown in Figure 11.3. However, q may be any value, not necessarily an integer.

The other thing these procedures all have is a general lack of statistical inferential procedures although inroads are beginning to be made. However, another procedure, MULTISCALE (Ramsay, 1977, 1978a, b, 1980a, b), provides maximum likelihood estimates based on certain distributional assumptions as well as confidence limits for the coordinates. Ramsay made the distinction between *exploratory* and *confirmatory* MDS, a distinction that has been made in Section 7.3 for PCA and will again for factor analysis in Chapter 17. Unlike

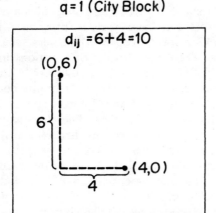
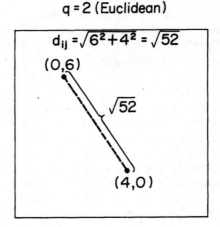

FIGURE 11.3. Minkowski metric.

the procedures previously discussed, MULTISCALE is primarily designed for confirmatory analysis and has a number of options that may be tested, such as independence of observations and homogeneity of variance. A description of MULTISCALE may also be found in Schiffman et al. (1981). MacKay and Zinnes (1986) gave a procedure for estimating both the location and variance parameters. Sibson et al. (1981) investigated the robustness of both metric and nonmetric MDS techniques. Spence and Lewandosky (1989) developed a median-based robust scaling procedure called TUFSCAL.

11.4.2 An Example

The use of nonmetric MDS of similarity data will be illustrated by another example taken from one of the Kodak Psychometric classes. This time, the task involved the judging of some color prints for exposure and color. A number of prints were made from a single negative, varying color and exposure by making adjustments to the printer. There were three levels of exposure and two variations in color, one being greener than the normal printing condition and the other less green than normal or magenta in appearance. This resulted in six different combinations as shown in Figure 11.4. All of these changes were fairly small and the task was designed to see how perceptive the respondents were of these changes. The photographic system included not only the printer but the paper on which the exposures were made and the process in which they were developed, but these other two factors were considered to be constant except for inherent variability. If this system was truly linear, then the difference between Prints #1 and #6 should appear to be the same as the difference between Prints #3 and #4. (See Table 11.2 for Print identification.) One class performed a more comprehensive experiment involving three exposure levels (Jackson, 1978) but this required over twice as many judgments.

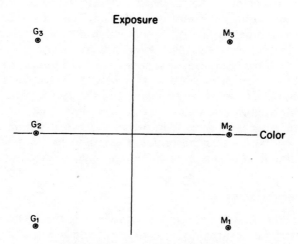

FIGURE 11.4. Color Prints: design space.

In order to reduce the size of the display of some of the intermediate results, one of these smaller experiments will be used here. It should be pointed out that MDS similarity procedures require a reasonable number of stimuli, this number increasing as the number of dimensions increases. The question of dimensionality will be discussed more fully in Section 11.4.4 but, for the moment, suffice to say that we have cut some corners for the sake of expediency in this example.

There are 15 different pairs of these prints. Five copies were made of each printing condition, allowing each pair to be mounted on a separate piece of cardboard. The method of *ranked* pairs was used in which each respondent ranked the 15 pairs in terms of appearance from least to greatest. This is a visual sorting task and although the differences were noticeable, they were not particularly large. This was not an easy task as can be seen from the variability displayed in the original data shown in Table 11.1. A rank of 1 corresponds to the pair the respondent perceived had the smallest difference and a rank of 15 had the largest difference.

In this example, the ranks for each pair were averaged over the respondents ($n = 17$ for this example) and these are also shown in Table 11.1. These data are unedited for reasons that will become apparent in Section 11.6. If the original data are in terms of similarities, dissimilarities are obtained by using their reciprocals, or, in the case of ranks, by subtracting the similarity rank from $[p(p - 1)/2] + 1$. They are then normalized so that the sum of squares is equal to $p(p - 1)$ and this is the dissimilarities matrix δ_{ij} shown in Table 11.2. (These preprocessing details will also vary from one procedure to another.) Note that $\delta_{1,6} = 1.677$ whereas $\delta_{3,4} = .953$, implying that this system may not be linear. Using (11.2.2), the matrix of scalar products, **B**, is obtained from the squares of the quantities in Table 11.2 and is displayed in Table 11.3. When rank data are used, Lingoes (1971) showed that the maximum number of dimensions required to fit p-dimensional data is $p - 2$. That means that this example would have, at most, four characteristic vectors. The first three **V**-vectors and their associated roots are shown in Table 11.4. The two-dimensional MDS representation would be the plot of v_1 vs. v_2 as shown by the points in Figure 11.5. (Not all scaling programs plot **V**-vectors. Some use **U**-vectors, which distorts the sense of the interpoint distances.) The third vector, had we chosen to use it, would have been primarily a contrast between M1 and M2 and another between G2 and G3. However, as we shall see in the next section, it is possible to obtain an improved representation in two dimensions that, for this example, one would hope to achieve since there were only two variables employed in producing the prints. Another reason for stopping at two dimensions is that, with only six stimuli, anything more would approach a forced fit.

11.4.3 The Enhancements

The first step in obtaining better estimates is to look at the *derived* or *estimated* interpoint distances d_{ij} using (11.4.1) with $k = 2$ and $q = 2$. These, compared with the original δ_{ij} are shown in Table 11.5 and are plotted in Figure 11.6.

Table 11.1. Color Prints. Original Rank Data

Print Pair										Respondent Number								Average
	1	2	3	4	5	6	7	8	9	10	11	12	13	14	15	16	17	
1-2	7	2	4	3	11	4	6	2	4	2	6	4	12	1	3	4	2	4.53
1-3	12	6	6	9	5	6	7	6	8	12	12	9	9	5	6	5	3	7.41
1-4	8	10	13	10	6	12	11	13	15	11	13	10	6	9	11	12	5	10.29
1-5	11	13	14	14	13	13	12	12	7	14	14	8	14	14	14	13	9	12.29
1-6	15	15	15	15	15	15	15	15	13	15	15	12	15	15	15	15	15	14.71
2-3	4	5	3	6	10	5	1	3	2	4	3	5	4	3	4	3	10	4.41
2-4	5	12	9	8	8	8	8	10	10	5	8	11	5	7	7	2	4	7.47
2-5	9	11	12	11	12	11	10	14	14	13	4	7	10	11	12	14	13	11.06
2-6	14	9	11	13	14	14	13	11	6	10	11	14	13	10	13	6	12	11.41
3-4	10	7	7	7	9	9	9	7	9	7	10	13	8	8	5	10	7	8.35
3-5	6	14	8	5	7	7	5	4	12	9	7	6	3	12	9	7	8	7.59
3-6	13	8	10	12	4	10	14	8	11	8	9	15	7	13	10	11	11	10.24
4-5	3	1	2	2	3	1	2	1	1	1	2	2	1	2	1	1	1	1.59
4-6	2	4	5	4	2	3	4	9	5	6	5	3	11	6	8	9	14	5.88
5-6	1	3	1	1	1	2	3	5	3	3	1	1	2	4	2	8	6	2.77

243

Table 11.2. Color Prints. Normalized Dissimilarity Matrix $[\delta_{ij}]$

	Print Number					
	1 M1	2 M2	3 M3	4 G1	5 G2	6 G3
1	0	.516	.845	1.174	1.402	1.677
2	.516	0	.503	.852	1.261	1.302
3	.845	.503	0	.953	.865	1.167
4	1.174	.852	.953	0	.181	.671
5	1.402	1.261	.865	.181	0	.315
6	1.677	1.301	1.167	.671	.315	0

Table 11.3. Color Prints. Scalar Products Matrix B

	Print Number					
	1 M1	2 M2	3 M3	4 G1	5 G2	6 G3
1	.77	.42	.15	−.22	−.43	−.69
2	.42	.34	.17	−.11	−.46	−.35
3	.15	.17	.25	−.25	−.09	−.23
4	−.22	−.11	−.25	.17	.23	.18
5	−.43	−.46	−.09	.23	.32	.44
6	−.69	−.35	−.23	.18	.44	.65

Table 11.4. Color Prints. Characteristic Vectors of Scalar Product Matrix

	v_1	v_2	v_3
M1	.839	−.184	.167
M2	.548	−.009	−.302
M3	.280	.432	.078
G1	−.302	−.316	−.005
G2	−.585	.017	.253
G3	−.781	.060	−.192
Characteristic Root	2.127	.324	.226

This is called a *Shepard's diagram*. One wishes to obtain a functional relationship between d_{ij} and δ_{ij}. In *metric* scaling this takes the form of linear or nonlinear regression. The purpose of this is to replace the d_{ij} by their estimates obtained from this equation. These estimates are then used in place of the original δ_{ij} to obtain a new **B**. From this a new set of vectors is obtained and a new Shepard's diagram (always using the original δ_{ij} for the ordinates). This is repeated until

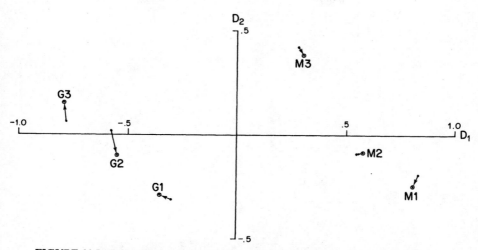

FIGURE 11.5. Color Prints: similarity plot. •, First approximation. ⊙, Final solution.

Table 11.5. Color Prints. Original Dissimilarities, δ_{ij}, and Derived Distances, d_{ij}

Print Pair	Original Dissimilarity δ_{ij}	Derived Distance d_{ij}
1–2	.516	.340
1–3	.845	.832
1–4	1.174	1.149
1–5	1.402	1.439
1–6	1.677	1.639
2–3	.503	.516
2–4	.852	.904
2–5	1.261	1.134
2–6	1.301	1.331
3–4	.953	.948
3–5	.865	.960
3–6	1.167	1.124
4–5	.181	.438
4–6	.671	.610
5–6	.315	.201

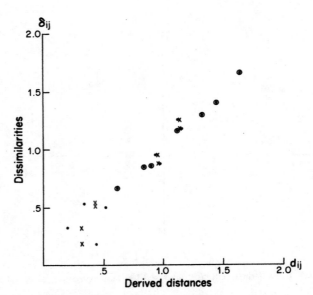

FIGURE 11.6. Color Prints: Shepard's diagram. $\odot$, d_{ij} in correct order ($\hat{d}_{ij} = d_{ij}$). $\bullet$, d_{ij} not in correct order ($\hat{d}_{ij}$ calculated). $\times$, calculated $\hat{d}_{ij}$.

Table 11.6. Color Prints. Monotone Regression

Print Pair	Original Dissimilarity δ_{ij}	Derived Distance d_{ij}	Disparity $\hat{d}_{ij}$	$d_{ij} - \hat{d}_{ij}$	Rank Image $d_{ij}^{\ast}$
4–5	.181	.438‡	.320	.118	.201
5–6	.315	.201	.320	−.119	.340
1–2	.503	.516‡	.428	.088	.438
2–3	.516	.340	.428	−.088	.516
4–6	.671	.610‡	.610	—	.610
2–4	.845	.832‡	.832	—	.832
3–5	.852	.904‡	.904	—	.904
1–3	.865	.960‡	.954	.006	.948
3–4	.953	.948	.954	−.006	.960
1–4	1.167	1.124‡	1.124	—	1.124
3–6	1.174	1.149‡	1.142	.007	1.134
2–5	1.261	1.134	1.142	−.008	1.149
2–6	1.301	1.331‡	1.331	—	1.331
1–5	1.402	1.439‡	1.439	—	1.439
1–6	1.677	1.639‡	1.639	—	1.639

‡ See text for explanation of figures marked with this symbol.

no further improvement is obtained. Most metric procedures have this option, of which COSCAL (Cooper, 1972) is typical.

The most widely used nonmetric procedure is monotone or isotonic regression (Kruskal, 1964a). This can best be illustrated by use of Table 11.6. The values in Table 11.5 have been reordered so that the original dissimilarities δ_{ij} are in ascending order. The next column, d_{ij} are the derived distances using (11.4.1). All of the values to which the symbol ($\ddagger$) is attached are also in ascending order; the others are not. For instance, between the first element, .438, and the third, .516 is a lesser quantity, .201. The monotone procedure obtains the average of the first two elements, .320, and replaces both elements by that value. This is the next column in Table 11.6, $\hat{d}_{ij}$, called *disparities*. The fourth element, .340, is less than the third, .516, and these two are also averaged. These two irregularities, incidentally, correspond to the contrasts represented by the third pc, which was not retained. This procedure is carried out for the rest of the table every time a value of d_{ij} is smaller than the one preceding it. The next column contains the differences between the derived distances, d_{ij}, and the disparities, $\hat{d}_{ij}$. This procedure is used both by the Bell Lab series such as KYST and ALSCAL, option CMDS. The Guttman–Lingoes series use *rank images* in which the original derived distances are rearranged in ascending order (Guttman, 1968). These are denoted by d_{ij}^* in Table 11.6. In this case the derived dissimilarity corresponding to the smallest original dissimilarity is replaced by the smallest derived dissimilarity, and so on.

The resultant disparities (or rank images) are now in the same rank order as the original dissimilarities, the goal of monotonic regression. The matrix of these disparities is then used in place of the original dissimilarity matrix from which are obtained a new set of vectors and a new set of derived distances. From these, a new set of disparities are obtained, always using the rank order of the original dissimilarities, δ_{ij}. This process continues until suitable convergence is obtained. Most procedures finish up with local explorations to avoid the problem of local minima.

The question of a convergence criterion is one on which the various procedures differ considerably and there are sometimes more than one within a procedure. The three widely used criteria are:

$$\text{STRESS1} = \sqrt{\sum\sum(d_{ij} - \hat{d}_{ij})^2 / \sum\sum d_{ij}^2} \qquad (11.4.2)$$

$$\text{SSTRESS1} = \sqrt{\sum\sum(d_{ij}^2 - \hat{d}_{ij}^2)^2 / \sum\sum d_{ij}^4} \qquad (11.4.3)$$

$$g = \sqrt{(1 - u^2)} \qquad (11.4.4)$$

where

$$u = \frac{\sum\sum d_{ij}\hat{d}_{ij}}{[\sum\sum d_{ij}^2][\sum\sum \hat{d}_{ij}^2]}$$

All summations are over i and j from 1 to p and $\hat{d}_{ij}$ may be replaced with d_{ij}^* if rank images are used. STRESS1 (Kruskal and Carroll, 1969) is used in both the Bell Lab and Guttman–Lingoes series. For the present example, STRESS1 = .055. There is no test for this quantity (nor for any of the others, for that matter) but Kruskal used guidelines such as .2 is poor, .1 is fair, .05 is good, .025 is excellent, and, of course, 0 is perfect. Note that all of these formulas are measuring the fit of the disparities to the derived dissimilarities, not the original dissimilarities, δ_{ij}. They are measuring how well the monotonicity is being maintained. A similar stress formula, STRESS2, differs only in that the denominator is corrected for the mean d_{ij}.

SSTRESS1 is used in the ALSCAL series and differs from STRESS1 in operating with squared distances instead of distances (Takane et al., 1977) and for these data has a value of .033. There is also a SSTRESS2, in which the denominator is corrected for the mean d_{ij}^2. The quantity g is like a coefficient of alienation and is included as an option in the Guttman–Lingoes series (Guttman, 1968). For these data, $g = .088$ using rank images.

The KYST procedure terminated with STRESS1 = .007. The resulting configuration is also given in Figure 11.5 but represented as circles indicating the change from the first to last iteration. These changes are not such as to change one's appraisal of the situation but there are examples in the literature where the extra effort was worthwhile. The results in this color print example indicate that enough nonadditivity did exist between the effects of exposure and color changes that it could be perceived by the respondents in terms of these unedited data. ALSCAL, option CMDS, terminated with SSTRESS1 = .009 and with results very close to those of KYST.

For more details on the operations of these procedures, including MULTISCALE, a good place to start would be Schiffman et al. (1982), which includes brief descriptions of them written by the people with whom each of the procedures is associated. In metric scaling, part of the estimation procedure involves the estimation of an additive constant that is more easily obtained by regression than by the methods originally used (e.g., Torgerson, 1958).

One final remark with regard to nonmetric procedures. We have already noted that the concept was put on a solid theoretical foundation by Shepard (1962), who showed the conditions under which interval data may be replaced by ranks and still recover the metric information regarding the interpoint distances in the final configuration. This has been further amplified by Shepard (1966), Young (1970), and MacCallum and Cornelius (1977). Although one can recover metric information from rank data, this is true only when the original data are in interval form. When the original data are in the form of ranks, as was the color print example, the final configuration is not an interval scale since no information is available on the relative spacing of the stimuli. One stimulus pair could be just barely larger than the others or much different and its rank would be the same. However, if there were gross differences, one or the other of the stimuli probably should not have been included in the first place. The experimenter may know this but the analyst, having only rank data, has no way of knowing. With interval data, this would not be the case. On the other hand, it might be difficult to obtain interval information for this kind of experiment.

11.4.4 The Assessment of Dimensionality

The question arises, as it has throughout this book: "How many dimensions are required to adequately represent the data?" In Chapter 2, a number of tests were proposed, some statistical and some graphical. Unlike the many procedures discussed in Chapters 2 and 4 for covariance and correlation matrices, graphical techniques are about all that are available for MDS. The customary procedure is to use an equivalent of the SCREE plot in which some measure of stress is plotted as a function of number of dimensions. Klahr (1969), Spence and Ogilvie (1973), and Spence (1979), among others, have given some guidelines for the sort of stress values one might expect to get using random numbers. These may be used in somewhat the same fashion as the work, in Section 2.8.5, of Horn (1965) and Farmer (1971) in relation to regular SCREE plots.

There is an additional problem, however. The Color Print example had $p = 6$ stimuli represented in $k = 2$ dimensions. That is a total of 12 coefficients to be estimated from 15 interpoint distances, nearly a forced fit. If random numbers were employed for interpoint distances, as the number of dimensions increases for fixed p, the stress values decrease to very small values. Figure 11.7 is taken from Klahr's paper and shows, based on simulations, the average stress expected from random numbers as a function of number of stimuli and the number of dimensions fitted. This would indicate, for the Color Print example, an expected stress of about .07 if random numbers were used. Fortunately, our results are much less than that. Young and Sarle (1983) recommended from three to four times as much data coming in as parameters to be estimated when the data are in the form of ranks but somewhat less for interval data. The original experiment, from which this one was adapted, had $p = 9$ prints, resulting in 36 pairs, still to produce 12 estimates, so this was a marked improvement. However, to do that, each respondent had to sort through 36 pairs of prints, requiring a large work space and considerable patience. Considering the difficulty in making the judgments, the results were most impressive (Jackson, 1978). Young (1970) and MacCallum and Cornelius (1977) investigated the effect of number of stimuli, number of dimensions, amount of error and, in the latter paper, sample size and the effect on recovery of metric information.

A point to be made here is that the majority of similarity MDS examples in the literature usually are displayed as two- or at the most, three-dimensional representations. Very few of them give much justification for the number of dimensions chosen. There also appears to be little regard for number of stimuli used in relation to the dimensionality. In the Color Print example, the number of stimuli was kept low in order to make the experiment practical from a classroom sense and the analysis was shown only for the two-dimensional case

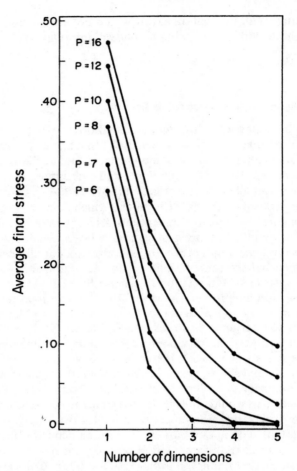

FIGURE 11.7. Klahr's STRESS simulation. Reproduced from Klahr (1969) with permission of Psychometrika and the author.

because there were only two experimental variables. However, in real-world situations, one should make sure that the third dimension does not include additional information because its existence would indicate that there was more variability in the experiment than was hypothesized.

11.5   DO HORSESHOES BRING GOOD LUCK?

The answer is *maybe*. In similarities scaling, there are certain patterns for which one should be on the lookout. Consider the similarity matrix in Table 11.7, obtained from ranked pairs. The two-dimensional representation of these data

Table 11.7. Horseshoe Example. Original Ranks

	Stimuli				
	A	B	C	D	E
A	—	3	6	9	10
B	3	—	2	7	8
C	6	2	—	4	5
D	9	7	4	—	1
E	10	8	5	1	—

is shown in Figure 11.8. Note that the stimulus points form somewhat of a semicircle. This phenomenon has often been referred to as a simplex pattern or a "horseshoe" although the ends do not actually curve back in. If each of these points is projected onto the horizontal axis and all of the interpoint distances are obtained among these projected points, it turns out that all but one of these projected distances are in the same rank order as the original data in Table 11.7. With a little effort, a straight line can be found below this axis and rotated slightly clockwise on whose projected points the ranks of the interpoint distances will exactly match the starting data. This means that the relationships among these five stimuli can be expressed in one dimension. Whenever the simuli appear in this fashion, be on the watch for a one-dimensional situation. The mathematical condition for this to occur is that the original similarity matrix can be row- and column-permuted in such a way that the similarities will decrease towards the main diagonal from both directions. Horseshoes of this type are lucky to the extent that, if detected, they can protect one from attempting to interpret a second dimension that exists only artificially.

There are some more exotic patterns also caused by relatively simple relationships among the data. One is a *circumplex*, which is a p-sided polygon, although in practice they look more like lumpy circles or ellipses. Another is

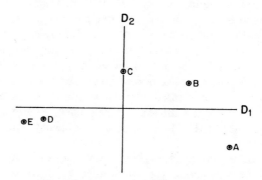

FIGURE 11.8. Horseshoe Example: similarity plot.

the *radex* which appears as spokes of a wheel. A discussion of these and other phenomena and the conditions that produce them may be found in Jöreskog (1974), Lingoes and Borg (1979), and Lingoes (1980).

11.6 SCALING INDIVIDUAL DIFFERENCES

The Color Print example given in the previous section was based on the data of 17 respondents but the analysis only dealt with their averages. It was assumed the respondents were somewhat homogeneous in the perceptions of color and exposure differences. In this section, methods of examining the individual responses will be considered. Again, there are a number of techniques available and unlike the techniques discussed in the previous section, which generally produce similar results, the individual scaling procedures have some marked differences. We shall reconsider the Color Print data by means of one of these techniques, INDSCAL (Carroll and Chang, 1970a; Carroll, 1972) and then discuss what the competition has to offer.

With the original data, one can actually obtain a similarity matrix for each respondent, so this becomes a three-mode analysis similar to the procedures discussed in Section 10.9. What this procedure does is change (11.2.1) to

$$\delta_{ijh}^2 = \sum_{m=1}^{k} w_{hm}(v_{im} - v_{jm})^2 \tag{11.6.1}$$

where δ_{ijh} is the interpoint distance between stimuli i and j for the hth respondent and w_{hm} is the weight for the hth respondent on the mth dimension. These weights are sometimes known as *saliences*. INDSCAL not only estimates **V** but the weights as well, so there are a large number of parameters, $m(p + n)$, to estimate. In the Color Print example with $p = 6$, $m = 2$, and $n = 17$, the number is 46. The basic assumption is that all of the respondents have the same perceptual space but differ in the weights they place on each of the dimensions. While this is probably a realistic assumption for many physical problems such as this one, there may be other situations where it may not hold.

The INDSCAL plot for these data is shown in Figure 11.9. While procedures such as KYST and SSA-1 have the option for rotating the final perceptual map, INDSCAL plots cannot be rotated because weights must be estimated as well as the coordinates for the stimuli. As can be seen in Figure 11.9, this is not a problem for this example. The first dimension is clearly color and the second is clearly exposure.

The weights, or saliences, are given in Table 11.8 and are plotted in Figure 11.10. (These weights are sometimes referred to as "stretching" coefficients although "contracting" might be a better term.) Note that the

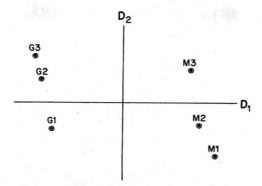

FIGURE 11.9. Color Prints: INDSCAL.

Table 11.8. Color Prints. INDSCAL Saliences

Respondent #	w_1	w_2	Proportion Explained
1	.67	.44	.64
2	.84	.19	.74
3	.90	.21	.85
4	.78	.39	.76
5	.61	.50	.62
6	.88	.28	.85
7	.86	.20	.80
8	.76	.33	.69
9	.81	.00	.66
10	.68	.49	.70
11	.72	.43	.70
12	.80	.17	.67
13	.43	.63	.58
14	.89	.20	.83
15	.82	.34	.79
16	.68	.30	.55
17	.49	.52	.51

weights are all positive and restricted to a quarter-circle of unit radius. The closer a respondent is to the perimeter of the circle, the better this model explains his or her data. Table 11.8 also includes the sums of squares of the weights for each respondent, which can be regarded as a coefficient of determination or proportion of explained variability. The closer a point is to the horizontal axis, the greater the weight on the first axis relative to the second; in other words, the more heavily their judgments were affected by color differences. Those closer to the vertical axis were more influenced by exposure differences.

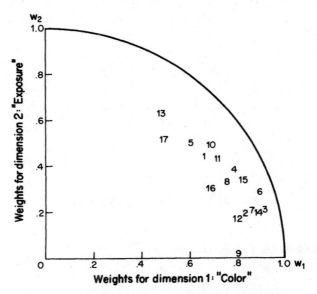

FIGURE 11.10. Color Prints: INDSCAL saliences.

Examining these coefficients and the corresponding measure of explained variability, we find that:

1. Respondents #3 and #6 had the highest explained variability, .85.

2. Respondent #17 had the best balance between the effect of color and exposure ($w_1 = .49$, $w_2 = .52$).

3. Respondent #9 was the most influenced by color relative to exposure ($w_1 = .81$, $w_2 = .00$).

4. Other then #17, respondent #13 was the only one influenced more by exposure than color ($w_1 = .43$, $w_2 = .63$) and was the only one to show a marked difference in that direction.

It appears that respondent #9 was marching to a different drummer. As it happened, respondent #9 was absent from class the day this exercise was carried out and performed the task at a later date. The instructor inadvertently told this individual to rank the pairs of prints on the basis of "color" rather than "overall" difference. This person did exactly that and hence appears on the abscissa of Figure 11.10.

An INDSCAL dissimilarity plot essentially represents a respondent who has equal weights for each axis. To the extent that the majority of respondents vary from this condition, one should guard against interpreting this plot too literally.

Of course, if one were doing serious research with a problem of this kind, more respondents would be used and outliers such as #9 would be deleted.

The color examples of Helm and of Indow and Kanazawa, referenced in Section 11.3, were revisited in terms of INDSCAL by Carroll and Chang (1970b) and by Wish and Carroll (1974). An example involving automobiles due to Rao (1972) is interesting in that the original data were obtained from reports published by Consumer's Union along with measures of respondents' familiarity with these models. Another example involving automobiles was given by Green et al. (1972). In both of these, the INDSCAL technique was used for market segmentation.

For another example, this class ($n = 18$ this time) also performed the ranked-pairs exercise with the same presidential hopefuls for whom they had earlier made preference judgments (Section 10.6). The INDSCAL plot is in Figure 11.11. The abscissa appears to represent conservative vs. liberal and the ordinate apparently Democrat vs. Republican. Remember that this plot represents *overall similarities*, whereas Figure 10.3 represented *similarity of preference* only. The saliences are displayed in Figure 11.12. This time there are more respondents associated with one dimension or the other. Apparently those with large values of w_1 were concerned about issues while those with large w_2 were more party-oriented. Respondent #14 had an explained variability of .12.

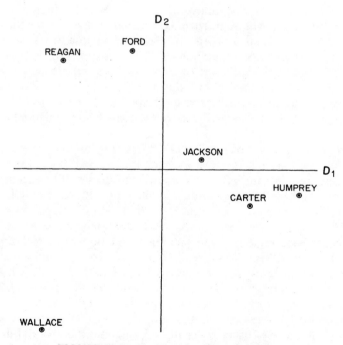

FIGURE 11.11. Presidential Hopefuls: INDSCAL.

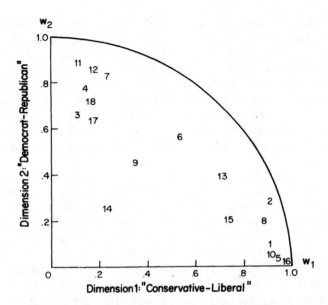

FIGURE 11.12. Presidential Hopefuls: INDSCAL saliences.

The use of salience plots does allow one to compare individuals and provides a basis for editing data. Although one might be tempted to cluster the saliences or carry out other formal analysis on them, these quantities do not lend themselves to that purpose (MacCallum, 1977; Coxon, 1982.). Jones (1983) advocated converting the saliences to vectors of unit length and performing significance tests on the angles, but this is only part of the story and this procedure would not distinguish between Respondents #6 and #14 in Figure 11.12, who have nearly the same direction but considerably different explained variability.

Some other comments about INDSCAL. First, the solution is metric. Second, because of all the parameters to be estimated, there are some computational considerations, both dealing with convergence and the resulting execution time, this problem affecting other individual scaling procedures as well. A faster version of INDSCAL using a different algorithm, was developed by DeLeeuw and Pruzansky (1978). Weinberg et al. (1984) developed some procedures for obtaining confidence regions using jackknife and bootstrap techniques and compared these with MULTISCALE, which also has an individual scaling option. Finally, the Shepard's diagrams associated with INDSCAL will generally exhibit much more variability because there will now be $np(p - 1)/2$ points, $p(p - 1)/2$ for each observer. If there are not too many observers, it might be useful to symbol-code them on the diagram.

There are a number of other procedures for handling individual differences and the models of most of these are less restrictive than INDSCAL. Among the pioneers were Tucker and Messick (1963), McGee (1968), and Horan (1969). Horan's work was developed into COSPA (COmmon SPace Analysis) by Schönemann et al. (1978, 1979). ALSCAL has some options to handle these problems including one which is nonmetric (Takane et al. 1977, 1980). PINDIS (Procrustean INdividual DIfferences Scaling) is a fairly extensive set of procedures ranging from one similar to INDSCAL to others allowing for differences in location and/or rotation (Lingoes and Borg, 1978; Borg and Lingoes, 1978; Borg, 1979; Langeheine, 1982). For a good introduction to PINDIS, see Coxon (1982). Vani and Raghavachari (1985) developed a model that obtains a solution by means of quadratic programming.

As mentioned earlier, the similarities procedures described in Section 11.4 such as KYST, ALSCAL, and SSA-1 will generally produce similar results. The same is not true of individual scaling procedures. There are a number of basic differences among them and anyone who plans to use these to any extent should look into what these differences are. For a comparison of ALSCAL and INDSCAL, see Takane et al. (1977) and MacCallum (1977). Ramsay (1977) made a comparison of MULTISCALE and INDSCAL. Comparisons of PINDIS and INDSCAL may be found in Borg and Lingoes (1978), Lingoes and Borg (1978), and Borg (1979). MacCallum (1976a) and Dunn and Harshman (1982) compared INDSCAL with Tucker's three-mode MDS. MacCallum (1976b) produced a transformation from Tucker's procedure to INDSCAL with an accompanying condition to test for the appropriateness of the INDSCAL model. A criticism of the CANDECOMP option of INDSCAL on the basis of a special case was given by TenBerge et al. (1988). In reading these articles, keep in mind that the authors are often fostering one method over another.

11.7   EXTERNAL ANALYSIS OF SIMILARITY SPACES

11.7.1   Introduction

This section considers the topic of *external* analysis of similarity spaces. It is assumed that a similarity space has already been established and that other information is to be added to the plot representing data other than that used to obtain that space. The most common application is to add preference data in the form of point–point or point–vector models. The difference is that this is not a joint space as it was in Sections 10.5–10.7. For instance, in the Presidential Hopefuls example displayed in Figures 10.3 and 10.6, the configuration of the candidates represents similarity of *preference*, not similarity of perception. In this section, the stimulus space represents similarity of perception and the preference data would be collected in a separate exercise. Given that one distinction, the interpretation would be the same as the dominance plots of Chapter 10. Not only may preference data be added but other types of information such as various perceptual properties.

The techniques discussed in this section do not use PCA and are included only for completeness with regard to the subject of multidimensional scaling. Without it, some confusion might exist between these techniques and those in Chapter 10, most of which do employ PCA.

11.7.2   Preference Mapping

By far the widest application of external analysis is preference mapping. This is a rather controversial subject as it is not clear in the minds of some whether this approach makes any sense or not. Its most popular application appears to be related to product positioning in market research (Urban and Hauser, 1980) and here the concept of an ideal point in a perceptual space does seem useful. Consider Figure 11.13, a fictitious preference map. Brands A, B,..., F are positioned on the basis of similarity of perception using one of the methods described in Section 11.3. Assume that Respondents 1, 2,..., 7 were asked, in a separate exercise, to score these brands on a scale of 0–100 based on preference. They appear as ideal points. Respondent #1 apparently prefers A, #2 prefers C or D, and so on. Respondent #6 appears to be halfway between C and D on one hand and B on the other. The implication here is that if a new product were designed that was perceived to be located at that point, that is the product #6 would prefer to any of the others. Note that Respondent #7 is denoted

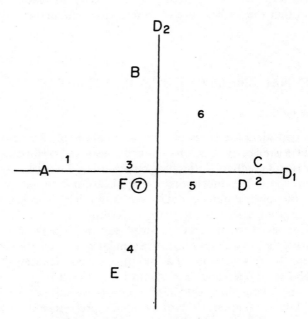

FIGURE 11.13. Fictitious data: PREFMAP III.

by a circle. This is a *negative* ideal point, the product that this individual would be least likely to buy. A negative ideal point can occur when a respondent such as #7 prefers B and E to the exclusion of the others, including F, even though B and E are conceptually different. The best an algorithm can do in that situation is produce this negative solution.

The algorithm that has received the most attention to date is PREFMAP (Carroll, 1972, 1980; Green and Rao, 1972). This program has four phases, each of which has a metric and a nonmetric form. We shall start by taking these phases in reverse order, beginning with the most simple.

Phase IV. This is a vector model that essentially fits each respondent's preference data to the similarity space by least-squares (metric) or isotonic regression (nonmetric) and is represented by a vector.

Phase III. This is a point–point model with one set of points, the stimuli, fixed. Each respondent is represented by an ideal point that minimizes the sum of squares of deviations or a nonmetric equivalent of each stimulus preference judgment to the stimulus itself.

Phase II. This is an extension of Phase III that allows, for each respondent, differential stretching of the axes in similarity space, similar to the notion of INDSCAL.

Phase I. This carries the process one step further and allows for individual rotation of the axes as well and is sometimes called the *interaction* model.

In practice, PREFMAP starts with the most complicated model and works through Phase IV, using the results of one phase as the starting approximation for the next. (It is not necessary to start with Phase I; many situations can be adequately handled with just Phases III and IV, or just Phase IV if all one wants is a point-vector representation.) The amount of information that the more complex model provides over a more simple one is evaluated by an F-test. (How well the F-distribution fits this ratio has not been established.) Since this procedure is carried out for each respondent separately, it is possible to have some respondents represented by ideal points and the rest by vectors. For a worked example, see Green and Rao (1972).

Returning to the Presidential Hopefuls, the preference data from Table 10.1 is fitted to the similarities configuration in Figure 11.11 using PREFMAP III. The results are shown in Figure 11.14. The stimuli are exactly the same as they were in Figure 11.10. The respondents marked with a * are represented by negative ideal points. Respondent #1 is off the page with a negative ideal point at $(-9.5, -2.1)$. Compare this plot with Figure 10.6, where the stimuli are part of the solution. There are some differences between overall similarity and similarity of preference.

PREFMAP seems to work best when the axes of the similarity space represent continuous variables. A case in point was a famous study of the Chicago beer

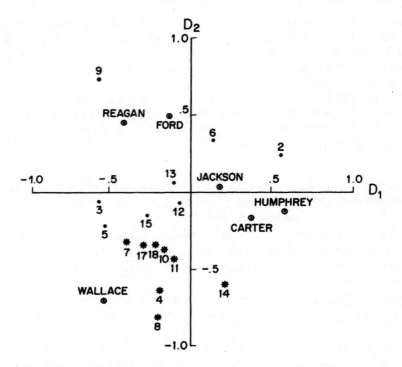

COORDINATES FOR:

RESP #1 (-9.47 -2.07)*

RESP #16 (5.89,-1.14)

* : NEGATIVE IDEAL POINT

FIGURE 11.14 Presidential Hopefuls: PREFMAP III.

market (Johnson, 1971) involving 500 male beer drinkers and 8 brands of beer. The first axis was related to price and the second axis was light vs. heavy. Johnson clustered the respondents and obtained an ideal point for each cluster. Market researchers would then probably try to characterize each cluster by studying the demographics associated with each. Both axes of this example represented conceptually continuous variables. [A similar study involving soft drinks by Best (1976) required three dimensions: calories, cola and noncola, and various fruit flavors.] When the stimulus space represents discontinuous perceptions, too often the only Phase III points that are significant are negative ideal points. A similar procedure by Davidson (1988) can restrict ideal points to being positive, although Davidson recommended that this should be used only for small negative ideal points; if they are substantial, they should be left that way.

11.7.3 Property Fitting

In a like manner, one can also fit other characteristics to a similarity configuration. In the beer study, the respondents were asked to rate each brand of beer on 35 attributes. Each of these attributes was then fitted to the similarity space as though it were a Phase IV PREFMAP. (In such examples, the same data could be used to obtain the similarity map in the first place, although it could be obtained as a joint space directly using the methods discussed in Chapter 10.) Two examples involving cigarettes were given by Smith and Lusch (1976), dealing with the effect of advertising and by Hooley (1984) on brand perception. Another example of such a study is shown in Figure 11.15. This comes from another psychometrics class in another election year, 1980. In that year the students were asked to rate each candidate on a bipolar scale of -7 to $+7$ with regard to a number of issues that were then fitted to the similarity space. The positive direction of each vector represents support for the particular program or strength in that characteristic. Quite obviously, the horizontal axis is liberal–conservative. The interpretation of the vertical axis might have been a challenge had not the vectors been included since Senator Edward Kennedy and Reagan were on one side of the origin and everyone else on the other. The "charisma" vector is the clue. The Bell Lab series actually has a separate program called PROFIT (PROperty FITting) to do this but it is really PREFMAP IV designed for property-fitting output.

Johnson (1971) also had a politician study using similar bipolar information for the year 1968. The first axis was also liberal–conservative and the second

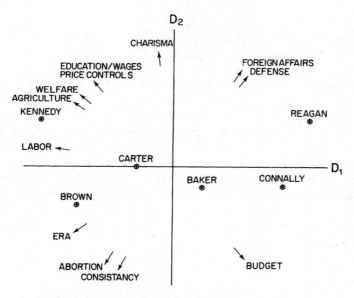

FIGURE 11.15. 1980 Presidential Hopefuls: property fitting.

was pro- or anti-administration. Instead of asking for direct preferences, the respondents were also asked some questions about how they, themselves, stood on these issues. These were then fitted as cluster points to the similarity data, making it a preference map by inference rather than direct estimation. A similar procedure was used in the 1976 election to help develop campaign strategy (Perry, 1977).

11.7.4 Other Algorithms

Although PREFMAP has had the most exposure of any external fitting algorithm, it is not the only one. ALSCAL, option CMDU, is similar to, Phase III; Option WMDU is like Phase II. Phase I may be handled by SAS PROC RSREG. Van der Kloot and Kroonenberg (1985) proposed an external version of the three-mode PCA model called TUCKALS.

The linear programming procedure LINMAP, referred to in Section 10.7, has also been used to obtain point–point representations (Srinivasan and Shocker, 1973a; Shocker and Srinivasan, 1975; Huber, 1976). For better or for worse, LINMAP does not produce negative ideal points because it works with absolute differences rather than quadratic regression.

Cooper and Nakanishi (1983) suggested logit models for both point–point and point–vector external analysis. The Bradley–Terry–Luce pair comparison model (Bradley and Terry, 1952; Luce, 1959; Bock and Jones, 1968, p. 132) is based on odds-ratios, producing a logit model which Cooper and Nakanishi felt might be better at producing improved fits for the point–point model when compared to the point–vector model.

11.8   OTHER SCALING TECHNIQUES, INCLUDING ONE-DIMENSIONAL SCALES

Before leaving the topic of multidimensional scaling, it should be pointed out that PCA has been applied to one-dimensional scaling problems as well. Some of these applications are primarily of academic interest, such as that of Gulliksen (1975), who described a number of one-dimensional scaling techniques in a unified matrix form. The method of successive categories with unequal category variances obtains scale estimates using the first characteristic root and vector (Gulliksen, 1954; Torgerson, 1958, p. 216). A variant of paired comparisons using ratios (Comrey, 1950; Torgerson, 1958, p. 105) was used by Hannan (1983) in the evaluation of contestants using an estimation procedure requiring PCA. *Cross-impact analysis*, a method of evaluating scenarios (Section 2.10.6), also uses odds-ratios.

CHAPTER 12

Linear Models I: Regression; PCA of Predictor Variables

12.1 INTRODUCTION

The next two chapters will deal with the use of principal components in linear models. Although regression and the analysis of variance are merely two alternative forms of linear models and it is possible to obtain most ANOVA results using regression, the two forms are generally treated separately and will be so dealt with here. The regression techniques will concentrate on the use of PCA to reduce the number of *predictor* variables while the subsequent chapter on the analysis of variance will concentrate on the use of PCA for the *response* variables. We will find, in both cases, that there are a number of alternatives for handling multivariate data, of which PCA is only one. The purpose of these two chapters will be to give the mechanics of principal components regression (PCR) and principal components analysis of variance (pc-ANOVA) and compare these with some of the alternatives available.

This chapter will focus on the topic of principal components regression (PCR), which concentrates on the use of PCA with the predictor variables. We will have p *predictor* variables $x_1, x_2, \ldots, x_p$ and one or more *response* variables $y_1, y_2, \ldots, y_q$. To put this in the proper perspective, the next section will contain a review of multiple regression or ordinary least squares (OLS). A numerical example will be introduced that will be used throughout the chapter and some modifications of OLS will be discussed. In addition, a number of properties of OLS may be described in terms of the principal components of the predictor variables. In discussing principal components regression, we shall see that while it may be quite useful, it is not a panacea and that it will be worth our while to investigate some alternative procedures. To this end, a number of similar techniques will be discussed and the chapter will conclude with some related techniques that are designed specifically for use when there is more than one response variable.

Presently, there is considerable interest in PCR and related techniques as applied to calibration problems. In some of the examples cited in the literature, in particular in connection with the partial least-squares techniques described in Section 12.5, it may seem that the true response and predictor variables have been reversed—the inverse calibration problem. This arises, for instance, when one initially predicts a response from a dose of some drug but later wishes to estimate what dose level would produce a certain response level. Bioassay applications generally regress the response to the dose level and then invert the relationship, contending that the dose levels are known and the responses are measured with error. The chemometricians on the other hand tend to regress the concentration of a solution on the measurements made from that solution, based on the philosophy that this is the way the equation will be used after the calibration phase. The question of which method to use has generated considerable controversy over the years and is a controversy into which this book shall not be drawn.

To simplify many of the expressions in this chapter, we will assume that all response and predictor variables are in terms of deviations from their respective means and, therefore, the regression equations or their equivalents will not have an intercept term. Also, *for this chapter only*, the symbol y will refer to response variables, not principal component scores.

12.2   CLASSICAL LEAST SQUARES

12.2.1   Multiple Regression

There are a number of texts devoted to multiple regression, among them, Draper and Smith (1980) and Freund and Minton (1979). It is not the purpose, here, to repeat a lot of this material but rather to outline the basic principles of least-squares regression, which will establish a framework from which the other sections may be presented.

Initially, we will consider the case of *one* response variable and p predictor variables. We will define X to be an $n \times p$ matrix of observations on the predictor variables and y to be an $n \times 1$ vector of response measurements. The traditional regression model is

$$y = Xb \tag{12.2.1}$$

where b is a $p \times 1$ vector of regression coefficients. The least-squares solution for b is

$$b = [X'X]^{-1}X'y \tag{12.2.2}$$

If the total sum of squares of the response is $\mathbf{y'y}$, then the amount explained by the OLS relationship is $\mathbf{b'X'y}$ and the standard error of estimate will be

$$s_{y.x} = \sqrt{\frac{[\mathbf{y'y} - \mathbf{b'X'y}]}{n - p - 1}} \tag{12.2.3}$$

If there are q responses rather than just one, $\mathbf{y}$ will be replaced by an $n \times q$ matrix $\mathbf{Y}$ and $\mathbf{B}$ will now be a $p \times q$ matrix of regression coefficients. $\mathbf{X'X}$ does not change, hence the operation really consists of q *separate* regression estimates since the relationship among the response variables is not taken into account.

A number of problems can occur when the predictor variables are highly correlated. This situation is called *multicollinearity*. (This term is also commonly used to indicate the existence of a linear constraint among the variables.) Some measures of this multicollinearity may be found in Stewart (1987) and Walker (1989) in addition to the measures introduced in Section 2.8.2. Multicollinearity causes a number of problems for OLS. Among them:

1. Obtaining a good inverse of $\mathbf{X'X}$ may be difficult. (The extreme case would occur when two predictors are perfectly correlated or some linear constraint exists among the predictors, in which case the inverse does not exist.) Although modern computational procedures have improved this situation considerably, it is still a legitimate problem.

2. The standard errors of the regression coefficients are equal to the square root of the product of the residual variance and their corresponding diagonal elements in the matrix $[\mathbf{X'X}]^{-1}$. These elements in the inverse matrix are equal to $1/[(n-1)s_i^2(1-R_i^2)]$, where R_i is the multiple correlation coefficient between x_i and the other predictors and s_i^2 is the variance of x_i. The quantity $1/(1-R_i^2)$ is called the *variance inflation factor*; the larger R_i, the larger will be the standard error of b_i. A near-singular case would inflate these estimates considerably and cast doubt on the interpretability of these coefficients.

Mandel (1982) showed that these standard errors could be expressed in terms of the characteristic roots and vectors of $\mathbf{X'X}$:

$$\text{St. error } (b_i) = s_{y.x} \sqrt{\sum_{j=1}^{p} \left(\frac{u_{ij}^2}{l_j} \right)} \tag{12.2.4}$$

where the quantity inside the square root is equal to the variance inflation factor. This shows that the principal components associated with the smaller roots will generally exert an undue influence on this quantity. Note, by (1.6.2), that (12.2.4) may also be expressed in terms of the sums of squares of the coefficients of the $\mathbf{W}$-vectors.

3. The correlation matrix of the regression coefficients is a function of $\mathbf{b'X'Xb}$, which says that as the predictor variables become more and more correlated, the same thing will happen to the regression coefficients themselves. This, like the inflated standard errors of the regression coefficients, will also cast doubt on the interpretability of these coefficients.

When one is analyzing response data from designed experiments, the predictors are usually uncorrelated, or nearly so, so that the above problems are not so prevalent. However, many other regression studies, in particular those from such fields as economics or the social sciences, do not have this luxury and the problem of predictor multicollinearity can become very real. Nelder (1985) used SVD to demonstrate various properties of OLS, particularly in the case of multicollinearity of which equation (12.2.4) is an example. One possible solution might be to transform the predictor variables into principal components and regress the responses on these. This will be taken up in Section 12.3. In Section 12.2.3 we will review some other alternative regression procedures currently in use. First, though, we need a numerical example to carry us through the rest of the chapter.

12.2.2   Numerical Example

The numerical example is found in Young and Sarle (1983) and is due to Dr. A.C. Linnerud of North Carolina State University. This example was chosen primarily because it is in a manual published by SAS Institute, Inc. and therefore readily accessible and because it is not a "pat" example from the standpoint of structure, as some of the intercorrelations are fairly low. It is a manageable data set and will serve to illustrate some of the strengths and weaknesses of PCR. The data consist of three physiological and three exercise variables measured on 20 middle-aged men in a fitness club. The physiological variables will be the predictors and the exercise variables the responses. The data are given in Table 12.1.

Some of the procedures to follow require these data to be in standard units, that is, each observation is corrected for its mean and this deviation is divided by its standard deviation. For simplicity in comparing all of these procedures, the standardized data given in Table 12.2 will be used for all of the examples. The first three columns of this matrix are the predictor variables and will be represented by the 20×3 matrix X while the remaining three columns are the responses and will be denoted by Y.

Because the data are now in standard units, the covariance and correlation matrices are the same and are displayed in Table 12.3. Some of the correlations among the predictors and among the responses are fairly strong but a glance at the intercorrelations between the predictors and responses should serve to prepare us not to expect too much from either the pulse measurements, x_3, or the number of jumps, y_3.

Using the relationships displayed in the previous section, the standardized regression coefficients, B, of Weight, Waist, and Pulse, on Chins, Situps, and Jumps along with their standard errors of estimate are given in Table 12.4.

Table 12.1. Linnerud Data

ID #	Weight	Waist	Pulse	Chins	Situps	Jumps
1	191	36	50	5	162	60
2	189	37	52	2	110	60
3	193	38	58	12	101	101
4	162	35	62	12	105	37
5	189	35	46	13	155	58
6	182	36	56	4	101	42
7	211	38	56	8	101	38
8	167	34	60	6	125	40
9	176	31	74	15	200	40
10	154	33	56	17	251	250
11	169	34	50	17	120	38
12	166	33	52	13	210	115
13	154	34	64	14	215	105
14	247	46	50	1	50	50
15	193	36	46	6	70	31
16	202	37	62	12	210	120
17	176	37	54	4	60	25
18	157	32	52	11	230	80
19	156	33	54	15	225	73
20	138	33	68	2	110	43
Mean	178.60	35.40	56.10	9.45	145.55	70.30
SD	24.69	3.20	7.21	5.29	62.57	51.28

Source: Young and Sarle (1983) with permission of SAS Institute.

Other information that will be used subsequently are the total sums of squares and crossproducts matrix of the responses:

$$Y'Y = \begin{bmatrix} 19.0000 & 13.2188 & 9.4194 \\ 13.2188 & 19.0000 & 12.7149 \\ 9.4194 & 12.7149 & 19.0000 \end{bmatrix}$$

and the residual sums of squares and crossproducts unexplained by the regression of the physical measurements on the exercises:

$$Y'Y - Y'XB = \begin{bmatrix} 12.5481 & 5.9626 & 7.8114 \\ 5.9626 & 10.7066 & 10.7012 \\ 7.8114 & 10.7012 & 17.9758 \end{bmatrix}$$

Table 12.2. Linnerud Data—Standardized

	Weight	Waist	Pulse	Chins	Situps	Jumps
ID #	x_1	x_2	x_3	y_1	y_2	y_3
1	.50	.19	−.85	−.84	.26	−.20
2	.42	.50	−.57	−1.41	−.57	−.20
3	.58	.81	.26	.48	−.71	.60
4	−.67	−.12	.82	.48	−.65	−.65
5	.42	−.12	−1.40	.67	.15	−.24
6	.14	.19	−.01	−1.04	−.71	−.55
7	1.31	.81	−.01	−.27	−.71	−.63
8	−.47	−.44	.54	−.65	−.33	−.59
9	−.11	−1.37	2.48	1.05	.87	−.59
10	−1.00	−.75	−.01	1.43	1.69	3.50
11	−.39	−.44	−.85	1.43	−.41	−.63
12	−.51	−.75	−.57	.67	1.03	.87
13	−1.00	−.44	1.10	.86	1.11	.68
14	2.77	3.31	−.85	−1.60	−1.53	−.40
15	.58	.19	−1.40	−.65	−1.21	−.77
16	.95	.50	.82	.48	1.03	.97
17	−.11	.50	−.29	−1.03	−1.37	−.88
18	−.87	−1.06	−.57	.29	1.35	.19
19	−.92	−.75	−.29	1.05	1.27	.05
20	−1.64	−.75	1.65	−1.41	−.57	−.53

Table 12.3. Linnerud Data. Correlation Matrix

	Weight x_1	Waist x_2	Pulse x_3	Chins y_1	Situps y_2	Jumps y_3
Weight	1.0000	.8702	−.3658	−.3897	−.4931	−.2263
Waist	.8702	1.0000	−.3529	−.5522	−.6456	−.1915
Pulse	−.3658	−.3529	1.0000	.1506	.2250	.0349
Chins	−.3897	−.5522	.1506	1.0000	.6957	.4958
Situps	−.4931	−.6456	.2250	.6957	1.0000	.6692
Jumps	−.2263	−.1915	.0349	.4958	.6692	1.0000

Table 12.4. Linnerud Data. Regression Analysis

	Chins	Situps	Jumps
Weight	.3683	.2872	−.2590
Waist	−.8818	−.8898	.0146
Pulse	−.0258	.0161	−.0546
Standard Error of Estimate	.8856	.8180	1.0599
Multiple Correlation Coefficient R	.5827	.6607	.2322

12.2.3 Enhancements and Alternatives for OLS

While this section concerns methods of dealing with collinearity, some quick measures and indications of collinearity would also be welcome and they involve principal components. We have alluded to these in Section 12.2.1. One, in particular, is the index of the matrix $\mathbf{X}$, $\sqrt{l_1/l_p}$. If one performs a PCA on the predictor variables, the more separated the characteristic roots, the more correlation there will be between the predictors and this will be reflected in this index. In particular, if one or more roots are equal to zero, it indicates that some of the predictors are linear combinations of some of the others. The solution might be to eliminate some of the predictor variables. An examination of the vectors associated with the smallest roots may be useful (Mason and Gunst, 1985). Variate reduction will be discussed in Sections 12.3.4 and 14.4. Mason and Gunst also suggested scatter plots of the pc's associated with the larger roots, looking for clusters and extreme values in two or more dimensions.

In addition to using PCA as a guide to eliminating one or more redundant predictor variables, a number of other techniques have been proposed to simplify regression models when collinearity exists among the predictors. Details may be found in many regression texts. The most common are sequential procedures, which will be found to some extent in all of the common statistical computer packages. *Forward selection* consists of determining that predictor that, by itself, has the highest correlation with the response. This is put into the model. Then the predictor that, with the predictor already chosen, has the highest correlation with the response is selected, and so on. *Backward elimination*, on the other hand, starts with the full regression equation and eliminates that predictor that contributes the least with the others still in the model; then the next weakest contributor is eliminated, and so on. Both of these procedures employ tests of significance to indicate when the sequential process should be terminated. These two procedures, in many cases, will not produce the same set of predictors. *Stepwise regression* is a combination of these; it is, essentially, a forward selection procedure with the option at each step of eliminating any predictor so far selected that at this point is no longer making a significant contribution. For details see Efroymson (1960). Since these procedures are somewhat automatic, they can, and have been, abused considerably.

Somewhat different is the so-called "all possible combinations" technique. The analyst may have already decided that some subset, k, of the set of p predictors should be optimal, possibly from a PCA or some other data analysis. These procedures would then evaluate all combinations of k predictors and select the one that produced the smallest standard error of estimate. Of particular utility in this procedure is the C_p-statistic due to Mallows (1973). Since these procedures can be lengthy for a large number of predictors, the use of the QR algorithm, proposed by Smith and Bremner (1989) may be worthwhile.

Another development is *ridge regression* (Hoerl and Kennard, 1970; McDonald, 1980; Hoerl, 1985). In this technique, the estimator for the regression coefficients given in (12.2.2) are replaced by the equation

$$\mathbf{b}_r = [\mathbf{X}'\mathbf{X} + c\mathbf{I}]^{-1}\mathbf{X}'\mathbf{y} \qquad (12.2.5)$$

where c is a small positive number. This procedure adds a small constant to $X'X$ before the inversion process which reduces the variance inflation factors and the mean square error. This enhances the chances of getting a better inverse at the cost of obtaining biased estimates of the regression coefficients. The chief problem related to ridge regression is the choice of c. The next section will deal with principal components regression, which is also a biased procedure. However, PCR has a sharp cut-off (i.e., a pc is either included in the model or it isn't) while ridge regression merely reduces the weights in general.

Since both principal components and ridge regression are concerned with multicollinearity, it is not surprising that the techniques are somewhat related. It can be shown, for instance, that

$$\mathbf{b}_r = \mathbf{U}[\mathbf{L} + c\mathbf{I}]^{-1}\mathbf{L}\mathbf{U}'\mathbf{b}_{OLS} \tag{12.2.6}$$

where $\mathbf{U}$ and $\mathbf{L}$ are the characteristic roots and vectors of $\mathbf{X}'\mathbf{X}$ (Hsuan, 1981). Hoerl and Kennard showed that the expected value of the mean square error is

$$E(\mathbf{b} - \boldsymbol{\beta})'(\mathbf{b} - \boldsymbol{\beta}) = \sigma^2 \sum_{i=1}^{p} (1/\lambda_i) \tag{12.2.7}$$

where $\boldsymbol{\beta}$ are the population regression coefficients and σ^2 is the residual variance. They also showed that σ^2/λ_p is a lower bound.

Marquardt and Snee (1975) pointed out that the expected improvement in ridge regression over OLS depends on the orientation of the true regression function relative to the principal axes defined by the characteristic vectors of $\mathbf{X}'\mathbf{X}$ and will be best when it coincides with the first vector. McDonald and Galarneau (1975) performed a simulation in which the regression models they created were based on either the first principal component, which would minimize the mean square error, or the last principal component, which would maximize it. They found that the models dominated by the largest pc would always produce a smaller mean square error than OLS while for models dominated by the smallest pc this was not always the case.

The sensitivity of PCA estimates to individual observations will be taken up in Section 16.4. For ridge regression, Walker and Birch (1988) showed that the ridge *leverage*, or effect, of the ith observation is equal to

$$h_i^* = \sum_{j=1}^{p} \frac{z_{ij}^2}{l_j + c} \tag{12.2.8}$$

and showed that h_i^* is bounded by

$$\left(\frac{l_p}{l_p + c}\right)h_i \leqslant h_i^* \leqslant \left(\frac{l_1}{l_1 + c}\right)h_i \tag{12.2.9}$$

where h_i is the ith diagonal element of $\mathbf{X}[\mathbf{X}'\mathbf{X}]^{-1}\mathbf{X}'$. They also gave an expression for the influence on the residuals and illustrated these quantities with the Longley data. Steece (1986) found that leverage was a monotonically decreasing function of the ridge parameter c. Reduction was greatest for observations having large scores for the pc's associated with the smaller roots so that ridge regression downweighted their influence.

Vinod (1976) proposed a method of determining c using a "multicollinearity allowance" that is a function of the characteristic roots and the number of pc's that would have been used in a PCA model. Hsuan (1981) established some bounds on the difference between ridge regression and PCR and proposed two estimators of c, both based on the characteristic root associated with the last pc included in a PCA model and the next smaller root. An adaptive estimation procedure involving the characteristic roots was proposed by Wang and Chow (1990) which also had a smaller mean square error then OLS under certain conditions.

There have been a number of empirical studies comparing one or more of these various methods with OLS as well as with each other. Hoerl et al. (1986) compared eight techniques (OLS, PCR, stepwise regression, two ridge regression, and three ridge selection schemes) in three data sets with responses simulated to reflect various conditions. They also compared their conclusions with those of many of the previous studies. Trenkler and Trenkler (1984) also compared a number of estimates.

None of these are cure-alls. If multicollinearity exists, there is a problem and although these techniques may alleviate the problem they will not make it go away. If there is more than one response, there is the additional problem of either using one or more of these techniques on each response separately or attempting to find a single solution that will suffice for all of the responses.

12.3 PRINCIPAL COMPONENTS REGRESSION

12.3.1 The Method of PCR

It may appear in this section that you are being shown how to do something and then are told not to use it. The details of PCR are necessary because it is a widely used technique—for better or for worse. PCR may be quite useful but it also has some serious drawbacks, which will be pointed out. To correct some of these drawbacks, some newer techniques have been developed. These include latent root regression (Section 12.3.4), partial least-squares regression (Section 12.5), and maximum redundancy (Section 12.6.4).

In principal components regression, one must first transform the predictor variables to principal components and then regress these against the original responses. Any scaling options for the pc's may be employed but for this chapter we shall use the relation:

$$\mathbf{z} = \mathbf{x}\mathbf{U} \qquad (12.3.1)$$

where $U'U = I$. This scales all of the vector coefficients to ± 1. The principal component scores for the Linnerud data will be the $20 \times k$ matrix Z and, initially, all three pc's will be used, so $k = 3$. To do *principal component regression* (PCR), one first transforms the original data into pc's and then obtains

$$y = zb_z \tag{12.3.2}$$

where b_z denotes the regression coefficients obtained by using principal components.

The rationale for principal components regression relates to the problems of multicollinearity of the predictors with its effect on the inverse of $X'X$, the standard errors of the regression coefficients and the correlations among these coefficients. The use of PCR avoids most of this. The matrix $Z'Z$ is diagonal, so the elements of the inverse are merely the reciprocals of the diagonal elements. The regression coefficients relating the pc's to the responses will have minimum standard errors since the predictors are uncorrelated and, for the same reason, the regression coefficients will be uncorrelated. If all of the pc's are used, PCR will predict the responses with exactly the same precision as OLS. (For numerical precision, PCR may be superior to OLS because it does not require the inverse of $X'X$.) However, a further possible benefit is that of model simplification due to the use of PCA particularly if fewer than a full set of pc's are used.

Other than dealing with multicollinearity, what does PCR have going for it? At the very best, the pc's are so readily interpretable that they become the new variables in the prediction model. At the very worst, they are not interpretable at all but one can still relate the responses to the original predictors as follows:

$$y = zb_z = xUb_z \tag{12.3.3}$$

since $z = xU$ and therefore:

$$b = Ub_z \tag{12.3.4}$$

This holds whether or not a full set of pc's is used, although the results will be the same as OLS only if the full set is used. (The relation is reversible: $b_z = U'b$.)

One need not compute the pc scores at all if they are only required as a means to an end. One can replace equation (12.2.2) with

$$b_z = [U'X'XU]^{-1}U'X'y \tag{12.3.5}$$

or, if none of the intermediate results is relevant,

$$b = U[U'X'XU]^{-1}U'X'y \tag{12.3.6}$$

For the case of the Linnerud data, all three characteristic vectors were obtained. The characteristic roots were $l_1 = 2.1041$, $l_2 = .7662$, and $l_3 = .1296$. (These pc's accounted for 70.1%, 25.5%, and 4.3%, respectively, of the total variability of the *normalized* predictors.) The characteristic vectors, normalized to 1.0, are

$$\mathbf{U} = \begin{bmatrix} .6437 & .2864 & .7096 \\ .6410 & .3048 & -.7045 \\ -.4181 & .9083 & .0126 \end{bmatrix}$$

The percent variability of each predictor accounted for by each pc is given in Table 12.5. The first pc accounts for most of the first two predictors while the second pc accounts for most of the third.

The matrix of regression coefficients, $\mathbf{B}_z$, using three pc's is given in Table 12.6 and the residual sums of squares and crossproducts are

$$\mathbf{Y'Y} - \mathbf{Y'ZB}_z = \begin{bmatrix} 12.5481 & 5.9626 & 7.8114 \\ 5.9626 & 10.7066 & 10.7013 \\ 7.8114 & 10.7013 & 17.9758 \end{bmatrix}$$

which is identical to $\mathbf{Y'Y} - \mathbf{Y'XB}$ obtained from the OLS solution in Section 12.2.2.

Although the correlations in this example are modest at best, we should constantly remind ourselves that high correlations may be caused by just one or two observations in a sample and we will have an example of this in Section 12.6.2. Dudzinski (1975) obtained scatter plots of pc's vs. responses for some sheep and cattle grazing data and found a number of cases of this phenomenon. With the facilities now available for graphics techniques, one should consider

Table 12.5. Linnerud Data. Variability of Each Variable Accounted for by Each Principal Component

	Weight	Waist	Pulse
z_1	87	86	37
z_2	6	7	63
z_3	7	7	0

Table 12.6. Linnerud Data. PCR Regression Coefficients

	Chins	Situps	Jumps
z_1	$-.3173$	$-.3922$	$-.1345$
z_2	$-.1868$	$-.1744$	$-.1194$
z_3	$.8822$	$.8308$	$-.1948$

doing this whenever it seems appropriate as long as it does not become a substitute for inferential techniques. The effect of individual observations on PCR estimates was investigated by Naes (1989).

Since the principal components are uncorrelated, the standard error of the ith regression coefficient is $s_{y \cdot x}/\sqrt{l_i}$; hence, the coefficients relating to the smaller pc's will have the larger standard errors. For this example, the standard error for the regression of z_1 on Chins is $.8856/\sqrt{2.1041} = .61$. The other two for Chins are 1.01 and 2.46. While this may cast doubt on the coefficients related to the third pc, z_3, there are some other considerations, which will be discussed in the next section.

12.3.2 Deletion of Principal Components

One of the advantages of PCA is its potential to represent a set of variables in a lower-dimensional space. However, all is not as it may seem when one uses this technique in regression. A PCA of the Linnerud data shows that the first pc accounted for roughly 70% of the total variability of the predictors, the second pc accounted for about 26% and the last pc, 4%. One could be tempted to drop the last pc and use just the first two as predictors. However, the manner in which the pc's are obtained for the predictors is unaffected by the responses. It is conceivable, for some pathological case, that the last pc could be perfectly correlated with the response and the remaining ones completely uncorrelated.

What happens if just the first two pc's are used as predictors? Or just the first pc? First, let us drop z_3, the third pc, as a predictor. In PCR this is easy; just drop it out of the equation. Since the pc's are uncorrelated, the deletion of z_3 will leave the other regression coefficients unchanged. The residual sum of squares and products using only two pc's as predictors is

$$\begin{bmatrix} 14.4652 & 7.7680 & 7.3882 \\ 7.7680 & 12.4069 & 10.3027 \\ 7.3882 & 10.3027 & 18.0693 \end{bmatrix}$$

As one would expect, these are larger than the ones obtained using all three pc's as predictors but not by very large amounts. (Note that the two crossproduct terms related to y_3 have actually increased a bit!) To obtain regression coefficients in terms of the original predictors, employ equation (12.3.4):

$$\mathbf{B} = \mathbf{U}\mathbf{B}_z = \begin{bmatrix} .6437 & .2864 \\ .6410 & .3048 \\ -.4181 & .9083 \end{bmatrix} \begin{bmatrix} -.3173 & -.3922 & -.1345 \\ -.1868 & -.1744 & -.1194 \end{bmatrix}$$

$$= \begin{bmatrix} -.2577 & -.3024 & -.1208 \\ -.2603 & -.3046 & -.1226 \\ -.0370 & .0056 & -.0522 \end{bmatrix}$$

Table 12.7. Linnerud Data. Proportion of Response Variability Explained by Principal Components

	Chins	Situps	Jumps
Explained by			
z_1	21.2	32.4	3.8
z_2	2.7	2.3	1.1
z_3	10.1	8.9	.5
Unexplained	66.0	56.4	94.6

which does not bear much resemblance to the original **B**. However, its predictive power is almost as good, as has already been seen.

If only the first pc, z_1, is used as a predictor, the residual matrix is

$$\begin{bmatrix} 14.9731 & 8.2421 & 7.7128 \\ 8.2421 & 12.8496 & 10.6058 \\ 7.7128 & 10.6058 & 18.2767 \end{bmatrix}$$

which is not an appreciable difference. An accounting for the variability of the responses explained by the principal components is revealed in Table 12.7.

The third pc explains more than the second for two of the responses. Granted this is not a very strong relationship in the first place, but the point is that one should not expect pc's to explain variability in responses in the same order that they do for the variables from which they were formed. Section (12.3.4) will touch on deleting variables, but from what has been seen so far, it would seem that pulse rate is doing very little in these regression models and could safely be eliminated.

The PCR coefficients in terms of the original variables may also be obtained by the expression

$$\mathbf{B}_{pc,k} = [\mathbf{I} - \mathbf{U}_k \mathbf{U}_k'] \mathbf{B}_{OLS} \tag{12.3.7}$$

where $\mathbf{U}_k$ represent the set of characteristic vectors with the first k vectors deleted (Hocking, 1976). For example, if the PCR model were based on the first two pc's, $k = 2$ so only the third vector would be included in (12.3.7), viz.,

$$\mathbf{B}_{pc,2} = \left\{ \begin{bmatrix} 1 & 0 & 0 \\ 0 & 1 & 0 \\ 0 & 0 & 1 \end{bmatrix} - \begin{bmatrix} .7096 \\ -.7045 \\ .0126 \end{bmatrix} [.7096 \quad -.7045 \quad .0126] \right\} \mathbf{B}_{OLS}$$

$$= \begin{bmatrix} -.2577 & -.3024 & -.1208 \\ -.2603 & -.3046 & -.1226 \\ -.0370 & .0056 & -.0522 \end{bmatrix}$$

which is the same result as obtained above. In this context, one needs only the set of characteristic vectors and the original OLS solution to produce the PCR solution.

Finally, what is the effect on the standard errors of the regression coefficients? Recall from (12.2.4) that the standard errors of the OLS regression coefficients may be obtained using the characteristic roots and vectors. For this example using Chins as the response, the standard errors of the regression coefficients for Weight, Waist, and Pulse are 1.81, 1.80, and .95, respectively, using OLS (or PCR with all three components). If the third component is dropped, the summation in (12.2.4) will go from 1 to $k = 2$, there will be a new standard error of estimate, and the standard errors of the regression coefficients are now .51, .52, and .99. If only the first pc is used, these become .40, .40, and .36. These are rather dramatic reductions caused, primarily, by the deletion of the third component, which has such a small root. Examples of this phenomenon were first pointed out by Massey (1965) using Census data. Although these standard errors are reduced by eliminating pc's in the model, the estimates are biased and this bias increases as more pc's are dropped. These two quantities make up the mean square error and the choice of the number of components is a trade-off between them. For this example, the choice seems clear.

Naes and Martens (1988) used cross-validation as a stopping rule but in this case it was performed on the principal components rather than the original variables.

12.3.3 Cautions in Using PCR

History has not recorded the first use of PCR but one would suspect that it did not take long for someone to conclude that this technique might be their salvation in some thorny regression problem. However, Hotelling (1957) pointed out, as we have already noted, that there is no reason for the pc's with the large roots to be the best predictors and demonstrated geometrically what deleting a pc would do to the model. However, even now, the practice is widespread of automatically dropping, as predictors, those pc's with small roots. Jolliffe (1982), in addition to compiling a list of statisticians who had advocated this practice in the past, gave four examples where this would have been unwise. A PCR of data representing chemical mixtures of two or more pure components by Martens et al. (1980) produced a case where the second pc was a better predictor than the first and Jolliffe (1986, p. 135) gave an example where two of the smaller components were among the better predictors.

A case could be made for dropping the small ones if they accounted only for the inherent variability. In this case, the regression estimates could be thought of as being unbiased in a practical rather than a mathematical sense (Basilevsky, 1981). (Factor analysis practitioners advocate a similar philosophy.) One would recommend exercising caution with this concept but it could make sense in applications, such as quality control, where the sources of inherent variability may be identified.

The answer to these problems would appear to be to obtain all of the pc's and regress the response on all of them. Since they are independent of each other, the amounts of the response variability accounted for by each of the pc's are also independent and hence the pc's could be ranked in order by that criterion and a cutoff employed when the desired residual has been obtained. Doing it this way minimizes the mean square error of the regression coefficients. This will also minimize the MSE for fixed k and would seem to be the recommended approach unless p is very large. (Keep in mind, however, that some of the smaller pc's may not be very stable or that one or more of them may represent near singularities.) Rather than use residual variability as a criterion, Gunst and Mason (1977b) considered the effect of dropping pc's on the mean square error and stated that the MSE will be reduced by including each pc for which

$$(\mathbf{u}_i'\mathbf{b}_{OLS})^2 < (\text{OLS Residual MS})/l_i$$

In a similar manner, Lott (1973) employed the maximum of the multiple correlation coefficient using a stepwise procedure with the pc's. Some other decision rules for use with PCR and their effect on the MSE were given by Belinfante and Coxe (1986). Procedures for comparing PCR models were developed by Hill et al. (1977).

Although PCR may present some operational difficulties, it does possess a number of desirable properties:

1. Using a complete set of pc's, PCR will produce the same results as the original OLS but with possibly more accuracy if the original $\mathbf{X}'\mathbf{X}$ matrix has inversion problems.

2. If $\mathbf{X}'\mathbf{X}$ is nearly singular, using a reduced set of pc's will produce a viable solution that OLS may not because of the variance inflation factor. If $\mathbf{X}'\mathbf{X}$ is singular, the vector associated with the zero root may furnish a suggestion about deletion of one or more of the original variables. (See Section 12.2.4.)

3. By virtue of the fact that the pc's are uncorrelated, straightforward significance tests may be employed that do not need be concerned with the order in which the pc's were entered into the regression model. The regression coefficients will be uncorrelated and the amounts explained by each pc are independent and hence additive so that the results may be reported in the form of an analysis of variance.

4. If the pc's can be easily interpreted, the resultant regression equations may be more meaningful.

5. There have been some cases where weighted PCR has been employed. Iglarsh and Cheng (1980) gave a form of weighted PCR that was a weighted sum of OLS and PCR and, for some econometric studies, had a smaller MSE than did OLS. Lee and Birch (1988) also used weighted PCR, showing that this was essentially the same as ridge regression.

6. Although PCR estimates will be biased for $k < p$, Greenberg (1975) and Fomby et al. (1978) showed that $\text{Tr}[\mathbf{b} - \boldsymbol{\beta}]'[\mathbf{b} - \boldsymbol{\beta}]$ will be less using PCR with the last $p - k$ pc's deleted than any other restricted least-squares estimator with $p - k$ restrictions.

7. Another, possibly dubious, claim for PCR (as well as most of the remaining methods to be discussed in this chapter) is that it can be performed when the number of predictor variables is greater than the number of observations since there cannot be more than $\min(p, n)$ pc's anyhow.

Examples of PCR in the physical sciences are probably less susceptible to many of the problems alluded to above. Naes (1985) used PCR for multivariate calibration where the covariances of the residuals are structured. Delaney (1988), using PCR with gas chromotography, obtained a single-component model that would determine the minimum concentration that could be detected. Mager (1980) used PCR for multivariate bioassays and included the relevant computational details.

12.3.4 Use of PCR in Variate Reduction

The use of PCA for help in deleting variables from a set will be discussed in Section 14.4. In this section, this topic will be discussed from the standpoint of regression analysis. The techniques to be discussed use principal components solely as a means to an end and assume, apparently, that a regression equation in terms of the pc's is not relevant and that the real goal is to delete as many predictor variables as seems prudent. Sometimes this can be done by examining the pc's, particularly some of the smaller ones that are suspected of representing multicollinearities. Wallis (1965) and Daling and Tamura (1970) obtained the pc's in the normal manner and then subjected them to a varimax rotation to obtain simple structure, concluding that this might facilitate making judgments about variables. (The principles of Section 8.4.2 still hold. Components obtained by rotation will account for exactly the same amount of variability of the responses as will the original pc's.) Mansfield et al. (1977) used backward elimination in connection with the pc's having the largest roots. Mason and Gunst (1985) gave a procedure for detecting outliers that cause collinearities. These procedures, however, still do not take into account the relationship between the pc's and the response variable.

To that end, Hawkins (1973) and Webster et al. (1974) proposed a technique called *latent root regression analysis* (LRRA). In this scheme, the response variable is added to the covariance or correlation matrix of predictor variables for a principal component analysis. The idea here is that any pc (here called *LRRA component*) that has a large coefficient corresponding to the response variable is important from the standpoint of the predictor variables that also have large coefficients. Gunst et al. (1976) concluded that LRRA is preferable to OLS when collinearities exist because the results are more stable even though they are biased. Webster et al. and Gunst et al. combined this with a backward selection technique to determine the predictor variables that remain in the model. This was enhanced with some asymptotic results by White and Gunst (1979), which gave more credence to these decision rules and also those used in identifying multicollinearities, the focal point of a paper by Gunst and Mason (1977a). For a review of these techniques, see Gunst (1983) and Mason (1986). In particular, any LRRA component that has a zero root indicates a linear restriction among the variables. If the vector has a sizeable coefficient for the response variable, it has the potential for a useable regression equation all by itself. While those with small roots may represent inherent variability, they may possibly be collinearities with a little error sprinkled in. So much for throwing the little ones away!

A latent root regression analysis of the Linnerud data using Situps, y_2, as the response variable, yields the following set of characteristic vectors, where the first element in each column vector corresponds to the response:

$$
\mathbf{U} = \begin{bmatrix}
-.47 & -.36 & .78 & .20 \\
\hline
.56 & .07 & .54 & -.63 \\
.59 & .17 & .24 & .75 \\
-.34 & .92 & .22 & -.01
\end{bmatrix}
$$

The characteristic roots are 2.55, .82, .52, and .11. For this example, nothing new emerges: $\mathbf{u}_1$, $\mathbf{u}_2$, and $\mathbf{u}_4$ essentially mirror the three original characteristic vectors of $\mathbf{X}'\mathbf{X}$; $\mathbf{u}_3$ accounts for the unexplained variability in the response. If there is a clue about deleting x_3 from the model, it should come from $\mathbf{u}_2$ but it is not very convincing. Another worked example may be found in Draper and Smith (1980, p. 332).

In his original paper, Hawkins used varimax rotation on the LRRA components. Interpretation to some of the matrix relationships was added by Eplett (1978) and some examples were given by Jeffers (1981). Hawkins later proposed a variation of LRRA called the *Choleskey inverse root* method (CIR), which was seen to be an improvement both in the ease of handling and interpretation (Hawkins and Eplett, 1982; Hawkins and Fatti, 1984).

In most of these procedures, the predictor variables are generally deleted one at a time, although sometimes it is possible to delete two or more in one step. That done, the remaining variables, again along with the response variable, are subjected to another PCR, and so on until the final solution is obtained. For another discussion of selection rules for PCR *and* LRRA, see Coxe (1982).

The use of pc-scatterplots has previously been suggested to discover patterns of nonlinearity and to study the effect of certain observations on the estimates of the characteristic roots and vectors. The use of these plots has also been suggested in connection with LRRA, not only for these reasons but as a procedure for detecting outliers (Hocking, 1984; Cerdan, 1989).

12.3.5   Other Comments with Regard to PCR

a. Singular Value Decomposition
Recall from Section 10.5 that the JK-biplot transformation, using SVD was
$X = ZU'$. Substituting into (12.3.6), along with the relation $Z'Z = L$, the
estimate for β becomes

$$b = UL^{-1}Z'y \qquad (12.3.8)$$

so the OLS regression coefficients are a function of the characteristic roots and
vectors and the corresponding z-scores of X. Since $Z = U*L^{1/2}$, (12.3.8) can be
rewritten as

$$b = UL^{-1/2}U*y \qquad (12.3.9)$$

showing that OLS may be carried out directly using SVD. (This result can also
be obtained directly from $y = Xb$ by using the Moore–Penrose generalized
inverse of X, $X^+ = UL^{-1/2}U*'$, Equation 10.3.7.)

If the rank of X is less than p, say r, then (12.3.8) or (12.3.9) still hold except
that U, $U*$, and Z will only have r columns and L will be $(r \times r)$, so this is a
viable option for OLS, particularly if X is ill-conditioned. This also means that
if one knew how many principal components, k, were to be used (assuming the
first k components were the correct ones to use) one could obtain b as modified
by PCR directly and obtain the pc-scores at the same time.

Eubank and Webster (1985) used SVD to show that certain linear combinations are estimable. Good (1969) and Nelder (1985) investigated some information
considerations of OLS using SVD. Mandel (1982, 1989) gave material on
computing standard errors and studying the nature of multicollinearities
using SVD, particularly with use of what he defined to be the *effective prediction
domain*.

b. Ridge Regression
As we have seen in Section 12.2.3, not only can PCA be used as a guide to the
choice of c in ridge regression but it may be used in conjunction with it. Baye
and Parker (1984) introduced the *k-c class estimator* of β:

$$b_k(c) = U(U'X'XU + cI)^{-1}U'X'y$$
$$= U[L + cI]^{-1}Z'y \qquad (12.3.10)$$

where k is the number of pc's employed, U is $(p \times k)$ and c is the constant
associated with ridge regression. This combines PCR and ridge regression into
a single form that Baye and Parker claimed should be better than either one
of them. Hawkins (1975) had previously obtained a similar expression in terms
of latent root regression and noted that in this case the inclusion of c in the

expression changed only the roots, not the vectors. Hawkins suggested that it might be possible for each pc to have its own value of c. Hocking (1976) and Hocking et al. (1976) showed that the Stein and Sclove estimators are modifications of PCR and that ridge regression includes all of these, including PCR, as special cases, but stated that PCR should still be employed to identify multicollinearities.

c. Partial Least Squares
Another attempt to obtain components in a manner that directly reflects the relationship between the predictors and the response is a technique called *partial least squares regression*. In this procedure, as each "component" is obtained, the $X'X$ matrix is corrected by the relationship of that component and the response. As this technique, in a more expanded version, is also applicable when there are multiple response variables, it will be deferred until Section 12.5.

d. Triangularization Methods
This technique, a substitute for PCA, is discussed in Section 18.3 and consists of transforming the original variables into a set of new variables in which the first involves one of the original variables, the second, two, and so on. These "components" may be used for predictors as a substitute for PCR. They have one of the PCA advantages of being uncorrelated but they are not orthogonal and hence some of the computational advantages of PCR is lost. It has no advantage over OLS unless the components can be interpreted to the extent that they may be used for predictors in their own right.

e. Bayesian Methods
We shall find, in Section 16.9.3, that very little has been accomplished so far with regard to Bayesian PCA, so it is not surprising that even less has been done with regard to PCR. Leamer and Chamberlain (1976) showed that the mean or mode of k-dimensional regression coefficients is a weighted average of $k + 1$ pc-scores and also showed that a spherical prior can be associated with methods that drop pc's. Oman (1978) obtained a modified Stein estimate that was smoother than PCR. Some other properties of Bayesian PCR were investigated by Chen (1974) and Whittle and Adelman (1982).

12.4 METHODS INVOLVING MULTIPLE RESPONSES

The techniques described so far in this chapter do not take into account whether there is a single response or more than one. OLS and PCR both operate on the $X'X$ matrix before taking into consideration the relationship between the predictors and the responses. The set of characteristic vectors of the predictors will be exactly the same whether there is one response or twenty and, further, this set does not take into account the relationship of the predictors to those responses nor of the responses among themselves. The methods given in the remainder of this chapter address both of these problems. This situation is sometimes referred to as the *two-block model*. A unifying discussion of several such models, including some to be discussed in this chapter and in Section 17.8, may be found in a book on partial least squares by Lohmöller (1989) which contains approximately 500 references.

There are occasions where PCA has been used for both prediction and response variables. In this case, the pc's for the predictors are obtained in one operation, the pc's for the responses in another, and the one set of pc's regresssed against the other (Schall and Chandra, 1986; Jolliffe, 1986, p. 142). One reason for doing this is to obtain model simplification. Since the pc's within each set of pc's are independent, the regression of the one set on the other would have the advantage that the regression coefficients would be independent and interpretable, if the original pc's were interpretable. Further, if one wished to relate out-of-control responses back to the predictors, this also might be more interpretable. However, this is not an optimal solution, as we shall see in Section 12.6, and may compound the problem of discarding relevant components.

It would be advisable to obtain the first canonical correlation coefficient (Section 12.6.2) and/or a measure of redundancy (Section 12.6.3) between the original predictor and response variables so that one would know how well the double-PCA model compared with the optimum. Obviously, there is little advantage to this method if the pc's are not interpretable, but if they are interpretable and recognized as such by those involved, the method might have merit. Schall and Chandra employed this technique for a process control application similar to the calibration problem.

For any multiple regression using multiple responses, such as OLS or PCR, where the predictors do not take into account the correlations among the responses, Gnanadesikan and Kettenring (1972) suggested performing a PCA on the response residuals to see whether they exhibit any interesting structure. Naes (1985) gave a procedure for estimation when the residual structure is a function of the order in which the observations are obtained.

12.5 PARTIAL LEAST-SQUARES REGRESSION

12.5.1 The Method of PLSR

The use of *partial least-squares regression* (PLSR) as a regression technique has been promoted primarily within the area of chemometrics although it would be equally useful in any application that had multiple predictors (Wold, 1982, 1985). Among some applications of PLSR are ones by Lindberg et al. (1983) involving impurities in water samples, Martens et al. (1984) dealing with the colors of some samples of jam, Frank et al. (1984) in a quality control application, Martens and Naes (1985) on infrared analysis of wheat samples, Lindberg et al. (1986) involving liquid chromatography, Brakstad et al. (1988) involving gas chromotography, and Persson et al. (1986) on the analysis of peat.

Although the material on PLSR has been left until this part of the chapter because it does take into account the case of multiple responses, it will also handle a single-response variable. Most of the development of this technique has been done in the Scandinavian countries and has been built on the NIPALS procedure, an alternating least-squares algorithm for obtaining principal components. (Recall from Section 10.3, that for NIPALS, "least squares" referred to the regression of the original variables on the principal components producing the characteristic vectors, not the relationship between predictors and responses, which is what PLSR will do.) The PLSR technique operates in somewhat the same way as PCR in that a set of vectors are obtained from the predictor variables. What makes PLSR different is that as each vector is obtained, it is immediately related to the responses and the reduction in variability among the predictors. The estimation of the next vector takes that relationship into account. Simultaneously, a set of vectors for the responses is also being obtained that also takes this relationship into account. For this reason, PLSR has received the nickname of "criss-cross" regression (Wold, 1966a).

The PLSR technique has often been presented by its proponents as an algorithm rather than as a formal linear model. The procedure is considerably more complicated than PCR and although software is available for both mini- and microcomputers, many users are inclined to think of it as a "black box."

In the account that is to follow, the notation will, as much as possible, follow that which appears in the applications literature, in particular, a paper by Geladi and Kowalski (1986). X and Y, are, as before, data matrices of size $n \times p$ and $n \times q$ respectively. The principal feature of this technique is that *two* operations will be carried out together:

$$X = TP + E \qquad (T, E \text{ are } n \times k, P \text{ is } k \times p) \qquad (12.5.1)$$

$$Y = UQ + F^* \qquad (U, F^* \text{ are } n \times k, Q \text{ is } k \times q) \qquad (12.5.2)$$

$k \leqslant p$ is the number of vectors associated with X. E is the matrix of residuals of X at the kth stage. When $k = p$, $E = 0$. F^* is an intermediate step in obtaining the residuals for Y at the kth stage.

In the SVD associated with PCA, the matrices Q and P would be the characteristic vectors and the matrices T and U would be the principal component scores. Although they do not enjoy the same properties in PLSR, they may still be thought of in that vein; the matrices T and U are referred to as "X-scores" and "Y-scores," respectively. In PLSR, a prediction equation is formed by replacing $\mathbf{U}$ by $\mathbf{TB}$ ($\mathbf{B}$ is $k \times k$), thus producing:

$$\mathbf{Y} = \mathbf{TBQ} + \mathbf{F} \qquad (12.5.3)$$

This $\mathbf{F}$ is the actual matrix of residuals for $\mathbf{Y}$ at the kth stage. The substitution takes place after each new component is added to the model rather than at the end of the process.

The algorithm required to perform these operations, sometimes referred to as PLS2 (PLS1, is the abbreviated version when there is only one response variable), is described in Wold et al. (1983), Lindberg et al. (1983), Martens and Naes (1985), and the aforementioned paper of Geladi and Kowalski. The paper by Martens and Naes compares various methods such as PCR, PLS, and others, contains material on sensitivity analysis of predictor variables, and also contains some material on various measures of inherent variability useful for detecting outliers and abnormal predictor variables. Other comparisons have been made by Naes and Martens (1985) and Naes et al. (1986). Hui and Wold (1982) showed the conditions under which PLSR estimates are consistent. Some more recent theoretical foundations for PLSR were given by Helland (1988), Hosküldsson (1988), and Lohmöller (1989, Chap. 3).

Stone and Brooks (1990) proposed a technique called *continuum regression*. Intrinsic in this scheme is the number of predictors and a parameter bounded between 0 and 1. They solve for both quantities by use of cross-validation. For parameter values of 0, .5 and 1, respectively, OLS, PLSR and PCR fall out as special cases.

Of course, none of these techniques is a substitute for common sense. One should always study the results carefully, including the residuals. Phelan et al. (1988) gave an example using infrared spectroscopy where, because of an interaction, stepwise regression with two variables did as well as PLSR with three components.

12.5.2 The PLS2 Algorithm

As before, there will be n observations on p predictor variables and q response variables. (Unfortunately, the letters p and q will also be used to denote some of the vectors but in that case they will always be subscripted and in boldface.) The subscript "h" will refer to the hth stage in the process of adding dimensions to the model, $1 \leqslant h \leqslant k \leqslant p$.

The vectors $\mathbf{u}_h$ and $\mathbf{t}_h$ are both of size $n \times 1$, the vectors $\mathbf{w}_h$ and $\mathbf{p}_h$ are both $p \times 1$, the vector $\mathbf{q}_h$ is $q \times 1$, and the matrices $\mathbf{E}_h$ and $\mathbf{F}_h$ are $n \times p$ and $n \times q$ respectively.

The first step is to normalize both $\mathbf{X}$ and $\mathbf{Y}$ by subtracting the mean from each variable and dividing by its standard deviation. These matrices will be denoted by E_o and F_o. Then for each dimension, h, the following operations are performed:

(1) $w_h' = \dfrac{u_h' E_{h-1}}{\|u_h' u_h\|}$

where $\|a\| = \sqrt{a'a}$. This is the least-squares solution of $u_h = w_h' E_{h-1}$. When obtaining the first estimates for the first dimension (i.e. $h = 1$), E_{h-1} will be E_o, the original data in normalized form. The first approximation to u_h is often chosen from one of the columns of F_o or merely a vector of 1's.

(2) $t_h = E_{h-1} w_h$

This is the least-squares solution of $E_{h-1} = t_h w_h'$.

(3) $q_h' = \dfrac{t_h' F_{h-1}}{\|t_h' t_h\|}$

This is the least-squares solution for $F_{h-1} = t_h q_h'$.

(4) $u_h = F_{h-1} q_h$

(5) Check the convergence of t_h. If t_h has converged, go to step (6); if not, return to step (1) except that a trial value of u_h will no longer be necessary. (If there is only one response variable, steps (3–5) are omitted. This is the PLS1 procedure).

(6) $p_h' = \dfrac{t_h' E_{h-1}}{t_h' t_h}$

(7) Renormalize p_h', t_h, and w_h':

$$p_{h,\,new}' = \frac{p_{h,\,old}'}{\|p_{h,\,old}'\|}$$

$$t_{h,\,new} = t_{h,\,old} \|p_{h,\,old}'\|$$

$$w_{h,\,new}' = w_{h,\,old}' \|p_{h,\,old}'\|$$

(8) $b_h = \dfrac{u'_h t_h}{t'_h t_h}$

(9) $E_h = E_{h-1} - t_h p'_h$ and $F_h = F_{h-1} - b_h t_h q'_h$

(10) The vectors p_h, w_h, and q_h are saved to begin building up the matrices **P**, **W**, and **Q**. **B** is a diagonal matrix with b_h forming the diagonal elements. E_h and F_h are the residual **X** and **Y** matrices after stage h and will be used in stage $h + 1$ should the decision rule dictate that to be the appropriate action.

(11) A number of techniques have been suggested for evaluating the adequacy of the model as a function of h. These will be discussed in Section 12.5.5.

A few properties are worth noting. The matrices **P** and **Q** are normalized to unit length but are not orthogonal. The matrices **W** and **T** are orthogonal but are not normalized to unit length. The matrix **U** has neither of these properties.

12.5.3 Prediction with PLSR

For most of the methods discussed in this chapter, once the initial relational estimates have been obtained, using them for predictions on a new set of data is quite straightforward. Not so for PLSR. Here one must go through some steps similar to those of the original estimation procedure. Assume that there are k "components" in the model. For $h = 1$ to k, perform the following operations:

$$t_h^* = E_{h-1}^* w_h \qquad (12.5.4)$$

$$E_h^* = E_{h-1}^* - t_h^* p'_h \qquad (12.5.5)$$

When the kth step has been completed, the predicted responses are:

$$Y^* = F_k^* + T^* BQ \qquad (12.5.6)$$

so the purpose of this operation is to produce **T***, which replaces the original **T** in (12.5.3). The distinction to be kept in mind is that the starred items, **T***, **E***, and **F***, all come from the new data set while **W**, **P**, and **B**, all come from the original data set.

12.5.4 Application of PLSR to the Linnerud Data

For the Linnerud data, all three components were obtained, producing the following set of matrices:

$$
T = \begin{bmatrix}
-.65 & .74 & .13 \\
-.78 & .21 & -.13 \\
-.92 & -.65 & -.05 \\
.70 & -.85 & -.35 \\
-.49 & 1.41 & .18 \\
-.23 & -.09 & -.02 \\
-1.42 & -.10 & .57 \\
.75 & -.26 & .03 \\
1.74 & -.82 & 1.56 \\
1.18 & .21 & -.33 \\
.37 & .87 & -.20 \\
.75 & .87 & .00 \\
1.20 & -.94 & -.34 \\
-4.45 & -.95 & -.26 \\
-.84 & 1.21 & .08 \\
-.76 & -.65 & .67 \\
-.40 & -.25 & -.56 \\
1.22 & .98 & -.09 \\
1.06 & .46 & -.32 \\
1.97 & -1.41 & -.57
\end{bmatrix}
\qquad
U = \begin{bmatrix}
-.37 & -.14 & -.28 \\
-1.34 & -1.03 & -.99 \\
-.08 & .48 & .31 \\
-.36 & -.53 & .06 \\
.46 & .83 & .41 \\
-1.31 & -1.19 & -.87 \\
-.86 & .01 & .22 \\
-.80 & -1.19 & -.74 \\
1.14 & .33 & .94 \\
3.03 & 2.03 & .41 \\
.41 & .51 & .46 \\
1.41 & .89 & .25 \\
1.53 & .81 & .87 \\
-2.22 & .15 & .36 \\
-1.50 & -.93 & -1.02 \\
1.31 & 1.58 & 1.27 \\
-1.88 & -1.57 & -1.07 \\
1.24 & .47 & .16 \\
1.61 & 1.04 & .90 \\
-1.43 & -2.56 & -1.63
\end{bmatrix}
$$

$$
Q = \begin{bmatrix}
.613 & .747 & .257 \\
.748 & .647 & .145 \\
.689 & .657 & -.307
\end{bmatrix}
\qquad
P = \begin{bmatrix}
-.656 & -.666 & .354 \\
-.016 & -.285 & -.958 \\
.658 & -.287 & .697
\end{bmatrix}
$$

$$
W = \begin{bmatrix}
-.598 & .584 & .658 \\
-.782 & -.708 & -.287 \\
.242 & -.843 & .697
\end{bmatrix}
\qquad
B = \begin{bmatrix}
.549 & 0 & 0 \\
0 & .361 & 0 \\
0 & 0 & .693
\end{bmatrix}
$$

To see how well PLSR has done, an analysis of variance can be performed on the response data. The total sums of squares and crossproducts is, as before,

$$\mathbf{Y'Y} = \begin{bmatrix} 19.00 & 13.22 & 9.42 \\ 13.22 & 19.00 & 12.71 \\ 9.42 & 12.71 & 19.00 \end{bmatrix}$$

the explained variability of $\mathbf{Y^{*\prime}Y^*}$, where $\mathbf{Y^*} = \mathbf{TBQ'}$, is

$$\mathbf{Y^{*\prime}Y^*} = \begin{bmatrix} 6.45 & 7.26 & 1.61 \\ 7.26 & 8.29 & 2.01 \\ 1.61 & 2.01 & 1.02 \end{bmatrix}$$

and the residual is

$$\mathbf{F'_3 F_3} = \begin{bmatrix} 12.55 & 5.96 & 7.81 \\ 5.96 & 10.71 & 10.70 \\ 7.81 & 10.71 & 17.98 \end{bmatrix}$$

This is the same as obtained by OLS or PCR using all three components.

If one wanted to stop when $h = 2$, the operations would be the same except that $\mathbf{T}$ and $\mathbf{F_2}$ would both be of size (20×2), $\mathbf{B}$ would be (2×2), and $\mathbf{Q}$ would be (3×2). The residual sum of squares from that operation would be

$$\begin{bmatrix} 13.57 & 6.94 & 7.36 \\ 6.94 & 11.64 & 10.27 \\ 7.36 & 10.27 & 18.18 \end{bmatrix}$$

These residuals are generally smaller than the comparable PCR case in Section 12.3.2. This is to be expected since PLSR takes the responses into consideration. Naes and Martens (1985) compared these two methods on a somewhat more theoretical basis.

Wold et al. (1987) extended these techniques to three-way and higher PLSR. An example of a third mode would be the introduction of more than one replication in a typical chemical absorption vs. concentration problem or the addition of varying experimental conditions in that type of experiment.

12.5.5 Stopping Rules for PLSR

As stated above, while there are no formal significance tests for PLSR, there are a number of stopping rules for deciding on a reduced dimensionality. Some of these are:

1. *Plotting* $\|\mathbf{F}_h\|$ *as a function of h.* This amounts to obtaining the total sum of squares of all the entries in $\mathbf{F}_h$, obtaining its square root and plotting this as a function of h. This is like a SCREE plot and would be interpreted in the same manner. For the Linnerud example, this would produce the following:

h	$\|\mathbf{F}_h\|$
0	7.55
1	6.71
2	6.59
3	6.42

which, since $p = 3$, is not very exciting but does show that the reduction in prediction variability is not monotonic, since in this case there is a greater reduction from the second to the third component than between the first and the second; this was the same experience for this set of data for two out of the three responses using PCR.

2. *Analysis of variance on the " Y-scores",* $\mathbf{u}_h$. At any stage, h, let $\mathbf{d}_h = \mathbf{u}_h - b_h \mathbf{t}_h$. Then the analysis of variance table would be:

Source	SS	d.f.
Explained	$b_h^2 \mathbf{t}_h' \mathbf{t}_h$	1
Residual	$\mathbf{d}_h' \mathbf{d}_h$	$n - 2$
Total	$\mathbf{u}_h' \mathbf{u}_h$	$n - 1$

For the case $h = 1$, this would be:

Source	SS	d.f.	MS	F
Explained	11.94	1	11.94	7.96
Residual	27.02	18	1.50	
Total	38.96	19		

This is significant at the $\alpha = .025$ level. For the case $h = 2$, $F = 1.31$, which is not significant, and for $h = 3$, $F = 3.85$, which has a probability of about .08.

3. *Analysis of variance of response data.* Although no formal procedures have been developed, something along the lines of the previous section might be employed.

4. *Cross-validation.* This is essentially a jackknife and is an extension by Lindberg et al. (1983) of the cross-validation method of Wold (1978) advocated for PCA in Section 16.3. In this method, the sample of n observations is divided into g groups. For each level of h, g PLSR solutions are obtained, each one with a different group deleted and the total prediction of that deleted group computed. These are then summed over the g groups and divided by their appropriate degrees of freedom to form the PRESS-statistic, which is plotted as a function of h. Lindberg et al. (1983) concluded that, for their particular example, the value of k tended to be understated and they considered the result obtained by cross-validation to be a lower bound on k, which they then increased by two to four components. They felt this should be stable as long as the final $k \leqslant p/3$.

12.6 REDUNDANCY ANALYSIS

12.6.1 Introduction

Maximum redundancy is another alternative technique for predicting a set of responses from a set of predictors. It is similar to partial least-squares regression in that it also takes into account the relationship between the two sets of variables in establishing the prediction equations. Like PLSR, this technique is also fairly new and is only beginning to be applied. Unlike PLSR, which has been mostly the domain of the chemists, redundancy analysis was developed by psychometricians. In order to appreciate how this technique operates, we shall first digress and introduce another useful relational technique known as *canonical correlation*, from which the concept of redundancy analysis is derived. We shall return to maximum redundancy in Section 12.6.4.

12.6.2 Canonical Correlation

The method of canonical correlations (CC) is due to Hotelling (1935, 1936b) and is customarily given a full chapter in most multivariate texts. Unlike PCR and PLSR, CC is not a prediction technique but rather a technique for portraying the relationship between two *sets* of multivariate data. In describing this method, we shall depart from our usual custom of using greek letters exclusively for population parameters. This is done primarily to distinguish the quantities used in this section from the rest of this chapter as well as to employ symbols customarily used for these quantities.

In the canonical correlation technique, one is looking for linear combinations of the predictors and of the responses which, themselves, have maximum correlation. The quantities:

$$\xi_i = a_{1i}x_1 + \cdots + a_{pi}x_p \tag{12.6.1}$$

and

$$\eta_i = a_{1i}^*y_1 + \cdots + a_{qi}^*y_q \tag{12.6.2}$$

are called the ith *canonical variates*, the a_{ij} and a_{ij}^* are called *canonical coefficients* and the correlation between them, r_i, is called the ith *canonical correlation* *coefficient.* ξ_1 and η_1 will be that pair of canonical variates having the maximum correlation; ξ_2 and η_2 will be that pair of canonical variates, independent of the first pair, which have maximum correlation, and so on. Assuming that both S_{xx} and S_{yy} are of full rank, the number of pairs of canonical variates will be the minimum of p and q. The canonical correlations are ordered in the same manner as characteristic roots: $r_1 \geqslant r_2 \geqslant r_3$, and so on, and in fact are obtained in essentially the same manner, being solutions of the determinental equation:

$$\begin{vmatrix} -rS_{xx} & S_{xy} \\ \hline S_{yx} & -rS_{yy} \end{vmatrix} = 0 \tag{12.6.3}$$

where S_{xy} is a $p \times q$ matrix of the covariances between the predictors and the responses and S_{yx} is its transpose. These correlations may, with less effort, be obtained from the solution of

$$|S_{xx}^{-1}S_{xy}S_{yy}^{-1}S_{yx} - r^2I| = 0 \tag{12.6.4}$$

or

$$|S_{yy}^{-1}S_{yx}S_{xx}^{-1}S_{xy} - r^2I| = 0 \tag{12.6.5}$$

The first determinant is of order q and the second of order p, so the choice would normally depend on which of p and q is the smaller. As is the case for multiple correlation coefficients, canonical correlations are biased for small sample sizes. A correction for this bias was given by Thompson (1990).

The vectors used in defining the canonical variates, the canonical coefficients, are obtained by the solution of the following pairs of homogeneous linear equations:

$$[S_{xx}^{-1}S_{xy}S_{yy}^{-1}S_{yx} - r_i^2I]a_i = 0 \tag{12.6.6}$$

and

$$[S_{yy}^{-1}S_{yx}S_{xx}^{-1}S_{xy} - r_i^2I]a_i^* = 0 \tag{12.6.7}$$

However, if a_i has already been obtained, then a_i^* may be obtained from the relation:

$$a_i^* = \frac{1}{r_i}S_{yy}^{-1}S_{yx}a_i \tag{12.6.8}$$

The canonical variates, $\xi = a'x$ and $\eta = a^{*'}y$ may be thought of somewhat in the same context as principal components. The vectors a and a^* may be considered as the characteristic vectors of X and Y rotated to obtain maximum correlation. The relationship between these canonical variates and the separate sets of pc's was given by Muller (1982). The correlation between any ξ_i and the corresponding η_i is r_i and the correlation between ξ_i and any other η_j $(j \neq i)$ is zero. Any two ξ_i are uncorrelated as are any pair of η_i.

Note that the nature of expressions (12.6.3) through (12.6.8) is such that if covariance matrices S are used, the units associated with them cancel out and the canonical correlations will be the same as if correlation matrices **R** had been used. This is unlike the case of conventional PCA, where the two matrices will yield different characteristic roots and vectors. In computing the canonical variates themselves, however, it is still important to keep track of the units. It should also be noted that the matrix products in (12.6.4) and (12.6.5) are *asymmetric* matrices so that the usual methods of obtaining characteristic vectors, which are described in Appendix C, would have to be modified.

The canonical correlation technique will be illustrated with the Linnerud data, which by the nature of it being put into standard units for these examples leaves us with a covariance matrix that is also the correlation matrix. This matrix was displayed in Table 12.3 and will now be partitioned to obtain the matrices required for the canonical correlation operations:

$$
S_{xx} = \begin{bmatrix} 1.0000 & .8702 & -.3658 \\ .8702 & 1.0000 & -.3529 \\ -.3658 & -.3529 & 1.0000 \end{bmatrix}
$$

$$
S_{xy} = \begin{bmatrix} -.3897 & -.4931 & -.2263 \\ -.5522 & -.6456 & -.1915 \\ .1506 & .2250 & .0349 \end{bmatrix}
$$

$$
S_{yy} = \begin{bmatrix} 1.0000 & .6957 & .4958 \\ .6957 & 1.0000 & .6692 \\ .4958 & .6692 & 1.0000 \end{bmatrix}
$$

Note that $S_{yy} = Y'Y/19$ since the data are normalized and similarly with the other matrices. Substituting these matrices into (12.6.4) or (12.6.5) yields canonical correlation coefficients .7956, .2006, and .0726. Substituting these coefficients into (12.6.6) and (12.6.7) yield the following sets of canonical coefficients:

$$
A = \begin{bmatrix} -.7754 & -1.8844 & -.1910 \\ 1.5793 & 1.1806 & .5060 \\ -.0591 & -.2311 & 1.0508 \end{bmatrix}
$$

$$
A^* = \begin{bmatrix} -.3495 & -.3755 & -1.2966 \\ -1.0540 & .1235 & 1.2368 \\ .7164 & 1.0622 & -.4188 \end{bmatrix}
$$

There are, as was the case with principal component characteristic vectors, a number of different scalings for these vectors. Those displayed above correspond to the **W**-vectors in PCA and have the property that $\mathbf{A'S}_{xx}\mathbf{A} = \mathbf{I}$, $\mathbf{A^{*'}S}_{yy}\mathbf{A^*} = \mathbf{I}$, and $\mathbf{A'S}_{xy}\mathbf{A^*}$ is equal to a diagonal matrix whose diagonal elements are the canonical correlation coefficients. Vectors scaled in this manner are sometimes known as *standardized* canonical coefficients.

The first canonical correlation coefficient is .7956. Considering that the highest of the three OLS multiple correlation coefficients is .6607, it indicates that the relationships among the responses do add some additional information. (The first canonical correlation coefficient must be *at least* as large as the largest multiple correlation coefficient since it represents the strongest linear relation between the two sets.) The first pair of canonical variates would appear to represent the comparison of Weight vs. Waist measurement for the physical measurements and a comparison of Jumps vs. Chins and Situps for the exercises. When these vectors are difficult to interpret, sometimes some plots may be of help. Figure 12.1 shows the plot of $\xi_1 = \mathbf{Xa}_1$ vs. $\eta_1 = \mathbf{Ya}_1^*$. It can be seen that this strong correlation is due primarily to just two observations. One of these is individual #9, who had the narrowest waist and highest pulse rate of the entire sample and who had one of the larger number of situps while producing one of the smaller number of jumps. The other, #14, was by far the heaviest individual, with the largest waist size, and could do only one chin and 50 situps. This might indicate that one or both of these men were outliers and without them this relationship (ξ_1 vs. η_1) might not exist. One should always keep in mind that outliers can cause effects like these and be on the lookout for them. We have previously suggested that pc-plots might be quite useful in detecting cases where correlations are primarily the result of one or two observations

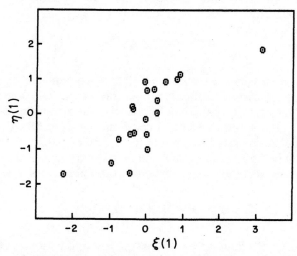

FIGURE 12.1. Linnerud Data: scatterplot of ξ_1 vs. η_1.

and the same holds for these plots. The other two sets of canonical variates have much smaller correlation coefficients and hence do not contribute much to the overall relationship.

As a descriptive statistical tool, canonical correlation analysis is a very powerful yet compact multivariate procedure. For a number of years, however, it appeared to be a technique in search of an application and was used primarily as a data-analytic technique. Most published examples would list the sort of information that has been displayed above but not much more. Hotelling (1957) indicated that perhaps its greatest use was as a preliminary test before embarking on some linear modeling. If the first canonical correlation coefficient was low, there was no need to go on and try anything else. Cliff and Krus (1976) and Röhr (1985) suggested that rotation might aid in interpretation. Bagozzi et al. (1981) felt that canonical correlation was most useful in confirmatory structure models. There have been some extensions of the canonical correlation technique to the case of three or more sets of variables (see, for example, Horst, 1961; McDonald, 1968; Kettenring, 1971; Shine, 1972; Van de Geer, 1984 and TenBerge, 1988).

It would seem natural that one would like to make use of the canonical correlation framework to predict one set of original variables from the other. Keep in mind, however, that canonical correlation analysis does not recognize predictor and response variables as such. The technique merely allows for a partition of the variables into two sets and examines the relationship between these sets. In many applications, nevertheless, the two sets have been predictor and response variables. Those using canonical correlation analysis for prediction purposes are, then, using the technique for something it was not intended to do. If one wishes to do this anyway, the following procedure found in Glahn (1968) may be used.

Let C be the diagonal matrix made up of the squares of the canonical correlation coefficients. Then

$$\mathbf{Y} = \mathbf{XACA^{*-1}} \tag{12.6.9}$$

If all of the canonical variates are used, $\mathbf{ACA^{*-1}}$ will be equal to the least-squares estimates of the regression coefficients, the same as was the case for PCR and PLSR. If only the first two pairs of predictor canonical variates are to be used, simply set $c_{33} = 0$ (denote by $\mathbf{C_2}$) and continue using (12.6.9). If only the first pair of canonical variates are used, set both c_{22} and c_{33} equal to zero. Equation (12.6.9) holds *only* if $q \leqslant p$ so that $\mathbf{A^*}$ will be a square matrix. This is generally the case, but when there are more responses than predictors, one must use

$$\mathbf{Y} = \mathbf{XACA^{*\prime}S}_{yy} \tag{12.6.10}$$

where one column of $\mathbf{A}$ and $\mathbf{A^*}$ as well as one row and column of $\mathbf{C}$ are deleted for each pair of canonical variates omitted from the model.

Using two pairs of canonical variates yields regression coefficients

$$\mathbf{ACA^{*-1}} = \begin{bmatrix} .3593 & .2879 & -.2623 \\ -.8581 & -.8917 & .0233 \\ .0232 & .0119 & -.0368 \end{bmatrix}$$

and the corresponding residual matrix

$$\begin{bmatrix} 12.5896 & 5.9591 & 7.8265 \\ 5.9591 & 10.7069 & 10.7000 \\ 7.8265 & 10.7000 & 17.9813 \end{bmatrix}$$

which, for the Linnerud data, is not only smaller than the residual for PCR with two components but also smaller than PLSR with two components.

Before leaving canonical correlation, it may be of use to have a single number to describe the relationship between two sets of variables. Among a number of proposals is the *vector correlation* or RV-coefficient of Escoufier (1973), a generalization of the Pearsonian coefficient of determination, r^2.

$$RV = \frac{\text{Tr}(\mathbf{S}_{xy}\mathbf{S}_{yx})}{\sqrt{\text{Tr}(\mathbf{S}_{xx}^2)\text{Tr}(\mathbf{S}_{yy}^2)}} \qquad (12.6.11)$$

Like r^2, RV is bounded between 0 and 1. For the Linnerud data,

$$RV = \frac{1.2792}{\sqrt{(5.0313)(5.3553)}} = .25$$

Some properties and distributional results associated with the RV-coefficient were given by Robert and Escoufier (1976) and Robert et al. (1985).

The RV-coefficient, like the Pearsonian coefficient of correlation, does not distinguish between predictor and response variables. When this distinction exists, as it does in this chapter, some generalization of the multiple correlation coefficient may be of use. Two such possibilities are

$$R_S = \sqrt{\frac{\text{Tr}(\mathbf{S}_{yx}\mathbf{S}_{xx}^{-1}\mathbf{S}_{xy})}{\text{Tr}(\mathbf{S}_{yy})}} \qquad (12.6.12)$$

due to Sampson (1984) and

$$R_{KG} = \sqrt{\frac{|\mathbf{S}_{yx}\mathbf{S}_{xx}^{-1}\mathbf{S}_{xy}|}{|\mathbf{S}_{yy}|}} \qquad (12.6.13)$$

due to Kabe and Gupta (1990).

12.6.3 Redundancy

Over the years, a number of people have proposed various indices to indicate how much of the variability of one set of variables is explained by the other using the canonical correlation structure. A short history of this activity is given by Levine (1977). Although the first proposal goes back to Hooper (1959), the index most commonly in use today is the *redundancy* index of Stewart and Love (1968), which is the trace of the explained covariance matrix of the responses divided by the trace of the covariance matrix of the responses. [There was some controversy about this, initially, but the matter has been put rest; see Gleason (1976) for a summary.] If the variables are in standard units (i.e., using correlation matrices), this index amounts to obtaining the square of the multiple correlation coefficient associated with each of the responses and averaging them. This measure, while proposed for canonical correlations, could be used for any linear model involving a multiple response. For example, from Table 12.4, the squared multiple correlation coefficients for the original least-squares solution (as well as PCR, PLSR, and canonical correlation, using all of the components) are

$$\begin{array}{lll} \text{Chins} & R_1^2 = .3396 \\ \text{Situps} & R_2^2 = .4365 \\ \text{Jumps} & R_3^2 = .0539 \end{array}$$

$$\text{Average} \quad .2767 = \text{Redundancy Index}$$

Eliminating the third pair of canonical variates drops this index to .2758. A comparison of all of the techniques discussed in this chapter in terms of this index will be given in Section 12.7. The redundancy index also has a bias for small sample sizes. A correction for this bias was given by Dawson-Saunders (1982), who also showed that the addition of more canonical variables does not increase the proportion of bias.

   An important point to keep in mind is that redundancy is asymmetric, unlike canonical correlation. This means that the redundancy index obtained by predicting y from x will not be the same as that obtained when predicting x from y.

12.6.4 Maximum Redundancy

Van den Wollenberg (1977) proposed a linear model that would depart from the canonical correlation model by *maximizing* the redundancy and hence could properly be used for prediction purposes. The new predictors were called *redundancy variates*. The redundancy vectors, a_i, used to obtain them are defined as the solution to the equation

$$[S_{xy}S_{yx} - r_d^2 S_{xx}]a = 0 \qquad (12.6.14)$$

where the subscript on r^2 is to distinguish it from canonical correlation coefficients. The vectors a_i are of unit length. Optimality conditions for redundancy analysis were given by Tyler (1982). In the unlikely event that all of the nonzero characteristic roots of the correlation matrix of the responses are equal, the results will be the same as canonical correlations (Buzas et al., 1989).

Several alternative computational procedures for maximum redundancy (MR) have since been introduced. Among them are:

1. The redundancy variates **a** are the right-hand vectors of the matrix product $S_{xx}^{-1}S_{xy}S_{yx}$. This is an asymmetric matrix. Israels (1984) and TenBerge (1985) pointed out that this form was equivalent to (12.6.14) and had been derived by Rao (1964) and Fortier (1966) for other purposes. Rao referred to the method as *principal components of instrumental variables*.

2. Obtain the unit characteristic vectors of $S_{yx}S_{xx}^{-1}S_{xy}$, U_d, and the corresponding roots r_d. Then

$$A = S_{xx}^{-1}S_{xy}U_d R_d^{-1/2} \qquad (12.6.15)$$

which has the advantage of obtaining the roots and vectors from a symmetric matrix. This form was suggested by Tyler (1982).

3. If one has already obtained the characteristic vectors from S_{xx} in the W-vector form, an identity of Muller (1981) may be of use and will work when S_{xx} is singular. Form the matrix:

$$M = W'S_{xy}S_{yx}W \qquad (12.6.16)$$

and from this obtain the unit characteristic vectors U_M. Then:

$$A = WU_M \qquad (12.6.17)$$

The U_M-vectors represent the rotation of the original characteristic vectors of the predictors required to obtain maximum redundancy.

4. For those who are working with correlation matrices and who have access to a canonical correlation program, Young and Sarle (1983) suggested a short-cut by using that program after first setting all of the off-diagonal elements in R_{yy} equal to zero. This has the effect of changing the matrix product in (12.6.4) to conform to the Rao–Fortier solution. The results of this "canonical correlation" analysis will then be the redundancy variates.

Once **A** has been obtained, by whatever method, the regression coefficients, B_R, are obtained by

$$B_R = AA'S_{xy} \qquad (12.6.18)$$

For the Linnerud data, the characteristic roots are $r_{d1} = .8931$, $r_{d2} = .1752$, and $r_{d3} = .0400$.

$$
\mathbf{A}' = \begin{bmatrix}
.4534 & -1.3744 & -.0165 \\
-1.9860 & 1.4227 & -.2548 \\
.2076 & -.4824 & -1.0469
\end{bmatrix}
$$

If all three columns of $\mathbf{A}$ are used in (12.6.18), the results will be the same as OLS. If only two columns are used, the regression coefficients will be

$$
\mathbf{B}_R = \begin{bmatrix}
.3625 & .2927 & -.2608 \\
-.8684 & -.9026 & .0189 \\
.0032 & -.0117 & -.0454
\end{bmatrix}
$$

with corresponding residual matrix

$$
\begin{bmatrix}
12.5627 & 5.9486 & 7.8161 \\
5.9486 & 10.7200 & 10.6968 \\
7.8161 & 10.6968 & 17.9773
\end{bmatrix}
$$

which is the smallest of any two-component model discussed in this chapter *for this numerical example*. The redundancy index is .2761.

Fornell (1979) and Israels (1986) suggested rotating the redundancy vectors to simple structure for better interpretation. Israels (1984) extended maximum redundancy to the case of qualitative variables. It is possible, using a procedure due to Johansson (1981), to obtain a set of transformed response variates, one corresponding to each of the redundancy variates in a manner similar to the pairing of canonical variates. These may be useful in the interpretation of the redundancy variates. Fornell et al. (1988) showed how this can be done using PLS. DeSarbo (1981) developed a procedure called *canonical/redundancy factoring analysis*, which simultaneously maximizes specific combinations of both canonical correlation and redundancy measures.

12.7 SUMMARY

In this chapter, five different linear models have been considered for use in predicting a set of related responses from a set of related predictors. These are ordinary least-squares (OLS), principal components regression (PCR), partial least-squares regression (PLSR), canonical correlations (CC) and maximum redundancy (MR). None of the alternatives to OLS will do better than OLS with regard to the residual sum of squares, although some of them may produce a smaller mean square error. All of them will be identical to OLS if all components are used. (Only CC may produce stronger relationships but those will be in terms of functions of both sets of variates.) However, the alternatives may be attractive from the standpoint of interpretation and/or model simplification. When using a reduced set of components, PCR may fare rather poorly since there is no guarantee that the largest components will necessarily be the best predictors. Although CC variates may be used for prediction, these are not optimal either, since they were derived for a different purpose, namely, to produce the best relation between two *sets* of variates. This leaves PLSR and MR as the two alternatives to OLS that do attempt to optimize the prediction of the response. Both of these techniques are fairly new and at this point in time it has not been established whether either of these methods has any theoretical advantage over the other. Certainly the operations for MR are more simple than PLSR. On the other hand, if $p > q$, MR is restricted to, at most, q components while PLSR can have as many as $\min(p, n)$, which may allow for more flexibility in modeling.

Listed in Table 12.8, for the Linnerud data, are the redundancy indices for OLS and for the all of the alternatives using either two components or one component. (Remember, the redundancy index for the three-component models will be the same as OLS for all of the alternatives.) Also included in the table is the square of the multiple correlation coefficient associated with each response. It is the average of these values that produces the redundancy index.

Even though maximum redundancy obtained the best results for one- and two-component models, this is just one set of data and no generalizations should be made from these results except that PCR does come out in last place as might be expected. Given its present popularity, PCR will undoubtedly enjoy continued use for some years to come, so it is well to know what its limitations

Table 12.8. Linnerud Data. Redundancy Index

Model[a]	Squared Multiple Correlation Coefficients			Redundancy Index
	y_1	y_2	y_3	
OLS	.3396	.4365	.0539	.2767
PCR(2)	.2387	.3470	.0490	.2116
PCR(1)	.2119	.3237	.0381	.1912
PLSR(2)	.2859	.3877	.0433	.2390
PLSR(1)	.2363	.3506	.0414	.2094
CC(2)	.3374	.4365	.0536	.2758
CC(1)	.3347	.4232	.0166	.2582
MR(2)	.3388	.4358	.0538	.2761
MR(1)	.3363	.4358	.0258	.2660

[a] Figures in parentheses indicate number of components in model.

are and how best to cope with them. It should be pointed out that if the first and third pc's were used instead of the first and second, the redundancy index would have been *increased* to .256.

There are a number of other linear models used with both multiple predictors and multiple responses that, while also useful, were not discussed here because their solutions do not belong to the class of solutions given in this chapter. These include the multivariate calibration methods of P.J. Brown (1982) and Naes (1985) and the systems of simultaneous equation models frequently employed in econometric models (Schmidt, 1976) and market research (Bagozzi, 1977). The latter are sometime referred to as *linear functional* or *structural* relations for which the LISREL® technique of Jöreskog is often employed, although solutions may also be obtained by PLS. We shall return to this topic in connection with confirmatory factor analysis in Section 17.8.

CHAPTER 13

Linear Models II: Analysis of Variance; PCA of Response Variables

13.1 INTRODUCTION

Chapter 12 was concerned with the use of PCA and some alternative procedures in regression analysis. The emphasis was on the reduction of the number of predictor variables. The use of pc's as predictors had the advantage that the transformed predictors were uncorrelated but had the potential disadvantage that there might well be no relationship between the amount of variability of the predictors explained by a particular pc and its worth as a predictor, in addition to the fact that the pc's might not be interpretable. Some techniques, such as partial least-squares regression and maximum redundancy, were designed to alleviate that situation to some extent. In this chapter, the emphasis will be on the *response* variables. Again, there will be no guarantee that pc's obtained from response variables will be interpretable either, but to the extent that they are, they may be used effectively in the analysis of variance.

The motivation for multivariate analysis will be familiar by now. If one has q response variables and conducts q univariate ANOVAs, the overall Type I error will be $1 - (1 - \alpha)^q$. Further, the results may conflict; a single answer is desirable. On the other hand, the majority of ANOVA data come from designed experiments where the predictor variables are either uncorrelated or nearly so; in this case PCA of the predictor variables is generally unwarranted.

Most multivariate techniques are extensions of similar univariate procedures. Although PCA is an exception, being a truly multivariate technique, multivariate analysis of variance (MANOVA) is an extension of univariate techniques. In this chapter, a set of data will be evaluated using univariate ANOVAs, multivariate analysis of variance (MANOVA), and pc-ANOVA. Although not involving a multivariate response, Section 13.7 will show how PCA may be employed in analyzing a univariate two-way ANOVA using singular value decomposition.

301

13.2   UNIVARIATE ANALYSIS OF VARIANCE

A new numerical example will be introduced for this chapter. It is similar to the chemical example used in Chapter 1 but we will now assume that it is in the form of an interlaboratory comparison. Samples of the same chemical solution are sent to each of $t = 3$ laboratories. Each of the labs makes four analyses of the sample sent them, using both methods. The data are shown in Table 13.1. From these data, one can perform two univariate analyses of variance, one for each test procedure. The results for these ANOVAs are shown in Table 13.2.

Table 13.1. Grouped Chemical Example. Grouped Data

Laboratory		Method 1	Method 2
1		10.1	10.5
		9.3	9.5
		9.7	10.0
		10.9	11.4
	Average	10.00	10.35
2		10.0	9.8
		9.5	9.7
		9.7	9.8
		10.8	10.7
	Average	10.00	10.00
3		11.3	10.1
		10.7	9.8
		10.8	10.1
		10.5	9.6
	Average	10.82	9.90

Table 13.2. Grouped Chemical Data. Univariate ANOVAs—Original Data

Source of Var.	Sum of Squares	d.f.	Mean Square
	Method 1		
Among labs	1.8150	2	.91
Within labs	2.7275	9	.30
Total	4.5425	11	
	Method 2		
Among labs	.4467	2	.22
Within labs	2.8100	9	.31
Total	3.2567	11	

The F-ratio for Method 1 is 3.0, which is significant at about the $\alpha = .10$ level, and the F-ratio for Method 2 is less than 1.0. Not a very exciting set of data.

Since these are multiple responses with correlated variables, we already know that univariate ANOVAs are not appropriate. Will a multivariate treatment of these data produce any different results?

13.3 MANOVA

In the *multivariate analysis of variance* (MANOVA), both methods are analyzed at the same time by replacing sums of squares and mean squares with matrices that are computed in the same manner. The resultant MANOVA table is shown in Table 13.3. Note that all of the diagonal elements in these matrices are found in Table 13.2. (The total mean square is not generally used in either the univariate or multivariate analysis of variance but use will be made of this matrix in the next section.) Obviously, an F-test cannot be performed on these matrices since they cannot be divided. What one does is to multiply the Among Mean Square Matrix $\mathbf{A}$ by the inverse of the Within Mean Square Matrix $\mathbf{E}^{-1}$ to form the product

$$\mathbf{AE}^{-1} = \begin{bmatrix} 40.20 & -38.59 \\ -17.27 & 16.88 \end{bmatrix}$$

Many of the MANOVA tests employ roots of the determinental equation

$$|\mathbf{A} - \lambda\mathbf{E}| = |\mathbf{AE}^{-1} - \lambda\mathbf{I}| = 0 \qquad (13.3.1)$$

so that the operation reduces to obtaining the characteristic roots of $\mathbf{AE}^{-1}$. Note that this matrix is asymmetric even though both $\mathbf{A}$ and $\mathbf{E}$ are symmetric. This means that the customary methods of obtaining characteristic roots and vectors cannot be employed here but the desired results can be obtained using an extension of the methods discussed in Appendix C, by the use of singular

Table 13.3. Grouped Chemical Data. MANOVA

Source of Variation	Sum of Squares		d.f.	Mean Square	
Among labs	$\begin{bmatrix} 1.8150 & -.6050 \\ -.6050 & .4467 \end{bmatrix}$		$2 = m_H$	$\begin{bmatrix} .9075 & -.3025 \\ -.3025 & .2234 \end{bmatrix} = \mathbf{A}$	
Within labs	$\begin{bmatrix} 2.7275 & 2.6300 \\ 2.6300 & 2.8100 \end{bmatrix}$		$9 = m_E$	$\begin{bmatrix} .3031 & .2922 \\ .2922 & .3122 \end{bmatrix} = \mathbf{E}$	
Total	$\begin{bmatrix} 4.5425 & 2.0250 \\ 2.0250 & 3.2567 \end{bmatrix}$		11	$\begin{bmatrix} .4130 & .1841 \\ .1841 & .2961 \end{bmatrix}$	

value decomposition, or a two-stage process similar to that discussed for discriminant analysis in Section 14.5.2. In this case, the row vectors by which AE^{-1} is premultiplied will not be the transpose of the column vectors by which it is postmultiplied.

One of the problems associated with MANOVA is that no single significance test corresponding to the univariate F-test exists. There are a number of tests, of which the most common, in order of their historical introduction are: (1) the ratio of the determinant of the Among SS matrix to the determinant of the Total SS matrix (Wilks, 1932b; Hsu, 1940); (2) the trace of AE^{-1}, which has the Lawley–Hotelling distribution introduced in Section 6.2, and (3) the maximum root of AE^{-1} (Roy, 1953). For a comparison of these methods and others, see Olson (1974), Obrien et al. (1982), Muller and Peterson (1984), and Seber (1984, p. 414). Generally, for most cases, these tests should all produce the same conclusions, although there will be differences in power against certain alternatives and in the effect of nonnormality. Since the Lawley–Hotelling statistic was used in Chapter 6, it will be used here also. The Lawley–Hotelling statistic in this case is

$$T_0^2 = (m_E/m_H)\text{Tr}(AE^{-1}) \tag{13.3.2}$$
$$= (9/2)(57.08) = 256.86$$

Referring to Appendix G.5 with $d = 2$, $m_H = 2$ and $m_E = 9$, $\alpha = .05$, $T_{0,.05}^2 = 8.55$, so T^2 is significant.

The trace of AE^{-1} is, of course, equal to the sum of its characteristic roots, which, in this example, are 56.86 and .22. Each of these roots has a column vector associated with it. If the amounts explained by these vectors were tested separately, the first would be significant; the second would not. The vector associated with the first root is

$$\mathbf{u}_1 = \begin{bmatrix} .918 \\ -.396 \end{bmatrix}$$

This is the linear combination of the two methods that maximizes the difference among labs. The interpretation of this vector would seem to be that this phenomenon is related to the difference between the two methods but also weights the first method more heavily. From Table 13.2, we already know that there is more variability in the first method and that the differences between the methods are not consistent from lab to lab.

If there are a larger number of variables, it is possible that AE^{-1} may produce more than one relevant vector. If two vectors were sufficient, some two-dimensional plots of their associated "scores" might be useful. Krzanowski (1989) gave expressions for the confidence ellipse for each group mean for the two-vector case.

There are some operational difficulties associated with MANOVA. There is the aforementioned problem of competing test statistics. Most of the commonly used computer packages have some MANOVA routines, but for more complex designs such as central composite designs, for example, the analysis could become complicated. Of more importance is the fact that the vectors associated with the major characteristic roots may not be as clear-cut as the one in this example. Because of these reasons, MANOVA has many critics. Finney (1956), for example, stated that in many agricultural experiments, the significance tests were secondary to the main goal of obtaining a measure of experimental error. He preferred inexact probability statements about univariate variables to exact statements about MANOVA that were not interpretable.

13.4 ALTERNATIVE MANOVA USING PCA

There are procedures that perform univariate ANOVAs in a sequential manner, most notably the step-down procedures of J. Roy (1958). In these procedures, one variable is selected for the first ANOVA. A second variable is then selected, corrected for what was explained by the first variable and run through a second ANOVA, and so on. (For a review of these procedures, see Mudholkar and Subbaiah, 1980.) Step-down procedures may have better interpretability and do have a guaranteed Type I error. The drawback to them is that the order in which the variables are entered into the sequential scheme is arbitrary. One possibility is to start with the variable having the largest variance, then the one with the largest residual variance, and so on.

An alternative is to obtain the principal components for the data set as a whole and then do univariate ANOVAs on as many of these pc's as seems relevant. An advantage of this procedure is that it may better aid in the interpretation of the data if the pc's themselves are interpretable. A second advantage is the relative computational ease in handling rather complex designs. Bratchell (1989) used pc-ANOVA for some response surface designs and Gemperline (1989) used it in some mixture experiments. The disadvantage of pc-ANOVA is that there is still more than one ANOVA and, as we shall see, the ANOVAs are not independent, even though the pc's are.

To perform pc-ANOVA for the data at hand, first obtain the characteristic roots and vectors from the Total Mean Square Matrix, which is displayed in Table 13.3. For this example, the characteristic roots are $l_1 = .5477$ and $l_2 = .1614$. The characteristic vectors are

$$\mathbf{U} = \begin{bmatrix} .8070 & -.5905 \\ .5905 & .8070 \end{bmatrix}$$

or

$$\mathbf{W} = \begin{bmatrix} 1.0905 & -1.4700 \\ .7979 & 2.0091 \end{bmatrix}$$

Table 13.4. Grouped Chemical Data. y-Scores

Laboratory		y_1	y_2
1		.14	1.09
		−1.53	.26
		−.69	.68
		1.73	1.73
	Average	−.09	.94
2		−.53	−.16
		−1.15	.37
		−.85	.28
		1.06	.47
	Average	−.37	.24
3		1.13	−1.47
		.24	−1.19
		.59	−.74
		−.14	−1.30
	Average	.46	−1.18

The vectors are about what one would expect; the first one represents level differences primarily and accounts for 77% while the second, representing method differences accounts for the remaining 23%. The W-vectors are used to obtain y-scores, which are displayed in Table 13.4. These scores, in turn, are used in the univariate analyses of variance, which are displayed in Table 13.5. (Both of the total sum of squares equal 11.0 since the principal components both have unit variance.) The F-test for y_1 is less than unity. For the second pc—method difference—the test is highly significant, so the conclusion would be that the labs do not differ in overall level but do differ in method bias.

Table 13.5. Grouped Chemical Data. Univariate pc-ANOVAs

Source of Var.	Sum of Squares	d.f.	Mean Square
	y_1		
Among labs	1.39	2	.69
Within labs	9.61	9	1.07
Total	11.00	11	
	y_2		
Among labs	9.30	2	4.65
Within labs	1.70	9	.19
Total	11.00	11	

Note that if one pre- and postmultiplies the Among and With Mean Square Matrices, **A** and **E** from Table 13.3 by the characteristic vectors used in the pc-ANOVAs, viz.,

$$\mathbf{W'AW} = \begin{bmatrix} .69 & -1.40 \\ -1.40 & 4.65 \end{bmatrix}$$

and

$$\mathbf{W'EW} = \begin{bmatrix} 1.07 & .31 \\ .31 & .19 \end{bmatrix}$$

the diagonal elements are the same as those given in Table 13.5. This points out that:

1. If one has already obtained the Among, Within and Total Mean Square Matrices, then the pc-ANOVAs may be obtained directly without having to obtain the y-scores for all of the original data.
2. Although the pc's, y_1 and y_2, are uncorrelated, the variability accounted for among and within labs is correlated, implying that the F-tests are not independent.

To take care of this correlation problem, Dempster (1963) proposed the following step-down procedure:

1. Obtain the characteristic roots l_i and vectors **U** from the Total SS matrix.
2. Obtain $\mathbf{G} = m_E \mathbf{U'EU}$.
3. Diagonalize **G** as follows:
   a. Stage 1

   $$g_{ij.1} = g_{ij} - g_{1i}g_{1j}/g_{11} \qquad i, j = 2, \ldots, p \tag{13.4.1}$$

   b. Stage 2

   $$g_{ij.12} = g_{ij.1} - g_{2i.1}g_{2j.1}/g_{22.1} \qquad i, j = 3, \ldots, p \tag{13.4.2}$$

   c. Continue through stage $p - 1$.
   d. Let $d_1 = g_{11}$, $d_2 = g_{22.1}$, $d_3 = g_{33.12}$, and so on.
4. The sequential test is of the form

   $$F_i = \frac{(1 - P_i)}{P_i}\left[\frac{(n - t - i + 1)}{(t - 1)}\right] \qquad i = 1, \ldots, p \tag{13.4.3}$$

where i is the component in the sequence, $P_i = d_i/l_i$, and F_i has $(t - 1)$ and $(n - t - i + 1)$ degrees of freedom.

Notice that at each stage, the within components, d_i are being corrected for this correlation and the error degrees of freedom are being reduced. Notice also that Dempster used the ratio of the *Within to the Total Sum of Squares*, reflecting the philosophy of the Wilks MANOVA technique, instead of the Among to Within Mean Squares.

For the Chemical example, $l_1 = 6.0247$ and $l_2 = 1.7754$. These are the same as before except that they have now been multiplied by the total degrees of freedom, 11. The characteristic vectors **U** are the same as before.

$$\mathbf{G} = \begin{bmatrix} 5.2627 & .8350 \\ .8350 & .2745 \end{bmatrix}$$

The within components are $d_1 = 5.2627$ and $d_2 = .1420$. Then $P_l = .8736$ with $F_{2,9} = .6511$, which is not significant. $P_2 = .08$ and $F_{2,8} = 46.00$, which is significant. The conclusions are the same as with the uncorrected pc-ANOVA, but for a larger number of variables this procedure could make a difference. Dempster also gave some options for varying the order in which the components were tested. Although the natural approach would be to test them in order of the amount of total variation explained, there is no guarantee (as we have seen in this example) that the components with the largest roots are the most discriminating among groups.

13.5 COMPARISON OF METHODS

These data have now been analyzed by three different procedures, univariate ANOVAs, MANOVA, and univariate ANOVAs of the principal components. Which one should be employed? Definitely not the univariate ANOVAs of the original data, since the variables are correlated. The multivariate procedures both detected the problem of method bias but using pc's probably produced more directly usable results—for this example, at least. There is no clear-cut answer as to which method will be the more useful. Some of my colleagues have found pc-ANOVAs quite helpful when dealing with response variables that represent points on a continuous curve.

People have been known to get rather emotional in debating the relative merits of MANOVA and pc-ANOVA. The merits of MANOVA include the always desirable control of the Type I error; there is a single answer to the question: "Is there a difference among the labs?" On the other hand, complex and/or unbalanced designs can produce estimation and computational problems. pc-ANOVA is more flexible, may provide more insight and may be worth the correlated F-tests. If one wished an overall test based on the retained pc's, a large-sample procedure due to Carter (1980) might be of use. Better yet, when appropriate, one may use Dempster's step-down procedure, which in this day and age should be fairly easy to employ. pc-ANOVA may also be considered as an adjunct tool when the MANOVA test indicates a significant result.

There have been a number of other suggestions that employ PCA within the framework of the multivariate analysis of variance. pc-ANOVA has been criticized because the PCA is based on the total variability of the data and does not take into account the structure of the experiment. Pearce and Holland (1960) obtained pc's separately from the Within and Among Mean Square matrices. Corsten and Gabriel (1976) used *h-plots*, a technique in which these two sets of vectors are then superimposed on the same plot. *h*-Plots can also be used to compare two or more covariance matrices. Tukey (1951) suggested obtaining the Among component of variance matrix and performing PCA on that. Joe and Woodward (1976) worked with the composites of the component of variance matrices. If a large number of covariance matrices are being compared, Gnanadesikan and Lee (1970) suggest making a gamma plot of either the trace of each matrix or the geometric mean of their characteristic roots.

13.6   EXTENSION TO OTHER DESIGNS

The data set used in Sections 13.2–13.4 is an example of one-way classification with equal subgroups, about the most simple of all designs. Extension of the multivariate procedures to other designs is straightforward. In the case of pc-ANOVA's, the analysis would be the same as for the original data. For MANOVA, it would also follow the same pattern, with the individual sums of squares being replaced with matrices. For an example of a two-way MANOVA as well as a multivariate analysis of covariance, see Finn (1977). Some special designs such as central composite designs might, however, produce some analytical problems related to expressions for curvature, since one is now working with matrices rather than scalar quantities. Another problem that can arise in MANOVA deals with random effects models and variance components. The univariate variance components are replaced by matrices. In the univariate case, some of these components may be negative; in MANOVA, not only may some of the diagonal elements of these matrices become negative, but even if they are positive the matrix may be ill-conditioned and/or negative-definite. However, that is not the fault of the method but of the data being analyzed by it, the same as in the univariate case.

Any univariate multiple comparison procedures used in conjunction with ANOVA may also be employed with pc-ANOVA. For MANOVA, the choice is somewhat restricted but a number of techniques are available. See, for instance, Gabriel (1968, 1969), Krishnaiah (1969, 1979), and Mudholkar et al. (1974).

13.7   AN APPLICATION OF PCA TO UNIVARIATE ANOVA

13.7.1   Introduction

PCA may be employed in univariate ANOVA. This application is related to tests for nonadditivity in two-way ANOVAs when there is only one observation per cell. The problem, in this case, is that there is no way to test for the interaction since there is no independent measure of error. The interaction mean square is made up of both the inherent variability and the true interaction effect, if any. Attempts have been made to establish models to partition the interaction sum of squares into something related to interactions and something more associated with inherent variability. The first attempt to do this was due to Tukey (1949) who split out a single degree of freedom associated with a test for nonadditivity. An alternative test for nonadditivity may be obtained using PCA.

To state the problem more formally, assume that one has a randomized block design with b blocks, t treatments, and one observation, x_{ij}, per cell. The linear model associated with this is

$$x_{ij} = \mu + \beta_i + \tau_j + \gamma_{ij} + e_{ij} \tag{13.7.1}$$

where μ is the overall population mean, β_i is the true effect of the ith block, τ_j is the effect of the jth treatment, γ_{ij} is the effect of the interaction for the ith block and jth treatment, and e_{ij} is the experimental error associated with that same combination. (This equation will hold for any two-way classification but the randomized block is one of the most common.) If there is no interaction, then γ_{ij} will be equal to zero for all i and j. In that case, this will be an *additive* model and γ_{ij} will drop out of (13.7.1). If there is an interaction, then two or more of the γ_{ij} are unequal to zero and the model is no longer additive.

In the analysis of interactions, one needs to double-center $\mathbf{X}$, that is, subtract both the row and column effects as well as the grand mean. In Chapter 10 this was referred to as M-analysis. Restating (10.3.6),

$$d_{ij} = [x_{ij} - \bar{x}_{i.} - \bar{x}_{.j} + \bar{x}_{..}] \tag{13.7.2}$$

(The resultant matrix $\mathbf{D}$ is a distance matrix in the context of Chapter 11 rather than a diagonal matrix that the symbol $\mathbf{D}$ often implies.) Then

$$\sum_{i=1}^{b} \sum_{j=1}^{t} d_{ij}^2 = \text{Interaction Sum of Squares}$$

13.7.2 Nonadditivity Models

Attempts to model nonadditivity focus on γ_{ij}. A common practice is to assume that the model is multiplicative. Tukey's model, also called the *concurrent* model, is

$$x_{ij} = \mu + \beta_i + \tau_j + \zeta_i \tau_j + e_{ij} \tag{13.7.3}$$

Mandel (1961) proposed a linear model of the form

$$x_{ij} = \mu + \beta_i + \tau_j + (v_i - 1)\tau_j + e_{ij} \tag{13.7.4}$$

where the interaction is a linear function of the treatments, or

$$x_{ij} = \mu + \beta_i + \tau_j + (v_j - 1)\beta_i + e_{ij} \tag{13.7.5}$$

a linear function of the blocks.

Bradu and Gabriel (1978) advocated the use of biplots on **D**. If the points (or endpoints of the vectors) representing the columns are collinear, it would suggest (13.7.4). If the points representing the rows are collinear, it would suggest (13.7.5). If both sets were collinear, it would suggest the concurrent model (13.7.3). The additive model would be indicated by both sets being collinear but at right angles to each other. An extension of these diagnostics to three-dimensional biplots may be found in Gabriel and Odoroff (1986a).

13.7.3 A SVD Model

Another multiplicative model is

$$x_{ij} = \mu + \beta_i + \tau_j + \sum_{h=1}^{g} \xi_{hi}\theta_h\eta_{hj} + e_{ij} \tag{13.7.6}$$

where $g = \min(b - 1, t - 1)$. ξ_{hi} is the ith element of a $b \times 1$ vector and η_{hj} is the jth element of a $1 \times t$ vector. This model is more appropriate when an interaction is present but there are no significant differences among the blocks, treatments, or both. The model is also appropriate when the interaction is present in only one cell or a single row or column.

The interaction term in (13.7.6), now in the form of a summation, looks suspiciously like a singular value decomposition and, in fact, the estimates of these quantities come from a SVD of the matrix **D**. There will be a maximum of g characteristic roots and the sum of these roots equals the interaction sum of squares. The estimate of θ_h is $l_h^{1/2}$, the estimate of the vector ξ_h is $\mathbf{u}_h^{*\prime}$, and the estimate of η_h is $\mathbf{u}_h$. A procedure for handling this problem is represented by the forthcoming equation (13.7.7) and the reader may skip to that point. However, it may be worthwhile to trace its development because it has attracted considerable interest and this section represents work drawn from a number of different journals.

The concept of using PCA for these models was first applied by Fisher and MacKenzie (1923) in a potato experiment in which the first vector was used to obtain deviations from the product formula. It was next suggested nearly 30 years later by Williams (1952) and it was another 16 years before it was rediscovered (Gollob, 1968a,b) and employed in the form as we know it today. Gollob called this FANOVA, being a combination of factor analysis (PCA, actually) and the analysis of variance. The interaction sum of squares was partitioned into the g terms $l_1, \ldots, l_g$. Generally, the last few roots were pooled as an error term and the others were tested by the means of F-tests.

Since the roots are not independent and are not distributed as central chi-squares, these tests are not independent. To overcome this, Mandel (1969, 1970, 1971, 1972) suggested adjusting the degrees of freedom. These degrees of freedom were based on the expected values of the roots, which were tabulated, using simulation, along with their standard deviations for the first three roots. Some improvements on these tables were given by Johnson and Graybill (1972). Approximations for both of these quantities were given by Cornelius (1980). Both Gollob and Mandel also suggested the use of the *vacuum cleaner*, a technique developed by Tukey (1962) to identify outlying residuals in a two-way table, before performing the SVD and gave adjustments to the degrees of freedom based on the number of sweeps. Freeman (1975) developed a procedure for handling missing data.

The practice today is to avoid the degrees of freedom problem entirely by using tests for the roots themselves. Some of the same procedures discussed in Sections 4.5.2 and 4.5.3 in connection with covariance matrices, may be employed with D as well. Johnson and Graybill (1972) showed that the first root could be tested by use of the statistic $l_1/(l_1 + \ldots + l_g)$ and prepared some tables for this statistic. [The tables of Schuurmann et al. (1973) described in Section 4.5.4, are also applicable.] The power of this test was studied by Hegemann and Johnson (1976a), Carter and Srivastava (1980), and Srivastava and Carter (1980). Hegemann and Johnson concluded that if the interaction was such that it could be eliminated with a transformation of the data, Tukey's test had a better chance of detecting it. If, however, a transformation was not required, the test for the first root would do better. Corsten and van Eijbergen (1972) developed a likelihood ratio statistic for testing a group of components for equality of the form $(l_m + \ldots + l_g)/(l_2 + \ldots + l_{m-1})$. Boik (1986) extended the Johnson–Graybill test to the case where there was more than one observation per cell and hence a separate estimate of error was available. In this case, the Error SS is added to the denominator of the Johnson–Graybill statistic. Boik called this the *studentized maximum root* test and included tables for its use.

Hegemann and Johnson (1976b) derived a test for the second component of the form $l_2/(l_2 + \ldots + l_g)$ and included some tables for its use. Finally, Yochmowitz and Cornell (1978) showed that this can be generalized so that the test for the kth root is

$$L_k^* = \frac{l_k}{l_k + \ldots + l_g} \tag{13.7.7}$$

and that the distribution for these tests are all the same as the test for the first root except that the number of "variables" in the various tables changes with each stage of the testing process. The assumption made for this procedure is that as each component is found to be significant, it is assumed to be included in the model when testing for the next component. The tables of Schuurmann et al. (1973) may now be used for any stage. Let $c_1 = \min(b, t)$ and $c_2 = \max(b, t)$. Then the critical value in their tables for testing

$l_k/(l_k + \cdots + l_g)$ is found by letting their $p = c_1 - k$ and $r = (c_2 - c_1 - 2 + k)/2$. These tables are somewhat limited in scope and for the cases $k = 1, 2$ the aforementioned tables of Johnson and Graybill (1972) and Hegemann and Johnson (1976b) may be useful although the latter need to be converted to use directly. Both of these tables were prepared for the most part by simulation, the values for $k = 1, c_1 = 3$ being exact. Krzanowski (1979a) gave exact values for $k = 1, c_1 = 3, 4$ for a wide range of significance levels.

13.7.4 What *is* the Inherent Variability?

If there is no interaction, then the interaction mean square is the best estimate of the inherent variability. However, if nonadditivity is detected, this implies that the interaction sum of squares is made up both of inherent variability and a true interaction. The sum of squares associated with the inherent variability is equal to the roots whose components are not significant but the degrees of freedom problem surfaces once more. For the case of only one significant component, Johnson and Graybill suggested

$$\hat{\sigma}^2 = \frac{l_2 + l_3 + \cdots + l_g}{(b-1)(t-1) - E(l_1)} \tag{13.7.8}$$

based on the suggestions of Mandel (1969, 1971) regarding degrees of freedom. Both of Mandel's papers contain tables of $E(l_1)$. One might conjecture that additional expected values could be subtracted from the denominator as additional components are added to the model, although no theoretical or empirical evidence is currently available to support this. Hegemann and Johnson (1976b) showed that there is an upward bias in (13.7.8) as a function of the size of the interaction. For this reason, Carter and Srivastava (1980) proposed, as an alternative for the one-component model, the estimator

$$\hat{\hat{\sigma}}^2 = \frac{R}{(n-1)(p-2)}\left[1 - \frac{R}{(n-1)[(p-2)l_1 - R]}\right] \tag{13.7.9}$$

where $R = l_2 + \cdots + l_g$. Marasinghe and Johnson (1982a) gave an estimate for the case where a subset of the data matrix could be shown to be free of any interactions.

13.7.5 Other Accessories

There are also a few results for simultaneous interval estimation and multiple comparisons. Johnson (1976) gave a procedure for linear contrasts which are of the form

$$\sum_{i=1}^{b} \sum_{j=1}^{t} c_i d_j \gamma_{ij} \tag{13.7.10}$$

where the c_i's define a contrast of the blocks (rows) and the d_j's define a contrast of the treatments (columns). The procedure assumes that $l_1, \ldots, l_g$ are independent of the inherent variability σ^2. Let $G_{\alpha, m, n, f}$ satisfy

$$P\{l_1/\hat{\sigma}^2 > G\} = \alpha \qquad (13.7.11)$$

where $m = \min(b - 1, t - 1)$, $n = \max(b - 1, t - 1)$, and f is the degrees of freedom associated with σ^2. If f is large, G may be obtained using Appendix G.6 where $p = m$ and $n_1 = n$. Then the $(1 - \alpha)\%$ confidence limits for

$$\left|\sum c_i d_j \bar{x}_{ij} - \sum c_i d_j \gamma_{ij}\right| < \sqrt{\sum c_i^2 \sum d_j^2 G_\alpha \hat{\sigma}^2 / r} \qquad (13.7.12)$$

where r is the sample size associated with the $\bar{x}_{ij}$. For the interaction problem under discussion (i.e., no subgroups), $r = 1$. If f is not large enough to use Appendix G.6, let $f_1 = 2E(l_1)/V(l_1)$ and $f_2 = 2[E(l_1)]^2/V(l_1)$. Then $f_1 l_1/(f_2 \hat{\sigma}^2)$ is distributed approximately as $F_{f_1, f}$ and $G_{\alpha, m, n, f} \simeq f_1 F_{\alpha, f_1, f}$.

Bradu and Gabriel (1974) also studied this problem but without building on models such as (13.7.4), (13.7.5), or (13.7.6). Instead they looked for the maxima of various types of contrasts in more of an exploratory vein. Marasinghe and Johnson (1981, 1982b) gave some procedures for testing that a subset of the data matrix has no interaction given that (13.7.6) holds for the whole matrix. These procedures involved testing either a subset of rows, columns, or both for lack of interaction. Cornelius et al. (1979) analyzed an agricultural experiment in which all of the factors were qualitative except for a time factor that was at 11 levels. They fitted all of the qualitative factors and then performed a SVD on the residuals whose variability was accounted for primarily by the time factor. This allowed them to model the time factor, which they smoothed with cubic splines and used in testing the multiway interactions. Aastveit and Martens (1986) used partial least-squares regression to relate additional variables to the interaction sum of squares.

13.7.6 Numerical Example

Some of these procedures will now be illustrated with the Presidential Hopefuls example of Section 10.6. The data along with the row and column averages were displayed in Table 10.1. The analysis of variance for these data is shown in Table 13.6 and the double-centered matrix $\mathbf{D}$ using (13.7.2) is shown in Table 13.7. The sum of squares of these values is equal to the interaction sum of squares, 38 939. From the ANOVA table we see that no significant difference was found among respondents, not surprisingly, since they were restricted to using values between 0 and 100. A SVD of $\mathbf{D} = \mathbf{U^{*'}L^{1/2}U}$ produced the characteristic roots $l_1 = 17684$, $l_2 = 13012$, $l_3 = 4650$, $l_4 = 2367$, $l_5 = 1226$, and $l_6 = 0$. The sum of these roots is equal to the interaction sum of squares.

First, we test the first component using (13.7.7) with $k = 1$, $L_1^* = l_1/(l_1 + \cdots + l_5) = .454$. Using linear interpolation in the tables by Schuurmann

Table 13.6. Presidential Hopefuls Data. Analysis of Variance

	Sum of Squares	d.f.	Mean Square
Among respondents	6066	17	356.8
Among candidates	9877	5	1975.4
Respondent × Candidate	38939	85	458.1
Total	54882	107	

Table 13.7. Presidential Hopefuls Data. Double-Centered Matrix D

Respondent	Carter	Ford	Humphrey	Jackson	Reagan	Wallace
1	19.3	−18.1	27.3	7.0	−19.3	−16.3
2	27.7	−4.7	5.7	.3	−15.9	−13.0
3	1.0	3.6	−21.0	−1.3	−7.6	25.3
4	−22.7	4.9	10.3	5.0	5.7	−3.3
5	−1.5	−28.9	−53.5	6.2	24.9	52.8
6	35.2	12.8	−11.8	−7.2	−8.4	−20.5
7	−15.7	21.9	37.3	−3.0	−29.3	−11.3
8	−11.5	21.1	1.5	−3.8	−.1	−7.2
9	−6.5	16.1	−23.5	1.2	14.9	−2.2
10	−14.0	3.6	9.0	−6.3	2.4	5.3
11	−31.5	1.1	11.5	1.2	14.9	2.8
12	19.3	−18.1	−7.7	7.0	−4.3	3.7
13	29.3	1.9	−22.7	7.0	.7	−16.3
14	−28.0	−2.4	−14.0	−13.3	52.4	5.3
15	−2.3	−29.7	−24.3	25.3	9.1	22.0
16	34.8	−27.6	27.8	2.5	−29.8	−7.8
17	8.5	11.1	16.5	−23.8	−.1	−12.2
18	−41.5	31.1	31.5	−3.8	−10.1	−7.2

et al. 1973) with $p = t - k = 5$ and $r = (b - t - 2 + k)/2 = 5.5$, the critical value for $\alpha = .05$ is .462. Bradu and Gabriel (1974) mentioned interactions of rank 2 or greater and suggested that if this situation existed, the first root might not be significant and therefore one should always look at more than the first one. $L_2^* = l_2/(l_2 + \cdots + l_5) = .612$, which, compared with the tabular value for $p = 4$ and $r = 6$, .525, is significant. $L_3^* = .564$ and the tabular value for $p = 3$ and $r = 6.5$ is .620, which is not significant. Apparently, we have a "rank two" interaction and the size of l_2 was such that it precluded l_1 from appearing significant. l_1 accounts for 45%, l_2 for 33%.

The first two characteristic vectors u_1 and u_2 are given in Table 13.8 and the corresponding principal components from U^* multiplied by $L^{1/2}$ to form Z are found in Table 13.9. Alternatively, they may be found by multiplying the double-centered matrix D in Table 13.7 by U. The first component appears to

Table 13.8. Presidential Hopefuls. U-Vectors from D

	u_1	u_2
Carter	−.089	.825
Ford	−.268	−.409
Humphrey	−.684	−.180
Jackson	.102	.133
Reagan	.469	−.314
Wallace	.471	−.056
Characteristic Root	17684	13012
% Explained	45.4	33.4

Table 13.9. Presidential Hopefuls. z-Scores

Respondent	z_1	z_2
1	−31.6	26.3
2	−18.6	29.5
3	21.5	3.9
4	−4.8	−23.5
5	81.7	10.3
6	−12.8	28.8
7	−49.4	−19.2
8	−9.5	−18.5
9	18.4	−12.1
10	−2.9	−16.5
11	3.1	−33.2
12	8.8	26.8
13	5.7	29.1
14	38.4	−38.1
15	42.0	13.9
16	−32.1	45.1
17	−23.2	−3.0
18	−34.7	−49.6

be a liberal–conservative contrast and the second component is a contrast between Carter and the two Republicans, Ford and Reagan. A two-dimensional MDPREF plot is shown in Figure 13.1. This is not much different from the results in Chapter 10. Reagan and Wallace are now closer together and Respondent # 13 has moved much closer to Carter, reflecting that this individual had a strong preference for Carter over Humphrey, but the averages for these candidates over the 18 respondents were about equal.

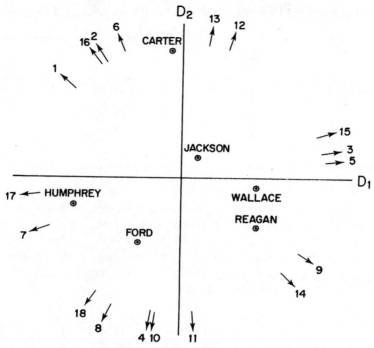

FIGURE 13.1 Presidential Hopefuls: MDPREF of double-centered data.

Next, what is the inherent variability? Assuming a two-component model and expanding on (13.7.8),

$$\hat{\sigma}^2 = \frac{l_3 + l_4 + l_5}{(b-1)(t-1) - E(l_1) - E(l_2)} = \frac{8243}{85 - 32.0 - 22.0}$$

$$= 266$$

where $E(l_1)$ and $E(l_2)$ were obtained from tables in Mandel (1971) using his $s = t - 1 = 5$ and $r = b - 1 = 17$. If we had stopped after a single component, the estimate for the variance would have been 401 and the Carter–Srivastava estimate, using (13.7.9) would have reduced it to 305. Quite a difference. Apparently the final chapter on this estimation procedure has yet to be written.

Finally, suppose we wish to find out whether the Carter–Ford difference was significantly different between respondents #1 and #3. The d_j, representing the Carter–Ford difference, are 1, −1, 0, 0, 0, 0. The c_j, representing the respondents, are all zero except $c_1 = 1$ and $c_3 = -1$. (This is a special type of contrast called a *tetrad*.) The computations for this contrast, using the original data from Table 10.1, are shown in Table 13.10 and show that this particular

Table 13.10. Presidential Hopefuls. Contrast between Carter and Ford for Respondents #1 and #3

Respondent		Candidate			
i	c_i	j	d_j	x_{ij}	$c_i d_j x_{ij}$
1	1	1	1	60	60
1	1	2	−1	40	−40
3	−1	1	1	50	−50
3	−1	2	−1	70	70
				Contrast	40

contrast had a value of 40. Assume that an independent estimate of inherent variability was available, say, $\sigma^2 = 200$. Then, with $f = \infty$, the tables in Appendix G.6 can be employed. $G_{.05, 5, 15, \infty} = 40.15$. $\sum c_i^2 = 2$, $\sum d_j^2 = 2$, and $r = 1$. Substituting in (13.7.12) results in $\sqrt{(2)(2)(40.15)(200)/1} = 179.22$. This means that any contrast smaller than ± 179.22 is not significant so, in the present case, the contrast of 40 is not significant.

CHAPTER 14

Other Applications of PCA

14.1 MISSING DATA

14.1.1 Introduction

One is not always so fortunate as to have a complete set of data with which to work. For any number of reasons, an observation vector may be missing a measurement on one or more variables. The largest amount of effort in coping with this problem has been expended in the area of the analysis of sample survey data where the response to each question would represent one variable. This problem is generally referred to as the *item nonresponse* problem. (This should not be confused with nonresponse caused by the inability of the researcher to contact the respondent in the first place or by the refusal of the respondent to participate in the survey. These conditions can produce serious biases if not detected and corrected.) Extremely complex imputation procedures have been employed for item nonresponse, in particular, for government surveys. A discussion of all these problems may be found in Madow, Nisselson and Olkin (1983), Madow and Olkin (1983), and Madow, Olkin and Rubin (1983). The most widely used techniques for multivariate data, in general, involve some sort of regression technique although there are instances where the use of the sample average for each variable may suffice. (For general articles on this subject, see, for instance, Beale and Little, 1975; Gleason and Staelin, 1975; Frane, 1976; and the book by Little and Rubin, 1987.)

It is not the purpose of this section to discuss all of these techniques but rather to focus on the relationship of this problem to principal component analysis. This will take two directions: (1) techniques for performing PCA in the presence of incomplete data and (2) techniques for estimating missing data using PCA.

In what is to follow, it is assumed that the cause of the missing data is unrelated to the relationships among the variables; to do otherwise would leave any results open to serious bias. An example of such a bias would be a sensitive question on a questionnaire for which the probability of nonresponse would be higher for certain identifiable groups of respondents than others.

(For a discussion of methods of testing whether or not missing data occur randomly or not, see Little, 1988b.) It is also well to keep in mind that if only a few elements are missing from a data matrix, most any of the recognized estimation procedures will produce satisfactory results, but if a lot of elements are missing, *none* of them will.

14.1.2   PCA in the Presence of Incomplete Data

Very little has been done in the nature of direct methods of obtaining characteristic roots and vectors when there are missing data. It is much more common to obtain the estimate of either the covariance or correlation matrix using missing data techniques. One proposal by deLigny et al. (1981) was motivated by a chemistry problem in which they wished to determine the solubility of $n = 49$ gases in each of $p = 6$ solvents and they had 167 missing pieces of data. Starting with the relationship $X = VY$, an initial estimate of V was obtained based on whatever chemical information might have been available. They would alternatively obtain new estimates of Y and V by regression. The number of vectors was fixed in advance by their knowledge of the chemical system. While this might work for some chemical examples, there is no record of it being used elsewhere. Let us now see what the alternatives might be.

14.1.3   Estimation of Correlation Matrices

The most popular techniques for estimating correlation matrices in the presence of incomplete data are:

1. Replace each incidence of a missing value with the average of all available data in the sample for that particular variable and obtain the correlation matrix for this adjusted set of data (Wilks, 1932a).
2. Obtain each correlation coefficient in the matrix on the basis of all data vectors in the data set for which neither value is missing for that particular pair (Glasser, 1964).

Gleason and Staelin (1975) concluded from some empirical work that the second method will be superior for values of φ (as defined in Section 2.8.2) greater than $.5323 - .0722 \ln(n)$ where n is the total number of data vectors with or without missing data. If φ is less than this, a state of poor correlation is considered to exist and the method of averages is actually superior.

Unfortunately, the correlation matrices obtained by either method may be ill-conditioned; in particular, they may not be positive semidefinite. It may be possible to improve these estimates either: (1) by use of the technique to be discussed in Section 14.1.5 to estimate the missing observations themselves or

(2) by a direct method proposed by Frane (1975) as follows:

1. Denote the initial estimate of R by R_0.
2. Obtain the characteristic roots and vectors of R_0:

$$R_0 = ULU'$$

3. If none of the characteristic roots of R_0 are negative, then R_0 is not ill-conditioned.
4. If any of the characteristic roots are negative, then R needs to be reestimated. Delete all of the rows and columns of L as well as the columns of U associated with the negative roots. Denote the remaining submatrices by L_1 and U_1, respectively.
5. Let $A = U_1 L_1 U'_1$.
6. Let F be a diagonal matrix whose elements are the square roots of the diagonal elements of A. Then the adjusted correlation matrix will be

$$R_1 = F^{-1}AF^{-1}$$

Knol and TenBerge (1989) proposed a least-squares approximation employing oblique Procrustes rotation. This scheme allows the correlations that were based on the complete data set to remain unchanged.

14.1.4 Estimation of Covariance Matrices

If one wishes to adjust the corresponding covariance matrix, this can be done by the operation

$$S_1 = CR_1 C$$

where C is a diagonal matrix made up of the standard deviations of each variable obtained from all data available from the sample for that variable. Note that this is done in conjunction with the new correlation matrix rather than performing Steps 1–5 on the initial covariance matrix directly. The rationale for this is that the initial standard deviations are estimated with more precision than the initial correlations and hence they should not be adjusted.

14.1.5 Estimation of Missing Data

The majority of techniques for estimating the data themselves are regression techniques which are summarized, for instance, in Beale and Little (1975), Gleason and Staelin (1975) and Frane (1976). However, Gleason and Staelin also included a technique that involves PCA as follows:

1. Assume all of the data have zero mean.
2. Form an estimated covariance matrix using the methods described in Sections 14.1.3 and 14.1.4.
3. Obtain the characteristic roots and vectors of this matrix. In particular, obtain the U-vectors and, using one of the methods described in Section 2.8, choose a relevant subset of k vectors.
4. For any row of data with one or more missing values, let the number of these missing values be m. Reorder the variables in this data vector so that the first m elements represent the variables with missing data. This partitioning will be:

$$[\mathbf{x}^{(1)} \vdots \mathbf{x}^{(2)}]$$

where $\mathbf{x}^{(1)}$ has m elements (all missing) and $\mathbf{x}^{(2)}$ has $p - m$ elements. In a similar manner, reorder the rows in $\mathbf{U}$ and partition it in the form

$$\mathbf{U}_m = \begin{bmatrix} \mathbf{U}^{(1)} \\ \hdashline \mathbf{U}^{(2)} \end{bmatrix}$$

where $\mathbf{U}^{(1)}$ is $m \times k$ and $\mathbf{U}^{(2)}$ is $(p - m) \times k$.
5. Form the product $\mathbf{U}_m \mathbf{U}'_m = \mathbf{M}$, which is partitioned as follows:

$$\mathbf{M} = \begin{bmatrix} \mathbf{U}^{(1)}\mathbf{U}^{(1)\prime} & \mathbf{U}^{(1)}\mathbf{U}^{(2)\prime} \\ \mathbf{U}^{(2)}\mathbf{U}^{(1)\prime} & \mathbf{U}^{(2)}\mathbf{U}^{(2)\prime} \end{bmatrix} = \begin{bmatrix} \mathbf{M}_{11} & \mathbf{M}_{12} \\ \mathbf{M}_{21} & \mathbf{M}_{22} \end{bmatrix}$$

Only $\mathbf{M}_{11}$ and $\mathbf{M}_{21}$ are required for this procedure.
6. The estimates for the missing elements $\hat{\mathbf{x}}^{(1)}$ are

$$\hat{\mathbf{x}}^{(1)} = \mathbf{x}^{(2)} \mathbf{M}_{21} [\mathbf{I} - \mathbf{M}_{11}]^{-1} \tag{14.1.1}$$

where $\mathbf{M}_{21}$ is $(p - m) \times m$ and $\mathbf{I}$ and $\mathbf{M}_{11}$ are $m \times m$.

Steps 4–6 would be repeated for each configuration of missing data. If there are more than one observation, say g observations, all having missing elements in the same variables, these operations may be performed on all of them at once. $\mathbf{X}^{(1)}$ will be $g \times m$ and $\mathbf{X}^{(2)}$ will be $g \times (p - m)$. Gleason and Staelin suggested that this process may be iterated and that it would be profitable to do so if $\varphi > .3$. After all of the missing values have been estimated, a new covariance matrix is obtained and steps 3–6 are repeated.

The related regression procedure would consist of reordering the covariance matrix in the same manner as steps 4 and 5. Let the covariance matrix be rearranged so that the first m rows and columns are related to the variables with missing values. Then the matrix is partitioned

$$S = \begin{bmatrix} S_{11} & \vdots & S_{12} \\ \hline S_{21} & \vdots & S_{22} \end{bmatrix}$$

where S_{11} is $m \times m$, S_{22} is $(p - m) \times (p - m)$, and so on. The regression estimates of $x^{(1)}$ are

$$\hat{x}_R^{(1)} = x^{(2)} S_{22}^{-1} S_{21} \tag{14.1.2}$$

The extent to which $\hat{x}_R^{(1)}$ and $\hat{x}^{(1)}$ agree is a function of the deleted pc's. However, the more pc's retained, the closer $[I - M_{11}]$ in (14.1.1) will approach singularity.

14.1.6   Example

The following example will illustrate these methods and also illustrate the risks in using them. Let us assume that the Audiometric data in Table 5.1 have been replaced by a similar table in standard units but that somewhere in the process the bottom right-hand corner of the table has been torn off so that no information is available for the last three observations on the seventh and eighth variables, the right ear at 2000 Hz and 4000 Hz. These values are, in fact,

$$X^{(1)} = \begin{bmatrix} .37 & 2.78 \\ .37 & .96 \\ 1.44 & -.59 \end{bmatrix}$$

For this correlation matrix (which is also the covariance matrix since the data are now in standard units), $\varphi = .45$, the Gleason–Staelin criterion from Section 14.1.3 with $n = 100$ is

$$.5323 - .0722 ln(100) = .20$$

Since $\varphi > .2$, a regression or PCA solution would be preferable to replacing the missing values with the variable averages, which would be zero or close to it for x_7 and x_8. The regression estimate is

$$\hat{X}_R^{(1)} = \begin{bmatrix} .58 & 1.06 \\ .42 & .02 \\ -.70 & -1.07 \end{bmatrix}$$

which is generally better than using averages except for $x_{100,\,7}$.

For this example, the PCA solution will do no better even after iteration, which would be recommended since $\varphi > .3$. Recall from Chapter 5 that four pc's were retained. Using four pc's in this solution, after two iterations,

$$\hat{X}^{(1)} = \begin{bmatrix} -.21 & 1.06 \\ .78 & -.11 \\ -.76 & -1.50 \end{bmatrix}$$

subsequent iterations being of little effect. Note from Table 5.13 that the right–left ear difference for the 100th observation is almost significant at the .05 level and is also rather strong for the 98th observation. Since both missing variables are for the right ear, this would contribute to the poor results by both methods. These missing data did not occur in a random pattern and when a situation as this exists the results should be used with extreme caution.

14.2   USING PCA TO IMPROVE DATA QUALITY

It has already been demonstrated that PCA can be used to enhance the quality of a set of data by using the Q-statistic or the Hawkins statistic to test for outliers. However, Gleason and Staelin (1973) used PCA to directly enhance the quality of a certain type of data set.

The type of data Gleason and Staelin had in mind is the sort that often appears on questionnaires. The respondent is either asked to rate something on, say, a scale of 1–10 or is presented with a set of ordered categories, which either are already identified by integers, say, 1–5 or have descriptive labels which the analyst will then convert into these numbers. The rationale for their procedure is that the true rating variable is continuous. If it were known for each respondent, it would produce, at the very least, interval data. However, by presenting the questions in a category form, the actual responses are in the form of ordinal data, or at best, quasiinterval data so that there has been some loss in quality by doing this. This is more than compensated by the increase in quality achieved by presenting the questions in this manner. Nevertheless, it may be possible to improve the quality of the response data and this is what they proposed to do.

It is assumed that some redundancy has been built into the questionnaire so that not only are the results from some of the questions highly correlated but the rank of the true data matrix, were it obtainable, would be $g < p$ where p is the number of questions. Gleason and Staelin suggested that the use of ordinal responses would add some measurement error and produce a data matrix of rank p with the last $p - g$ pc's representing this error. They concluded that their procedure would not produce a worthwhile improvement unless $g < (p/2)$.

Since the number of categories may differ from question to question, it is best to transform the response data into standard units. If there are p questions and n respondents, let the $n \times p$ matrix of standardized data be $\mathbf{X}$. In Chapter 10, it was shown that $\mathbf{X}$ could be decomposed using the SVD technique, viz.,

$$\mathbf{X} = \mathbf{U}^{*\prime}\mathbf{L}^{1/2}\mathbf{U} \tag{14.2.1}$$

where $\mathbf{U}^*$ are the scores of the pc's and $\mathbf{L}$ and $\mathbf{U}$ are the characteristic roots and vectors of the correlation matrix $\mathbf{X}'\mathbf{X}/n$. Since it is hypothesized that the true rank of the interval data is g, an estimate of g is obtained, the first g roots and vectors are retained (denoted by the subscript g), and an estimate of the "true" response data is obtained by the relationship

$$\hat{\mathbf{X}} = \mathbf{U}_g^{*\prime}\mathbf{L}_g^{1/2}\mathbf{U}_g \tag{14.2.2}$$

It can be shown this is equivalent to

$$\hat{\mathbf{X}} = \mathbf{X}\mathbf{U}_g'\mathbf{U}_g \tag{14.2.3}$$

so that only $\mathbf{U}_g$ need be computed.

Gleason and Staelin performed a number of simulations to study the effect of sample size, number of variables, and true rank. One application they suggested for this technique was the matching of similar respondents, which they felt should be enhanced by using the adjusted data. Another suggested application would be to perform the same process on $\mathbf{X}'$, which would have the effect of removing the idiosyncrasies of the respondents.

Young and some of his colleagues produced a similar procedure using alternating least squares (DeLeeuw et al., 1976; Young et al., 1978; Young, 1981) and extended this to individual difference scaling (Takane et al., 1980). Didow et al. (1985) illustrated this with two examples where the correlations among the $\hat{\mathbf{X}}$ were larger than among the $\mathbf{X}$, confirming the conjecture of Gleason and Staelin.

14.3 TESTS FOR MULTIVARIATE NORMALITY

It is convenient in many cases to assume that a sample is drawn from a population that is normally distributed. This assumption is useful in obtaining the distribution of test statistics, in constructing confidence limits, and in estimating the proportion of a population that falls within certain limits. This is also an important property in multivariate analysis, more so in fact, because the majority of multivariate techniques available are large sample procedures whose properties depend on multivariate normality and on which little is known about the effect of nonnormality.

Standard univariate goodness-of-fit tests do not generalize very well. One cannot merely test the marginal distributions because that does not take into account the correlations among the variables. It is possible to have nonnormal multivariate distributions with normal marginals. Consider the example in Table 14.1. All of the cells save the four that are more heavily bordered contain the densities one would expect for a bivariate normal distribution with uncorrelated variables. The densities for these four cells should all be equal to .0084. However, the cells in the northeast and southwest quadrants have had .0084 added to them and the other two cells have had that quantity subtracted from them. The marginal distributions for x_1 and x_2 are unchanged but the joint distribution is no longer bivariate normal. Thus, testing only the marginal distributions would have produced an incorrect conclusion.

The time-honored chi-square goodness-of-fit test can be expanded, in theory, to any number of variables but is not practical for more than two or three variables, partly because of the labor of computing all of the expected frequencies and because, with the increasing number of variables, the number of cells increases exponentially and with it the probability of cells with very small expected frequencies. For bivariate distributions, a two-way array of frequencies is required; the trivariate case would require a three-way array, and so on. A number of procedures for testing multivariate normality have been proposed, mostly involving tests on the third and fourth moments or plotting procedures (Seber, 1984, p. 184). Other techniques include a multivariate extension of the Shapiro–Wilk test (Royston, 1983) and some tests of multivariate normality using the Anderson–Darling and Cramer–von Mises procedures (Paulson et al., 1987). For a review paper on the subject, see Koziol (1986).

The method of principal components has been used for an ad hoc test of multivariate normality. This procedure is based on the fact that if a set of variables has a joint multivariate distribution, then any linear combination of these variables will also be normally distributed. Brown et al. (1956) were investigating the distribution of color matches, a trivariate problem. The three marginal distributions were all normal. They then obtained the three principal components and found that their marginal distributions were also normal. They concluded that this was "strong evidence" that the matches were trivariate normal. Of course, it was just that—strong evidence—a necessary but still not sufficient condition for multivariate normality. Nevertheless, the gates were beginning to close. If the original variables are well correlated, as they were in this case, then the axes represented by the principal components will not lie close to the original coordinate axes and if these marginals are also normal, it will remove that much more possibility that nonnormality could exist. Of course, if the original marginal distributions are nonnormal, there is no point going any further. In fact, Gnanadesikan and Kettenring (1972) pointed out that because the pc's are linear combinations of the original variables they would tend towards normality even if the original variables were nonnormal.

Brown et al. used classical chi-square goodness-of-fit test on the pc's. Since that time, alternative tests of normality have become available that may be adapted for use with principal components. In particular, Srivastava and Carter (1983, p. 286) recommended the use of the Shapiro–Wilk (1965) statistic on

Table 14.1. Bivariate Nonnormal Distribution with Normal Marginals. Proportions × 10⁴

		Cell Midpoints for x_1												
		Under −2.50	−2.25	−1.75	−1.25	−.75	−.25	.25	.75	1.25	1.75	2.25	Over 2.50	Marginal Totals
Over	2.50	0	1	3	6	9	12	12	9	6	3	1	0	62
	2.25	1	3	7	15	25	32	32	25	15	7	3	1	166
	1.75	3	7	19	40	66	84	84	66	40	19	7	3	438
	1.25	6	15	40	0	138	176	176	138	168	40	15	6	918
	.75	9	25	66	138	225	287	287	225	138	66	25	9	1500
Cell	.25	12	32	84	176	287	367	367	287	176	84	32	12	1916
midpoints	−.25	12	32	84	176	287	367	367	287	176	84	32	12	1916
for x_2	−.75	9	25	66	138	225	287	287	225	138	66	25	9	1500
	−1.25	6	15	40	168	138	176	176	138	0	40	15	6	918
	−1.75	3	7	19	40	66	84	84	66	40	19	7	3	438
	−2.25	1	3	7	15	25	32	32	25	15	7	3	1	166
Under	−2.50	0	1	3	6	9	12	12	9	6	3	1	0	62
	Marginal totals	62	166	438	918	1500	1916	1916	1500	918	438	166	62	10000

327

the pc's and proposed the following form:

$$C_1 = \sum_{i=1}^{p} \frac{1}{l_i} \left[\sum_{j=1}^{n} a_j(z_{i[j]} - \bar{z}_i) \right]^2 \tag{14.3.1}$$

where $z_{i[j]}$ are the order statistics of the ith pc with z_i as defined in (1.4.1) using U-vectors and a_j are tabulated in Shapiro and Wilk. However, if one used W-vectors to obtain the pc-scores, then (14.3.1) may be simplified to

$$C_1 = \sum_{i=1}^{p} \left[\sum_{j=1}^{n} a_j(y_{i[j]} - \bar{y}_i) \right]^2 \tag{14.3.2}$$

The distribution of C_1 is not known at this point but one could test each component separately, keeping in mind the cumulative effect of these on the Type I error. Matthews (1984) used *robust* PCA (a topic to be discussed in Section 16.5) to investigate multivariate normality and, for his examples, suggested that the extension of the Shapiro–Wilk test was not very powerful against the alternative of a normal distribution contaminated by a small number of outliers.

The example in Table 14.1 is not a strong aberration, the correlation between x_1 and x_2 being only .05. Since both x_1 and x_2 are in standard units, the covariance and correlation matrices are the same and, as we shall see in Section 15.3, this particular situation will produce principal components whose coordinate axes fall along the 45° lines because there are only two variables in standard units. Both of these pc's would have nonnormal distributions. This is a very special case. In general, however, the higher the intercorrelations, the more likely are these new axes to be rotated significantly from the original and make one more confident that the demonstration of normality for the pc's would, indeed, produce the "strong evidence" desired. Cases where the new axes lie close to the original would not be very convincing.

If the new axes have a sufficiently different orientation, then, at the very least, the use of PCA to test the necessity condition would appear to have some merit. Admittedly, this is not an elegant solution but one which, in the many cases, will get the job done.

PCA has also been applied to univariate goodness-of-fit procedures. Durban and Knott (1972) and Durban et al. (1975) decomposed the Cramer–von Mises statistic into principal components of the differences between the actual and hypothetical cumulative distributions and showed that these pc's were also equal to the coefficients of the Fourier expansion of this difference.

14.4 VARIATE SELECTION

14.4.1 Introduction

In Section 12.3.4, some techniques were discussed for employing PCA to reduce the number of predictor variables used in regression analysis. In this section, a similar problem will be dealt with but in this case there is but a single set of variables and the rationale for deletion of variables is merely the detection of variables that are redundant within the set.

The key to these techniques lies in the vectors associated with the smallest characteristic roots. If one of the roots is equal to zero, as was the case in Section 2.4, this indicates that two or more variables are linearly related. The nature of this relationship is portrayed by the associated characteristic vector. In that example, the vector associated with the zero root was $u' = [-.577 \ -.577 \ .577]$, implying that $x_1 + x_2 = x_3$, the constraint that was introduced in that example. This situation can exist when the unwary analyst has been presented with some data that contains one or more variables of this type, included by a researcher on the philosophy that "more is better," and this can happen with predictor variables in regression analysis as well. Zero roots also occur when there are more variables than observations (the crime data described in Appendix E.2.2, for example) but that would be known ahead of time and the linear combinations will be ones that are forced on the data by that situation.

More common will be the situation where two or more variables are highly correlated, resulting in one or more roots being positive but quite small. This will be the major subject of this section.

14.4.2 Use of Vector Coefficients

These techniques deal, for the most part, with the coefficients of the vectors associated with the smallest roots. The large coefficients in these vectors may relate to redundancies and these techniques are designed with this in mind.

Jolliffe (1972, 1973) compared a number of variable-deletion techniques, several of them involving PCA. One of them, which he called B_2, consisted of performing a PCA on the correlation matrix. One would select all of the W-vectors associated with characteristic roots that are less than some arbitrary value λ_0. (On the basis of some simulation studies, he recommended setting $\lambda_0 = .7$. He also advocated using stopping rules to determine the number of vectors to examine. Starting with the vector having the smallest root, w_p, the variable is deleted for which w_p has the largest coefficient in absolute value. Next, w_{p-1} is examined and the variable having the largest coefficient of those variables remaining is deleted. The process is repeated until all of these vectors have been examined.

Consider, in Table 14.2, the W-vectors and their associated characteristic roots of the correlation matrix for the Black-and-White Film examples. This correlation matrix was displayed in Table 7.5. Note that only the first three roots exceed .7 so this procedure would imply that only three variables be retained. (Recall in Section 2.9, when working with a covariance matrix, that three pc's were retained.) Consider, first, w_{14}. The largest element, in absolute value, is $w_{5,14} = -12.06$ indicating that x_5 should be deleted. Next, for w_{13}, the largest element for those variables remaining is $w_{3,13}$ which eliminates x_3,

Table 14.2. B&W Film Example. W-Vectors from Correlation Matrix.

	w_1	w_2	w_3	w_4	w_5	w_6	w_7	w_8	w_9	w_{10}	w_{11}	w_{12}	w_{13}	w_{14}
x_1	.08	-.11	.43	-.83	.33	.51	.69	.67	-1.46	-4.21*	-4.26	2.15	2.72	-1.24
x_2	.09	-.11	.35	-.40	.20	.36	.24	.30	-.68	1.14	5.34*	-4.08	-7.85	2.26
x_3	.10	-.11	.22	.01	.05	-.04	-.14	-.57	1.68	3.88	3.47	1.57	9.20*	-3.23
x_4	.11	-.09	.08	.44	-.20	-.44	-.42	-.85	2.26	2.55	-5.25	3.53	-4.16	7.89
x_5	.12	-.07	-.05	.65	-.36	-.74	-.82	-.75	.56	-1.16	-2.48	-4.17	-2.67	-12.06*
x_6	.12	-.05	-.14	.73	-.53	-.57	-.74	-.09	-.76	-4.19	2.23	-3.48	4.55	8.74
x_7	.12	-.01	-.22	.46	-.37	-.06	.04	.72	-2.89	-.35	3.09	8.45*	-2.81	-3.06
x_8	.11	.04	-.23	.20	.01	.80	1.12	2.03	-2.33	3.78	-3.37	-4.26	1.97	.92
x_9	.11	.08	-.24	-.15	.39	.97	1.31	.81	4.93*	-2.52	1.50	.63	-1.16	-1.00
x_{10}	.10	.12	-.19	-.83	.83	.20	.55	-3.69*	-1.79	.17	-.13	-.55	.13	.72
x_{11}	.08	.18	-.00	-.99	.63	.07	-3.01*	1.43	.27	.43	-.26	.25	-.01	.16
x_{12}	.05	.21	.14	-.98	-1.94*	-1.26	1.00	.20	.39	.43	.14	-.22	-.03	-.06
x_{13}	.03	.21	.28	1.08*	1.52	-1.55	.91	.65	-.24	-.09	.29	.24	.07	.08
x_{14}	.01	.22	.33	1.18	-.80	1.98*	-.54	-.97	-.30	-.25	-.18	.19	.02	-.37
Characteristic Root	8.206	4.020	1.278	.134	.115	.091	.061	.041	.019	.012	.008	.006	.005	.003

330

and so on. The elements that specified the variables to be deleted are indicated with asterisks and by the time each of the last eleven vectors had been examined, only x_4, x_6, and x_8 remain. Note, for $\mathbf{w}_4$, that the largest element was $w_{14,4}$ but x_{14} had already been eliminated by $\mathbf{w}_6$ and the largest element corresponding to a variable still in the running was $w_{13,4}$. Although the $\mathbf{W}$-vectors were used for this exercise, either the $\mathbf{U}$- or $\mathbf{V}$-vectors could have been used since $\mathbf{u}_i$, $\mathbf{v}_i$, and $\mathbf{w}_i$ are all proportionate to one another and the elimination is done one vector at a time. Tortora (1980) used this technique when investigating a number of weighting techniques for a sample survey designed to measure the attitude and understanding of some farmers and ranchers in North and South Dakota with regard to surveys run on behalf of the USDA.

A variant of this, B_1, would then perform a PCA on the correlation matrix of the remaining variables and keep repeating this procedure until none of the remaining roots qualify (Beale et al., 1967). Another variant, B_4, was a backward version of B_2 in that it started with the largest root and retained the variable with the largest coefficient, and so on. These roots were compared with a larger value of λ_0. Jolliffe also examined a number of other techniques that did not use PCA, some involving examination of the correlation matrix and some involving clustering of the variables. These will be listed in Section 18.5. One of these, A_2, a stepwise procedure, consisted of deleting that variable whose multiple correlation with the remaining $p - 1$ variables was a maximum. The procedure is repeated with the remaining $p - 1$ variables and so on until some prescribed number of variables remain or the remaining variables have multiple correlation coefficient no larger than some prescribed amount (Jolliffe used .15 for his studies).

Jolliffe did a number of simulation studies as well as examining some real data sets. For the simulation data, A_2 appeared to be the most consistent in that it retained the "best" or "good" subsets of variables, never "moderate" or "bad" sets. B_4 retained best sets more than A_2 but also retained some moderate or bad ones. B_2 was in the middle. Generally speaking, when compared with A_2, the PCA procedures had smaller Type II errors and larger Type I errors. Jolliffe recommended retaining at least four variables using either B_2 or B_4

Hawkins and Fatti (1984) used the $\mathbf{W}$-vectors in a different manner. Suppose that the kth vector, $\mathbf{w}_k$ had a large element, say w_{jk}. Then one gets x_j in terms of the others by the relationship

$$x_j = -\sum_{i \neq k}^{p} \frac{w_{ik} x_i}{w_{jk}} \qquad (14.4.1)$$

and the residual variance for x_j will be $1/(w_{jk})^2$. Illustrating this with the last vector from the covariance matrix of Ballistic Missile example,

$$\mathbf{w}_4 = \begin{bmatrix} -.0643 \\ -.0364 \\ .2126 \\ -.1013 \end{bmatrix}$$

$w_{43} = .2126$ is the largest value in the entire **W**-matrix. Solving in (14.4.1),

$$x_3 = -(-.0643x_1 - .0364x_2 - .1013x_4)/(.2126)$$
$$= .302x_1 + .171x_2 + .476x_4$$

These results are in terms of deviations from their respective means. The residual variance for x_3 is $1/(.2126)^2 = 22.12$. It is interesting to note that if one had used least squares to predict x_3, the corresponding result would have been

$$x_3 = .245x_1 + .249x_2 + .366x_4$$

with a residual variance equal to 21.19. (Neither of these quantities is corrected for degrees of freedom.) Kane et al. (1985) used PCA to study dependency structure as a guide to deletion of variables and included several classic examples.

14.4.3   Principal Variables

This technique is due to McCabe (1984) and is to PCA what the "all possible combinations" is to regression analysis. The term *principal variables* refers to that subset of variables which optimizes one or more criteria suggested by McCabe. For an arbitrary number of retained variables, k, let S_M be the residual covariance (or correlation, whichever is appropriate) matrix of the deleted variables with the effect of the retained variables removed. S_M is of order $p - k$. Let $l_{M,1}, l_{M,2}, \ldots, l_{M,p-k}$ be the characteristic roots of S_M. Compute the quantity

$$M_1 = \sum_{i=1}^{p-k} l_{M,i} \tag{14.4.2}$$

The set of k variables for which M_1 is a minimum are the principal variables. McCabe suggested several other criteria including the product of the roots (M_2), the sums of squares of the roots (M_3), and a fourth involving the canonical correlations between the retained and deleted variables (M_4). M_4 is equal to the sum of squares of the canonical correlation coefficients and, in this case, the principal variables will be that set for which M_4 is a *maximum*.

This procedure will also be illustrated with the Ballistic Missile data, this time using the covariance matrix as that was the form in which the original problem was handled. One would not ordinarily use these methods for such a small example but its size makes it useful for illustrative purposes. For this example, the task will be to find the combination of *two* retained variables that defines the principal variables. Consider one of the subsets of retained variables to be x_1 and x_2, which means that x_3 and x_4 would be deleted. Partition the covariance matrix (Table 2.1) so that the covariance matrix of the retained variables is

$$S_R = \begin{bmatrix} 102.74 & 88.67 \\ 88.67 & 142.74 \end{bmatrix}$$

the matrix of the deleted variables is

$$S_D = \begin{bmatrix} 84.57 & 69.42 \\ 69.42 & 99.06 \end{bmatrix}$$

and the covariance between them is

$$S_{RD} = \begin{bmatrix} 67.04 & 54.06 \\ 86.56 & 80.03 \end{bmatrix}$$

Using the relationships given in (12.2.2) and (12.2.3), the residual covariance of the deleted variables given the retained variables is

$$S_{D,R} = S_D - S'_{RD}S_R^{-1}S_{RD} \tag{14.4.3}$$
$$= \begin{bmatrix} 28.38 & 19.68 \\ 19.68 & 53.79 \end{bmatrix}$$

The characteristic roots of this matrix are 64.51 and 17.67, whence $M_1 = 1140$, $M_2 = 82.18$, and $M_3 = 4474$. To obtain M_4, use (12.6.4), from which the squared canonical correlation coefficients are .6702 and .0290, making $M_4 = .6992$. The results from all possible combinations of two retained and two deleted variables are given in Table 14.3. Retaining x_2 and x_4 produces the lowest values of M_1, M_2, and M_3 and the maximum value of M_4. M_4 is a function only of the split between the groups, not which is retained, so that the winner, x_2 and x_4, produce the same M_4 as x_1 and x_3. Jolliffe (1986, p. 110) compared M_1 with his B_2 and B_4 on some real data including the crime data described in Appendix E.2.2. For examples with a larger number of variables than the example given above,

Table 14.3. Ballistics Missile Example. McCabe Procedure

Retained Variables	Deleted Variables	M_1	M_2	M_3	M_4
1, 2	3, 4	1140	82.18	4474	.6992
1, 3	2, 4	1848	88.09	4064	.8282
1, 4	2, 3	1069	74.76	3451	.7854
2, 3	1, 4	1693	82.76	3463	.7854
2, 4	1, 3	1002	71.34	3084	.8282
3, 4	1, 2	2179	101.80	6006	.6992

McCabe concluded that to account for the same amount of variability as a fixed number of pc's, one or two more principal variables would be required.

Krzanowski (1987a) gave a procedure in which k pc's $\mathbf{Y}_p$ were obtained from the full set of variables and k pc's $\mathbf{Y}_q$ from a reduced set of $q \geqslant k$ variables. The two sets of pc's were compared using generalized procrustes rotation (Section 8.6) with $\mathbf{Y}_p$ as the target matrix. He combined this with a backward elimination scheme to obtain the optimum subset based on the sums of squared differences between $\mathbf{Y}_p$ and $\mathbf{Y}_q$,

$$\mathbf{Y}_p\mathbf{Y}_p' + \mathbf{Y}_q\mathbf{Y}_q' - 2\mathbf{Y}_q\mathbf{U}\mathbf{U}^{*\prime}\mathbf{Y}_p \tag{14.4.4}$$

where $\mathbf{U}$ and $\mathbf{U}^*$ are obtained from the singular value decomposition of $\mathbf{Y}_p'\mathbf{Y}_p$.

14.5 DISCRIMINANT ANALYSIS AND CLUSTER ANALYSIS

14.5.1 Introduction

In Chapter 13 on MANOVA, we were interested in whether or not a significant difference existed among two or more groups when the response was multivariate in nature. Given that a significant difference does occur, additional techniques may be utilized just as they are in univariate analysis of variance. Much has been written about both discriminant analysis and cluster analysis and it is not the intention of this section to describe either of these procedures in any detail. (Of interest may be a 1989 article by the Panel on Discriminant Analysis, Classification, and Clustering, a review of the state of the art at that time along with a discussion of some unsolved problems and a description of available software.) The purpose of this section is to point out what contributions, if any, PCA can make towards the fulfillment of the aims of these procedures. Compared to the other techniques described in this chapter, we shall see that these contributions are rather limited.

14.5.2 Discriminant Analysis

The purpose of *discriminant analysis* is to produce one or more functions that will permit one to assign an individual observation to one of two or more groups of individuals such that this individual will be more like the group to which it has been assigned than any other. If successful, these functions will be said to be able to *discriminate* among groups. For these functions to be successful, there must be a significant difference among the groups in the first place. The applications of discriminant analysis are widespread, including chemistry (where the item to be classified might be an unknown substance), anthropology (skulls of an unknown individual to be classified as belonging to a group such as a tribe or caste), market research (market segmentation), and so forth. Most multivariate texts will have a chapter on this subject.

The general procedure is to collect a data set in which the identification of the individuals as to group is known and establish the discriminant functions on the basis of these data. There are two main methods used to obtain these functions (see for instance, Rao, 1952, Chapter 8).

1. Use of likelihood functions in which such a function is established for each group. (This requires an assumption about the distribution within the group.) Then each unknown is evaluated in terms of each likelihood function and assigned to that group whose likelihood is the largest.

2. An extension of the multivariate analysis of variance that was discussed in Section 13.3. In that context, the discriminant functions are the characteristic vectors associated with the matrix AE^{-1}.

To illustrate this second method, we shall return to the Chemical data in Table 13.1, which has been reproduced in Table 14.4. In the MANOVA example in Section 13.3, there was only one significant root from AE^{-1}. The corresponding vector was

$$\mathbf{u}_1 = \begin{bmatrix} .918 \\ -.396 \end{bmatrix}$$

The discriminant function will be

$$u_{11}x_1 + u_{21}x_2 = .918x_1 - .396x_2$$

One does not need to correct for the means in this case. First, all three averages are evaluated in terms of this function and they are 5.08 for Lab 1, 5.22 for

Table 14.4. Grouped Chemical Data. Discriminant Analysis

Laboratory		Method 1	Method 2	Value of Discriminant Function	Classification
1		10.1	10.5	5.11	Lab 1
		9.3	9.5	4.78	Lab 1
		9.7	10.0	4.94	Lab 1
		10.9	11.4	5.49	Lab 2
	Average	10.00	10.35	5.08	
2		10.0	9.8	5.30	Lab 2
		9.5	9.7	4.88	Lab 1
		9.7	9.8	5.02	Lab 1
		10.8	10.7	5.68	Lab 3
	Average	10.00	10.00	5.22	
3		11.3	10.1	6.37	Lab 3
		10.7	9.8	5.94	Lab 3
		10.8	10.1	5.91	Lab 3
		10.5	9.6	5.84	Lab 3
	Average	10.82	9.90	6.01	

Lab 2 and 6.01 for Lab 3. Since the characteristic root associated with u_1 is significant, it means that these three values are significantly different. Now, each individual observation is evaluated. The first observation for Lab 1 produces a value of 5.11. This is closest to the Lab 1 mean so that observation would be classified as Lab 1. The fourth observation, however, is classified as Lab 2 and for the data set as a whole there are four misclassifications. This might be construed to imply that if an unidentified result were found, there would be about one chance in three that it would not be classified correctly. The problem is primarily Lab 2, whose mean lies between the other two; even though the means are significantly different, there is some overlap among the individuals.

This example does not make use of PCA as such nor would there be any point in it since there are only two variables. How might PCA be used to enhance discriminant analysis? The most obvious way is to perform a PCA on the data and do the analysis in terms of the pc-scores. As in the case of principal components regression, if all the pc's are used, the answers will be the same and the only reason to use PCA is to reduce the size of the computing problem plus, possibly, the fact that the pc's are uncorrelated. However, as in the case of PCR, there is no guarantee that the larger pc's will necessarily produce the best discriminant functions [Kshirsager et al. (1990) and Dillon et al. (1989), the latter including an example]. If one has to obtain all of the pc's, there is probably no advantage in using PCA at all. If near singularity is suspected, then PCA has the advantage of obtaining a solution that otherwise might not be obtainable (see, for instance, Loh and Vanichsetakul, 1988). For the case where $p > n$, as may occur in some chemical applications, PCA may be a practical way to obtain a solution. Heesacker and Heppner (1983) used the pc's themselves as discriminators. A single pc cannot discriminate better than the first discriminant function but sometimes it may be good enough that it may be of use, particularly if the same pc has been used elsewhere and is interpretable.

Multiple regression can be used to obtain discriminant functions by using dummy variables to identify the groups. In the most simple case—two groups—one group is assigned the value 1 and the other 0 or -1. These dummy variables are then regressed against the variable set to obtain the discriminant function. Stahle and Wold (1987) and Vong et al. (1988) used partial least-squares regression to obtain discriminant functions in this manner.

One of the problems with using the vectors of AE^{-1} is that, like MANOVA, it assumes homogeneity of the covariance matrices among groups. There are ways to handle this, one being to use likelihood functions instead. Another proposal, using PCA, is the SIMCA (soft independent modeling of class analogy) procedure (Wold, 1976). In this procedure, a PCA is performed separately for each group. Each unknown observation is then transformed into pc's for each group and assigned to that group for which the residual sum of squares, Q, is the smallest. If each group produces about the same value of Q, the observation is considered an outlier and is not assigned to any group. The SIMCA procedure is claimed to be able to handle the case where one group is quite diverse. Most of the published applications of SIMCA are related to chemistry. Some examples of chemical structure have been given by Dunn et al. (1978), Dunn and Wold (1978) and Cartier and Rivail (1987). Luotamo et al. (1988) used SIMCA in analyzing blood samples for PCB's. Derde and Massart (1988) compared SIMCA with two other pattern recognition techniques, UNEQ and PRIMA.

Principal components are used directly in an alternative procedure known by various names including *canonical* or *two-stage* discriminant analysis. This notion goes back at least to Rao (1952, p. 367). Here, the characteristic vectors **W** are obtained initially from the pooled within-group covariance matrix **E**. The group means are obtained in terms of the *y*-scores and a weighted among-group covariance matrix is obtained from these results. This is equal to **W'AW** where **A** is the original among-group matrix. The characteristic roots of **W'AW** are the same as of AE^{-1}. The final discriminant function requires an extra step. The advantages of this method are that one may use conventional PCA procedures and that it can be used when there are more variables than observations, in which case **E** cannot be inverted. Krzanowski (1990) recommended using this technique when heterogeneity exists among the group covariances. For a tutorial on this topic, including some geometrical interpretation, see Campbell and Atchley (1981). Campbell (1980a) used this framework in obtaining robust estimators of the discriminant functions. Yendle and MacFie (1989) suggested standardizing the data based on the within-group standard deviations and obtaining the characteristic vectors of the among-group matrix obtained from these data. Gower (1966b) suggested another alternative in the form of *Q*-analysis and distances.

14.5.3 Cluster Analysis

Cluster analysis is the reverse of discriminant analysis. In this case, there are no predefined groups and the clustering techniques are designed to break the data set into two or more homogeneous groups. There have been a large number of algorithms designed to accomplish this. Whole books have been written on this subject, among them, Hartigan (1975), Everitt (1980), and Aldenderfer and Blashfield (1984). PCA is often used to reduce the data to a smaller number of transformed variables before the clustering is carried out. This is done partly to reduce the size of the computing problem (generally more complex than discriminant analysis) and because the use of pc's may help identify the characteristics of the clusters. If two pc's are sufficient, PCA would also allow a scatter plot of the two pc's showing the nature of the clusters. The caveat is the same as for regression and discriminant analysis: there is no guarantee that the pc's with the larger roots will do a better job of clustering, and if all of the pc's are required one might as well perform the clustering on the original variables.

There is, however, a case for selective use of pc's. Dietz et al. (1985) did a study on the performance of 99 mutual funds over a period of 8 years. The first pc, accounting for 80% of the variability related to the level of the overall market. The second pc related to small vs. high capitalization stocks, and the third pc represented high-growth stocks vs. high-yield stocks. They used y_2 and y_3, but not y_1, to cluster portfolio managers. Zani (1980) suggested clustering each pc separately. A relationship between the groups of a minimum spanning tree and the characteristic roots of a dissimilarity matrix of the data was given by Gondran (1977).

A decision often faced in cluster analysis is how the data should be normalized before clustering. In pc-cluster analysis, this would include the choice between covariance and correlation matrices and the choice between euclidean distances (z-scores) and Mahalanobis distances (y-scores).

14.6 TIME SERIES

14.6.1 Introduction

We shall deal only briefly with the topic of time series analysis as there have been a number of texts written on the subject, among them, Anderson (1971), Jenkins and Watts (1968), Box and Jenkins (1970), and Brillinger (1975). There are a few occasions when one might wish to use PCA in connection with a time series and it will be the purpose of this section to list what these possibilities might be. Some situations, such as chemical reaction curves or growth curves may technically be considered time series, but the nature of the functional form is the main purpose of the analysis. These constitute a special case and the use of PCA in this context will be deferred until Section 16.7.

For the purpose of statistical inference, it is assumed that the observation vectors are sampled independently. For such applications as process control, this assumption would imply that although the time between observations is a constant, the observations should be independent under the null hypothesis that the mean of the process has not shifted. For many time series, however, one knows that the mean not only has shifted but may shift in a predictable manner. PCA may still be used with these data in an exploratory manner.

14.6.2 Use of PCA in Time Series Analysis

There have been some examples in this book where the data were ordered in time as in the cases of some of the quality control examples. In these cases, the PCA did not reflect the effect of time directly but was used to study its effect. On the other hand, the Assault Data analyzed in Section 10.8.3 employed time as one of the two classifications in a contingency table. A cancer study in England and Wales by Osmond (1985) used biplots with time and age at death as the two factors. Although one could have used counts here also, Osmond used death rates/million persons at risk instead. Bozik and Bell (1987) suggested using PCA for modeling and forecasting of age-specific fertility curves rather than the traditional time series methodology commonly employed.

Principal components may also be used to analyze seasonality in a time series. Craddock (1966) analyzed the monthly temperatures of central England from 1680 to 1963 and performed a PCA using the months as variables. Rather than work with a covariance matrix of months, he subtracted the overall mean from all of the data and then obtained a product matrix of these differences. As one might expect, the first pc, representing average seasonal effect (i.e., cold winters and warm summers) explained over 90%. The second pc accounted for year-to-year differences, while the next two pc's represented different patterns in winter temperature. A discussion of the pros and cons of the use of PCA in meteorology representing different points of view at that time may be found in Craddock (1973a,b) and Jenkins (1973).

One may wish to carry out forecasting and/or control functions using procedures such as the Box–Jenkins model but where the time series has a multiple response. In this case, it may be advantageous to perform a PCA on the original data and then model the pc's. The success of this strategy will depend on how interpretable the pc's are. An example of the use of PCA in a distributed lag model was given by Doran (1976).

14.6.3 Power Spectrum Analysis

More complicated is the classical time series analysis involving power spectrum analysis. For a univariate time series $x(t)$, one obtains the autocovariance function made up of the covariance between $x(t)$ and $x(t-1)$, $x(t)$ and $x(t-2)$, and so on, where $x(t-g)$ is the same series displaced by g periods in time. This is sometimes called a *lag* function and is expressed in the *time domain*. Alternatively, one may obtain the Fourier transform of the autocovariance function. This is called the *power spectrum* and is in the *frequency domain*. If the original data were taken every hour for a week, the time domain would be in terms of hours. The frequency domain would be in terms of number of observations per unit of time, customarily denoted by the wavelength symbol λ, this symbol having nothing to do with population characteristic roots. For instance $\lambda = \pi$ corresponds to a period of 2 hours. This is the *Nyquist frequency*; no cycles of period less than twice the sampling frequency may be resolved. $\lambda = \pi/2$ would represent 4-hour cycles, and so on. The power spectrum may be thought of as the decomposition of the variability of the entire time series as a function of frequency. The portion of the power spectrum from $\pi/2$ to π represents that portion of the variability made up of cycles between 2 and 4 hours. An analog of the ANOVA variance components called *harmonic* variance components was given by Jackson and Lawton (1969).

If a time series extends from $t = -T$ to $t = T$ and its autocovariance function is denoted by $C(t)$, the relationships between it and the power spectrum $g(\lambda)$ are

$$g(\lambda) = \frac{1}{2\pi} \sum_{t=-T}^{T} C(t)e^{-i\lambda t} \qquad 0 \leqslant \lambda \leqslant \pi$$

$$C(t) = \int_{-\pi}^{\pi} g(z)e^{itz}\,dz \qquad \text{for } t = 0, \pm 1, \pm 2, \ldots$$

(14.6.1)

where $i = \sqrt{-1}$. Although this pair of expressions appears very neat and compact, there are many operational details to be dealt with in getting from one to the other.

The autocovariance matrix of this series will be $(2T + 1) \times (2T + 1)$. Its characteristic roots are

$$l_h = \frac{1}{2T + 1} \sum_{t=-T}^{T} C(t)e^{-2i\pi ht} \qquad h = -T, \dots, T \qquad (14.6.2)$$

Recall that

$$\exp(-2i\pi ht) = \cos(-2\pi ht) + i\sin(-2\pi ht)$$
$$= \cos(2\pi ht) - i\sin(2\pi ht)$$

Since the time intervals are evenly spaced, h and t are integers, so $\sin(2\pi ht) = 0$ and the characteristic roots will be real. These characteristic roots are sometimes called the *spectrum* of the autocovariance function because it can be shown that (14.6.2) is proportionate to the power spectrum. It is for this reason that the largest root is sometimes referred to as the *spectral radius*.

The elements of the corresponding characteristic vectors are

$$u_{ht} = [1/\sqrt{\pi(2T + 1)}] \exp\left(\frac{-2i\pi ht}{2T + 1}\right) \qquad t, h = -T, \dots, T \qquad (14.6.3)$$

There will be $2T + 1$ vectors, each with $2T + 1$ elements. The reason this comes out so neatly is that the autocovariance matrix is a *circulant* matrix. Once the first row has been specified, all of the rest of the matrix is determined. The principal components $u_h' x(t)$ are finite Fourier transforms and have variance l_h.

More detail in these results may be found in Brillinger (1969, 1975). Proofs of (14.6.2) and (14.6.3) may be found in Good (1950) and some other results may be found in Wahba (1968) and Taniguchi and Krishnaiah (1987), the latter also having investigated the distributional aspects. Brillinger (1975) also discussed the case where $x(t)$ is, itself, complex and where $x(t)$ has a multiple response. Similar results may be found in Chow (1975), a text dealing with dynamic economic systems. This book also discusses the characteristic roots and vectors of systems of difference equations.

One may prefer to obtain the characteristic roots and vectors of the power spectrum $g(\lambda)$. For p responses, this becomes a $p \times p$ matrix. Brillinger's book contains an entire chapter on these operations, including an example of monthly temperature for 13 European cities over a period of 150–250 years, depending on the city. Corresponding data were also included for New Haven, Connecticut. Because of the different time spans involved, a multivariate analysis of the original time series would have many blocks of missing data with which to contend but the power spectra may be compared directly given the assumption that the period from 1755 to 1962 was fairly stationary for each city. The first pc corresponded to the average of the European cities. The second pc appeared, primarily, to be a contrast between New Haven and the rest. Because the characteristic vectors may be complex, the principal components may be decomposed into *gain* and *phase*. The first phase component was relatively stable with the exception of New Haven. Cohen and Jones (1969) used PCA of some meteorological data to predict temperature. Devaux et al. (1987) used these techniques on the chemical analysis of wheat and then used the results to study the relationship of wheat analysis to the bread made from it.

One may also compare two series simultaneously by means of *cross-spectral* analysis. These are established by means of a covariance function involving both series and the correlations between them. The use of principal components is then extended to an analog of canonical correlations that were introduced in Section 12.6.2. Brillinger (1975) has a second chapter on this topic, which was also investigated by Chow (1975) and Tiao and Tsay (1989).

CHAPTER 15

Flatland: Special Procedures for Two Dimensions

The title of this chapter was inspired by the book of the same title by E.A. Abbott (1884) describing life in a two-dimensional world. [Hopefully, the other material in this book will not be confused with—*And He Built a Crooked House* by R.A. Heinlien (1943), the trials and tribulations of life in a four-dimensional house.] This chapter will describe a number of special techniques designed for PCA in two dimensions. While two-dimensional PCA may seem to be of little importance other than for demonstrations as in Chapter 1, there are some useful applications for this special case where PCA may be used to enhance situations involving the analysis of two related variables.

15.1 CONSTRUCTION OF A PROBABILITY ELLIPSE

One of the advantages of flatland is the ability to plot the data and study the relationship between the two variables. If these variables have a bivariate normal distribution, then it is possible to construct probability ellipses for them. Their use in quality control was illustrated in Section 1.7.5. Mandel and Lashof (1974) used ellipses for interlaboratory comparisons where n laboratories each analyzed the same two compounds. The results were plotted as a scattergram, each lab being represented by one point. Mandel and Lashof developed a rather elaborate series of hypotheses for additive and concurrent models and used the ratio l_1/l_2 as an alternative test for sphericity, for which they included a short table.

In all that is to follow in this section, the horizontal axis will represent x_1 and the vertical axis x_2. The equation for the $100(1-\alpha)\%$ ellipse such as the one for the Chemical data in Figure 1.6 is

$$\frac{s_1^2 s_2^2}{(s_1^2 s_2^2 - s_{12}^2)}\left[\frac{(x_1 - \bar{x}_1)^2}{s_1^2} + \frac{(x_2 - \bar{x}_2)^2}{s_2^2} - \frac{2s_{12}(x_1 - \bar{x}_1)(x_2 - \bar{x}_2)}{s_1^2 s_2^2}\right] = T_\alpha^2 \quad (15.1.1)$$

A unique ellipse is defined for given values of $\bar{x}_1$, $\bar{x}_2$, s_1^2, s_2^2, s_{12}, n, and α. Some computer packages have a subroutine to perform this operation. However, it should be worthwhile to present a procedure for constructing an ellipse because it gives additional insight into the method of principal components as well as providing an alternative if no computer is available.

It is possible to work directly with equation (15.1.1) without using PCA at all. For the chemical example, with $\alpha = .05$, this becomes

$$\frac{(.7986)(.7343)}{[(.7986)(.7343) - (.6793)^2]}$$
$$\times \left[\frac{(x_1 - 10)^2}{.7986} + \frac{(x_2 - 10)^2}{.7343} - \frac{2(.6793)(x_1 - 10)(x_2 - 10)}{(.7986)(.7343)} \right] = 8.21$$

which reduces to

$$.72(x_1 - 10)^2 + .78(x_2 - 10)^2 - 1.32(x_1 - 10)(x_2 - 10) - 1 = 0$$

Points on the perimeter of the ellipse may be determined by setting x_1 equal to some constant and solving the resulting quadratic equation in x_2. For instance, corresponding to $x_1 = 10$, the solutions for x_2 are 11.13 and 8.87. Each arbitrary value of x_1 will yield two similar values of x_2. This procedure involves rather tedious algebra but for the few values required to obtain a rough sketch of the ellipse it will probably suffice, particularly in this day of pocket calculators.

The use of characteristic vectors simplifies this procedure considerably in addition to supplying the major and minor axes of the ellipse. The quantities required for this are $\bar{x}_1$, $\bar{x}_2$, n, α, and the V-vectors. The major and minor axes of the ellipse intersect at the point $x_1 = \bar{x}_1$, $x_2 = \bar{x}_2$. The slope of the major axis, the orthogonal regression line, is $b_{or} = v_{21}/v_{11}$, involving the elements of the first characteristic vector and the slope of the minor axis is $v_{22}/v_{12} = -v_{11}/v_{21}$. For the Chemical example, the slope of the major axis is $(.8301)/(.8703) = .9538$ and the slope of the minor axis is -1.0484. (The slope of the major axis is also equal to $(l_1 - s_1^2)/s_{12}$, which does not involve the vectors at all and is also equal to

$$b_{or} = \frac{s_{22} - s_{11} + \sqrt{(s_{22} - s_{11})^2 + 4s_{12}^2}}{2s_{12}} \tag{15.1.2}$$

which involves only elements of the covariance matrix.)

The length of the semimajor axis is

$$\sqrt{l_1 T_\alpha^2} = \sqrt{(1.4465)(8.21)} = 3.45$$

Table 15.1. Ellipse Computations

g	h	x_1	x_2
2.685	1	12.13	12.44
−2.685	1	7.46	7.98
2.685	−1	12.54	12.02
−2.685	−1	7.87	7.56
1	2.685	10.33	11.40
1	−2.685	11.42	10.26
−1	2.685	8.58	9.74
−1	−2.685	9.67	8.60

and the length of the semiminor axis is

$$\sqrt{l_2 T_\alpha^2} = \sqrt{(.0864)(8.21)} = .84$$

With a knowledge of the length of these axes, their orientation, and their intersection, one has a fair idea of the shape of the ellipse (and could actually draw one with the time-honored method using a piece of string and three pencils).

To obtain the actual coordinates, the value of T_α^2 must be broken down into pairs of real numbers g and h such that $g^2 + h^2 = T_\alpha^2$. It is common to use 0, 1, 2,..., as the values of either g or h so that the corresponding values of the other would be T_α, $\sqrt{T_\alpha^2 - 1}$, $\sqrt{T_\alpha^2 - 4}$, ... The coordinates of the points would then be as follows:

$$\mathbf{x} = \bar{\mathbf{x}} + g\mathbf{v}_1 + h\mathbf{v}_2 \tag{15.1.3}$$

For instance, if $h = 0$, $g = T_\alpha = 2.865$, then

$$\begin{bmatrix} x_1 \\ x_2 \end{bmatrix} = \begin{bmatrix} 10 \\ 10 \end{bmatrix} + (2.865)\begin{bmatrix} .8703 \\ .8301 \end{bmatrix} + (0)\begin{bmatrix} -.2029 \\ .2127 \end{bmatrix} = \begin{bmatrix} 12.49 \\ 12.38 \end{bmatrix}$$

which is the upper end of the major axis. Letting $g = -T_\alpha$ would produce the lower end, and letting $g = 0$ with $h = \pm T_\alpha$ would produce the ends of the minor axis.

Using 1 and $\sqrt{T_\alpha^2 - 1} = 2.685$, another eight points may be obtained as shown in Table 15.1. The use of 2 and $\sqrt{T_\alpha^2 - 4} = 2.052$ will yield eight more points. These three sets would provide 20 points on the ellipse, which should be enough to ensure an adequate representation.

15.2 INFERENTIAL PROCEDURES FOR THE ORTHOGONAL REGRESSION LINE

As has been stated in the previous section and elsewhere, the first characteristic vector describes the *orthogonal regression line*, which minimizes the sums of squares of deviations *perpendicular* to the line itself. In Chapter 12 we said that we would not get drawn into the calibration controversy. This controversy involves the choice among various proposals to predict x_2 from x_1 when both are subject to error. One proposal is to regress x_2 on x_1 assuming that x_1 represents observations on an *instrumental* variable (i.e., $E(x_1)$ is unbiased). The other is to regress x_1 on x_2 and invert the equation.

PCA can provide an alternative by using the orthogonal regression line as a prediction equation. This may be particularly useful if one wishes, sometimes, to predict x_2 from x_1 but at other times predict x_1 from x_2. As an example, suppose one is monitoring a process using a reference material and when the supply of this material is exhausted a new batch is selected to replace it. Customarily, there is a *crossover* period in which samples from both batches are tested. However, the new batch may have a different sensitivity from the old one so that the regression line between them does not have a slope of 1.0. Since one may wish to relate test results before and after this crossover to each other, and in either order, the orthogonal regression line would seem a reasonable choice.

Before discussing orthogonal regression further, let us digress a bit and consider the topics *linear functional relationship* and *linear structural relationship*. Assume the observation x_i is made up of two components:

$$x_i = g_i + e_i \tag{15.2.1}$$

g_i is called the *systematic* component. It is equal to the true value associated with this observation (i.e., $E(x_i) = g_i$). e_i is called the *random* component and has an expected value of zero. This is the random error. Neither g_i nor e_i is observable, only x_i.

Consider observations on two variables such that

$$x_1 = g_1 + e_1 \qquad E(x_1) = g_1 \qquad E(e_1) = 0$$
$$x_2 = g_2 + e_2 \qquad E(x_2) = g_2 \qquad E(e_2) = 0$$

and that g_1 and g_2 are related in a linear fashion without error. The problem is that they cannot be measured directly to ascertain what this relationship might be. Instead, one observes x_1 and x_2 and, from that, attempts to estimate this relationship in the presence of the errors e_1 and e_2. If g_1 and g_2 are fixed as in the case of a Model 1 ANOVA, this is called a linear *functional* relationship. If g_1 and g_2 are random, as in a Model 2 ANOVA, this is called a linear *structural* relationship.

To estimate these relationships, certain assumptions must be made about the e_i. There are three possibilities:

1. $\qquad\qquad\qquad V(e_1) = V(e_2) \qquad \rho_{e_1 e_2} = 0$

2. $\qquad\qquad\qquad V(e_1) \neq V(e_2) \qquad \rho_{e_1 e_2} = 0 \qquad (15.2.2)$

3. $\qquad\qquad\qquad V(e_1) \neq V(e_2) \qquad \rho_{e_1 e_2} \neq 0$

with Case 3 being the general case. For Case 1, both the resultant functional and structural relationships are the orthogonal regression line. (This result was originally obtained by Adcock in 1878!) Orthogonal regression lines exist for any number of variables since they are represented by the first characteristic vector. Some of the factor analysis models discussed in Chapter 17 deal with Cases 2 and 3. For a full discussion of these functional and structural models, see Anderson (1984b).

The angle between the orthogonal regression line, b_{or} and the horizontal axis is $\hat{\theta} = \arctan(b_{or})$. A robust estimator for b_{or} was given by M.L. Brown (1982), Ammann and VanNess (1989), and Zamar (1989). A number of inferential procedures are relevant with respect to either b_{or} or $\hat{\theta}$ and will be illustrated by the Chemical example.

1. *Test of the hypothesis* H_0: $\theta = \theta_0$

$$t_{n-2} = \sqrt{\frac{n-2}{l_1 l_2}} \left[\frac{(l_1 - l_2)\sin(2\theta_0)}{2} \right] \tag{15.2.3}$$

This expression has been obtained in various forms by Creasy (1956), Williams (1959, p. 199), and Jolicouer (1968). For an example, let H_0: $\theta = 45°$. Then

$$t_{13} = \sqrt{\frac{13}{(1.4465)(.0864)}} \left[\frac{(1.4465 - .0864)\sin(90°)}{2} \right]$$

$$= 6.94$$

If $\alpha = .05$, $t_{13,.025} = 2.16$ so the hypothesis is rejected. Anderson (1980) obtained the distribution of b_{or} under the null hypothesis that the true slope is equal to zero and studied the effect on this of the increase in value of a subset of the observations in the sample.

2. *Confidence limits for* β_{or}. Let

$$\sin(2c) = \sqrt{\frac{l_1 l_2}{n-2}} \left[\frac{2t_{n-2,\alpha/2}}{(l_1 - l_2)} \right] \tag{15.2.4}$$

Then the $100(1 - \alpha)\%$ confidence limits for β_{or} will be $\tan(\hat{\theta} \pm c)$. For the Chemical data, $\hat{\theta} = 43.65$, $c = 9.07$ and the confidence limits are

$$.69 < \beta_{or} < 1.31$$

If the sample size had been $n = 100$ instead of $n = 15$,

$$.86 < \beta_{or} < 1.06$$

For very small correlations, these limits may become infinite or even imaginary. Imaginary limits are considered a special case of infinite limits. Jolicoeur (1973) investigated the conditions for which this might occur. It is not much of a problem because it will occur only for such low correlations that one should not be making inferences in the first place.

3. *Test of sphericity.* This is the same problem as (4.4.1) with $k = 0$. For the case $p = 2$, an exact test is available.

$$F_{2,n-2} = \frac{(n-2)(l_1 - l_2)^2}{8l_1 l_2} \tag{15.2.5}$$

This is a test of the hypothesis $\lambda_1 = \lambda_2$. If this hypothesis is rejected, it means that the ellipse is now a circle and that the characteristic vectors are undefined. This test should be run first since acceptance of this hypothesis would preclude any inferences about the slope. For this example,

$$F_{2,13} = \frac{(13)(1.8499)}{(8)(.1250)} = 24.05$$

For $\alpha = .05$, $F_{2,13} = 3.81$ so the hypothesis is rejected.

The square of the orthogonal regression coefficient, b_{or}^2, is an F-statistic with 1 and $n - 2$ degrees of freedom. James and Venables (1980) pointed out that, for a given sample size, there are certain combinations of l_1 and l_2 such that the hypothesis $\theta = 45°$ (along with other values of θ_0) could be *rejected* while the hypothesis of sphericity would be *accepted*, that is,

$$F_{1,n-2} < \frac{(n-2)(l_1 - l_2)^2}{4l_1 l_2} < 2F_{2,n-2} \tag{15.2.6}$$

for fixed α. Although this is an unhappy state of affairs, the chance of it happening is relatively small. For the Chemical example, with $l_1 + l_2 = 1.53$, $l_1 > l_2$,

$$.76 < l_1 < 1.16 \quad \text{both hypotheses are accepted}$$

$$1.16 < l_1 < 1.23 \quad \text{the paradox exists}$$

$$1.23 < l_1 < 1.53 \quad \text{both hypotheses are rejected}$$

This corresponds to the region of 76–80% explained by the first pc. If the sample size were $n = 100$ instead of $n = 15$, the paradox would occur for roughly $.92 < l_1 < .95$, the 60–62% range, so this problem is diminished with larger sample sizes. James and Venables did obtain expressions for confidence limits on the slope that overcame this. These limits will always be wider than the conventional limits, as much as 50% larger, but diminish with increasing correlation. The expressions are quite complicated, involving, among other things, the ratio of two $_2F_1$ hypergeometric functions. A short table of these limits as a function of correlation and sample size is included in their article.

15.3 CORRELATION MATRICES

For the two-dimensional case, working with correlation matrices may seem trivial because everything is a function only of the correlation coefficient and the slope of the major axis is always 1.0 (-1.0 if the correlation is negative). The fundamental quantities in general terms and for $r = .5$ and $.9$ are as follows:

	$r = .5$	$r = .9$
Characteristic roots		
$l = 1 \pm r$	1.5, .5	1.9, .1

$$U = \begin{bmatrix} \sqrt{.5} & -\sqrt{.5} \\ \sqrt{.5} & \sqrt{.5} \end{bmatrix} \qquad \begin{bmatrix} .71 & -.71 \\ .71 & .71 \end{bmatrix} \qquad \begin{bmatrix} .71 & -.71 \\ .71 & .71 \end{bmatrix}$$

$$V = \begin{bmatrix} \sqrt{(1+r)/2} & -\sqrt{(1-r)/2} \\ \sqrt{(1+r)/2} & \sqrt{(1-2)/2} \end{bmatrix} \qquad \begin{bmatrix} .87 & -.50 \\ .87 & .50 \end{bmatrix} \qquad \begin{bmatrix} .98 & -.22 \\ .98 & .22 \end{bmatrix}$$

$$W = \begin{bmatrix} 1/\sqrt{2(1+r)} & -1/\sqrt{2(1-r)} \\ 1/\sqrt{2(1+r)} & 1/\sqrt{2(1-r)} \end{bmatrix} \qquad \begin{bmatrix} .58 & -1.00 \\ .58 & 1.00 \end{bmatrix} \qquad \begin{bmatrix} .51 & -2.24 \\ .51 & 2.24 \end{bmatrix}$$

$$y_1 = \frac{x_1 + x_2}{\sqrt{2(1+r)}} \qquad .58(x_1 + x_2) \qquad .51(x_1 + x_2)$$

$$y_2 = \frac{-x_1 + x_2}{\sqrt{2(1-r)}} \qquad x_2 - x_1 \qquad 2.24(x_2 - x_1)$$

If the correlation coefficient is negative, all of the values will remain the same but the minus sign will be shifted to the first element of the first vector from the first element of the second vector for **U**, **V**, and **W**.

Another property of correlation matrices is that

$$r = \frac{l_1 - l_2}{l_1 + l_2} \tag{15.3.1}$$

so that the test for equality of roots is the same as the test for the correlation coefficient (Hotelling, 1933).

15.4 REDUCED MAJOR AXIS

The motivation for this technique stems from a desire to overcome the fact that the orthogonal regression line is not invariant under a change of scale. The *reduced major axis* technique (also called *organic correlation* or *geometric mean functional relationship*) consists of obtaining the slope of the major axis using the correlation matrix, which from the previous section is always 1.0 (or -1.0 for negative correlation) and then using the slope that results from transforming the two variables back to their original units. This slope turns out to be $b_m = s_2/s_1$, taking the sign of s_{12}. For the Chemical problem this is $b_m = (.8569)/(.8936) = .9589$ compared with the slope of the major axis using the covariance matrix, $b_{or} = .9538$. The difference between these two numbers is not very impressive in this case because of the high correlation between the variables, but for lower correlations the difference could be considerable. Sprent (1969) showed that for a large spread of values of x_1, b_m will be a reasonable estimate of the underlying functional relationship between x_2 and x_1 when both have measurement errors. This technique is popular in allometry.

There are some interesting relationships among the various regression coefficients. The regression of x_2 on x_1 is $b_{2,1} = .8506$ and the regression of x_1 on x_2 with respect to the x_1-axis is $b_{1,2} = 1.0810$. Recalling that the correlation between the two variables is $r = .887$, it turns out that $b_{2,1} = b_m r$ and $b_{1,2} = b_m/r$. Furthermore, the reduced major axis is the geometric mean of the two standard regression coefficients, that is, $b_m = \sqrt{(.8506)(1.0810)} = .9589$. Finally, let

$$a = [b_m - (1/b_m)]/(2r)$$

$$= (.9589 - 1.0429)/[(2)(.887)] = -.0473 \qquad (15.4.1)$$

Then,

$$b_{or} = a + \sqrt{(a^2 + 1)}$$

$$= -.0473 + \sqrt{(1.002237)} = .9538 \qquad (15.4.2)$$

Although b_m does not involve the correlation coefficient, its standard error does. The standard error of b_m is $b_m\sqrt{[(1 - r^2)/n]}$, which is equal to the standard error for $b_{2,1}$. The distribution of b_m is quite asymmetric so that confidence limits can be obtained only with some difficulty. However, to test the hypothesis $H_0: \beta_m = \beta_{m0}$, the test

$$t_v = |\ln(b_m) - \ln(\beta_{m0})|\sqrt{[(n - 1)/(1 - r^2)]} \qquad (15.4.3)$$

with $v = 2\{1 + [(n - 1)/(2 + r^2)]\}$ degrees of freedom is a good approximation, even for $n = 10$ (Clarke, 1980). Clarke also gave a procedure for a test of the hypothesis $H_0: \beta_{m1} = \beta_{m2}$, a likely situation in allometry. For other results related to the reduced major axis, see Kermack and Haldane (1950), Kruskal (1953), Carlson (1956), and Barker et al. (1988).

CHAPTER 16

Odds and Ends

16.1 INTRODUCTION

The topics contained in this chapter are made up of special situations in PCA that do not fit in elsewhere in this book. These topics include nonlinear PCA, cross-validation, sensitivity of estimates to aberrations in the data, robust estimates, procedures involving two or more samples, PCA with functional data, and PCA with discrete data.

16.2 GENERALIZED PCA

We shall now return from Flatland in the previous chapter to the more rugged terrain of p-dimensional PCA and venture into new territory where straight lines may not be the norm. There are a number of different definitions for generalized PCA. Among them are the notion of *weighted PCA* (Section 3.6), *k-mode PCA* (Sections 10.9 and 10.10), and *nonmetric multidimensional scaling* (Section 11.4.3). However, for now, we shall use the definition of Gnanadesikan and Wilk which deals with the extension of PCA to include nonlinear models (Gnanadesikan and Wilk, 1969; Gnanadesikan, 1977, p. 48).

Consider the data set in Table 16.1. These are the coordinates for a unit circle. If one were to perform a traditional PCA on these data, the following results would be obtained:

$$ S = \begin{bmatrix} .5217 & 0 \\ 0 & .5217 \end{bmatrix} = L $$

$$ U = \begin{bmatrix} 1 & 0 \\ 0 & 1 \end{bmatrix} $$

The analysis would conclude that x_1 and x_2 are uncorrelated, which they are *in a linear sense*. The fact that these variables are perfectly related in a nonlinear

Table 16.1. Circle Data

Observation No.	x_1	x_2
1	0	1
2	.259	.966
3	.500	.866
4	.707	.707
5	.866	.500
6	.966	.259
7	1	0
8	.966	−.259
9	.866	−.500
10	.707	−.707
11	.500	−.866
12	.259	−.966
13	0	−1
14	−.259	−.966
15	−.500	−.866
16	−.707	−.707
17	−.866	−.500
18	−.966	−.259
19	−1	0
20	−.966	.259
21	−.866	.500
22	−.707	.707
23	−.500	.866
24	−.259	.966

sense escapes unnoticed. An example of a generalized procedure, proposed by Gnanadesikan and Wilk (1969) to handle such situations, is as follows:

1. Add three more variables:

$$x_3 = x_1^2$$
$$x_4 = x_2^2$$
$$x_5 = x_1 x_2$$

2. The expanded covariance matrix is

$$S = \begin{bmatrix} .5217 & 0 & 0 & 0 & 0 \\ 0 & .5217 & 0 & 0 & 0 \\ 0 & 0 & .1304 & -.1304 & 0 \\ 0 & 0 & -.1304 & .1304 & 0 \\ 0 & 0 & 0 & 0 & .1304 \end{bmatrix}$$

The means are $\bar{x}_1 = \bar{x}_2 = \bar{x}_5 = 0$ and $\bar{x}_3 = \bar{x}_4 = .5$.

3. A PCA is carried out on this expanded covariance matrix. The characteristic roots are .5217, .5217, .2609, .1304 and zero. This means that **S** is only rank 4. However, four characteristic vectors will be required to obtain pc's that will adequately reproduce the original data. The characteristic vectors are

$$\mathbf{U} = \begin{bmatrix} 0 & -1 & 0 & 0 & 0 \\ 1 & 0 & 0 & 0 & 0 \\ 0 & 0 & .7071 & 0 & .7071 \\ 0 & 0 & -.7071 & 0 & .7071 \\ 0 & 0 & 0 & -1 & 0 \end{bmatrix}$$

The last vector, $\mathbf{u}_5$, is important even though its corresponding root is zero. As pointed out in Chapter 2, zero roots represent singularities in the data and the vectors associated with them may be used to characterize what these singularities may represent. In this example, the vector defines the nature of the nonlinearity. Using this vector, the pc, z_5, would be

$$z_5 = .7071(x_3 - .5) + .7071(x_4 - .5)$$
$$= .7071x_3 + .7071x_4 - .7071$$
$$= .7071x_1^2 + .7071x_2^2 - .7071$$

If this were set equal to zero and divided by .7071, the resultant equation would be the unit circle which these data represent.

4. To check this example further, the first two data rows produce the following y-scores.

$$\begin{bmatrix} 1.3844 & 0 & -1.3844 & 0 \\ 1.3373 & -.3586 & -1.1986 & -.6922 \end{bmatrix}$$

and these in turn, using (2.7.2) will reproduce the original data exactly:

$$\begin{bmatrix} 1 & 1 & 0 & 1 & 0 \\ .2590 & .9659 & .0671 & .9329 & .2502 \end{bmatrix}$$

The **V** and **W**-matrices are not displayed here but can be easily obtained using (1.6.1) and (1.6.2). Note that the last characteristic root in this example was zero only because there was no random variability associated with the data. This example used only second-order terms in the expansion of the covariance matrix but any transformation may be employed.

If one is working with only two variables, nonlinearity can be detected by a simple plot, but the generalized PCA technique might be useful for larger problems, at least to detect that nonlinearity exists. Nonlinearity problems may sometimes be resolved by simple transformations of the original data.

For another example, consider the Chemical example of Chapter 1. If one were to obtain all of the second-order elements and the resultant 5×5 covariance matrix, the characteristic roots would be 900.9185, 33.9509, .0060, .0027, and .0003. Because these data do include inherent variability, the last three roots are not equal to zero but they are close enough to indicate that the linear relationship assumed for this example is adequate.

Another concept of generalized PCA is the notion of *principal curves* (Hastie and Stuetzle, 1989). A principal curve bears somewhat the same relation to a spline function that the first characteristic vector (or orthogonal regression line) does to the regression of a response variable on one or more predictor variables; it is not a function of a predictor variable but rather a joint function of all the response variables. It is a nonlinear function obtained in such a way that the variability about that curve is minimized perpendicular to the curve. Hastie and Stuetzle also used a circle for an example and included two physical examples, one of these dealing with a linear collider. The other example dealt with pairs of gold assays, similar to the Chemical example in Chapter 1 except that what should have been a straight-line relationship had a buckle in the middle of it. An extension to principal surfaces was also suggested.

16.3 CROSS-VALIDATION

16.3.1 Motivation

The method of *cross-validation* was introduced in Section 2.8.10 in connection with stopping rules. The principal of cross-validation is due to Mosteller and Wallace (1963) and Stone (1974) among others. It is related, somewhat, to jackknifing and validation samples. In PCA it consists of randomly dividing the sample into g groups. The first group is deleted from the sample and a PCA is performed on the remaining sample. The vectors obtained from that reduced sample are used to obtain pc's and Q-statistics for the deleted group. That group is returned to the sample, the next group is deleted, and the procedure is repeated g times. The grand average of the Q-statistics, divided by p, is called a PRESS-statistic (PREdiction Sum of Squares). Its primary use in PCA is as a stopping rule (Wold, 1976, 1978; Eastment and Krzanowski, 1982; Krzanowski, 1983, 1987b). It differs from other stopping rules in that it is based on the Q-statistic rather than the characteristic roots. Krzanowski pointed out that it is possible to have different data sets produce the same covariance or correlation matrix but would probably produce different PRESS-statistics although their characteristic roots would be the same. Cross-validation also differs from other stopping rules in requiring the original data while the other procedures work directly from the covariance or correlation matrix. Although the procedure described below is not a significance test, it is more quantitative than most of the stopping rules given in Chapter 2.

16.3.2   A Simple Cross-validation Procedure

The principle of cross-validation as a stopping rule will be illustrated using the Audiometric data in Chapter 5, Table 5.1, where $p = 8$ and $n = 100$. In that chapter, the PCA was performed on a correlation matrix and so, for the purposes of cross-validation, these data were transformed into standard units by subtracting from each observation its mean and dividing by its standard deviation. This makes the covariance and correlation matrices the same. The transformed data set will be denoted by the 100×8 matrix **X**. In this example, the data set will be divided into $g = 5$ groups of 20 observations each. Although this subdivision is supposed to be carried out in a random fashion, in order for the reader to more easily follow this example, it will be assumed that the sample was obtained in a random fashion to start with so that the first group will be observations 1–20, the second group 21–40, and so on. The procedure is as follows:

1. Delete the first group from the sample. Perform a PCA on the remaining observations (i.e., 21–100). Obtain all eight vectors. Since this example used a correlation matrix, the data will be in standard units.

2. For the deleted sample, obtain all eight *y*-scores for each observation using the vectors obtained in step 1.

3. Using, in turn, the first pc, the first two pc's, and so on, obtain the predicted values of the deleted sample using (2.7.2). $\bar{x}$ will be equal to zero.

4. For each observation in the deleted sample, obtain Q (equation 2.7.4). For the first observation, $Q = 1.054$ for one pc, $Q = .639$ for two pc's, and so on.

5. Return the deleted group to the sample and remove the second group. Repeat steps 1–4. Do the same for the other three groups. This concluded, there will now be 100 values of Q for a one-pc model, another 100 for a two-pc model, and so on.

6. For each pc model, add up the 100 Q-statistics and divide each sum by $np = 800$. These are called PRESS-statistics and will be designated by PRESS(1), PRESS(2), and so on. It will also be necessary to obtain PRESS(0), the sum of squares of the original data, again assuming a mean of zero.

7. To determine whether the addition of another pc, say the kth pc, to the model is warranted, form the statistic

$$W = \frac{[\text{PRESS}(k-1) - \text{PRESS}(k)]/D_M}{\text{PRESS}(k)/D_R} \qquad (16.3.1)$$

where

$$D_M = n + p - 2k \qquad D_R = p(n-1) - \sum_{i=1}^{k}(n+p-2i)$$

Table 16.2. Audiometric Data. Cross-validation

k	PRESS(k)	D_M	D_R	W
0	.9900		792	
1	.5337	106	686	5.53
2	.3478	104	582	2.99
3	.2100	102	480	3.09
4	.1532	100	380	1.41
5	.1149	98	282	.96 < 1

If $W > 1$, then retain the kth pc in the model and test the $(k + 1)$st. For example, to test whether the first pc should be included, one would form

$$W = \frac{[\text{PRESS}(0) - \text{PRESS}(1)]/106}{\text{PRESS}(1)/686} = \frac{.004305}{.000778}$$

$$= 5.53.$$

and the first pc would be included in the model.

For the Audiometric data, this test procedure is shown in Table 16.2. The process terminated with the inclusion of the fourth pc, the same number as retained in Chapter 5, although the reasoning there was that the remaining variability represented inherent variability on the part of the subjects.

In practice, if one had a large number of variables and was confident that only a small number of pc's would be retained, a different strategy might be employed, in which each characteristic vector is obtained and tested sequentially before obtaining the next and, in that way, only one unwanted vector is obtained.

It is possible that if one were to continue this process beyond the first occurrence where $W < 1$, later values of k might produce one or more occurrences of $W > 1$ and in fact that happens with this data set. This may be due to the presence of outliers which, from Chapter 5, are known to exist in this set.

16.3.3 Enhancements

The example shown in the previous section is the most simple. The general model is

$$\hat{\mathbf{x}} = \bar{\mathbf{x}} + \mathbf{V}\mathbf{y}$$

In the previous section, $\bar{\mathbf{x}}$ was assumed to be zero since it was equal to zero for the entire sample. However, it may be that the mean is not equal to zero and the cross-validation technique for it is more complicated, involving the deletion of variables. Furthermore, both $\mathbf{V}$ and $\mathbf{y}$ are considered estimates and, as we now know from Chapter 10, may be estimated simultaneously using singular value decomposition. This, of course, is not possible here because the y-scores are obtained for the observations not included in the sample from which V is obtained. The solution to this problem is to use all n observations in each subsample but randomly delete elements from each data vector. The good way to do this is to randomly order the observations and use a cyclic deletion pattern given by Wold (1978). The estimation procedure will, of necessity, require SVD but the SVD algorithm employed must be able to handle missing data. Another alternative is to use procrustes fits as a cross-validation procedure (TenBerge, 1986). Cross-validation has also been employed to obtain bounds on the inherent variability (Wold, 1978).

Krzanowski (1983) carried out some simulations to compare cross-validation with some other stopping rules. This was summarized in the material on stopping rules in Section 2.8.12.

16.4 SENSITIVITY

16.4.1 Introduction

In Chapter 4, the distributions of characteristic roots and vectors were presented, making it possible to construct significance tests and confidence limits based on the variability that one would expect in estimates of these quantities based on chance alone. This section will be concerned with variations due to known causes and in the sensitivity to the estimates due to these changes. Specifically, we shall look at the effect of a change in a single characteristic root, the effect of a single observation on all of the roots and vectors, the effect of rounding, and the effect of changes in the data collection method with particular emphasis on stratified sample surveys.

16.4.2 Sensitivity of a Characteristic Vector to a Change in its Characteristic Root

Although the standard errors for the individual coefficients of the characteristic vectors gives one a fair idea of the precision one might expect, some other techniques may prove useful.

Krzanowski (1984b) considered the maximum possible change in a characteristic vector u_i, given a specified reduction in the value of its corresponding root l_i. The change in the ith vector is a function only of the ith and $(i + 1)$st roots and vectors. For the ith vector, this is

$$u_{(i)} = \frac{u_i \pm u_{i+1}\sqrt{l_i/[c(l_i - l_{i+1})]}}{\sqrt{1 + l_i/[c(l_i - l_{i+1})]}} \qquad (16.4.1)$$

where c is an arbitrary constant related to the desired variation in the ith characteristic root. To study the effect of an $e\%$ variation in a characteristic root, c is set equal to $100/e$. The reciprocal of the denominator of this expression is the cosine of the angle between $\mathbf{u}_{(i)}$ and $\mathbf{u}_i$.

Using the Ballistic Missile example of Section 2.3, the first two characteristic vectors are

$$\mathbf{u}_1 = \begin{bmatrix} .468 \\ .608 \\ .459 \\ .448 \end{bmatrix} \qquad \mathbf{u}_2 = \begin{bmatrix} -.622 \\ -.179 \\ .139 \\ .750 \end{bmatrix}$$

with characteristic roots $l_1 = 335.34$ and $l_2 = 48.03$. Suppose one was interested in a 10% change in the first characteristic root: $c = 100/10 = 10$ and from this (16.4.1) becomes

$$\mathbf{u}_{(1)} = \frac{\mathbf{u}_1 \pm \mathbf{u}_2(.3416)}{1.0567}$$

Taking the positive sign,

$$\mathbf{u}_{(1)} = \begin{bmatrix} .242 \\ .518 \\ .479 \\ .666 \end{bmatrix}$$

and for the negative sign,

$$\mathbf{u}_{(1)} = \begin{bmatrix} .644 \\ .633 \\ .389 \\ .182 \end{bmatrix}$$

The angle between $\mathbf{u}_{(1)}$ and $\mathbf{u}_1$ is $\arccos(1/1.0567) = 19°$. Although these perturbations may seem impressive, they represent only a 10% reduction in l_1 from 335.34 to 301.81, whereas the lower confidence limit for λ_1, using (4.2.1) is 233.16, since the sample size was only $n = 40$.

The angle between $\mathbf{u}_{(2)}$ and $\mathbf{u}_2$ is $27°$, reflecting the smaller difference between l_2 and l_3. However, the angle is actually a function of $l_i/(l_i - l_{i+1})$, not just $l_i - l_{i+1}$ and for $\mathbf{u}_3$, the angle between $\mathbf{u}_{(3)}$ and $\mathbf{u}_3$ is $25°$, smaller than $\mathbf{u}_2$ even though $(l_3 - l_4) < (l_2 - l_3)$.

Krzanowski's technique might be of use in situations where the pc's are fairly well defined and constant over time or some other kind of replication so that the pc model itself is of importance. If this were a quality control application, for instance, and one of the pc's was associated with a process variable on which some improvements might be made, this might reduce the variability of that pc, its characteristic root, and this technique could be used to estimate what this anticipated improvement would do to the associated characteristic vector. Krzanowski felt that this technique should be useful for the last few pc's, that may be used to delete variables, the subject of Section 14.4, or to look for functional relationships. His method was based on a procedure proposed by DeSarbo et al. (1982) to handle the same problem for canonical correlation. Tarumi (1986) presented a unified procedure for investigating sensitivity in PCA, biplots, correspondence analysis, and canonical correlations. Tanaka (1988) investigated the influence on the *subspace* spanned by the retained pc's. Keep in mind that a change in a root of a correlation matrix must must result in a change of at least one of the other roots.

16.4.3 Influence of a Single Observation

Probably of more immediate application in data analysis is a technique designed to measure the effect of an individual observation on the characteristic roots and vectors, the *influence function*. This technique has been used elsewhere, particularly in multiple regression (see for instance, Andrews, 1974).

Using the z-scores as defined by (1.4.1), $z = U'[x - \bar{x}]$, the influence function for the ith characteristic root, λ_i, is

$$I(x, \lambda_i) = z_i^2 - \lambda_i \qquad (16.4.2)$$

and for the corresponding characteristic vector

$$I(x, \Upsilon_i) = -z_i \sum_{j \neq i}^{p} z_j \Upsilon_j (\lambda_j - \lambda_i)^{-1} \qquad (16.4.3)$$

These results are due to Radhakrishnan and Kshirsagar (1981) and Critchley (1985). Note that while the influence function for λ_i involves only the ith root and vector, the influence function for Υ_i involves them all except for Υ_i itself. Further, any observation that influences one vector will have to influence at least one more since the vectors are orthogonal to each other. This is consistent with expressions for the standard errors associated with the roots and vectors. Note that these are theoretical influence functions. In practice, neither the population roots or vectors will be known and the sample estimates for them will be substituted in (16.4.2) and (16.4.3). These expressions will not be reproduced exactly if one actually deletes the observation and recalculates the sample root and vector. An example comparing the estimated with the actual quantities may be found in Jolliffe (1986, p. 190).

Influence functions are illustrated here with the Chemical data. One would ordinarily not need to use this technique with such a simple example and it is included here merely as a numerical example. The results are shown in Table 16.3 and, as is usually the case, the population values are not known and are replaced with sample estimates. The influence functions for the characteristic roots, $I(x, l_i)$ have been divided by their respective roots merely to put them on a relative rather than actual basis, making it easier to study the effect on smaller roots. When $p = 2$, the elements of u_1 and u_2 are the same except for position and the sign of one of them. For the first observation,

$$I(x, u_1) = \begin{bmatrix} .13 \\ .12 \end{bmatrix}$$

and

$$I(x, u_2) = \begin{bmatrix} -.12 \\ .13 \end{bmatrix}$$

For this special case, only a single characteristic vector is involved and is a constant. This being the case, the only quantity that varies with each observation is $-z_1z_2/(l_1 - l_2)$, which is also displayed in Table 16.3. For $p > 2$, Jolliffe (1986) recommended computing the sum of squares of these values, $I(x, u_i)'I(x, u_i)$.

Note that the influence on the roots of individual z-scores is a function of the *absolute* value of the score. Further, note that scores that are close to zero, such as z_1 for the 14th observation and z_2 for the 10th observation, have a

Table 16.3. Chemical Data. Influence Functions

Observation No.	z_1	z_2	$\dfrac{z_1^2 - l_1}{l_1}$	$\dfrac{z_2^2 - l_2}{l_2}$	$\dfrac{-z_1z_2}{(l_2 - l_1)}$
1	.48	.51	−.84	2.01	.18
2	.15	−.42	−.98	1.04	−.05
3	−.22	.21	−.97	−.49	−.03
4	−.15	.28	−.98	−.09	−.03
5	2.27	−.09	2.56	−.91	−.15
6	1.28	−.11	.13	−.86	−.10
7	−1.77	.03	1.17	−.99	−.04
8	−.84	−.16	−.51	−.70	.10
9	−.34	−.50	−.92	1.89	.12
10	−.57	−.01	−.78	−1.00	.00
11	.64	−.06	−.72	−.96	−.03
12	−1.27	−.17	.12	−.67	.16
13	2.04	.26	1.88	−.22	.39
14	−.07	−.21	−1.00	−.49	.01
15	−1.64	.46	.86	1.45	−.55

decided negative effect on their respective roots since their omission from the sample would result in marked increases in the sizes of these roots. z-Scores will have a positive influence only when they are greater, in absolute value, than the square root of their corresponding characteristic root. For normally distributed z-scores, one would expect 68% of the pc's to have a negative influence and for this small sample exactly two-thirds of the scores have this distinction.

Critchley (1985) proposed two different influence functions. One deletes the ith observation in estimating the roots and vectors and the other is based on the difference between these and (16.4.2) or (16.4.3). These may be of interest when one is working with small sample sizes. Emerson et al. (1984) developed a procedure for measuring the leverage of individual observations in a two-way table of residuals on nonadditivity. As an alternative procedure, Hadi (1987), Hadi and Wells (1990) suggested using the index $\sqrt{\lambda_1/\lambda_p}$. This index may be employed when the variances are nearly equal and will be a minimum when Σ is diagonal. Naes (1989) investigated influence functions for principal components regression.

If the PCA has been carried out on a correlation matrix, the influence functions become much more complicated. Let

$$I(\mathbf{x}, \rho_{jk}) = -\tfrac{1}{2}\rho_{jk}(x_j^2 + x_k^2) + x_j x_k \tag{16.4.4}$$

where the x's are all in standard units. This is the influence function for the correlation coefficient ρ_{ij} (Devlin et al., 1975). Then equation (16.4.2) becomes (Calder, 1986; Jolliffe, 1986, p. 189)

$$I(\mathbf{x}, \lambda_i) = \sum_{\substack{j=1 \\ j \neq k}}^{p} \sum_{k=1}^{p} \Upsilon_{ij}\Upsilon_{ik} I(\mathbf{x}, \rho_{jk}) \tag{16.4.5}$$

The influence function for the vectors themselves is (Pack et al., 1988)

$$I(\mathbf{x}, \Upsilon_i) = \left\{ \sum_{\substack{h=1 \\ h \neq i}}^{p} \Upsilon_h (\lambda_h - \lambda_i)^{-1} \right\} \left\{ \sum_{\substack{j=1 \\ j \neq k}}^{p} \sum_{k=1}^{p} \Upsilon_{hj}\Upsilon_{ik} I(\mathbf{x}; \rho_{jk}) \right\} \tag{16.4.6}$$

These complications arise, in part, because the sum of the roots of a correlation matrix is known in advance to be equal to p. Therefore, any observation that has influence on one root will, by necessity, also have to affect one or more other roots in the opposite direction. Pack et al. (1988) included an example consisting of seven physical measurements on a number of college students. They compared the influence function with the actual changes made by deleting an observation. In this set of data, two of the measurements for *one* student— wrist and hand circumference—had originally been reversed. This, by itself, became the second principal component. Talk about influence! Pack et al. also pointed out that if one pair of roots were close together, the pc's might become interchanged in order and, if that were so, the influence functions might be too large for both of them.

16.4.4 Effect of Rounding on PCA

In the Introduction, the reader was warned to expect small discrepancies when following through the examples in this book. These discrepancies could be the result of rounding carried on primarily to enhance the presentation of the results. Rounding may also occur during intermediate steps of some computations, either on the part of the author or those whom the author has cited (or the reader, for that matter). None of this should have any serious consequences, even with fairly pronounced rounding (e.g., one-digit vector coefficients) but it will certainly affect things such as orthogonality, unit length, and so on. (In Section 7.3, severe rounding was advocated for purposes of display to enhance the ability of the user to interpret the results.)

Let a_i be an approximation to u_i. The difference in the amount accounted for by that component is

$$\Delta_i = \frac{a_i'Sa_i}{a_i'a_i} - l_i \qquad (16.4.7)$$

As an example, consider the Ballistic Missile example of Section 2.3. Suppose that the second characteristic vector in Table 2.2

$$u_2 = \begin{bmatrix} -.622 \\ -.179 \\ .139 \\ .750 \end{bmatrix} \text{ was replaced by } a_2 = \begin{bmatrix} -.6 \\ -.2 \\ .1 \\ .8 \end{bmatrix}$$

$$\Delta_2 = (50.3137/1.05) - 48.03 = -.11$$

or, in relative terms, $\Delta_2/l_2 = -.11/48.03 = -.0023$. A similar calculation for the first vector, also rounded to one decimal place, yields $\Delta_1 = -1.49$ with $\Delta_1/l_1 = -.0044$. The correlation between these two "components" is $r = -.014$ so no real damage has been done by severely rounding these coefficients. Other types of simplification, such as replacing small coefficients with zeros, may be evaluated in a similar manner although the practice of zeroing will produce much more noticeable difference (Green, 1977).

If one is concerned only about rounding effects, bounds can be placed on the expected difference $E(\Delta_i)$. Under the assumption that rounding errors are uniformly distributed over the interval $-c$ to $+c$,

$$\tfrac{1}{3}pc^2(l_p - l_i) \leqslant E(\Delta_i) \leqslant \tfrac{1}{3}pc^2(l_1 - l_i) \qquad (16.4.8)$$

Again, taking the second vector rounded to the nearest .1, $c = .05$. With $p = 4$ variables, (16.4.8) becomes

$$(\tfrac{1}{3})(4)(.05)^2(16.41 - 50.32) \leqslant E(\Delta_2) \leqslant (\tfrac{1}{3})(4)(.05)^2(335.34 - 50.32)$$

or

$$-.11 \leqslant E(\Delta_2) \leqslant .95$$

These results are due to Bibby (1980) who also gave some other bounds for approximations. Green (1977) gave similar bounds in terms of expected squared loss for each parameter estimate.

On the other side of the coin, Box et al. (1973) pointed out that roots, which should be zero because of linear constraints among the variables, may be positive because of rounding errors. The usual rules of numerical analysis apply: Save the rounding until the end of the operations. In this day and age, with the computer facilities available, rounding is no longer the problem it once was.

16.4.5 Sensitivity Due to Nonhomogeneity of the Data

Most of the examples used in this book have assumed that the observations used to obtain the dispersion matrices, from which the characteristic roots and vectors are obtained, were drawn from a homogeneous population. There have been exceptions, particularly in Chapters 12 and 13 whose purpose was to introduce techniques in which PCA might be of use in relating prediction and response variables. In pc-ANOVA (Section 13.4), the PCA was based on the total variability of the responses from a designed experiment in which the only assumption of homogeneity dealt with the observations for each experimental condition. In the Color Film Process example from Section 6.6, it was known that the level of the process shifted from week to week and so the covariance matrix on which the PCA was performed was based on within-week variability and the assumption of homogeneity referred to within-week, not the population as a whole.

This section will discuss one such application, principal component analysis of a sample survey from a stratified population. Sample survey data usually involve a multiple response and PCA is frequently used as an aid in the analysis. Often, the population from which the data are obtained is stratified by such criteria as geography, income, sex, and so on. The sizes of these subpopulations are generally not equal and a simple random sample drawn from the entire population may not represent all of the strata equally. A PCA performed on such data may produce biased results. In addition, information is often available with regard to each respondent, which might produce better estimates of the predicted responses than would result merely from knowing which stratum he or she represented.

Skinner et al. (1986) investigated these problems with respect to PCA, citing evidence of similar situations with tests of independence (Rao and Scott, 1981) and regression (Nathan and Holt, 1980). They approached this problem by first examining the various covariance matrices that could be obtained by using or ignoring information available about the stratification. Assume that there are q response variables, y, and p predictor variables, x. "Response" variables in this case would be answers to various questions asked during the survey and "predictor" variables refer to such things as demographic information available for each respondent. This situation parallels those in Chapter 13 where the PCA would be performed on the response variables. Let the population covariance matrix of the responses be Σ_{yy}, of the predictors be Σ_{xx} and the covariance between them, Σ_{xy}. The total covariance matrix is:

$$\Sigma_T = \left[\begin{array}{c|c} \Sigma_{yy} & \Sigma'_{xy} \\ \hline \Sigma_{xy} & \Sigma_{xx} \end{array} \right]$$

Skinner et al. investigated three possible covariance matrices. These are the equivalent of simple random sampling, the regression estimate, and the Horvitz–Thompson estimate.

In the *simple random sampling* estimate, all information about stratification is ignored. A random sample is obtained from the entire population. The estimate of Σ_{yy} is simply S_{yy}, the sample covariance matrix. The estimate of the overall mean, μ_y is equal to the average of the sample. The expected value of S_{yy} is

$$\Sigma_{yy} + \beta(S_{xx} - \Sigma_{xx})\beta' \qquad (16.4.9)$$

where $\beta = \Sigma_{xy}\Sigma_{xx}^{-1}$. The second term represents the potential bias that may occur. This is made up of two parts, β, representing the regression of the predictors on the responses for the entire population and $S_{xx} - \Sigma_{xx}$ representing the difference between the simple random sampling covariance matrix and the population covariance matrix. For this second term to be of importance, *both* parts must contribute. If either of them is zero, the term will drop out.

With this in mind, Skinner et al. suggested as an alternative estimate:

$$S_{yy.ML} = S_{yy} + B[S_{xx} - \Sigma_{xx}]B' \qquad (16.4.10)$$

where $B = S_{xy}S_{xx}^{-1}$. Σ_{xx} is presumed to be known assuming that the predictor information is known for the entire universe. This is related to the *regression* estimate of μ_y (Cochran, 1977, p. 187) and as the subscript indicates, is a maximum likelihood estimator.

The third estimate, the so-called *design-unbiased* scheme due to Nathan and Holt (1980), is based on the probabilities associated with a particular observation being drawn and producing the Horvitz–Thompson (1952) estimate of μ_y.

What effect does simple random sampling have on the PCA? Although Skinner et al. (1986) studied this only for a single predictor variable ($p = 1$), their results should imply what one can expect for the general case. For simple random sampling, the expected value of the ith characteristic root is

$$E(l_i) = \lambda_i\left(1 + \rho_i^2 \frac{s_x^2 - \sigma_x^2}{\sigma_x^2}\right) \tag{16.4.11}$$

where ρ_i is the population correlation between the predictor variable and the ith pc. This is the same situation as the expected value for the covariance matrix; $E(l_i)$ will be affected only if ρ_i^2 and $(s_x^2 - \sigma_x^2)/\sigma_x^2$ contribute. If either quantity is zero, the expected value will not be inflated. Note that if $s_x^2 < \sigma_x^2$, the expected value will biased downwards and vice versa.

The expected value of the corresponding characteristic vector is

$$E(\mathbf{u}_i) = \Upsilon_i + \left\{\rho_i \frac{(s_x^2 - \sigma_x^2)}{\sigma_x^2}\right\}\left\{\sum_{k \neq i}^{q} \rho_k \Upsilon_k \frac{\sqrt{\lambda_i \lambda_k}}{\lambda_i - \lambda_k}\right\} \tag{16.4.12}$$

The second term here is a function not only of ρ_i and $(s_x^2 - \sigma_x^2)/\sigma_x^2$ but also $\lambda_i - \lambda_k$.

For this special case of one predictor variable, Skinner et al. performed some simulations using stratification. Merely by varying allocations among the strata, they were able to effect differences of 60% in the first characteristic root of the covariance matrix obtained using simple random sampling. The "design-unbiased" scheme had similar problems. On the other hand, their maximum likelihood estimator was insensitive to changes in allocation.

In an earlier simulation study using correlation matrices, Bebbington and Smith (1977) compared the results for four designs:

1. Simple random sampling.
2. Stratified sampling, samples within strata chosen by simple random sampling.
3. Single-stage cluster sampling where strata are chosen by simple random sampling. This technique is often used to choose which neighborhoods, blocks, and so forth will be used in personal interview surveys.
4. Single-stage cluster sampling with the strata having unequal probability of being selected.

The results produced estimates of the variance of the larger roots that were markedly smaller than one would expect from (4.7.1) when using simple random sampling. Even smaller estimates were obtained using stratified sampling. However, cluster sampling produced results larger than (4.7.1) and, as one might expect, unequal probabilities of selection only made things worse. Similar results were obtained for the first characteristic vector; simple random and stratified sampling produced vectors that were more highly correlated with the population vectors than did the cluster sampling results. These results would suggest caution when applying PCA to survey data.

16.5 ROBUST PCA

16.5.1 Introduction

The majority of techniques discussed in this book, as well as the examples used to illustrate them, have assumed that the data with which we are working are basically "good." A number of problems may occur. First, the assumptions regarding the underlying distributions used in inferential procedures may be incorrect. This was discussed in Section 4.8. Fortunately, the estimates of the characteristic vectors, themselves, are distribution-free although inferential procedures associated with them are not. Second, there are assumptions of independence of the sample observations. If the observations are obtained as a function of time, it is assumed that the covariance structure does not change over this period of time. These assumptions may be checked by examining the individual pc's and residuals. In fact, PCA is a good method for discovering changes in structure.

But what about outliers? What effect may they have? In some examples, such as the Chemical example in Section 1.7, these were detected with the T^2-test. Others were detected with residual tests such as the Q-test in Section 2.7. Both of these examples assumed that the estimates of the characteristic roots and vectors were adequate. It is possible that outliers can affect the roots and vectors themselves and for that, robust estimation procedures will be required. The material in the previous section on sensitivity could intensify a feeling of insecurity about one's results.

Unlike the majority of techniques described in this book, robust estimation procedures tend to become quite complicated—not the sort of thing one would do on a pocket calculator! It is the old case of getting what you pay for; if one really wants to feel secure about the results, one must be prepared to work for them.

In the case of PCA, there are three ways of obtaining robust estimates:

1. Obtain a robust estimate of the covariance or correlation matrix used in the PCA. From these, a conventional PCA would be carried out.
2. Obtain robust estimates of the characteristic roots and vectors themselves. Some of these results may, in turn, be used to obtain robust estimates of the covariance or correlation matrix.
3. Carry out some sort of data analysis on the starting data and perform conventional PCA on this edited data set.

Most procedures identified as *robust PCA* employ strategy (1) or (2). It is conceivable that one could merely carry out a PCA on data that has been screened by some data analysis procedures and call that robust PCA but, given the other procedures now available, this would not seem to be a viable alternative.

Many of these techniques use one form or another of the *Mahalanobis distance function*

$$d_i = \{(\mathbf{x}_i - \mathbf{a})'\mathbf{S}^{*-1}(\mathbf{x}_i - \mathbf{a})\}^{1/2} \qquad (16.5.1)$$

where $\mathbf{x}_i$ is an observation vector, $\mathbf{a}$ is an estimate of the mean, and $\mathbf{S}^*$ is an estimate of the covariance matrix. This is the same form as Hotelling's T^2 for individual observation vectors, the only difference being that $\mathbf{a}$ and $\mathbf{S}^*$ are no longer the standard estimators but are part of the robust process. Most of these techniques are iterative so that $\mathbf{a}$ and $\mathbf{S}^*$ may change with each iteration. These distances are used to create weights $w(d_i)$ which, in turn, are used to produce M-estimators (Maronna, 1976) of $\mathbf{a}$ and $\mathbf{S}^*$, viz.,

$$\mathbf{a} = \sum_{i=1}^{n} w_a(d_i) \frac{\mathbf{x}_i}{\sum_{i=1}^{n} w_a(d_i)} \qquad (16.5.2)$$

$$\mathbf{S}^* = \frac{\sum_{i=1}^{h} w_s(d_i^2)[\mathbf{x}_i - \mathbf{a}][\mathbf{x}_i - \mathbf{a}]'}{f\{w_s(d_i^2)\}} \qquad (16.5.3)$$

The determination of $w_a(d)$, $w_s(d^2)$, and $f\{w_s(d_i^2)\}$ is a function of the technique involved and will be illustrated shortly. One way in which various robust techniques may be compared is in terms of their *breakpoint*, the maximum proportion of the data set that may be made up of outliers without effectively changing the result (Hampel, 1974). There is a lot of literature regarding robust estimation, including a book on the subject by Huber (1981) and two expository papers by Hogg (1979a, 1979b). Seber (1984, p. 165) has a section on multivariate robust estimation.

16.5.2 Robust Estimates of Σ and P

This section will cover methods of obtaining robust estimates of covariance or correlation matrices. Estimates of the characteristic roots and vectors would be obtained from these matrices by conventional PCA methods. Devlin et al. (1981) compared a number of these techniques and some of the names given here correspond to their nomenclature.

One group of techniques are the so-called "univariate" methods in which each element of the covariance or correlation matrix is estimated independently of the others. One of these, SSD, (Devlin et al., 1975) consists of obtaining a correlation matrix by estimating robustly each correlation coefficient separately using whatever information is available. For covariance matrices, trimmed or Winsorized estimates of each standard deviation are obtained, again from all available data, and these in conjunction with the robust correlations are used to obtain the estimate of the covariance matrix. The ability to work with missing data in this way is one of its strong points. Other features include its requiring the shortest computing time of all the methods and having among the highest breakpoints, equal in this case to the percentage of the sample trimmed in each iteration. The main drawback of SSD is that the resultant covariance matrix may not be positive-definite, in which case a shrunken estimate must be obtained to achieve that property. With this in mind, it is not surprising that SSD has the largest bias of any of the methods and does the poorest in producing estimates of characteristic roots and vectors corresponding to near singularities in the data. A technique for robust estimation of μ and Σ with missing data was given by Little (1988a).

A variant on this due to Ruymgaart (1981) also obtains independent robust estimates of the variances along with a linearization of the multivariate structure but, so far, has been worked out only for the case $p = 2$. A different approach, REG (Mosteller and Tukey, 1977), uses robust regression to obtain the regression of each variable on the rest and obtain the covariance matrix from that. Of all the techniques compared by Devlin et al. (1981) in their simulation studies, REG took the longest to perform in addition having some invariance problems and has consequently dropped from the scene.

All of the multivariate procedures to be discussed below are sensitive to starting values and usually work best with robust estimates at the beginning. Of the multivariate procedures, the most simple involves multivariate trimming, MVT (Gnanadesikan and Kettenring, 1972; Devlin et al., 1975). In this method, the Mahalanobis Distance Function, d_i, (equation 16.5.1), is evaluated for each observation. A fixed proportion of these observations are trimmed from the sample on the basis of the size of the d_i and the remainder are used to obtain robust estimates of the mean and the covariance matrix. The trimmed data are then replaced and the procedure is repeated until the correlation matrix converges. MVT is the fastest of all the multivariate methods and has the highest breakpoint among them, the breakpoint again being equal to the proportion of the sample trimmed. On the other side of the ledger, the MVT estimates of the characteristic roots and vectors are generally biased, though not as much as SSD. Coleman (1986) used this technique in a quality control application involving semiconductor wafers.

The final group of robust estimators of Σ and P are the M-estimators (not to be confused with *M-analysis*, the SVD of a double-centered matrix). M-estimators involve weighting each observation in obtaining estimates both of the mean using (16.5.2) and the covariance matrix using (16.5.3). Outlying observations get down-weighted rather than being trimmed. A number of choices of weights have been proposed of which we shall give three:

1. *MLT* (Maronna, 1976)

$$w_a(d_i) = \frac{p + v}{v + d_i^2} = w_s(d_i^2) \qquad (16.5.4)$$

where v are the degrees of freedom associated with the multiple t-distribution. Generally, v is set equal to 1, the Cauchy distribution. This is the value used by Devlin et al. (1981). $f\{w_s(d_i)\} = 1/n$. Boente (1987) obtained the asymptotic distribution of the principal components obtained from this matrix.

2. *HUB* (Huber, 1964)

$$
\begin{aligned}
w_a(d_i) &= 1 & d_i &\leqslant c_1 \\
&= c_1/d_i & d_i &> c_1
\end{aligned}
$$
(16.5.5)

$$
w_s(d_i^2) = \{w_a(d_i)\}^2/c_2
$$
(16.5.6)

Devlin et al. (1981) used $c_1 = \chi^2_{p,.10}$. c_2 is a correction to make the estimate of Σ unbiased. Again, $f\{w_s(d_i)\} = 1/n$.

3. Campbell (1980b). Let

$$
\begin{aligned}
w^*(d_i) &= d_i & d_i &\leqslant c_3 \\
&= c_3 \exp\{-\tfrac{1}{2}(d_i - c_3)^2/c_4^2\} & d_i &> c_3
\end{aligned}
$$

where $c_3 = \sqrt{p} + c_5/\sqrt{2}$. Then

$$
w_a(d_i) = \{w^*(d_i)\}/d_i
$$
(16.5.7)

$$
w_s(d_i^2) = \{w_a(d_i)\}^2
$$
(16.5.8)

$$
f\{w_s(d_i)\} = \sum_{i=1}^{n} \{w_s(d_i)\}^2 - 1
$$
(16.5.9)

Campbell suggested using $c_4 = 1.25$ and $c_5 = 2$. This produces weights that decrease at a faster rate than the corresponding Huber scheme, which in Campbell's format would have $c_4 = \infty$ and $c_5 = 2$. (If $c_5 = \infty$, the weights would all be equal to unity.) c_5 is equated with a percentage point of the normal distribution. In Campbell's case, c_5 is approximately $\alpha = .05$, which corresponds to the χ^2-level given above. Campbell also suggested using all three sets of weights and making probability plots of the associated d_i to detect possible outliers. For the case $c_4 = 1.25$, $c_5 = 2$, he recommended normal probability plots of $d_i^{2/3}$ and that weights of .3 or less indicated outliers. For the case $c_4 = \infty$, this would be increased to .6.

Devlin et al. (1981) concluded that MLT and HUB gave the best estimates of the characteristics roots and vectors (Campbell's scheme was not included in their study) with a slight edge to MLT somewhat compensated for by MLT's longer computational time. M-estimators generally have a breakpoint of roughly $1/p$, which could prove a problem for an application with a large number of variables. This and the shorter run time were the reasons that Coleman used MVT for his semiconductor problem. His problem had 25 variables so the breakpoint for the M-estimators would have been 4%.

16.5.3 Robust Estimates of the Characteristic Roots and Vectors

16.5.3.1 Introduction

The methods discussed in Section 16.5.2 produced robust estimates of Σ and P and the characteristic roots and vectors were obtained from them by conventional PCA. These results are called robust PCA because the starting matrices were robust. The three methods discussed in this section are used to obtain robust estimates of the characteristic roots and vectors directly. From these, robust estimates of Σ and P may be obtained by the relationship: $S = ULU'$. In addition, robust estimators of the orthogonal regression line for the case $p = 2$ has been given by M.L. Brown (1982), Ammann and van Ness (1989), and Zamar (1989).

Unlike the previous section, there have been no studies comparing the various procedures given below.

16.5.3.2 Campbell's Method

The procedure for this method (Campbell, 1980b) is as follows:

1. Obtain an initial estimate of the first characteristic vector u_1.
2. For each observation, obtain the first z-score, using u_1, that is, $z_1 = \acute{u}_1[x - a]$.
3. Obtain the M-estimators of the mean and variance of z_1 and the associated weights w_i for each observation. As M-estimators are sensitive to starting values, Campbell suggested starting with the median and $\{.74 (\text{interquartile range})\}^2$. After the first iteration, use for each observation, the minimum weight produced for it by that time.
4. Use these weights to obtain new estimates a and S^* employing (16.5.2) and (16.5.3).
5. Obtain the first characteristic root and vector from S^* and repeat steps 2–5 until the characteristic root converges sufficiently. The final value of u_1 is the robust estimate of the first vector, u_1^*.
6. Once the first characteristic root and vector have been determined, obtain the residuals $(x - \hat{x})$ unexplained by the first robust pc and repeat steps 1–5. This will produce the second characteristic root and vector. Continue the process until all of the characteristic roots, L^*, and vectors, U^*, have been obtained.

A robust estimate of Σ is now $S^* = U^*L^*U^{*'}$. If one is not interested in obtaining S^*, step 6 may be terminated when the appropriate number of roots and vectors have been determined. The plotting recommendations Campbell made with regard to the direct robust estimation of Σ in Section 16.5.2 are equally appropriate here.

A similar procedure using *projection pursuit* was proposed by Li and Chen (1985) in which a robust estimate of scale is maximized rather than the variance. This also uses M-estimation with weights similar to HUB. The Li and Chen procedure appears to provide similar estimates to other M-estimators described here with higher break points (25% or better) and higher computational costs.

16.5.3.3 The Method of Gabriel and Oderoff

A similar method, also using M-estimation, is due to Gabriel and Oderoff (1984) but operates within the operations of the singular value decomposition. Although this procedure also reweights the data with each iteration of the vector estimation procedure, unlike Campbell's procedure, it does not require successive estimates of Σ since SVD works with the data directly. In fact, this method is not designed to obtain robust estimates of Σ at all, seeking only to represent the data set with the smallest number of pc's possible consistent with the philosophy of PCA.

Gabriel and Oderoff made use of *iteratively reweighted least squares* or IRLS (Green, 1984). Unlike the methods described so far, which compute weights for each observation, this method computes a weight for each variable of each observation and is a function of the residual unexplained by the PCA model at that point. Let $d_{ij} = (x_{ij} - \hat{x}_{ij})/c_6$ where c_6 is an arbitrary scaling constant. Then, by use of the Beaton–Tukey (1974) bi-square weight function:

$$
\begin{aligned}
w_{ij} &= [1 - (d_{ij}/c_7)^2]^2 & |w_{ij}| \leqslant c_7 \\
&= 0 & |w_{ij}| > c_7
\end{aligned} \right\} \tag{16.5.10}
$$

The choice of the "tuning" constant, c_7 is also arbitrary. Gabriel and Oderoff used, for c_6, the median of the absolute residuals from the initial fit. The choice of c_7 is dependent on the type of outlier against which one is guarding. If there are a few large outliers, a value of $c_7 = 8$ might be suitable; $c_7 = 4$ would be appropriate for a larger number of smaller outliers, particularly along one row or column. They suggested trying more than one. A similar procedure was used by Cleveland and Guarino (1976) in the analysis of some air pollution data and by Ramsay (1984), who used weights of the form $\exp[-d^2/\chi_k^2]$.

As M-estimators are sensitive to starting values, Gabriel and Oderoff developed an algorithm called LORANK that contains an IRLS procedure due to Gabriel and Zamir (1979) preceded by a method of providing good initial estimates. LORANK includes several options for initial estimates but the one they preferred is a multiplicative analog of Tukey's *median polish* (Tukey, 1977). Gabriel and Oderoff illustrated this procedure with a subset of Fisher's iris data Fisher (1936) in which they had introduced a large outlier. LORANK produced essentially the same results as the standard SVD did on the unperturbed data.

16.5.3.4 The Method of Galpin and Hawkins

Entirely different is the method due to Galpin and Hawkins (1987). Although this procedure does obtain robust estimates of the characteristic roots and vectors, its purpose in doing so is to obtain a robust estimate of Σ. Their procedure involves either quadratic or linear programming, neither technique designed to work with negative values such as those which the principal components might take. To handle this difficulty, let the value of the jth principal component for the ith observation be written as

$$z_{ij} = \mathbf{u}'_j[\mathbf{x}_i - \bar{\mathbf{x}}] = P_{ij} - N_{ij} \qquad (16.5.11)$$

If, for the ith observation, z_{ij} is positive, then $P_{ij} = z_{ij}$ and $N_{ij} = 0$. If z_{ij} is negative, then $P_{ij} = 0$ and $N_{ij} = |z_{ij}|$. The quadratic programming operation is designed to optimize, with respect to $\mathbf{u}_j$, the quantity

$$\sum_{i=1}^{n} |z_{ij}| \qquad (16.5.12)$$

subject to

$$\sum_{i=1}^{n} (P_{ij} + N_{ij}) = 0$$

$$P_{ij} - N_{ij} = z_{ij}$$

$$P_{ij} \geqslant 0$$

$$N_{ij} \geqslant 0$$

$$\mathbf{u}'_j \mathbf{u}_j = 1$$

and
$$\mathbf{u}'_j \mathbf{u}_k = 0 \qquad (j \neq k)$$

Although the characteristic vectors $\mathbf{U}$ will still be orthonormal, the principal components will no longer be uncorrelated. This is not considered a problem since the primary goal of this procedure is to produce a robust estimate of Σ. In addition, the estimation of the corresponding characteristic root will require an additional operation. Galpin and Hawkins had two suggestions for this, one involving (16.5.12) and the other involving the projections of the data points onto the axes defined by the robust $\mathbf{u}_j$ and obtaining some measure of dispersion such as the mean absolute deviation to estimate the characteristic root λ_j. (In numerical work, the second option appeared to produce better results.) Once this has been done, a robust estimate of Σ may be obtained. Galpin and Hawkins found that improved results were obtained by using the MLT weight function (16.5.4). The median was used in place of the mean when computing the initial d_i^2.

To make things a little easier operationally, Galpin and Hawkins also formulated this as a linear programming problem by changing the constraint that the squares of the coefficients of $\mathbf{u}_j$ sum to 1 to a constraint that their absolute values sum to 1. The LP procedures produce a global optimum while the QP procedures provide only a local optimum. However, the LP estimates of the vectors may not be too good unless the characteristic roots are well separated. Again if one is carrying out this procedure only to obtain a robust estimate of Σ, this is not a problem since the vectors are only artifacts to the final end. For the five data sets with which Galpin and Hawkins worked, the QP procedure generally produced better estimates of Σ than the LP procedure, but not by a great amount.

16.6 g-GROUP PCA

16.6.1 g-Group Procedures

In a PCA of Scottish voting patterns in the late 1960s, Lincoln et al. (1971) noted that the correlation matrices, and hence the characteristic vectors obtained from them, were fairly consistent from year to year. This is a common situation and the purpose of this section is to present some PCA procedures involving two or more independent samples, or groups of data, representing different populations of individuals, methods, time periods, and so on. The question is: "Is the principal component model the same across populations?" If the answer is "yes," then one set of characteristic vectors will satisfy all of the samples.

At this point, it is important to specify which normalization of vectors is under consideration. If the hypothesis is in terms of V-vectors, it will specify both the orientation and length of the vectors. Since these two properties will uniquely define a covariance matrix, this type of hypothesis may be handled with conventional tests of g covariance matrices found in any multivariate analysis text. However, if the hypothesis is in terms of U-vectors, then it involves orientation only (i.e., the vectors but not the roots). We have already seen an example of this in Chapter 9 where the U-vectors related to audiometric examinations were relatively independent of age; it was the characteristic roots associated with these vectors that varied. Formally, the hypothesis for g-populations is

$$H_0: \Upsilon'\Sigma_i\Upsilon = \Lambda_i \qquad i = 1,\dots,g \tag{16.6.1}$$

where Λ_i is a diagonal matrix. This is called *common principal components* (CPC).

To test (16.6.1), one must obtain an estimate for Υ, U, and estimates of Λ_i, $F_i = U'S_iU$. Although Λ_i are hypothesized to be diagonal, the sample F_i will not ordinarily be so and the farther they are from being diagonal, the more likely that H_0 will be rejected. The large-sample test of H_0 (Flury, 1984) is

$$\chi_v^2 = \sum_{i=1}^{g} n_i \ln\left\{\frac{|\text{Diag }F_i|}{|F_i|}\right\} \tag{16.6.2}$$

where $v = (g - 1)p(p - 1)/2$ degrees of freedom. Multiplication of a covariance matrix by a constant, c, will not alter these results since the U-vectors of cS are the same as those of S (Theorem B.8).

How does one obtain U? Flury obtained maximum likelihood estimates by solving the system of equations:

$$\Upsilon_j'\left[\sum_{i=1}^{g} n_i\left(\frac{\lambda_{ij} - \lambda_{ih}}{\lambda_{ij}\lambda_{ih}}\right)S_i\right]\Upsilon_h = 0 \qquad \begin{matrix} j, h = 1,\dots,p \\ j \neq h \end{matrix} \tag{16.6.3}$$

subject to $\Upsilon'\Upsilon = I_p$ and $\lambda_{ij} = \Upsilon_j' S_i \Upsilon_j$. There will be $p(p-1)/2$ of these equations. A solution for (16.6.3) using the FG algorithm was produced by Flury and Gautschi (1986), whose computer code may be found in Flury and Constantine (1985) and Flury (1988, Appendix C).

Krzanowski (1984a) proposed a simplified alternative, which was to obtain the characteristic vectors of

$$S_p = \frac{\sum_{i=1}^{g} n_i S_i}{\sum_{i=1}^{g} n_i}$$

The properties of this procedure have not yet been established but for several examples examined by Krzanowski, including the Iris data, the two sets of estimates were quite similar. Two alternative procedures were given by Meredith and Tisak (1982).

Flury (1987) extended this to the case where only $(p-k)$ vectors are tested (*partial common principal components*). This is a practical modification since, most of the time, a number of the components are of little or no interest. In addition, the components associated with small and not significantly different roots may have an undue effect on the test. The test procedure is the same as (16.6.2) except that the degrees of freedom have now been reduced to $(g-1)k(p-k)$ and there are now many fewer equations in (16.6.3). The subset of components need not be the first $p-k$ components nor need they be the same pc's from each population (i.e., the first three pc's from one population, the first, second and fourth from another population, etc.) Flury (1986) also obtained some procedures for testing the equality of vectors with a hypothesized set of vectors and a likelihood ratio test for simultaneous sphericity of $p-k$ pc's in g populations.

Krzanowski (1979b) proposed another type of procedure based on comparing the U-vectors from each of the samples. This is a descriptive procedure involving the vectors only, and as such, may be used with correlation matrices as well as covariance matrices. An example for the case $g = 2$ will be given in Section 16.6.3.

Keramidas et al. (1987) suggested some graphical procedures for comparing g samples. In particular, they suggested producing box-plots (Tukey, 1977) for the characteristic roots. There would be one "box" for each root, the spread representing the g samples. Secondly, they suggested a gamma $Q-Q$ plot (Wilk and Gnanadesikan, 1968; Gnanadesikan, 1977, p. 198) of a dissimilarity measure that is essentially the squared euclidean distances between the jth vector for each sample and some reference vector. This reference vector may be either a prespecified vector or the average jth vector over the g samples. ("Average" in this case is such that it *maximizes* the sums of cosines of the angles between itself and the sample vectors.) Not surprisingly, the more separated the characteristic roots, the more effective the $Q-Q$ plots. By and large, these techniques require a large number of samples to be effective—probably more than are customarily available in practice. However, one of their examples, an evaluation of questionnaires for class instructors, had over 100 different samples.

Vogt (1988) suggested plotting the coefficients of U-vectors from one sample against their corresponding coefficients from another.

Another procedure, used in the analysis of longitudinal data is called *component analysis* (yet another use of this term!). This procedure maximizes the correlations between the principal components and the original variables (Meredith and Tisak, 1982; Meredith and Millsap, 1985; Millsap and Meredic 1988; Kiers and TenBerge, 1989).

Flury (1988) combined much of the material described in this section along with its theoretical background and many numerical examples. He established a hierarchy of significance tests:

a. Equality of covariance matrices
b. Proportionality of covariance matrices
c. CPC
d. Partial CPC

One would first test for the equality of the g covariance matrices. If that hypothesis is accepted, that is the end. If it is rejected, then one tests the hypothesis that the covariance matrices are proportionate, and so on.

16.6.2 Numerical Example

For a numerical example, we shall return to the analysis of variance data in Table 13.1. In practice, one would not ordinarily use this analysis on such a small data set but it serves the purpose here because it is small. There were $n = 4$ observations on each of $p = 2$ test methods for $g = 3$ labs. The covariance matrices for these groups and their related U-vectors are

$$\mathbf{S}_1 = \begin{bmatrix} .4667 & .5533 \\ .5533 & .6567 \end{bmatrix} \quad \mathbf{U}_1 = \begin{bmatrix} .6445 & -.7646 \\ .7646 & .6445 \end{bmatrix}$$

$$\mathbf{S}_2 = \begin{bmatrix} .3267 & .2567 \\ .2567 & .2200 \end{bmatrix} \quad \mathbf{U}_2 = \begin{bmatrix} .7757 & -.6311 \\ .6311 & .7757 \end{bmatrix}$$

$$\mathbf{S}_3 = \begin{bmatrix} .1158 & .0667 \\ .0667 & .0600 \end{bmatrix} \quad \mathbf{U}_3 = \begin{bmatrix} .8325 & -.5540 \\ .5540 & .8325 \end{bmatrix}$$

The U-vectors for the individual covariance matrices are not used for this test but are included for interest. We wish to ascertain whether one set of vectors will fit all three labs. Even though (16.6.3) consists only of one equation in this case, we will use Krzanowski's approximation and obtain the vectors from the pooled covariance matrix. This matrix and its associated U-vectors are

$$\mathbf{S}_p = \begin{bmatrix} .3031 & .2922 \\ .2922 & .3122 \end{bmatrix} \quad \mathbf{U}_p = \begin{bmatrix} .7015 & -.7126 \\ .7126 & .7015 \end{bmatrix}$$

Obtaining $F_i = U_p' S_i U_p$ produces:

$$F_1 = \begin{bmatrix} 1.1164 & .0863 \\ .0863 & .0069 \end{bmatrix}$$

$$F_2 = \begin{bmatrix} .5291 & -.0574 \\ -.0574 & .0175 \end{bmatrix}$$

$$F_3 = \begin{bmatrix} .1541 & -.0290 \\ -.0290 & .0217 \end{bmatrix}$$

The ratio of the determinant of the diagonal elements of F_1 to the determinant of F_1 itself is 30.21 and $4 \ln(30.21) = 13.63$. For the other two groups, the results are 1.75 and 1.17. The sum of these three quantities is 16.55. For $g = 3$ and $p = 2$, there are 2 degrees of freedom so a value of 16.55 is highly significant. Remember that (16.6.2) is a large-sample test for which this example does not qualify. However, the result is so large that one would probably feel comfortable in concluding that U_p does not adequately represent all three labs. This is not surprising in view of the interaction found in Section 13.4.

16.6.3 Two-group Procedures

A few other techniques are available for the two-sample case. Krzanowski (1979b) gave a scheme for comparing the two subspaces represented by $k < p$ vectors. The method is to obtain the characteristic roots of $C = U_1' U_2 U_2' U_1$. The arccosines of the square roots of these characteristic roots give the maximum angles between these subspaces. For the Chemical example used in Section 15.7.2, we can only use the case $k = 1$ and in this case, C is a scalar and is its own characteristic root. For samples 1 and 2, using the U-vectors displayed in Section 16.6.2, the root is .9653 so that the corresponding angle is $10.7°$. The angle between samples 1 and 3 is $16.2°$ and between samples 2 and 3, $5.5°$. Because this procedure works solely with vectors, it may be used for correlation and product matrices as well as covariance matrices. Although there is no formal test for this procedure, Krzanowski (1982) did some simulations to provide a crude table for guidance both for the case where the underlying matrices are equal and for some cases where they were not.

Newcomb (1961) showed that for two symmetric positive-semidefinite matrices A and B (which would cover any sort of dispersion matrix), there exists a nonsingular matrix T such that

$$A = T' A_0 T \qquad B = T' B_0 T \qquad (16.6.4)$$

where A_0 and B_0 are both diagonal matrices. DeLeeuw (1982) gave conditions for all possible T that would satisfy (16.6.4).

Flury (1983) also used PCA to test whether or not two covariances matrices were proportionate, the first step in his hierarchy of tests.

16.7 PCA WHEN DATA ARE FUNCTIONS

16.7.1 Introduction

A number of examples in this book represent situations where an observation on p variables represents selected points from a continuous curve. These include the B&W Film example from Section 2.9 and the Audiometric example from Chapters 5 and 9. In both cases, the number and spacing of the variables was a function of the equipment involved. For the B&W Film example, there were exactly 14 exposure levels on the exposing tablet and an increase in the number of variables would have required a redesign of the equipment. On the other hand, there are many cases where the original data are in the form of a curve (spectrophotometric curves, spirometer curves, and EKG exams are examples) and the variables are defined arbitrarily when certain locations are selected and digitized.

PCA, being a mechanical operation, would treat these variables as any others and would find linear combinations of them that maximize explained variability. Among the consequences to be considered are the following:

a. Designed experiments whose measured response is a set of points from a curve are often analyzed with the analysis of variance technique. If PCA is employed, then one may also use pc-ANOVA, which was discussed in Section 13.4. Recall, from that section, that the ANOVA results of the pc's are correlated unless some type of step-down procedure is employed.

b. The residual variability may include variability in the level of the curve itself. This means that after k pc's have been retained in accordance with a stopping rule, the residuals may still be correlated. However, instead of that, this type of variability may have been buried in the first principal component (or one of the others for that matter), thus inflating the estimate of its variance and underestimating the true inherent variability.

c. As stated earlier, one has a better chance of obtaining interpretable vectors for this type of data than any other.

16.7.2 Growth and Learning Curves

A special case of functional data is the situation where the function is monotonically increasing (or at least nondecreasing). Each of the variables, in order, should be higher than the one before it. One would expect this type of result from animal feeding studies or chemical reactions where a reagent is being added as a function of time. The resultant responses are called *growth* curves and represent weight as a function of time or product as a function of concentration. Test results given to the same individuals over a period of time result in *learning* curves. The principle is the same and the use of PCA for these two situations was first advocated by Rao (1958) and Tucker (1958a). Ramsay and Abrahamowicz (1989) used splines to model test items in examinations as a function of the ability of the students and then used PCA to analyze the resultant curves. Although growth and learning curves have a "positive" aura to them, the reverse situation of decay or survival curves is equally applicable. An example of PCA with tumor data using proportional hazards or survival data regression may be found in Danielyan et al. (1986). A review article dealing with prediction of future observations in growth curve models by Rao (1987) made use of a number of techniques including both principal component regression and factor analysis, the subject of the next chapter. Tucker (1958a), Izenman and Williams (1989), and Meredith and Tisak (1990) also used factor analysis to study growth curves.

Because this is a very specialized situation, so are some of the anticipated results as well as the recommended operations required to obtain them. A 2^5 experiment reported by Church (1966) is typical. The coefficients for the first vector were all positive, the second was a contrast between the earlier and later measurements and the third represented a buckle in the curve orthogonal to the second. (The B&W Film example, which may be thought of as a growth curve relating density to exposure, also had these same three vectors.)

Snee (1972a) advocated analyzing growth curve data as a two-way table, using the techniques of Section 13.7, feeling that the interactions might be detected more readily in this form. (He referred to this as *modified* PCA.) Snee et al. (1979) compared several ANOVA models with some rat-feeding experiments. Snee also used this type of PCA to characterize shape of objects, one example dealing with carrots (Snee, 1972b).

Much of the time, the measured response will always be positive, as in the case of weights, concentrations, or many test results. In this case, use of the product matrix rather than the covariance or correlation matrix has been recommended (Rao, 1958 and Ross, 1964), as was done in Section 3.4. Weitzman (1963) also used product matrices in obtaining pc's to cluster observations. In this case, the first vector will often represent the mean curve and aids in model description. As we have seen in Section 3.4, product matrices are quite popular with analytical chemists for that reason. Rao also considered using first differences rather than the observations themselves.

In modeling growth or learning curves with PCA, the first pc ordinarily characterizes the growth and the succeeding pc's account for the nonlinearity of that function. In allometry, it is often assumed that the first component is the only "real" one, the other pc's representing inherent variability (Hopkins, 1966; Sprent, 1968). Jolicoeur (1963) suggested taking logarithms of the data first.

Whether there is only one pc of interest or more than one, these are sample estimates, subject to sampling and experimental variability, and the characteristic vectors may not produce as smooth functions as one might like. Arbuckle and Friendly (1977) suggested an orthogonal rotation procedure that would minimize the differences among successive times. Assuming that p, the number of variables, now represents the number of times at which data are collected, let $\mathbf{P}$ be a $(p-1) \times p$ differencing matrix of the form

$$\mathbf{P} = \begin{bmatrix} 1 & -1 & 0 & 0 & \cdots & 0 & 0 \\ 0 & 1 & -1 & 0 & \cdots & 0 & 0 \\ 0 & 0 & 1 & -1 & \cdots & 0 & 0 \\ \vdots & \vdots & \vdots & \vdots & & \vdots & \vdots \\ 0 & 0 & 0 & 0 & \cdots & 1 & -1 \end{bmatrix}$$

The function $\mathbf{t}$ that will minimize the sums of squares of the successive differences within a vector, that is, minimize

$$d^2 = \sum_{i=2}^{p} (u_i - u_{i-1})^2$$

is the vector associated with the *smallest* characteristic root of $\mathbf{C} = \mathbf{U}'\mathbf{P}'\mathbf{P}\mathbf{U}$ and d^2 is that root. $\mathbf{t}$ may be used to smooth vectors (i.e., $\mathbf{u}_1^s = \mathbf{u}_1\mathbf{t}$). Arbuckle and Friendly also generalized the procedures to smooth more than one vector, to use quadratic and higher-order differencing formulas, and to allow for uneven spacing. It should be emphasized that this technique is *only* for use with the growth or learning model. Using this for any other vector sets will make growth curves out of them. As an example of such misuse, Table 16.4 displays the first characteristic vector, $\mathbf{u}_1$, of the Black and White Film example from Section 2.7 and $\mathbf{u}_1^s$, that vector smoothed using this procedure. Note that the coefficients are now all positive and monotonically decreasing. $\mathbf{u}_1^s$ is still normalized to 1

Table 16.4. Black & White Film Example. Illustration of
the Arbuckle–Friendly Smoothing Algorithm

$\mathbf{u}_1$	$\mathbf{u}_1^s$
.292	.405
.319	.384
.356	.358
.397	.346
.416	.321
.398	.304
.332	.280
.244	.275
.146	.208
.077	.140
.031	.102
.009	.078
.000	.076
−.009	.070

and $\mathbf{Pu}_1^s = .0139$, the smallest root of $\mathbf{C}$, while $\mathbf{Pu}_1 = .0337$ so the algorithm does work and can be useful when applied properly.

Using spirometer data obtained at discrete time intervals, van Pelt and van Rijckevorsel (1986) fitted various nonlinear functions such as splines to the data and then performed a PCA on the results. In studying monthly temperature data for 32 French cities, Winsberg (1988) removed the main curve shape by filtering and then performed a PCA on the residuals. (This article includes the raw data.) On the other hand, Rossi and Warner (1986), working with some emission data, obtained a Fourier transform of the vectors. *Primary* vectors would be weighted towards lower frequency Fourier coefficients and *secondary* vectors, representing inherent variability, would be weighted toward the higher frequencies.

In another interesting technique, Cartwright (1987) had spectroscopic data representing n samples, each at p wavelengths. Each sample represented a different pH level and the data were ordered by pH. A PCA was obtained for $n = 1$ (a special case since there is only a single root), then $n = 2$ (two roots), and so on. Plots were obtained of $\ln(l_i)$ vs. pH that showed that as the number of observations (and hence the range of pH) increased, the apparent rank of the system went from one to two to three.

16.7.3 Eigenfunctions

In this section, the use of PCA with functional data will be discussed, particularly with regard to the questions:

1. In view of the fact that the data are actually functions, is PCA appropriate?
2. What is the effect on the PCA results of unequally spaced sampling locations on those functions?

The answers to these questions have been examined by Ramsay (1982), Besse and Ramsay (1986), Besse (1988), and Castro et al. (1986), who all treated the problem as a time series analysis, conducting their investigations in the time domain using functional analysis. The concepts are basically the same as PCA except that the term characteristic vector or eigenvector is replaced by *eigenfunction*.

Besse and Ramsay used *interpolating splines* (Winsberg and Ramsay, 1983) with either Green's function or reproducing kernels and established a class of eigenfunction models of which conventional PCA is a special case. Another special case corresponds to the PCA of residuals obtained from a regression analysis of multiple responses. A numerical example was included using data for the movement of the human tongue while pronouncing the syllable "Kah." Although the details for their method would have to be worked out by the user, Besse and Ramsay did list the reproducing kernels for about a dozen situations. Ramsay (1988) also discussed a SAS PROC that handles monotone spline principal components.

Castro et al. were concerned more directly with the analog of conventional PCA and showed that the eigenfunction procedure is nothing more than a weighted PCA of the form

$$\mathbf{S}_w = \mathbf{D}^{1/2}\mathbf{S}\mathbf{D}^{1/2} \qquad (16.7.1)$$

where $\mathbf{D}$ is now a diagonal matrix whose coefficients are obtained using the trapezoidal rule. If the data points were equally spaced, x_1 and x_p would have weights of .5 and $x_2, \ldots, x_{p-1}$ would have weights of 1.0 so that the results would differ little from conventional PCA. Table 16.5 compares these for the Black and White Film example. Note that the differences become more pronounced as the characteristic roots become smaller. The real merit of this technique is that the results are fairly robust when the sampling points are not evenly spaced and Castro et al. gave some rather dramatic examples to demonstrate this. Although they used the trapezoidal rule, other functions could have been used. A number of such functions were given by van Rijckevorsel (1988). Besse (1988) gave a procedure for evaluating various smoothing coefficients with the object of maximizing the "elbow" effect of the SCREE plot.

In using any of these functions, one might normalize the weights such that their sum is equal to the number of variables. In our example, with $p = 14$, the weights would be $p/[2(p - 1)] = .538$ and $p/(p - 1) = 1.077$ instead of .5 and 1.0. The U-vectors will be unaffected but the roots and other vector normalizations will change.

Anyone wishing to perform PCA using observations from continuous functions should seriously consider using one of these techniques, particularly if the data are unequally spaced.

Table 16.5. Black & White Film Example. Comparison of Conventional Vectors u_i and Eigenfunctions u_i^c

	u_1	u_1^c	u_2	u_2^c	u_3	u_3^c
	.29	.20	.49	.39	.35	.32
	.32	.32	.42	.47	.19	.28
	.36	.36	.27	.33	−.01	.05
	.40	.40	.10	.15	−.17	−.14
	.42	.43	−.09	−.05	−.29	−.29
	.40	.41	−.25	−.23	−.25	−.27
	.33	.34	−.36	−.35	−.06	−.09
	.24	.25	−.38	−.39	.26	.25
	.15	.15	−.31	−.32	.34	.34
	.08	.08	−.20	−.21	.31	.32
	.03	.03	−.11	−.12	.31	.32
	.01	.01	−.07	−.08	.29	.30
	.00	.00	−.05	−.06	.31	.32
	−.01	−.01	−.04	−.03	.32	.23
%Expl.	83.1	84.3	13.5	12.4	2.3	2.1

16.8   PCA WITH DISCRETE DATA

16.8.1   Binary Data

If one has p binary variables $x_1, \ldots, x_p$ such that x_i takes the value 1 with probability b_i and the value 0 with probability $1 - b_i$ and the x_i may or may not be correlated, this is a multivariate binomial distribution. Because the variables take only the values 0 and 1, people tend to be leery of using them directly in such things as ANOVA and PCA. However, the data could be arranged as a p-dimensional contingency table that could then be handled by multiple correspondence analysis (Section 10.8.5). Gower (1966a) justified the use of PCA on the covariance matrix of the binary data directly by showing an analogy between those results and distances obtained from the frequencies arranged similarly to a Burt matrix where each variable was represented by one row and column.

Cox (1972) and Bloomfield (1974) investigated the use of contrasts among the variables for use as new variables which, in turn, could be used for PCA. Multivariate logit analysis is another transformation possibility. It appears, at this point in time, that there has been very little in the line of potential applications to produce much experience in these areas.

Binary data can also be handled using *entropy*. Brambilla (1976) used this for some questionnaire data related to women's rights. There were 6 groups of 10 questions each. If the proportion of positive responses to the 10 questions in each of the 6 groups is p_i, then the entropy would be

$$- \sum_{i=1}^{6} p_i \ln(p_i)$$

Brambilla also analyzed these same data with PCA and pointed out that the PCA procedures produce a plot of scores while entropy produced only a single point, essentially an overall measure.

16.8.2   Compositional Data

Of more frequent occurrence is the extension to the *compositional* case in which, for an observation vector, the variables $m_1, \ldots, m_p$ are proportions that sum to 1.0. This constraint, in turn, places restrictions on the correlation matrix of these variables and produces pc's that are difficult to interpret as well as making it difficult to work with PCA models in this context. A discussion of these problems may be found in Aitchison (1983) along with a number of suggestions for alternative solutions. Although the most common occurrence of data of this sort is related to the multinomial distribution, data normalized to sum to 1 also occur in such instances as mixture and directional data, for which these procedures would also be applicable.

Aitchison's recommended procedure is to transform the original m_i as follows:

$$x_i^* = \ln(m_i) - \frac{1}{p} \sum_{j=1}^{p} \ln(m_j) \qquad i = 1, 2, \ldots, p \qquad (16.8.1)$$

The PCA is then carried out on the covariance matrix of the x_i^*. This matrix will have one characteristic root equal to zero and the corresponding U-vector with coefficients all equal to $1/\sqrt{p}$.

A discussion of the properties of this transformation may be found in Aitchison (1982, 1983) but it included the premise that subsets of these variables are easier to work with in this form and that interpoint distances between two observation vectors could be obtained directly. Watson (1984) gave an expression for the confidence limits for the first characteristic vector and included details for the case $p = 3$.

Table 16.6 contains a form of the United States government budget for the years 1967–1986. Although these data were obtained from various issues of the *Statistical Abstract* (Bureau of the Census, various years, 1967–1986), the reader should consider them to be *informal*. Many of these values had been restated in subsequent years and over a period of 20 years budget categories had been created or discarded so that certain categories had to be grouped. Other categories are grouped merely because of their smaller size and, finally, categories

Table 16.6. U.S. Budget. (in Billions of Dollars) 1967–1986

Year	x_1	x_2	x_3	x_4	x_5	x_6	x_7
1967	76	13	31	7	7	6	15
1968	88	14	34	10	8	7	19
1969	89	16	38	12	8	6	17
1970	89	18	44	13	9	7	19
1971	88	20	56	14	11	9	18
1972	89	21	65	17	11	10	23
1973	88	23	73	18	13	10	20
1974	92	28	84	22	13	12	24
1975	104	31	109	28	16	15	26
1976	108	35	127	33	17	18	28
1977	116	38	138	39	19	21	34
1978	124	44	146	44	24	26	43
1979	138	53	160	50	28	30	41
1980	155	52	205	55	40	32	46
1981	181	69	240	66	46	34	51
1982	209	85	263	74	41	27	53
1983	235	90	293	82	37	27	61
1984	253	111	291	88	38	28	53
1985	279	129	316	100	36	29	67
1986	299	136	319	106	37	30	73

such as revenue sharing were not included at all because that activity did not exist during all 20 years. The category groupings used for this example are the following:

x_1: Defense and Veterans Affairs

x_2: Interest on the National Debt

x_3: Income Security and Social Security not including Medicare

x_4: Health and Medicare

x_5: Commerce, Transportation and Energy

x_6: Education and Manpower

x_7: A composite of smaller items including International Affairs, Space and Technology, Agriculture, Natural Resources, Community Development, and General Government

A PCA of the correlation matrix for these data yielded a first component accounting for nearly 94% of the total variability and represented the general increase in the federal budget over that time. The second component, accounting for about 5%, represented a contrast primarily between x_1 and x_2, both of which had dramatic increases in the last few years of the period, and x_5 and x_6, which actually decreased during that same time. The last five pc's accounted for 1.2%. It is difficult to make many intercategory comparisons because the overall growth of the budget swamped everything else. A better way to do this might be to analyze the proportions of the budget for each category.

These proportions are shown in Table 16.7. The transformation of these data using (16.8.1) are shown in Table 16.8 and the covariance matrix of these data in Table 16.9. The characteristic roots of this matrix are .1557, .0626, .0109, .0037, .0032, .0015, and 0. There is a definite break after the third root, by which time 96.4% has been accounted for. The first three U-vectors are given in Table 16.10. When working with this kind of data, the first pc will not be the customary "overall" component because that effect has been removed by working with proportions. The first pc, accounting for 65.5% is primarily a contrast between x_1 and x_7, on one hand, and x_3 and x_4 (and to some extent x_2), on the other hand, and probably represents the fact that these latter three had increased during the time period. The second component (26.3%) appears to be a contrast of two categories that gained in their share of the budget in the last few years, x_1 and x_2, with two that lost share, x_5 and x_6. The third component (4.6%) appears to represent the contrasts in curve shape for the minor (in share) categories of the budget; x_7 had a steady decline while x_6 had all of its decline in the last half of the period and x_5 actually gained for a while before declining.

16.8.3   Other Discrete Conditions

Young et al. (1978) produced a technique for handling the case where variables are *mixed*, some being continuous while other variables may be discrete or

Table 16.7. U.S. Budget 1967–1986. Proportions

Year	m_1	m_2	m_3	m_4	m_5	m_6	m_7
1967	.490	.084	.200	.045	.045	.039	.097
1968	.489	.078	.189	.056	.044	.039	.106
1969	.478	.086	.204	.065	.043	.032	.091
1970	.447	.090	.221	.065	.045	.035	.095
1971	.407	.093	.259	.065	.051	.042	.083
1972	.377	.089	.275	.072	.047	.042	.097
1973	.359	.094	.298	.073	.053	.041	.082
1974	.335	.102	.305	.080	.047	.044	.087
1975	.316	.094	.331	.085	.049	.046	.079
1976	.295	.096	.347	.090	.046	.049	.077
1977	.286	.094	.341	.096	.047	.052	.084
1978	.275	.098	.324	.098	.053	.058	.095
1979	.276	.106	.320	.100	.056	.060	.082
1980	.265	.089	.350	.094	.068	.055	.079
1981	.263	.100	.349	.096	.067	.049	.074
1982	.278	.113	.350	.098	.055	.036	.070
1983	.285	.109	.355	.099	.045	.033	.074
1984	.294	.129	.338	.102	.044	.032	.061
1985	.292	.135	.331	.105	.038	.030	.070
1986	.299	.136	.319	.106	.037	.030	.073

Table 16.8. U.S. Budget 1967–1986. Aitchison Transformation (15.9.1)

Year	x_1^*	x_2^*	x_3^*	x_4^*	x_5^*	x_6^*	x_7^*
1967	1.66	−.11	.76	−.73	−.73	−.88	.03
1968	1.63	−.21	.68	−.54	−.77	−.90	.10
1969	1.62	−.10	.77	−.38	−.79	−1.08	−.04
1970	1.51	−.08	.81	−.41	−.78	−1.03	−.03
1971	1.39	−.09	.94	−.45	−.69	−.89	−.20
1972	1.29	−.15	.98	−.36	−.80	−.89	−.06
1973	1.24	−.10	1.05	−.35	−.67	−.93	−.24
1974	1.15	−.04	1.06	−.28	−.81	−.89	−.19
1975	1.10	−.11	1.14	−.22	−.78	−.84	−.29
1976	1.02	−.11	1.18	−.17	−.83	−.77	−.33
1977	.97	−.15	1.14	−.12	−.84	−.74	−.26
1978	.88	−.15	1.05	−.15	−.76	−.68	−.18
1979	.88	−.08	1.03	−.14	−.72	−.65	−.33
1980	.86	−.24	1.14	−.18	−.50	−.72	−.36
1981	.86	−.11	1.14	−.15	−.51	−.82	−.41
1982	.97	.07	1.19	−.07	−.66	−1.08	−.41
1983	1.02	.06	1.24	−.03	−.83	−1.14	−.33
1984	1.06	.23	1.20	.00	−.84	−1.14	−.51
1985	1.06	.29	1.18	.03	−.99	−1.21	−.37
1986	1.08	.29	1.14	.04	−1.01	−1.22	−.33

Table 16.9. U.S. Budget 1967–1986. Covariance Matrix of x_i^*

	x_1^*	x_2^*	x_3^*	x_4^*	x_5^*	x_6^*	x_7^*
x_1^*	.072	−.008	−.039	−.045	−.005	−.012	.037
x_2^*	−.008	.024	.012	.019	−.012	−.022	−.013
x_3^*	−.039	.012	.028	.030	−.002	−.003	−.025
x_4^*	−.045	.019	.030	.041	−.008	−.009	−.029
x_5^*	−.005	−.012	−.002	−.008	.016	.012	−.001
x_6^*	−.012	−.022	−.003	−.009	.012	.031	.004
x_7^*	.037	−.013	−.025	−.029	−.001	.004	.027

Table 16.10. U.S. Budget 1967–1986. U-Vectors

	u_1	u_2	u_3
x_1^*	.637	.358	.125
x_2^*	−.193	.509	.026
x_3^*	−.405	.012	.092
x_4^*	−.489	.176	−.227
x_5^*	.029	−.367	.779
x_6^*	.031	−.668	−.387
x_7^*	.390	−.019	−.408

consist of rank data. This is a general solution that essentially puts all of the variables in standard units and obtains, using alternating least squares, a solution that minimizes the sum of the Q-statistics.

Discreteness may also be the result of categorizing continuous data. An example of this is the grouping of data into class intervals. Another example replaces variables with quantities such as Likert scale values, bipolar scales, and so on. In these cases, traditional PCA may still be employed but is has been suggested that this type of discreteness may cause the apparent number of retained pc's to increase. Bernstein and Teng (1989) studied the effect of various categorizations on the characteristic roots for both PCA and factor analysis.

16.9 [ODDS AND ENDS]²

16.9.1 Introduction

This section contains a number of topics, most of which have attracted little attention with regard to PCA. The purpose for including them is to provide information on what has been done so that anyone wishing to pursue any of these topics will have a base, such as it may be, on which to build.

16.9.2   Minimax Estimation

PCA minimizes, for fixed k, the sums of squares of residuals $[\mathbf{X} - \hat{\mathbf{X}}]$. Other criteria are possible. Bargmann and Baker (1977) developed a minimax technique that, for fixed k, will obtain characteristic roots and vectors such that

$$\sum_{i=1}^{n} \sum_{j=1}^{p} (x_{ij} - \hat{x}_{ij})^{q}$$

is minimized. The x_{ij} are in quasistandard units being divided by the maximum permissible error for each variable rather than its standard deviation. q is an even positive integer. (This is a bit like the Minkowski metric in multidimensional scaling.) To obtain a solution, one must obtain not only the characteristic roots and vectors but also a vector of constants required to balance the sum of odd powers of errors. Since the pc-scores are also required to obtain the solution, Bargmann and Baker defined this problem in terms of singular value decomposition.

The minimax procedure of Bargmann and Baker is designed to require fewer pc's than would be required by conventional PCA. Their suggested stopping rule was the smallest value of k such that $\text{Max} |x_{ij} - \hat{x}_{ij}| \leq 1.0$. They suggested doing this sequentially, starting with $k = 1, 2, \ldots$ and for each value of k trying a sequence of values of q on a power of four scale (i.e., 4, 16, 64, 256, ...). In some of their examples, they ended up with quite large values of q but the resultant values of k were markedly less than would have been obtained with conventional PCA.

16.9.3   Bayes Estimation

Little has been done with regard to Bayesian PCA. Geisser (1965) obtained a general expression for the posterior distribution of the characteristic roots. This involves the classical distribution of the roots, but for some of the special cases discussed in Chapter 4 some usable results might be obtained. Tiao and Fienberg (1969) obtained, for the bivariate normal distribution, the posterior distribution of $l_1/(l_1 + l_2)$ and the angle associated with the principal axis rotation. Some optimality properties of Bayesian PCA have been investigated by Chen (1974). Some additional Bayesian procedures are listed in Section 12.3.5 in connection with principal components regression.

16.9.4   Populations of Variables

One problem related to PCA modeling is the choice of the original variables employed. Specifically: Are the variables employed representative of the class of variables for which inferences will be made? For some exploratory PCA, this is not a problem; one is furnished with a data set and the inferences are to be made with regard to those variables only. However, when building models or doing confirmatory PCA, the choice of variables does play a part. This problem surfaced originally in connection with the use of aptitude tests and similar exercises to characterize one's ability. To wit: "These are the tests we used but there are others we could have used, and if we had, would the results have been any different?" Put another way, how robust is the model to change of variable?

Take as an example the Seventh-Grade example with four variables, two having to do with reading and two with arithmetic. The first pc represented overall ability, accounting for 46% and the second pc represented the difference between reading and arithmetic, accounting for 37%. The hypothesis would be that the exchange of new variables for old or the addition of new variables would not change the interpretation of the first two pc's or the proportion of variability accounted for. Implied in this hypothesis is that only reading and arithmetic variables are of concern. If one were really trying to measure *overall* ability, other criteria besides these two would have to be considered. There are some mechanical problems here, particularly for small *p*. The inclusion of just one more reading variable would change all of the coefficients and the proportion just because there would now be three reading variables and only two for arithmetic.

Given that the problem has been well defined and there are no mechanical problems, the way to deal with this situation is to assume that the variables used are a *random sample* of the variables that could have been chosen. Hotelling (1933) showed, under this assumption, that the estimate of the proportion of the total variability explained by each pc was consistent and gave a procedure for specifying the number of variables chosen such that the difference between the true and sample proportion was less than some arbitrary quantity. Tucker (1958b) gave a procedure for comparing *two* test batteries on this basis.

16.9.5 Monte Carlo Operations

A number of articles referenced in this book used simulation or Monte Carlo studies to investigate the properties of various PCA procedures. These operations with principal components would follow the same general procedures as in any other field of statistical analysis. One can, for instance, generate sample covariance or correlation matrices having a certain structure by specifying the characteristic roots and characteristic vectors normalized to those roots. One can then randomly generate principal components, use (2.7.2) to obtain the corresponding values for **X**, and from these obtain the covariance and correlation matrices. One could also add a random error component and use (2.7.3). [The covariance matrix, without error, could be obtained simply by multiplying the characteristic vectors by their transpose as defined by equation (1.6.4)]. Some techniques especially designed for PCA were developed by Chalmers (1975) and Lin and Bendel (1985). Care should be exercised in the choice of random number generator. Some that appear to be adequate for univariate distributions do not hold up when used for multivariate analysis.

CHAPTER 17

What is Factor Analysis Anyhow?

17.1 INTRODUCTION

At last, we have come to the topic of *factor analysis*. What is it? The definition depends on the person with whom you are speaking as there are many opinions as to just what comprises this technique. These various points of view will be taken up in Section 17.10. One definition has, by now, become generally accepted and this will be described in the next section. For now, it will suffice to say that factor analysis (FA) is a technique similar in nature to principal component analysis and is its principal competitor, enough so that it merits this chapter of its own. Like PCA, FA also goes back to the turn of the century (Spearman, 1904). Most of the early work was done by psychologists and was, in general, long on intuition and short on inferential procedures. Unlike PCA, the solutions were not unique and the early reputation of FA was felt by many in the statistical community to be closely associated with witchcraft. However, modern FA techniques have changed that. Factor analysis is a viable solution for many problems and the purpose of this chapter will be to briefly describe these techniques and suggest where they might be used in place of PCA and vice versa.

There have been a number of books written about factor analysis, among them Harmon (1976), Lawley and Maxwell (1971), and Cureton and D'Agostino (1983), and there continue to be a great many papers on the subject, particularly in psychological journals. Because of all the available material, it is not the purpose in this chapter to cover the topic in that much detail. Rather, the basic principles will be discussed and it will be shown how PCA and FA coexist in the modern world. References to other material on factor analysis will be made where appropriate.

388

17.2 THE FACTOR ANALYSIS MODEL

17.2.1 Review of the PCA Model

Recall, in the development of the PCA model in Chapters 1 and 2, it was shown that given a full set of principal components, one could determine the original variables, that is,

$$x = \bar{x} + Vy \tag{17.2.1}$$

If less than a full set of pc's were used, the original variables could be estimated but there would be some discrepance between the predicted and actual values of the original variables. This would replace **x** in (17.2.1) with an estimate $\hat{x}$ and give the more general model:

$$x = \bar{x} + Vy + (x - \hat{x}) \tag{17.2.2}$$

In Chapter 2, the Q-statistic, $Q = (x - \hat{x})'(x - \hat{x})$ was developed to measure this discrepancy. Along with this was another expression in terms of the covariance (or correlation) matrix:

$$E = S - V'V \tag{17.2.3}$$

or, turned around,

$$S = V'V + E \tag{17.2.4}$$

where **E** is the residual covariance after k characteristic vectors had been estimated.

In PCA, equation (17.2.4) was somewhat of a by-product of the method. This equation says that the covariance matrix is made up of the part explained by the principal components, $V'V$ and the portion unexplained by them, **E**. In FA, a form of (17.2.4) is the starting point. Recall, in Section 2.8.5, that one of the stopping rules for the number of retained pc's was based on the size of the diagonal elements in **E** and one that we felt was the best one to use in many cases *if the information was available.* In the FA model, it assumed that **E** is a *diagonal* matrix, that is, the residuals are uncorrelated. More about this later. First, we shall need to define some terms.

17.2.2 Factor Analysis Terminology

Not only are there some differences between PCA and FA with regard to models and methodology but also to the names associated with many of the quantities involved. What in PCA are called principal components will now be referred to as *common factors*. There will be $k < p$ of these factors.

(The case $k = p$ would be the same as a full PCA.) To avoid confusion with PCA notation, we shall use the same symbols, in general, but will affix a superscript "f" to them, viz. $\mathbf{U}^f$, $\mathbf{V}^f$, and so on. The coefficients of these matrices are generally referred to as *loadings*.

Common factors refers to factors that account for variability in at least two of the original variables. Associated with this is another new term, *communality*. The communality, h_i^2, is the variability associated with each of the original variables explained by the common factors, that is,

$$h_i^2 = \sum_{j=1}^{k} (v_{ij}^f)^2 \qquad i = 1, 2, \dots, p \qquad (17.2.5)$$

This corresponds to the diagonal elements of the first term on the right-hand side of (17.2.4). The diagonal elements of the second term are called *unique* factors because they represent that variability in each of the original variables that has nothing in common with any of the other variables. There will be p of these, although it is possible that one or more of them may be equal to zero. The unique factor for each of the original variables includes its inherent variability. If this quantity were known, it would lead to the general FA model:

$$\mathbf{x} = \bar{\mathbf{x}} + \mathbf{V}^f \mathbf{f} + \mathbf{c}_1 + \mathbf{c}_2 \qquad (17.2.6)$$

where the four terms on the right-hand side represent the mean, the contribution of the common factors, the contribution of the *specific* factors, $\mathbf{c}_1$, and the inherent variability, $\mathbf{c}_2$. $\mathbf{c}_1$ represents real variability associated with each of the original variables that are uncorrelated (as opposed to $\mathbf{V}^f \mathbf{f}$, which represents variability that is correlated) and $\mathbf{c}_2$ represents the inherent variability associated with each of the original variables. If $\mathbf{c}_2$ is not known, then the last two terms must be combined and (17.2.6) becomes

$$\mathbf{x} = \bar{\mathbf{x}} + \mathbf{V}^f \mathbf{f} + \mathbf{e} \qquad (17.2.7)$$

This is the most common case and the one with which the majority of this chapter will be concerned. Either way, this implies that one could not have a common factor with zeros for every coefficient but one.

Both (17.2.6) and (17.2.7) look much like the linear model associated with multiple regression and this is what the originators of FA had in mind. Their motivation was that they wanted to obtain regression equations but, unlike their counterparts in the physical and engineering sciences, they could not run the designed experiments necessary to obtain estimates of the parameters. Unlike the predictor variables in regression, the vector $\mathbf{f}$ is unobservable and must be estimated from the structure of the observable $\mathbf{x}$. The true variables driving the system, of which the common factors, $\mathbf{f}$ are an estimate, are called *latent* variables. The true values of the observed variables, $\mathbf{x}$, before any error has been added in are called *manifest* variables.

17.2.3 Comments on Notation

As stated above, the same notation will be used in this chapter as for PCA for corresponding quantities except for the inclusion of the superscript "f". One reason for this is to present a unified treatment of the subject. The other is that symbols commonly used for certain quantities in FA are used for other quantities in PCA. In particular, the vectors associated with the common factors, v_i^f are often denoted by λ_i. The residual matrix is often denoted by $\boldsymbol{\Psi}$. Using this notation, (17.2.4), in terms of sample estimates, would then be

$$S = \hat{\boldsymbol{\Lambda}}\hat{\boldsymbol{\Lambda}}' + \hat{\boldsymbol{\Psi}} \qquad (17.2.8)$$

17.2.4 The Factor Analysis Model

The general factor analysis model is

$$S = V^f R^f V^{f'} + (D_s - H) \qquad (17.2.9)$$

R^f is a correlation matrix of the latent variables. Generally, the V^f are orthogonal so R^f will be an identity matrix and (17.2.9) simplifies to:

$$S = V^f V^{f'} + (D_s - H) \qquad (17.2.10)$$

D_s is a diagonal matrix whose elements are the diagonal elements of S, and H is another diagonal matrix with elements equal to the p communalities. The diagonal elements of $D_s - H$ are the unique factors. The off-diagonal elements of $D_s - H$ are all equal to zero, implying that once the relevant factors have been obtained, there are no residual covariances left. The *structure* has been completely explained. Therein lies the difference between PCA and FA.

PCA continues extracting characteristic vectors until the residual variance has been reduced to some prescribed amount. FA will perform a similar operation until the covariances have been reduced to zero—actually as close to zero as is possible. In other words, PCA explains *variability*, FA explains structure or *correlations*. PCA is attempting to reduce the diagonals of S, while FA is reducing the off-diagonals. Obviously, when one is reducing the diagonals, the off-diagonals are going to be reduced also and vice versa. As we shall see, each method does best what it is supposed to do but the results obtained by the two methods will generally be similar.

With its background in psychology and education, most FA studies are carried out on correlation matrices because the units of the original variables are different. In that case, (17.2.10) becomes

$$R = V^f V^{f'} + (I - H) \qquad (17.2.11)$$

so that the elements in $(I - H)$ are somewhat like alienation coefficients.

Recall from Chapter 8 that it is possible to perform a rotation on either principal components or factors. This practice is fairly common in factor analysis since the ultimate goal is not only estimating structure but being able to interpret it.

17.2.5 An Example

Before getting into the actual mechanics of factor analysis, let us compare the results of FA and PCA using the Audiometric data from Chapter 5. The starting point for both methods will be the correlation matrix displayed in Table 5.3. The four characteristic vectors obtained by PCA, and displayed in Table 5.6, are given again for comparison in Table 17.1 along with the vectors associated with the first four factors obtained by the method of principal factors, which will be described in the Section 17.5.2. (Both sets of vectors are normalized to 1.0.)

There is very little difference between the coefficients produced by these two methods. (This will generally be the case unless the residual variances differ substantially from one to another.) The largest differences occur in the fourth vectors and would not be considered significant when compared with the standard errors in Table 5.6. *Every* PCA characteristic root is larger than its FA counterpart. The consequence of this is evident in Table 17.2, which shows the residual correlation matrices for the two methods.

Note, first, that the diagonal elements in the FA matrix are all larger than any of the corresponding elements in the PCA matrix. The trace of the PCA matrix is 1.01, corresponding to about 13% unexplained; the trace of the FA matrix is 2.56 or about 32% unexplained. This confirms that FA does not explain as much of the total variability as PCA—not surprising since PCA is the optimal method for this goal and, in fact, FA does not even attempt to explain the inherent variability. On the other hand, a glance at the off-diagonal

Table 17.1. Audiometric Example. Coefficients for Principal Components and Common Factors

Frequency	Principal Components				Common Factors			
	u_1	u_2	u_3	u_4	u_1^f	u_2^f	u_3^f	u_4^f
500L	.40	−.32	.16	−.33	.41	−.33	.17	−.36
1000L	.42	−.23	−.05	−.48	.43	−.23	−.08	−.53
2000L	.37	.24	−.47	−.28	.36	.27	−.48	−.19
4000L	.28	.47	.43	−.16	.27	.47	.45	−.09
500R	.34	−.38	.26	.49	.34	−.37	.25	.48
1000R	.41	−.23	−.03	.37	.41	−.22	−.04	.48
2000R	.31	.31	−.56	.39	.30	.33	−.53	.30
4000R	.25	.51	.43	.16	.24	.50	.44	.07
Characteristic root	3.93	1.62	.98	.47	3.57	1.21	.55	.11

Table 17.2. Audiometric Example. Residual Correlation Matrices

Principal Components

.13	−.07	−.02	−.03	−.01	−.06	.06	.03
	.11	−.07	−.03	−.03	.02	.02	.04
		.13	.01	.08	−.02	−.08	−.02
			.13	.02	.03	−.01	−.13
				.12	−.11	−.01	−.03
					.18	−.05	−.01
						.08	.00
							.13

Common Factors

.23	.04	.01	.00	.04	−.03	−.01	.01
	.24	.02	.00	−.04	.04	−.02	.00
		.31	.04	−.01	−.02	.06	−.04
			.36	.00	.00	−.03	.08
				.36	.04	.00	.00
					.31	.03	.00
						.37	.03
							.38

elements, which are residual correlations, will show that the elements in the PCA matrix are generally larger than those in the FA matrix—this is the optimal property of factor analysis.

A further comparison may be made in terms of the U-vectors. Both U and U^f are orthonormal. We already know that $U'RU = L$ is a diagonal matrix. $U^{f'}RU^f$ will *not* be diagonal since the common factors are not uncorrelated but $U^{f'}[R − (I − H)]U^f$ should be diagonal. This implies that within the common factor space itself, unencumbered by the unique factors, the common factors are uncorrelated.

Since the corresponding characteristic roots for PCA are larger than those for FA, the associated V-vectors for PCA will have larger coefficients. Assuming that the underlying FA model holds, the PCA V-vectors would then be considered inflated by including some inherent variability.

Many numerical comparisons between PCA and FA have been carried out over the years. The simulation studies of Snook and Gorsuch (1989) are typical.

17.2.6 Identifiability and Indeterminacy

Another new term for use with factor analysis is *identifiability*. This has to do with whether or not there is a unique set of parameter values that are consistent with the data.

There are a lot of parameters to be estimated in (17.2.9). If there are k factors, there are $p \times k$ parameter estimates associated with V^f, $k(k − 1)/2$ with R^f, and p with $[\mathbf{D}_s - \mathbf{H}]$. However, if $\mathbf{V}^f$ is orthogonal so that $\mathbf{R}^f$ is an identity matrix, this eliminates $k(k-1)/2$ estimates. One may, however, wish to leave $\mathbf{R}^f$ unrestricted which will lead to an *oblique* solution. This may be desirable in some instances since one is searching for structure in factor analysis. That is, one wishes to obtain estimates of true underlying factors and the fact that two or more of these latent factors are correlated may be a small price to pay.

In some estimation procedures, other restrictions are added, such as making $\mathbf{V}^{f\prime}[\mathbf{D}_s - \mathbf{H}]^{-1}\mathbf{V}^f$ a diagonal matrix, which places $k(k-1)/2$ constraints on the remainder. (This restriction is equivalent to stating that $\mathbf{V}^f$ will reduce $\mathbf{S}$ to the difference of two diagonal matrices.)

How much information is available to estimate these parameters? There are $p(p+1)/2$ distinct elements in $\mathbf{S}$. This means that if that quantity is not smaller than $p + pk - k(k-1)/2$, that is,

$$p(p+1)/2 - [p + pk - k(k-1)/2] = \tfrac{1}{2}[(p-k)^2 - (p+k)] \geqslant 0$$

$$(17.2.12)$$

then a unique or determinate solution is possible up to rotation of the factors. As we shall see later, this quantity is also equal to the degrees of freedom for testing the adequacy of a model, so this cannot go negative either. This may seem rather restrictive in view of the fact that PCA will allow for as may as p pc's. However, to quote Mulaik (1986): "...the only fatal flaw that factor interdeterminancy reveals for the common factor model, is fatal only for the unreasonable expectations we have for this model, especially in its exploratory applications... What is unreasonable is to expect common factor analysis to produce from data unambiguous, self-evident insights into the workings of the world."

Define a *reduced* covariance matrix to be a covariance matrix with the variances replaced by their corresponding communalities. If m is the rank of the reduced covariance matrix, then the smallest number of linearly independent factors that will account for the correlations is m or, said another way, the common factor space is of m dimensions (Harmon, 1976). This value will be the smallest integer, k, for which

$$k \geqslant \tfrac{1}{2}[2p + 1 - \sqrt{8p+1}]$$

$$(17.2.13)$$

This is the Ledermann inequality (Ledermann, 1937). Table 17.3 gives the value of m for $3 < p < 20$ as well as the maximum value of k admissible under most estimation procedures and also gives the solutions of (17.2.12) for $k = 1, 2$ and 3. The solutions for $k = 1$ are the number of independent relationships that must exist among the correlations such that the rank of the reduced matrix would be 1. The solutions for $k = 2$ are the number required for a rank of 2, and so on.

Note that for $p = 3$, $m = 1$ and the solution for $k = 1$ is 0. This means that the minimum rank is 1 and that there are no special relationships required to attain a single-factor system.

Table 17.3. Properties Associated with Equation (17.2.12)

Number Variables p	Minimum Rank m	Solution of (16.2.12) for		
		$k = 1$	$k = 2$	$k = 3$
3	1	0	-2	[a]
4	2	2	-1	-3
5	3	5	1	-2
6	3	9	4	0
7	4	14	8	3
8	5	20	13	7
9	6	27	19	12
10	6	35	26	18
11	7	44	34	25
12	8	54	43	33
13	9	65	53	42
14	10	77	64	52
15	10	90	76	63
16	11	104	89	75
17	12	119	103	88
18	13	135	118	102
19	14	152	134	117
20	15	170	151	133

[a] Number of factors equals number of variables.

For $p = 4$, $m = 2$ so the minimum rank for the reduced covariance matrix is 2. The solution for $k = 1$ is 2. This means that for the reduced covariance matrix to be represented by one factor, there must be two linearly independent conditions on the correlations and these are

$$r_{12}r_{34} - r_{14}r_{23} = 0$$

and

$$r_{13}r_{24} - r_{14}r_{23} = 0$$

These quantities are called *tetrads* and these conditions are due to Spearman (1904).

For $p = 5$, the minimum rank $m = 3$, and five conditions are required to produce a rank of 1. Only one condition is required to produce a rank of 2 and this is the *pentad* criterion due to Kelley (1935), which is a function of 12 products of 5 correlations each. An expanded discussion of these relationships may be found in Cureton and D'Augustino (1983).

It would be an extraordinary situation that would produce reduced-rank situations such as these, but similar situations could produce some further reductions of rank beyond the value of m. The practical implication of this is that a lower bound is placed on the number of factors that one should expect to obtain *before* the analysis is carried out. If the apparent rank produced in the analysis appears to be less than that, it is evidence that some unusual relationships exist among the correlations. This bound also allows one to guard against trying to get by with too few factors. In the early days of factor analysis, great stress was placed on models hypothesized to have one common factor (e.g., general intelligence) and the preceding results would indicate that the existence of such models would be rare under normal circumstances. One of the reasons for all of the effort with regard to minimum rank was an attempt to bring some reality to these models. For a more detailed discussion of this topic, see Harmon (1976).

On the other side of the coin, (17.2.12) also indicates the most factors one can have and obtain a unique or determinate solution. The audiometric example has been used throughout this book with $p = 8$ variables and $k = 4$ pc's and was compared with factor analysis using four factors. By coincidence, $k = 4$ is the maximum number of factors possible without an indeterminate solution when $p = 8$. This is an important point because most of the estimation procedures associated with FA will produce more than that number if asked to and one should realize that such solutions are indeterminate, in that more than one result is possible. Furthermore, factor analysis methodology usually includes a rotation to simple structure. As we know from Chapter 8, these rotation procedures leave the explained variability unchanged. Therefore, even if a solution is determinate, it is so only up to a rotation. Although Algina (1980) gave some conditions for removing this restriction, there has been considerable furor about this practically since the beginning of FA and even at this point in time there is some controversy about the properties of this condition. Rather than dwell on this any further here, the reader may consult Steiger (1979) for more details of this problem. A pair of papers by Heermann (1964, 1966) are also of interest because they discuss both the geometric and algebraic concepts of indeterminacy.

Before leaving this topic, it is well to reiterate that PCA does have this problem with indeterminacy. By the very nature of its definition, PCA is a mechanical procedure and the solutions are always unique up to a change of scale unless there are multiple roots.

17.2.7 Why Would Anyone Use Factor Analysis?

The evidence from the previous section does not present a very convincing argument for factor analysis but there is more to the story and this relates to the residual variances and the communalities. The FA model defines the communalities directly. Furthermore, the FA procedure, one way or another, actually estimates the communalities rather than merely defining them to be the unexplained variability after the characteristic vectors have been obtained. In the principal factor method used in the previous section and described in

Section 17.5.2, the communalities are substituted for the diagonals of the covariance or correlation matrix *before* the vectors are calculated. The resultant matrix represents an estimate of all the variability that the original variables have in common due to their intercorrelations but none of the variability that is *unique* to each of the original variables. In that context, the factors *are* uncorrelated.

The choice between FA and PCA, then, is a function of the underlying model. Does one want to explain structure or variability? There are some other considerations but that should be the main one. Table 17.4 contains a list of properties for each method. Some of these topics have not yet been discussed and will be dealt with later in this chapter.

This comparison probably does little to add to the allure of factor analysis but the main decision should be on the basis of the underlying model. The other pros and cons are a consequence of that decision. As the results of PCA and FA are usually similar, it is common practice to use PCA since it is easier to use. However, as we have seen, each method optimizes different things.

That there is still some question in the minds of many, as to the choice of method to employ, is illustrated by an in-depth comparison by Velicer and Jackson (1990) which, along with the invited discussion, ran well over 100 pages.

Table 17.4. Comparison of Factor Analysis and Principal Component Analysis

Factor Analysis	Principal Component Analysis
Explains correlations.	Explains variability.
Factors are uncorrelated only within the common factor space. (i.e., $U^{f\prime}[S - D_s + H]U^f$ is diagonal. $U^{f\prime}SU^f$ is not).	Principal components are uncorrelated unconditionally. $US'U$ is diagonal.
Residuals are uncorrelated. (In theory, not in practice. They will be less correlated than the PCA residuals).	Residuals will usually be correlated.
Several estimation procedures. Estimates are not unique.	One estimation procedure. Solution is unique.
Adding another factor may change the earlier ones.	Adding another principal component will leave earlier pc's unchanged.
Some solutions are invariant with respect to scale change.	Solution will differ with respect to scale change.
For the lesser specified models, estimation of communalities may be a problem.	PCA does not have to estimate communalities.
For the more fully specified models, computations can become a problem.	Computations are relatively straightforward.

This article contained a number of references to numerical studies of the two procedures.

Although overshadowed by PCA over the years, there have been many examples of the use of factor analysis, primarily in the fields of psychology and education, for whose use it was originally intended. Examples in other, rather diverse, fields include vision testing (Zachert, 1951), measurements on termites (Stroud, 1953), ratings on wines (Baker, 1954), fossils (Reyment, 1963), and the estimation of the effect of R & D efforts (Blackman et al., 1973). Each of these examples includes either the original data or the correlation matrix obtained from it.

In a study of Hauser and Urban (1986) on consumer budget plans for durable goods, respondents were asked, for each of a number of products, questions about price, probability of purchase, and desirability and/or usefulness. Fact scores were used to predict the utility for each item for each individual. These utilities were then compared with the actual customer purchases of these items. (It is a known fact that stated buying intention is not as well correlated with actual purchase behavior as one might believe.)

17.3   ESTIMATION METHODS

It has already been implied that the estimation process for factor analysis, relative to PCA, is no simple matter. The unique factors or residual variances in FA are parameters, so that estimates must be obtained for these in addition to the coefficients in the vectors associated with the factors. We have already discussed the problems of identifiability in Section 17.2.6, but there are some additional complications. These have to do with the fact that it is not easy to obtain both of these sets of estimates.

Some of these techniques will simultaneously obtain both V^f and the diagonal elements of $[D_s - H]$ but, for these, the number of factors must be specified in advance. The most widely used technique of this class is *maximum likelihood estimation*. A second group of techniques estimate the communalities first and from those obtain the V^f. Many of these procedures then use V^f to obtain new estimates of the communalities. The principal factor method is one of these.

Until recently, most applications of FA have entailed the use of this second class of techniques because of their relative simplicity, there being tremendous numerical problems associated with the simultaneous estimation techniques. However, some major breakthroughs occurred in the late 1960s so that the standard against which the other methods are now compared is maximum likelihood and this technique is now beginning to appear in some of the computer packages.

For a number of these estimation procedures, the inclusion of an additional factor will change all of the ones already obtained. Adding another factor to the model may not be the straightforward "add-on" that it is in PCA.

Before we begin with these procedures, a little more notational housekeeping is in order. For simplification, the diagonal matrix of unique factors, $[\mathbf{D}_s - \mathbf{H}]$ will be denoted by $\mathbf{D}_U$ and the population parameters for which $\mathbf{V}^f$ and $\mathbf{D}_U$ are estimates will be denoted $\mathbf{V}^P$ and $\mathbf{D}_U^P$, respectively. Then the FA model (17.2.10) can be restated in terms of population parameters as:

$$\mathbf{\Sigma} = \mathbf{V}^P \mathbf{V}^{P\prime} + \mathbf{D}_U^P \tag{17.3.1}$$

That done, let us now describe some of the more widely used estimation procedures. There are a large number of methods available. It is not the purpose of this book to go into these in any great detail since that information is available elsewhere. Rather, the choice has been made primarily on the basis of the methods available in the more popular computer packages and the intent is to provide just enough detail to characterize each of them. A numerical comparison of some of these methods will be carried out with the Physical Measurements Data introduced in Section 8.4.1.

17.4 CLASS I ESTIMATION PROCEDURES

17.4.1 Introduction

Class I estimation procedures simultaneously obtain the estimates $\mathbf{V}^f$ and $\mathbf{D}_U$ subject to k, the number of factors, being specified in advance. Let $\hat{\mathbf{\Sigma}} = \mathbf{V}^f \mathbf{V}^{f\prime} + \mathbf{D}_U$ for fixed k. Three Class I procedures currently in use are

1. *Maximum likelihood* (MLE), which minimizes the quantity

$$F_1 = \mathrm{Tr}(\hat{\mathbf{\Sigma}}^{-1}\mathbf{S}) - \ln|\hat{\mathbf{\Sigma}}^{-1}\mathbf{S}| - p \tag{17.4.1}$$

2. *Unweighted least squares* (ULS), which minimizes the quantity

$$F_2 = \tfrac{1}{2}\mathrm{Tr}\,(\mathbf{S} - \hat{\mathbf{\Sigma}})^2 \tag{17.4.2}$$

3. *Generalized least squares* (GLS), which minimizes the quantity

$$F_3 = \tfrac{1}{2}\mathrm{Tr}\,(\mathbf{I} - \mathbf{S}^{-1}\hat{\mathbf{\Sigma}})^2 \tag{17.4.3}$$

These will be described in this section. Algorithms are available for all of them. Jöreskog (1977a) gave a procedure that handles all three.

17.4.2 Maximum Likelihood Estimation (MLE)

This method is due to Lawley (1940). If the manifest variables, $\mathbf{X}$, have a multivariate normal distribution so that the covariances follow the Wishart distribution, the estimates $\mathbf{V}^f$ and $\mathbf{D}_U$ may be obtained by maximizing the log-likelihood function with respect to $\mathbf{V}^P$ and $\mathbf{D}_U^P$ subject to the constraint that $\mathbf{V}^{P\prime}[\mathbf{D}_U^P]^{-1}\mathbf{V}^P$ be diagonal. A slight variant commonly used is to minimize the following form of (17.4.1):

$$F_1 = \ln |\Sigma| + \mathrm{Tr}\,(\mathbf{S}\Sigma^{-1}) - \ln |\mathbf{S}| - p \tag{17.4.4}$$

$\mathbf{V}^P$ is $p \times k$. The first algorithm for the solution of (17.4.4) was also due to Lawley (1942) and there have been many modifications and enhancements since then.

The literature is full of the travail associated with this task. Most early attempts at solution were plagued with convergence problems and also with a phenomenon known as the *Heywood* case, in which solutions lead to estimated communalities that are larger than the sample variances of which they are supposed to be a part. Finally, Jöreskog (1967) developed a two-stage procedure where F_1 is minimized over $\mathbf{V}^P$ for fixed $\mathbf{D}_U^P$ and then, in the second stage, minimized over $\mathbf{D}_U^P$ using an algorithm due to Fletcher and Powell (1963) as well as some constraints on the Heywood case. For details of these solutions, see Jöreskog and Lawley (1968) and Jöreskog (1977a). Even so, some computational problems still remain.

One desirable feature of the MLE solution is that it is invariant with respect to change in scale. The use of a covariance or correlation matrix will produce the same vectors except for a factor of s_i, the standard deviation of each variable. Another, not so desirable, feature shared with many of the other techniques, is that the vectors are not invariant with respect to the addition of factors. The solution for a one-factor model, $\mathbf{v}_1^f$, will not generally be the same as its counterpart for a two-factor model; the addition of a second factor changes the first one. This is because the MLE procedure finds those estimates that will maximize F_1, not explain variability as was the case with PCA. This is not unlike the situation in OLS, where the inclusion of a second predictor may change the regression coefficient for the first predictor.

For any choice of k, the number of factors in the model, it is possible to test the goodness of fit of the model using the minimum of (17.4.1), which when multiplied by

$$n - 1 - (2p + 5)/6 - 2k/3 \tag{17.4.5}$$

will have an asymptotic χ^2-distribution with

$$v = \tfrac{1}{2}[(p - k)^2 - (p + k)] \tag{17.4.6}$$

degrees of freedom (Bartlett, 1954). Note that this is the same as equation (17.2.12), so if there is an identifiability problem, v will be negative. If one solves (17.2.12) for the largest value of k for the inequality to hold, the *maximum* number of factors possible is equal to or less than the minimum rank of the reduced matrix. For $p = 4$, one factor is possible. For $p = 8$, four are possible.

Although the degrees of freedom problem does not carry over to other estimation procedures, it still provides a warning. A study of the power of this test was done by Geweke and Singleton (1980), who concluded that it was adequate unless V^f was not full column rank or that one or more of the unique factors was zero. A technique called AIC (Akaike's Information Criterion) has been used as a stopping rule (Akaike, 1974, 1977; Bozdogan and Ramirez, 1988). AIC is equal to $[-2\ln(\text{likelihood}) + 2(\text{no. of parameters})]$, which is essentially a χ^2-test corrected for bias.

Customarily, one will start with low values of k and continue with increasing values of k until a satisfactory fit is obtained. There is a problem, as in PCA, with the significance level, since successive and not independent tests are carried out, but Lawley and Maxwell (1971) concluded that this is not serious. For a discussion of these sequential tests and sequential difference tests, see Steiger et al. (1985). Expressions for the standard errors of the factor coefficients for the MLE case were given by Jennrich and Thayer (1973) and Jennrich (1974).

Two alternative solutions are of interest because both of them end up with the same solution as maximum likelihood. The first of these, due to Rao (1955), is called *canonical factor analysis*, whose solution maximizes the canonical correlation between the latent and manifest variables. The second, due to Howe (1955), maximizes the determinant of the matrix of partial correlations. The significant point of these results is that these additional properties may then also be associated with the MLE solutions, including the very important one that no distributional assumptions are required to carry them out. This also means that correlation matrices may be used even though they do not follow the Wishart distribution. Of course, one still needs to make distributional assumptions to perform the significance tests. This, again, has a parallel in OLS where the Gauss–Markov theorem produces the same results as MLE without any distributional assumptions.

The data are sometimes in discrete steps. If some variables are skewed in opposite directions and have high loadings, Olsson (1979) found that more factors may be required than for continuous data; this is independent of the number of steps. However, discretizing attenuates the loading estimates and this *is* increased with decreased number of steps. Boomsma (1982) conducted a similar investigation with regard to the LISREL solution.

As stated above, now that some of the major computational problems have been solved, the MLE technique is becoming the standard against which the other estimation procedures are judged.

17.4.3 Unweighted Least Squares (ULS)

Also present in some computer packages is an estimation procedure using *unweighted least squares*. This involves minimizing (17.4.2) subject to $V'V$ being diagonal. Probably the best known ULS procedure is MINRES (MINimum RESiduals), the name stemming from the fact that it is the residual covariances or correlations that are being minimized (Harmon and Jones, 1966;

Harmon, 1976, 1977). The Heywood case for ULS has been dealt with by Harmon and Fukuda (1966) and Hafner (1981).

The principal advantage of ULS is that it has a much more simple solution than MLE and consequently will produce results when MLE will not. ULS will also handle singular covariance and correlation matrices, which MLE will not. On the other hand, the results for ULS are not scale-invariant, so that the use of a covariance matrix will produce a different result from a correlation matrix. Inclusion or deletion of a factor will change the other factors in the model.

17.4.4 Generalized Least Squares (GLS)

The GLS method (Jöreskog and Goldberger, 1972) minimizes (16.4.3). The specific factors are not equal to the diagonal elements of $S - V^f V^{f'}$ and are generally smaller than their MLE counterparts. This procedure has not attracted as much attention, probably because of the availability of the other two. Like MLE, GLS differs from ULS in being scale-invariant and requiring that S or R must be positive-definite. Inclusion or deletion of a factor will change the other factors in the model. (This is not the same GLS procedure as the one given earlier by Jöreskog, 1963, which is related to image analysis.)

17.5 CLASS II ESTIMATION PROCEDURES

17.5.1 Introduction

The Class II estimation procedures require that the communalities be known (or at least estimated) before the process begins. On the other hand, the number of factors does not have to be specified in advance.

17.5.2 Principal Factor Analysis (PFA)

By far the most popular method of estimation has been the *principal factor* method (Thompson, 1934). This was a marriage of the philosophy of the factor structure model advocated by such people as Thurstone and the mechanics of Hotelling's PCA. The inclusion or deletion of a factor leaves the other factors in the model unchanged. In this method, the diagonal elements of the covariance or correlation matrix are replaced with estimates of the communalities. The characteristic roots and vectors are obtained from the resultant matrix. (If the residual variances are equal for all variables, then PFA and PCA will have the same solution.) This reduced matrix is not positive-definite and will have negative roots. Two problems exist: (1) stopping rules and (2) communality estimates.

The stopping rules here are roughly the same as for PCA. The most popular seem to be the retention of all factors having positive roots and the use of the SCREE plot. The so-called Kaiser–Guttman rule of using only factors with roots greater than unity (for correlation matrices) has been criticized by some as being too restrictive since the diagonal elements have been replaced by communalities (Yeomans and Golder, 1982). Cattell and Vogelmann (1977) compared this with the SCREE plot. An expression for the expected values of these roots, obtained in the same manner as those discussed in Section 2.8.7 for PCA, was obtained by Montanelli and Humphreys (1976). Influence functions associated with PFA were developed by Tanaka and Odaka (1989).

PFA is generally carried out iteratively. An initial estimate is obtained for the communalities and the characteristic roots and vectors are obtained on the basis of the stopping rule. From these, new estimates of the communalities are obtained and the process is repeated until convergence is attained. A number of initial estimates for the communalities have been proposed over the years but the squared multiple correlation of each variable with all the others is the most widely used at this point in time. These turn out to be lower bounds on the communalities (Roff, 1936; Guttman, 1956). Yanai and Ichikawa (1990) gave bounds for the communalities themselves. Cureton and D'Agostino (1983, p. 139) recommended stopping after two iterations to avoid overfitting but this is not the generally accepted practice.

The larger the communalities, the closer will be the results of PFA (or any other estimator, for that matter) to those of PCA since, in the limiting case, the communalities would be equal to the variances. In a study about college counselors, Heesacker and Heppner (1983) had very high communalities and used this as a justification for using PCA.

17.5.3 α-Factor Analysis

This Class II technique is due to Kaiser and Caffrey (1965) and is motivated by the use of factor analysis for test battery data. The object is to establish a model that would be representative of all test batteries, not just the ones used for producing the data being analyzed. This is similar to the problem discussed for PCA in Section 16.9.4. Rather than maximize the likelihood function, Kaiser and Caffrey maximized the generalized Kuder–Richardson reliability coefficient:

$$\alpha = \frac{p}{p-1}\left(1 - \frac{\mathbf{w'Hw}}{\mathbf{w'[R - D_s]w}}\right) \tag{17.5.1}$$

This is obtained by solving the characteristic equation

$$[(\mathbf{R} - \mathbf{D}_s) - l\mathbf{H}]\mathbf{w} = \mathbf{0} \tag{17.5.2}$$

which amounts to obtaining the characteristic roots and vectors of

$$[\mathbf{H}^{-1/2}(\mathbf{R} - \mathbf{D}_s)\mathbf{H}^{-1/2} - \mathbf{I}]$$

The successive values of α are $\{p/(p-1)\}\{1 - 1/l_i\}$. The corresponding α-factors are $\mathbf{H}^{-1/2}\mathbf{V}$, where $\mathbf{V}$ is the solution of (17.5.2).

This procedure is iterative, like many of the others, and customarily uses the squared multiple correlation coefficients as the first approximation for the communalities. Factors are retained for which the corresponding roots are greater than unity. The above relationships also hold for covariance matrices and this procedure is invariant with regard to change of scale. Inclusion or deletion of a factor will change the other factors in the model.

17.5.4 Centroid Method

An early method was the *centroid* method (Burt, 1917; Thurstone, 1931). This was a sequential method in that the first factor was obtained, then the second factor was obtained from the residual matrix of the first, and so on, much in the spirit of the method Hotelling recommended for PCA. The procedure was only approximate but was quite popular because it was fairly simple to use. With the advent of modern computers, the centroid method is now of historical interest only. It is mentioned here because of its widespread use in the 1930s and 1940s and that so much research carried on during that time, particularly in psychology and education, used that technique. Many people also used the centroid method in place of PCA because it was easier but probably did not appreciate the difference between the underlying models. More about that in Section 17.10. Typical of early applications of factor analysis are papers by Heath (1952) and Green et al. (1953). Both include correlation matrices for around 30 variables on which the centroid method was used followed by early versions of rotation. Heath's example is interesting for the fact that she used *partial* correlations, women's measurements corrected for age.

Anyone wishing to verify any results found in the literature may find details of the method in Harmon (1976) and Cureton and D'Agostino (1983).

17.5.5 Instrumental Variables

An approximate method involving the method of *instrumental variables* was proposed by Madansky (1964). The coefficients in $\mathbf{V}^f$ are fixed for a subset of the variables, called *reference* variables, as though each represented a single factor. Some of the remaining variables, called *instrumental* variables are used to estimate the coefficients corresponding to those variables still remaining. The resulting solution is direct, not iterative, and so is much faster than MLE. Hägglund (1982) proposed two models, FABIN2 and FABIN3, the latter comparable to the two-stage estimator used in econometrics. Jennrich (1987) developed new algorithms for both models. A similar procedure was developed by Bentler (1982a).

With improved algorithms for MLE, this method may have lost some of its appeal but is still useful for obtaining initial approximations. LISREL, for instance, uses FABIN3 to obtain its initial estimates.

17.5.6 Image Analysis

The *image analysis* technique was proposed by Guttman (1953) and is commonly thought of as an alternative estimation procedure for FA, many computer packages including this as an FA option. However, it does not conform to the fundamental FA model (17.2.9) because the residuals are correlated and, for that reason, a description of this technique will be deferred until Section 18.2 as another alternative to PCA. However, in the comparison of the FA estimation procedures in Section 17.6, image analysis will be included.

17.6 COMPARISON OF ESTIMATION PROCEDURES

17.6.1 Introduction

Needless to say, there has been much discussion regarding the merits of these various methods. Harris (1967) compared PCA, α-analysis, and image analysis on a number of examples, as did Velicer (1977) and Acito and Anderson (1980). Velicer noted that the largest differences were always in the last factor and concluded that perhaps too many factors had been extracted in some of the examples. Velicer (1974) compared these same methods on the basis of stability when only a subset of variables were employed. (Conclusion: a standoff.) McDonald (1975) compared these same three procedures on a theoretical basis. McDonald (1970a) also compared α-analysis, MLE, and PFA, while Derflinger (1984) compared α-analysis and MLE. Browne (1968) compared MLE, PFA, and the centroid method. To these, we shall add two comparisons of our own.

17.6.2 Physical Measurements Example

Table 17.5 contains, for the Physical Measurements data, the V-vectors or their equivalent for many of the techniques described in the previous two sections along with the PCA results for comparison. Table 17.6 shows the residuals for each method after two factors have been obtained. These results were obtained using SAS PROC FACTOR. The general nature of the vectors is the same for all methods. As one would expect, the residuals are generally smaller for the PCA solution. However, for x_5 (weight), PCA had the largest residual, every FA method having explained more of its variability. The image analysis method also had a smaller residual for x_2, arm span. The image analysis, overall, explained almost as much as PCA and, as will be shown in Section 18.2, this method has a lot in common with PCA.

17.6.3 Audiometric Example

The audiometric example was a different matter. This example is of interest here because of the degrees of freedom situation. If we set $k = 4$ factors, which seems reasonable in view of our knowledge of the system, there are $pk = 32$ common factor coefficients to estimate in addition to $p = 8$ unique factors for a total of 40. There are $p(p + 1)/2 = 36$ pieces of information available so, initially, we are 4 short. Operationally, this is of concern only for the MLE

Table 17.5. Physical Measurements. V-Vectors

| | Principal Components | | Factor Analysis | | | | | | | | Image Analysis | |
| | | | MLE | | ULS | | Principal Factor | | Alpha Analysis | | | |
	v_1	v_2	v_1^f	v_2^f	v_1^f	v_2^f	v_1^f	v_2^f	v_1^f	v_2^f	v_1^I	v_2^I
Height	.86	-.37	.88	-.24	.86	-.32	.86	-.32	.81	-.42	.85	-.27
Arm span	.84	-.44	.87	-.36	.85	-.41	.85	-.41	.80	-.50	.83	-.36
Length of forearm	.81	-.46	.85	-.34	.81	-.41	.81	-.41	.76	-.49	.81	-.36
Length of lower leg	.84	-.40	.86	-.26	.83	-.34	.83	-.34	.79	-.43	.82	-.30
Weight	.76	.52	.70	.64	.75	.57	.75	.57	.81	.48	.68	.47
Bitrochanteric diameter	.67	.53	.59	.54	.63	.49	.63	.49	.68	.42	.60	.47
Chest girth	.62	.58	.53	.55	.57	.51	.57	.51	.62	.44	.54	.49
Chest width	.67	.42	.57	.37	.61	.35	.61	.35	.64	.29	.58	.33
Amount explained	4.67	1.77	4.43	1.52	4.45	1.51	4.45	1.51	4.41	1.54	4.64	1.75

. Table 17.6. Physical Measurements. Residual Variability

	PCA	MLE	ULS	PFA	α	Image
Height	.12	.17	.16	.16	.16	.12
Arm span	.10	.11	.11	.11	.11	.08
Length of forearm	.13	.17	.18	.18	.18	.14
Length of lower leg	.14	.20	.19	.19	.19	.15
Weight	.15	.09	.11	.11	.12	.07
Bitrochanteric diameter	.26	.36	.36	.36	.36	.28
Chest girth	.28	.42	.42	.42	.42	.33
Chest width	.38	.54	.51	.51	.50	.44

solution and there it may be overcome by assuming that $\mathbf{V'D}_U\mathbf{V}$ is diagonal. (SAS PROC FACTOR automatically makes this assumption; LISREL does not.) This places $k(k-1)/2 = 6$ restrictions on the 40 estimates so now we are 2 on the plus side, agreeing with (17.4.6). However, this implies that we are on the ragged edge and it will be interesting to see how these estimation procedures handle it. Owing to space considerations, the results will be given only in general terms.

The principal factor solution has already been displayed in Table 17.1 and 17.2 and is fairly close to the PCA solution as has been noted. Interestingly, PFA produced just four positive roots. The unweighted least-squares procedure (ULS) produced a similar solution differing in two ways: (1) The final communalities were much larger, four of them leaving negative residual variances and (2) the coefficients for the fourth factor were, excepting for sign, much more heterogeneous than PFA. ULS obviously had some Heywood problems with which to contend. The α-analysis estimates were similar to ULS.

Maximum likelihood had great difficulty with this example. It was extremely sensitive to starting approximations for the communalities. It would not converge using either multiple correlations or any of the other "automatic" options of SAS PROC FACTOR. Solutions were obtained using extremely large estimates of the communalities. These produced solutions, likewise, with extremely high communalities and with factors that had little relation to reality.

The image analysis procedure was somewhat different in that the first three images were much like the other methods but the left–right difference did not show up until the sixth image. This is not surprising in view of some of the comparisons published in the literature where image analysis tended to require more "factors" than the others.

17.7 FACTOR SCORE ESTIMATES

Although we have covered a variety of topics in the first 16 chapters, much of the emphasis in this book has been on the use made of the principal components themselves. Is there a parallel for factor analysis? Well, sort of. The corresponding quantity in factor analysis is referred to as a *factor score*. One of the best ways to demonstrate the difference between the uses of PCA and FA is in the treatment of these respective transformations of the original variables. In contrast to PCA, factor scores appear to be little used, some multivariate texts omitting the topic completely while even the factor analysis texts devote relatively little space to it. The main reason for this appears to be the overriding interest in FA on models and structure. Data analysis appears to be of secondary interest while in PCA this is the end to which most of the means are put. Be that as it may, it would seem that factor scores *should* be of interest, if for no better reason than to screen the data used in the estimation procedure. We shall see, as in other aspects of factor analysis, that (1) the problem is more complicated in FA than it is in PCA and (2) there is more than one way to do it.

Pc's are obtained directly by multiplying the data by the characteristic vectors using either (1.4.1) or (1.6.9), or they may be obtained in a single operation using SVD. It is also relatively easy to invert these equations to obtain estimates of the original variables given a subset of the pc's (2.7.3). In factor analysis, the basic equation (17.2.7) already describes the original variables as a function of the factors, so to obtain factor scores the "inversion" goes the other way and therein lies the first problem. In FA there are k common factors and p unique factors or $p + k$ in all to be obtained from only p original variables, so there is not a unique solution to obtain the scores as there would be in PCA. This means that rather than inverting (17.2.7), one must use some sort of estimation procedure to obtain *factor score coefficients*, which are then multiplied by the original variables to obtain the factor score estimates. The situation is a bit like regression in that one estimates the regression coefficients and uses these to obtain estimates of the response for a given choice of levels of the original variables. The difference in the case of regression is that one does know the values of the response for the sample data set, while for factor analysis one does not. It is for this reason that FA practitioners emphasize that these are factor score *estimates*, not factor *scores*. Add to this the determinacy problem of the factors themselves, the number of estimation techniques for the factors, the number of rotation options, and the number of methods suggested for estimating the scores, and one can understand why the FA literature has been filled with articles on this subject. Green (1976) had a relatively nontechnical account of the problem while more extensive discussions may be found in Heerman (1964, 1966) and Williams (1978) among others. Green recommended obtaining, as a measure of factor score indeterminacy, the multiple correlation of each factor with the observed variables. With respect to rotation, Harris (1975, p. 220) recommended as a criterion for rotation, not simple structure in the rotated V^f but in the factor score coefficients. From the intensity and emotional level of the discussion appended to an article by Velicer and Jackson (1990), it is safe to say that it will be some time before these issues will be resolved to the satisfaction of the working majority.

In what is to follow, the factor scores estimates will be denoted by y^f and their "true" counterparts by y^p. When one is estimating anything, there is a goal to be achieved of maximizing or minimizing something. In the case of factor scores, it is a minimization problem and there are several possibilities of viable quantities to be minimized, each producing a different set of factor score coefficients, $\mathbf{W}$. Among them are

1. Thurstone (1935)

$$\mathbf{W}'_t = \mathbf{V}^{f\prime}\mathbf{S}^{-1} \tag{17.7.1}$$

which minimizes $\text{Tr}\{E[(\mathbf{y}^f - \mathbf{y}^P)(\mathbf{y}^f - \mathbf{y}^P)']\}$.
This was originally very popular.

2. Least squares

$$\mathbf{W}'_{ls} = [\mathbf{V}^{f\prime}\mathbf{V}^f]^{-1}\mathbf{V}^{f\prime} \tag{17.7.2}$$

which minimizes $\text{Tr}\{E[(\mathbf{x} - \mathbf{V}^f\mathbf{y}^f)(\mathbf{x} - \mathbf{V}^f\mathbf{y}^f)']\}$. Unlike $\mathbf{W}_t$, $\mathbf{W}_{ls}$ and the others to follow minimize functions of the original variables. Over the years, this may have been the most popular. See Horst (1965), among others, for details.

3. Bartlett (1937)

$$\mathbf{W}'_b = [\mathbf{V}^{f\prime}\mathbf{D}_U^{-1}\mathbf{V}^f]^{-1}\mathbf{V}^{f\prime}\mathbf{D}_U^{-1} \tag{17.7.3}$$

which minimizes

$$\text{Tr}\{E[\mathbf{D}_U^{-1/2}(\mathbf{x} - \mathbf{V}^f\mathbf{y}^f)(\mathbf{x} - \mathbf{V}^f\mathbf{y}^f)'\mathbf{D}_U^{-1/2}]\}$$

This is a weighted form of $\mathbf{W}_{ls}$. As we shall see, this one does rather well on a theoretical basis.

4. Anderson and Rubin (1956)

$$\mathbf{W}'_{ar} = [\mathbf{V}^{f\prime}\mathbf{D}_U^{-1}\mathbf{S}\mathbf{D}_U^{-1}\mathbf{V}^f]^{-1/2}\mathbf{V}^{f\prime}\mathbf{D}_U^{-1} \tag{17.7.4}$$

which is the same as Bartlett's with the additional restriction that $E(\mathbf{y}^f\mathbf{y}^{f\prime}) = \mathbf{I}$. However, this restriction makes it biased.

5. Thompson estimator, given by Mardia et al. (1979, p. 274)

$$\mathbf{W}'_{tp} = [\mathbf{I} + \mathbf{V}^{f\prime}\mathbf{D}_U^{-1}\mathbf{V}^f]^{-1}\mathbf{V}^{f\prime}\mathbf{D}_U^{-1} \tag{17.7.5}$$

This is a ridge estimator that will minimize the mean square error at the expense of being biased. These coefficients will be scalar multiples of the coefficients in $\mathbf{W}_b$ (Seber, 1984, p. 221). A favorite among Bayesians.

There are some others including some modifications due to Heermann (1963). Although $\mathbf{W}_{ls}$ is called the least-squares estimator, all of these are least-squares estimators of a sort and Mardia et al. (1979) showed that some of them are maximum likelihood estimators as well. McDonald (1979) developed a procedure for simultaneously estimating V^f and the factor scores using either MLE or ULS.

McDonald and Burr (1967) compared the first four of these estimators. Among the properties on which these methods were compared were:

1. Are the factor scores, y_i^f highly correlated with their true counterparts, y_i^P for $i = 1, \ldots, k$? All four of these methods conform.
2. Are the factor scores, y_i^f uncorrelated with their noncorresponding y_j^P for $i \neq j$? This property is called *univocal* (Guilford and Michael, 1948). W_t and W_{ar} are not univocal unless canonical factor analysis is employed.
3. Are the factor scores uncorrelated with each other as the principal components are? In this case W_t and W_b fulfill this condition only with canonical factor analysis estimators and W_{ls} does not conform at all.
4. Are the factor scores unbiased? Yes: W_{ls} and W_b. No: W_t and W_{ar}.

It can be seen that it is possible to have properties (2) or (3) but not both. The only exception to this occurs when the V^f are obtained by Rao's canonical factor method (producing estimates identical to MLE) in which case W_t, W_b, and W_{ar} satisfy both conditions. Not surprisingly, McDonald and Burr expressed a preference for W_b not only because of these properties but because its generalized variance of errors will be equal to or less than that of W_{ls}. As one would surmise, rotation does nothing to enhance any of these properties.

To illustrate these various estimators, we shall return to the Physical Measurements example. The factor coefficients or loadings using PFA were given in Table 17.5. Note that the last three residuals (Table 17.6) are fairly large. Table 17.7 gives the factor score coefficients using these methods as well the PCA W-vectors. (Remember that the factors scores all come from the same factor vectors obtained by PFA. PCA will be different.) Note that the coefficients within columns of W are more homogeneous than any of the factor score estimators. The pc's are, of course, uncorrelated and have unit variance as do the Anderson–Rubin estimates. None of the others have correlations as large as .03. Note that with the exception of the least-squares estimator, all factor score estimators have a large coefficient for the fifth coefficient of the second factor. Although the sixth, seventh, and eighth coefficients of the second vector in Table 17.5 are nearly as large as the fifth coefficient, this does not follow through to the factor score coefficients, possibly because these three have such large residuals.

Velicer (1976a) did a comparison of FA scores with image- and pc-scores and found little difference among them. Acito and Anderson (1986) did a simulation study of factor score indeterminacy and felt that image analysis recovered the original data structure better than did either PCA or PFA.

Table 17.7. Physical Measurements. Factor Score Coefficients

	Thurstone W_t		Least Squares W_{ls}		Bartlett W_b		Anderson–Rubin W_{ar}		Thompson W_{th}		PCA W	
x_1	.16	−.24	.19	−.21	.19	−.19	.18	−.18	.18	−.17	.18	−.21
x_2	.32	−.24	.19	−.27	.26	−.38	.26	−.36	.25	−.34	.18	−.25
x_3	.11	−.25	.18	−.27	.15	−.24	.15	−.22	.15	−.21	.17	−.26
x_4	.15	−.17	.19	−.23	.15	−.18	.15	−.16	.15	−.16	.18	−.23
x_5	.32	.65	.17	.38	.32	.72	.31	.67	.30	.63	.16	.29
x_6	.08	.16	.14	.33	.09	.19	.08	.18	.08	.17	.14	.30
x_7	.05	.13	.13	.34	.07	.17	.07	.16	.06	.15	.13	.33
x_8	.05	.11	.14	.23	.06	.10	.05	.09	.05	.09	.14	.24

17.8 CONFIRMATORY FACTOR ANALYSIS

17.8.1 Introduction

Section 7.4 considered a tests of hypotheses regarding the coefficients of one or more characteristic vectors and it was remarked that this could be thought of as *confirmatory* PCA. There is a corresponding *confirmatory factor analysis* (CFA), a technique that is currently receiving quite a bit of attention. Keeping in mind that factor analysis is concerned with structure, not variability, it is not surprising that those engaged in FA are also concerned with model development and this is where confirmatory FA is generally put to use. In a philosophical paper about the misuse or overuse of factor analysis (or PCA, for that matter), Cliff (1983) suggested that the use of confirmatory procedures would guard against these excesses, and van der Linde (1988) felt that this was the *only* use to which factor analysis should be put.

Recall, from (17.2.9), the general FA model, in terms of the sample covariance matrix, was

$$S = V^f R^f V^{f'} + [D_s - H]$$

In CFA, the same model is used except that it is now in the form of a hypothesis and some of the parameters are specified. Most commonly, some of the elements of V^f are set equal to zero, implying that some of the factors, at least, are a function of a subset of the original variables. The maximum likelihood method is used to estimate the remaining "free" parameters and test whether or not the combination of fixed and estimated values of the parameters fit the data. Since the fixed parameters usually involve V^f, one cannot always assume that R^f, the correlation matrix of the vectors, can any longer remain an identity matrix. If it cannot, then the elements in R^f are now free also. Some of the problems of factor indeterminacy hold for CFA as well (Vittadini, 1988). One of the applications of CFA is in testing the factor analysis framework, the simple structure models that were discussed in Chapter 8. Practitioners often find the use of *path diagrams* useful in developing these models. These diagrams show, visually, which factors are related to which variables.

Although the concept of CFA goes back at least to the 1950s (Howe, 1955; Anderson and Rubin, 1956; Lawley, 1958), the primary development of this technique is due to Jöreskog (1966, 1969b). The current success of CFA is due in part to the development of LISREL (Jöreskog and Sörbom, 1984), a software package that handles both exploratory and confirmatory FA as well as more general models. A general review article on structural modeling by Bentler (1986) includes both CFA and *covariance analysis*, a topic to be discussed briefly in Section 17.8.3. Anderson and Amemiya (1988) showed that under most conditions, the standard asymptotic formulas for standard errors of V^f used in exploratory MLE FA also hold for confirmatory FA.

Although these procedures generally use maximum likelihood estimation, partial least squares may also be employed (Jöreskog and Wold, 1982; Dupačová and Wold, 1982; Wold, 1982). These people feel that PLS is complementary, not competitive. Whereas MLE employs likelihood ratio tests, PLS uses cross-validation. Qualls (1987) used PLS in a study on the effect of the husband–wife relationship on purchase decisions because PLS estimates were distribution free and because PLS could handle large problems more easily. Examples comparing LISREL and PLS may be found in Areskoug (1982), Bookstein (1982), Noonan and Wold (1982), and Lohmöller (1989, Chapter 5).

17.8.2 Numerical Example

For an example, let us return to Chapter 9, the Audiometric case study. Recall that we had obtained some simplified characteristic vectors whose orientation, but not length, was assumed constant over the age span 17–64. Let us now see whether this "hypothetical" model fits the sample data of 100 39-year-old males from Chapter 5 that have been used throughout this chapter. Consider, again, the general FA model (using a correlation matrix):

$$\mathbf{R} = \mathbf{V}^f \mathbf{R}^f \mathbf{V}^{f\prime} + \mathbf{D}_U \tag{17.8.1}$$

The correlation matrix was originally taken from Table 5.3. The hypothetical vectors, which now become fixed parameters in the model, are taken from Table 9.6. This leaves as the only free parameters to be estimated, the off-diagonal elements of $\mathbf{R}^f$, the correlation matrix of the vectors, and the diagonal elements of $\mathbf{D}_U$ representing the specific factors. Since $p = 8$ and $k = 4$, there will be $k(k - 1)/2 = 6$ elements from $\mathbf{R}^f$ and $p = 8$ specific factors, or 14 free elements in all.

This example was run on LISREL VI. The vectors in Table 9.6 are normalized to unit length and in the estimation procedure LISREL rescales these to V-vectors although they are still considered to be fixed. These rescaled vectors are shown in Table 17.8. The correlations of these vectors are also shown along with the estimates of the unique factors. On substituting these values into the right-hand side of (17.7.1) and subtracting it from the original correlation matrix $\mathbf{R}$, the maximum discrepancy is .12 for $r_{2,6}$. However, of the 28 off-diagonals, only four differ by more than .05. The diagonal elements all differ by less than .001. Because we are using a correlation matrix, strictly speaking, we cannot use the likelihood ratio test that accompanies this technique. Jöreskog and Sörbom (1984) recommended using it anyway as a goodness-of-fit indicator. This test is

$$\chi^2 = (n - 1)\{\ln |\mathbf{\Sigma}| - \ln |\mathbf{S}| + \mathrm{Tr}[\mathbf{\Sigma}^{-1}\mathbf{S}] - p\} \tag{17.8.2}$$

where $\mathbf{\Sigma}$ is equal to the right-hand side of (17.3.1). The degrees of freedom are equal to $p(p + 1)/2$ less the number of free parameters, 14 in this case, so the degrees of freedom are $36 - 14 = 22$. Substituting the sample correlation matrix for $\mathbf{S}$, for this example, $\chi^2 = 24.71$. With 22 degrees of freedom, this should make one feel fairly comfortable even though it is only a descriptive quantity.

Table 17.8. Audiometric Case History. Confirmatory Factor Analysis. Hypothetical V-Vectors

Frequency	v_1^f	v_2^f	v_3^f	v_4^f	Unique Factor
500L	.683	−.393	.165	−.160	.170
1000L	.683	−.262	0	−.160	.251
2000L	.683	.262	−.495	−.160	.154
4000L	.683	.655	.330	−.160	.239
500R	.683	−.393	.165	.160	.356
1000R	.683	−.262	0	.160	.333
2000R	.683	.262	−.495	.160	.307
4000R	.683	.655	.330	.160	.273

$$R^f = \begin{bmatrix} 1 & -.230 & -.070 & -.278 \\ -.230 & 1 & -.077 & .114 \\ -.070 & -.077 & 1 & .070 \\ -.278 & .114 & .070 & 1 \end{bmatrix}$$

17.8.3 Covariance Analysis

Confirmatory factor analysis is but a subset of an even larger field of endeavor, the *analysis of covariance structures* (Bock and Bargmann, 1966; Jöreskog, 1970, 1977b, 1978; Graff and Schmidt, 1982; Bentler, 1982b, 1986; Jöreskog and Sörbom, 1984). For instance, (17.2.9) may be generalized to

$$S = B[V^f R^f V^{f'} + D_1]B' + D_2 \tag{17.8.3}$$

where D_1 refers to the matrix of specific factors in (17.2.9) and D_2 is a similar diagonal matrix relating to this generalized model. There may even be two sets of variables such as equations arising in econometrics:

$$\beta_1 \eta = \beta_2 \xi + \zeta \tag{17.8.4}$$

where η are latent endogenous variables, ξ are latent exdogenous variables, β_1 are the regressions of the endogenous variables on themselves, and β_2 are the regressions of the endogenous variables on the exdogenous variables. ζ are residuals uncorrelated with ξ. Both of these sets of latent variables are essentially factors as defined in (17.2.7). Applications include analysis of longitudinal data (repeated measurements, often as a function of time) and congeneric measurements (variables assumed to represent measurements on the same thing, such as the Chemical example in Chapter 1). The technique may also be used as a generalization of the analysis of covariance (Sörbom, 1978). A nonlinear approach was proposed by Lee and Jennrich (1984). The problem of heterogeneity was discussed by Muthén (1989). Pudney (1982) used a generalized PCA model in dealing with the identification problem. The effect of nonnormality was investigated by Sharma et al. (1989).

It is not the intent to go into any detail with these techniques. Introductory material may be found in Everitt (1984, Chap. 3), Everitt and Dunn (1983, Chap. 12), and Dillon and Goldstein (1984, Chap. 12). Cudeck and Browne (1983) discussed cross-validation of covariance structures, remarking that although fully parameterized models may fit better, more simple models may be preferred because of repeatability from sample to sample. Mukherjee and Maity (1988) gave conditions under which OLS, GLS, and MLE produced the same results. Covariance analysis has come into its own with the development of LISREL. For a "how-to-do-it" text on the operations of LISREL with regard to covariance analysis, see Hayduk (1987). There is at least one newer version of LISREL (Jöreskog and Sörbom, 1989) than the one used in this book. As with CFA in Section 17.8.1, solutions may also be obtained for covariance analysis using PLS.

A major application of covariance analysis appears to be in the field of market research and will be discussed in Section 17.8.4. However, some applications in psychology may be found in Jöreskog (1974) and Breckler (1990), alienation and peer influences in Jöreskog (1982), and political behavior in Rossa (1982).

17.8.4 Market Research Applications

Nowhere has a technique been adopted so quickly and enthusiastically as has been the use of covariance analysis or the study of structural equations by the marketing community. The reasons for this would appear to be that the technique is model-driven and the fact that one has to hypothesize a model to get started is appealing to developers of market research projects. Furthermore, the path diagrams that accompany this technique are a useful tool both in the development of the model and in the description of it later. Almost any article using this technique will include the path diagram, not only because it looks impressive but because it aids the writer in the description of the problem and provides the reader with a map to follow the confirmation process.

One application that has made good use of covariance analysis is the study of marketing channels and the factors, both internal and external, that affect them. These include such noneconomic issues as the power wielded by a large supplier (Gaski, 1986, 1987; Howell, 1987), bureaucracy in dominated channels (Dwyer and Oh, 1987), and diversity among consumers, as well as some economic issues such as capacity (Achrol and Stern, 1988). In using this technique to model the motivation of sales people, Sujan (1986) examined the factors that affect the dual components of motivation, "working harder" and "working smarter." Crosby and Stephens (1987) investigated the effect of client relationship and service in the life insurance field.

Advertising applications include a study on the effect of attitude towards an advertisement on its general effectiveness (MacKenzie et al. 1986) and a study of print advertisement recognition (Finn, 1988). The effect of price promotions in supermarkets such as loss leaders, coupons, and in-store specials on profit, sales, and store traffic was investigated by Walters and MacKenzie (1988).

Another field of application deals with consumer behavior. Oliver and Bearden (1985) modeled buying patterns including brand switching. Cadotte et al. (1987) investigated models of consumer satisfaction and Malhotra (1986) investigated customer preference behavior when supplied with less than full information on some choices. Qualls (1987) studied household purchase decisions, the proverbial battle of the sexes. While most marketing studies employ LISREL, Qualls used PLS for the reasons given in Section 17.8.1—size of the study and distribution-free estimates.

17.9 OTHER FACTOR ANALYSIS TECHNIQUES

17.9.1 Introduction

As stated earlier, it is not the intent to go into great detail about factor analysis as many texts on the subject are already available. The purpose of this section is to briefly comment on some other topics to include, primarily, a statement of the problem and a few relevant references.

17.9.2 Discrete FA

Section 16.8 dealt with the use of PCA for discrete variables, the problem being one or more constraints among the variables. The situation is the same in FA. An additional distinction is made as to whether the factors themselves are continuous or discrete. Bartholomew (1987) presented the following classification:

1. Original variables and factors are both continuous: Factor Analysis.
2. Original variables are discrete, factors are continuous: Latent Trait Analysis or Factor Analysis of Categorical Data.
3. Original variables are continuous, factors are discrete: Latent Profile Analysis.
4. Original variables and factors are both discrete: Latent Class Analysis.

For case 2, methods of handling binary data were given by Bartholomew (1984, 1987), Christoffersson (1975, 1977), Muthén (1978), Takane and deLeeuw (1987), and, for multiple classes, Muthén and Christoffersson (1981). For multicategorical data, see Bartholomew (1980, 1987). For case 4, latent class analysis, see Formann (1988), Rost (1988), and Duncan et al. (1982), this last presenting a method of examining for patterns of consistency in responses to questionnaires. Espeland and Handelman (1989) employed latent class analysis for some dental x-ray data in the form of a 2^5 pattern of diagnoses. The response was binomial; either a tooth was sound or it was not. For two-way contingency tables, latent class analysis encounters indentifiability problems, which Evans et al. (1989) overcame by the use of Bayesian methods. A general model for multilevel data including some latent variable models was given by Goldstein and McDonald (1988). The effect of categorization of continuous variables, discussed in Section 16.8.3 for PCA, was also addressed by Bernstein and Teng (1989) for factor analysis.

The term *latent structure* analysis has come to be applied to the use of discrete variables and is a technique developed somewhat apart from factor analysis (Lazersfeld and Henry, 1968; Everitt, 1984, p. 76). For a survey article on latent structure analysis, see Fielding (1977) or Dillon and Goldstein (1984, Chap.13), who have a chapter on the subject. Dillon et al. (1983) illustrated this technique with some housing satisfaction data as a function of neighborhood and landlord. The use of latent structure analysis for contingency tables was discussed by Clogg and Goodman (1984) and Mooijaart (1982), the latter handling up to third-order cross-products with OLS or generalized least squares.

17.9.3   *m*-Sample FA

Section 16.6 examined PCA procedures when k different populations were involved. These "populations" could represent such things as different groups of respondents, test conditions, or times. The same situations may occur in factor analysis. Early attempts to obtain invariant factor patterns involved the rotation of the estimates for each population to a common estimate (Meredith, 1964a,b). Multiple population FA was also investigated by Tisak and Meredith (1989). Levin (1966) showed, for fixed k, that the least-squares fit to all matrices by one factor matrix is the PFA solution to the average of the matrices. However, before performing these operations, it would be well to ascertain how alike the populations are.

In 1963, Lawley and Maxwell devised, in the first edition of their book, a more formal procedure for $m = 2$ populations. With the advent of an operationally feasible MLE, this approach was possible for $m > 2$. Jöreskog (1971) suggested the following hierarchy of tests:.

1. Test

$$H_{01}: \Sigma_1 = \cdots = \Sigma_m$$

If this hypothesis is accepted, one may then obtain an FA solution for the average of the sample covariance matrices.

2. If H_{01} is not accepted, then test

$$H_{02}: k_1 = \cdots = k_m$$

This hypothesis implies that each population requires the same number of factors.

3. If H_{02} is accepted, then test

$$H_{03}: \mathbf{V}_1^P = \cdots = \mathbf{V}_m^P = \mathbf{V}^P$$

To do this, the estimates $\mathbf{V}^f$, $\mathbf{D}_{U1}, \ldots, \mathbf{D}_{Um}$, and $\mathbf{R}_1^f, \ldots, \mathbf{R}_m^f$ must be estimated simultaneously using MLE.

4. If H_{03} is accepted, then test

$$H_{04}: \mathbf{V}_1^P = \cdots = \mathbf{V}_m^P = \mathbf{V}^P$$
$$\mathbf{D}_{U1}^P = \cdots = \mathbf{D}_{Um}^P = \mathbf{D}_U^P$$

this time, obtaining estimates $\mathbf{V}^f$, $\mathbf{D}_U$ and $\mathbf{R}_1^f, \ldots, \mathbf{R}_m^f$.

5. Finally, if H_{04} is accepted, test

$$H_{05}: \mathbf{V}_1^P = \cdots = \mathbf{V}_m^P = \mathbf{V}^P$$
$$\mathbf{D}_{U1}^P = \cdots = \mathbf{D}_{Um}^P = \mathbf{D}_U^P$$
$$\mathbf{R}_1^P = \cdots = \mathbf{R}_m^P = \mathbf{R}^P$$

obtaining estimates $\mathbf{V}^f$, $\mathbf{D}_U$, and $\mathbf{R}^f$.

Sörbom (1974) added the case where the population means may differ.

There may be some problems with the Type I error associated with this series of procedures and with identifiability. This latter situation, as well as one where the populations may have different numbers of observable variables, may be dealt with by fixing some of the parameters as one would do in confirmatory factor analysis.

Corballis and Traub (1970) developed a paired-sample model in which each individual was measured by certain criteria both before and after some event. Their model allowed for a change in the factors and factor score coefficients between samples.

17.9.4   FA Regression

The use of factor analysis in regression does not seem to have attracted much attention. The method, generally, is to obtain factor scores from the original predictor variables and regress the responses on these (Scott, 1966; Basilevsky (1981). (It could, however, be the other way around: Regress the factor scores derived from the response variables on the original predictor variables.) The main problem would appear to be the number of possible methods of estimating factor scores and the cloud of doubt surrounding them. In addition, the problems associated with principal component regression would hold here: there is no guarantee that the major factors are the best predictors.

The properties of MLE factor analysis regression were given by Browne (1988). Lawley and Maxwell (1973) proposed a technique in which the response variable was included in the matrix to be factored—a method similar to the latent root regression discussed in Section 12.3.4. Isogawa and Okamoto (1980) showed the interrelations among several of these methods. Muthén (1982) described a method much like multivariate probit analysis using factors as predictors of a categorical response.

17.9.5 Missing Data

The common procedures for handling missing data in FA are essentially the same as given in Section 14.1 for PCA: obtain an estimate of the covariance or correlation matrix using missing data techniques. Finkbeiner (1979) compared a number of these on some factor analysis models and concluded that the maximum likelihood estimation did the best, albeit much more costly. Conventional regression models and the method of Gleason and Staelin fared less well because they optimize a model different from the FA model. Brown (1983) also considered several of these methods in terms of efficiency as a function of communalities, whether or not a particular variable had any missing data and the pattern of missing data. ML did the best, followed, in some cases by a pairwise ML procedure that he devised.

LISREL handles missing data by using all possible pairs for each correlation coeficient. As we saw in Section 14.1.3, these matrices may be ill-conditioned and, therefore, ULS must be used rather than maximum likelihood to obtain factor estimates.

17.9.6 Three-mode FA

Multimode PCA, the analysis of three- or higher-order arrays, was discussed in Section 10.9. A similar set of procedures, in theory, can also exist for factor analysis. A three-mode FA model, the equivalent of (10.9.4) was developed by Tucker (1966) and Bloxom (1968). Bloxom also considered the case of stability over m populations. As one can imagine, this could lead to rather intricate analyses given the problems of identification and estimation associated with the ordinary FA model. However, Bentler and Lee (1978) simplified the three-mode model by assuming that the factors (there are now two sets of them) are orthogonal and produced the following extension of (17.2.10):

$$S = [AA' \otimes BB'] + D_U \qquad (17.9.1)$$

where S and D_U are both $(pq \times pq)$, the factor matrix A is $p \times k$ and factor matrix B is $q \times r$. They were then able to make use of some analysis of covariance structure techniques to obtain a solution.

McDonald (1970b) employed a technique of partitioning the correlation matrix into two or three parts representing different types of variables and obtained an FA solution for this using Rao's canonical factor technique, previously noted to provide the same estimates as MLE. A PLS solution for three-mode factor analysis was given by Lohmöller (1989, p. 232) while Apel and Wold (1982) gave a procedure for PLS estimation and testing in structures with up to four sets of factors.

17.9.7 Bayesian FA

There has been little activity in this field. Press (1972, p. 317) suggested using priors on some of the elements of V^f in order to remove the ambiguity of the solution during the rotation phase. Martin and McDonald (1975) implemented such a method and also dealt with the Heywood case. Koopman (1978) expanded on this, suggesting some empirical Bayes procedures and also considered the problem of missing data. Lee (1981) gave a similar procedure for confirmatory FA in which priors might be used for the coefficients of the vectors, the correlations among them and/or the specific factors. Akaike (1987) proposed the use of a form of his AIC statistic for use in evaluating Bayesian models. A Bayesian approach to covariance analysis was given by Fornell and Rust (1989).

17.9.8 Multidimensional Scaling

We saw, in Chapters 10 and 11, the relationship of PCA to multidimensional scaling. Some of these techniques have been modified to utilize the factor analysis model. These include asymmetric similarity matrices (TenBerge and Kiers, 1989) and dominance models including ideal point models (Brady, 1989).

17.10 JUST WHAT IS FACTOR ANALYSIS ANYHOW?

There has been considerable confusion over the years as to just what constitutes factor analysis and what its relation is to PCA. The reader should be aware that there are many articles, including some listed in the bibliography of this book, where PCA has been employed even though it has been called factor analysis. This is particularly true in journals related to analytical chemistry, business and, in earlier days, psychology. Do not be misled by titles. This confusion on the part of the casual practitioner may be understandable, since the techniques are similar, particularly PCA and principal factor analysis. However, in light of the material in Section 17.2, the difference seems rather clear-cut and in this enlightened age one would wonder why any confusion should remain at all. A lot of this stems from the historical background of both methods.

As we have seen, both PCA and FA date back to the early 1900s. Karl Pearson was a biometrician, normally publishing in journals such as *Biometrika*.

Spearman was a psychologist publishing in the psychological journals, so there was probably little chance for interchange of ideas in those early days. The initial uses made of these procedures were, understandably, fairly simple; PCA was used for orthogonal regression and the axes for bivariate ellipses, while Spearman's model assumed that there was but one common factor related to ability or intelligence. There the matter rested for over 25 years. In the late 1920s, people such as Thurstone, Holzinger, and Kelley began to formulate the FA problem within the framework we have today. This culminated in a general treatment of the multiple factor problem, an approximate solution for it, the centroid method (Thurstone, 1931) and the principle of rotation and simple structure (Thurstone, 1932). Although the concepts had been introduced by then, the literature did not include explicit models such as (17.2.6), (17.2.7), or (17.2.9). The world was clearly ripe for a rigorous analytical tool and Hotelling furnished one complete with statistical inferential procedures. Hotelling credits Kelley with the motivation for his 1933 paper, which was published in the *Journal of Educational Psychology*. The model Hotelling used defined the original variables as function of the pc's rather than vice versa. That Hotelling saw no conflict between PCA and the FA model is due to the fact that the FA model had not yet fully evolved and PCA was the only "unique" solution available. (Hotelling called his transformed variables "principal components" rather than "factors", as he put it, "...in view of the prospect of application of these ideas outside of psychology and the conflicting usage attaching to the word 'factor' in mathematics...".) However, the following year, Thompson (1934) published a paper in the same journal, proposing the principal factor method, which modified Hotelling's PCA method by using communalities, thus formally pointing out the difference in philosophy as we know it today.

The FA community, which already had a problem with the number of procedures available, now had added to that the problems of estimating communalities and rotation. The psychometricians of the day had remarkable insight but the many different models and estimation procedures proposed for FA were developed, for the most part, without benefit of statistical inference. These, together with the myriad of rotation procedures, gave FA a reputation from which it took another 20 years to escape. To quote Hotelling (1942), "Factor analysis persists in the old habit of regarding sampling distributions as luxuries to be considered after everything else is done, if at all." Mulaik (1986), in tracing the history of factor analysis, was of the opinion that American psychometricians were overly preoccupied during the 1930s and 1940s with the concepts of rotation and simple structure and left the problems of estimation to their colleagues in Europe.

The psychological literature was full of articles on FA, not only because everyone was going off in different directions but because the FA principle was something that practitioners needed and were using in their daily research. At the same time, little was going on in PCA because, for one thing, the PCA model was fairly straightforward and required little elaboration other than inferential procedures, many of which would have to wait for another generation of statisticians. Second, there did not appear to be interest in it, outside of the psychological and educational communities, and these people were more inclined to use the centroid method of factor analysis because it was easier.

Although the confusion surrounding these two methods probably dates from the publication of Hotelling's 1933 paper, there were enough people who did perceive the difference in the beginning that a controversy began almost immediately over which method, PCA or centroid, should be used. This pitted a rigorous, unique, method (PCA) that did not fit the philosophy of (17.2.9) against an approximation technique (centroid) that did. (Principal factor analysis seems to have gotten lost in the shuffle, possibly because Thompson used it to get a single common factor.) This controversy would seem to have been much ado about nothing in light of subsequent developments but at the time it was very real. (See, for instance, Thurstone, 1935; Girshick, 1936, and McCloy et al., 1938.) Although that confusion is now gone, the controversy about the appropriateness of one procedure or the other for a particular application still exists. However, the proponents are now much more knowledgeable about both procedures and concede the right for both to exist (Velicer and Jackson, 1990).

With the beginning of the 1950s, computers had been developed to the point where PCA (and principal factor analysis, for that matter) was no longer a formidable task. By now, mathematical statisticians had become interested in both PCA and factor analysis. At the same time, the next generation of psychometricians had appeared on the scene. These two groups of people combined to provide such advances as the maximum likelihood method combined with some inferential procedures associated with stopping rules and interval estimates, thereby putting FA on a much more solid foundation. By the mid 1960s, the differences between PCA and FA were well recognized, multivariate texts being written at that time customarily devoting a separate chapter to each.

Even so, some confusion with PCA remained. One school of thought stated that any technique that tried to obtain transformations of the original variables in this general manner should be considered factor analysis, of which PCA was a special case. Others went even further and said that PCA referred only to the specific case where *all p* components were obtained, this possibly because in his original paper, Hotelling did get them all in his numerical example. Some others proclaimed that rotation constituted factor analysis, even if performed on principal components. These definitions may still be found in the literature occasionally.

A product of the 1960s was *Kaiser's Little Jiffy* (Kaiser, 1960). This was a computer package designed to help graduate students with little mathematical background do research that required the use of these techniques without their having to spend a lot of time learning how to use them. The motivation for this was that it was more important for the student to get on with his or her project without getting bogged down with a lot of details of the analysis. In the "Little Jiffy", the user had no options. The analysis was done with a correlation matrix, a PCA was carried out retaining all pc's whose characteristic roots were greater than unity, and the characteristic vectors were then rotated using varimax rotation. This was a very popular procedure but I suspect many of the users thought they were doing factor analysis. (This procedure was modified to a "second generation" Little Jiffy (Kaiser, 1970; Kaiser and Rice, 1974), which replaced PCA with image analysis and varimax with orthoblique rotation.)

The latest in this continuing saga of confusion came with the introduction of the mainframe statistical packages such as SPSS®, SAS®, and BMDP®. Many of these packages included PCA in their factor analysis procedure, often as a default option. Later releases of SAS included a separate PROC PRINCOMP but the PCA option in PROC FACTOR remained along with the remark: "The most important type of analysis performed by the FACTOR procedure is principal component analysis." This, after stating in the preceding section: "You should not use any type of component analysis if you really want a common factor analysis..." (SAS Institute Inc., 1989). As another example, the book *Analyzing Multivariate Data* (Green, 1978) had a chapter entitled: "Introductory Aspects of Factor Analysis" but covered only PCA and rotation. Malinowski and Howery (1980, p. 38) stated that PFA and PCA were the same. There are a number of others.

In summary, at this point in time, it is clearly established that PCA and FA are two separate and distinct models but a lot of users are still not aware of these differences. There are still a great many people who are using PCA thinking that they are performing FA and to the extent that this situation persists, both procedures are being done a disservice.

CHAPTER 18

Other Competitors

18.1 INTRODUCTION

The previous chapter was devoted to factor analysis, the chief competitor of PCA. A discussion of the strengths and weaknesses of PCA and FA was given in Section 17.2.7 and it was shown that the choice between the methods should relate to the choice of the underlying model. Because the two methods are quite similar in many respects, there has been quite a bit of confusion regarding them, as was pointed out in Section 17.10.

The main purpose of PCA (or FA) is to reduce a system of correlated variables to a smaller number of new variables which, one way or another, will be of use in dealing with a multivariate problem. Chapter 7, on simplification, and Chapter 8, on rotation, offered alternatives of a sort to PCA but both required PCA as a first step. The simplified pc's would be much like the original pc's, as they would be so designed and should have only low correlations with each other. The rotated components, on the other hand, would *not* resemble the original pc's nor would they be uncorrelated, but they produce results that may be useful in interpreting a set of data.

There are a number of other methods of reducing the complexity of a multivariate problem that do not directly involve PCA and it will be the purpose of this chapter to describe a few of these procedures. These will not yield pc's and hence will give up one or more of the optimal properties associated with PCA but may, nevertheless, produce some useful results. In particular, we shall revisit image analysis, which was mentioned briefly in Chapter 17, and follow this with discussions of triangularization methods, subjective procedures, variate deletion, and Andrews plots.

PCA may be combined with some of the procedures described here. For instance, one may wish to obtain the first pc to remove some variability common to all of the variables (if that is what the first pc represents) and then do an analysis on the residuals using one of these techniques.

18.2 IMAGE ANALYSIS

Image analysis was introduced in Section 17.5.6 because it is generally associated with factor analysis but, in fact, it does not make use of the fundamental model of FA (17.2.9) and for that reason is included in this chapter rather than as an alternative FA estimation procedure. However, because it is commonly thought of as a factor-analytic technique, the transformed variables associated with the method are commonly referred to as "factors." Unlike some of the other techniques to be described in this chapter, image analysis contains the same complexity as PCA and, in fact, may be considered a *weighted* PCA (Meredith and Millsap, 1985). It has enjoyed some popularity having replaced PCA in Kaiser's second-generation Little Jiffy (Kaiser, 1970). Although this presentation, in keeping with most of the literature on image analysis, will be in terms of correlation matrices, it applies equally well to covariance matrices. Moreover, the image analysis procedure is scale invariant (Velicer and Jackson, 1990).

The rationale for this method, due to Guttman (1953), is that the matrix from which the "factor" vectors are obtained not only has the diagonal terms adjusted to become communalities but the off-diagonal terms are adjusted as well. In this procedure, the multiple correlation coefficients are used to determine both. In terms of correlation matrices, let

$$D_R = [\text{Diag}(R^{-1})]^{-1} \tag{18.2.1}$$

and

$$B^I = R^{-1}[R - D_R] \tag{18.2.2}$$

The coefficients of B^I, called *image coefficients*, are the regression coefficients for predicting each of the original variables from all of the others. (The diagonal elements are zero.) From this is obtained

$$G = B^{I\prime}RB^I \tag{18.2.3}$$

The characteristic vectors of G, V^I, then determine the components except that they are called *images* (Kaiser, 1963). (The unique "factors" are called *anti-images*.) V^I is called the *image factor matrix*. The matrix decomposition of R becomes

$$R = [B^{I\prime}]^{-1}V^I V^{I\prime}[B^I]^{-1} \tag{18.2.4}$$

$[B^{I\prime}]^{-1}V^I$ is called the *image pattern*. Nicewander et al. (1984) recommended that, in obtaining simple structure using image analysis, it is the image pattern that should be rotated, not the image factor matrix.

In conventional factor analysis, the unique factors are defined to be independent of each other and also uncorrelated with the common factors, but in image analysis, neither of these properties hold. The anti-images are correlated with each other and are also correlated with the images.

The operations associated with image analysis have more in common with PCA than do most FA techniques. There is a single step to prepare the reduced correlation (or covariance) matrix and another one to obtain the final image pattern; in between is the characteristic vector operation and, as in the case of PCA, this sequence is only performed once. There is no iterative procedure to obtain new communalities, such as in PFA, and hence, the results are unique. The components of the image pattern are unchanged by the addition or deletion of images—the same as PCA. Furthermore, the image score coefficients are obtained directly as is the W-matrix in PCA, rather than having to go through one of the FA estimation procedures given in Section 17.7.

Kaiser (1970) recommended the average root stopping rule for image analysis but in terms of the roots of the original correlation or covariance matrix, not the image covariance matrix. There is some controversy about this (Velicer and Jackson, 1990).

For the Seventh-Grade example, whose correlation matrix is given in Table 3.1 and whose first two roots are greater than 1.0, the image coefficients are

$$
\mathbf{B}^I = \begin{bmatrix} 0 & .7913 & .5784 & -.4008 \\ .7510 & 0 & -.5194 & .4160 \\ .4678 & -.4427 & 0 & .7252 \\ -.2660 & .2908 & .5949 & 0 \end{bmatrix}
$$

and the image covariance matrix is

$$
\mathbf{G} = \begin{bmatrix} .6262 & .4022 & .0478 & .2308 \\ .4022 & .6061 & .1436 & -.0719 \\ .0478 & .1436 & .5378 & .2588 \\ .2308 & -.0719 & .2588 & .4366 \end{bmatrix}
$$

The first two image "factors" are

$$
\mathbf{V}^I = \begin{bmatrix} .7041 & -.1558 \\ .6215 & -.3712 \\ .3576 & .5072 \\ .3134 & .5075 \end{bmatrix}
$$

with corresponding image pattern

$$[\mathbf{B}^{I\prime}]^{-1}\mathbf{V}^I = \begin{bmatrix} .8974 & -.2684 \\ .7890 & -.4698 \\ .4751 & .8210 \\ .4182 & .7033 \end{bmatrix}$$

Finally, the residual matrix after two images have been extracted is

$$\begin{bmatrix} .1226 & -.1368 & .0580 & -.1056 \\ -.1368 & .1556 & -.0505 & .0921 \\ .0580 & -.0505 & .1003 & -.1821 \\ -.1056 & .0921 & -.1821 & .3305 \end{bmatrix}$$

The diagonal elements are the anti-images. Note that the off-diagonals are relatively large, demonstrating the correlations among the anti-images. The image analysis results for the Physical Measurements Data were given in Table 17.5 and 17.6.

Yanai and Mukherjee (1987) obtained a solution for a singular correlation matrix. Jöreskog (1969a) proposed a slightly different image model in which the residual variances were proportional to $[\text{Diag}(\Sigma^{-1})]^{-1}$ and both the vectors and the proportionality constant were estimated by maximum likelihood.

Since image analysis is usually described along with other factor-analytic techniques, it is often included in the controversies comparing them and, in this, seems to have had more than its share, both pro and con (see, for instance, Kaiser, 1970, and McDonald, 1975). Some of this may be because it does not conform to the FA analysis model and is being compared with estimation procedures that do. In many of the published comparisons, image analysis appears to require more components than either PCA or FA. However, as observed above, image analysis is the recommended procedure for the second-generation Little Jiffy.

18.3 TRIANGULARIZATION METHODS

Triangularization methods are somewhat analogous to the step-down procedures mentioned in Section 13.4. One "component" is defined to be equal to one of the original variables, a second component to be a function of this first variable and one other, the third to be a function of these two and one other, viz.,

$$y_1 = a_{11}x_1$$
$$y_2 = a_{12}x_1 + a_{22}x_2$$
$$y_3 = a_{13}x_1 + a_{23}x_2 + a_{33}x_3$$

and so on. Note that the matrix of coefficients, **A**, will be triangular with zero elements below the diagonal, that is, $a_{ij} = 0$, for $i > j$. In many of these procedures, it is conventional to normalize the columns of this matrix so that all of the diagonal elements, a_{ii}, are equal to unity although other normalizations, particularly unit length may be of use. Strictly speaking, these arrays will be triangular only if the variables are entered into the model in order. Some methods use other criteria for choosing these variables.

One way of obtaining triangular transformations is to use "sweep-out" or similar methods as are employed in regression analysis (see for instance, Rao, 1952, p. 345). If this method is employed with the Ballistic Missile data, the transformations are

$$y_1 = x_1$$
$$y_2 = x_2 - .8631x_1$$
$$y_3 = x_3 - .4335x_2 - .2783x_1$$
$$y_4 = x_4 - .6933x_3 - .2035x_2 + .1017x_1$$

These components are still uncorrelated but they no longer account for variability in descending order. The matrix made up of these coefficients is not orthogonal.

For this four-variable case, there are 24 permutations of the ordering of the variables so there are 24 possible solutions of this type. Sometimes the variables may be entered on the basis of some a priori information. If not, a procedure called *regression components* (Ottestad, 1975) may be employed. This method also produces a triangular array but starts with all p variables and drops them one at a time. As the name implies, multiple regression is used to make these decisions but does not involve a separate response variable. One first obtains the multiple correlations of each variable with the rest. For the Ballistic Missile example,

$$R^2_{1.234} = .5940$$
$$R^2_{2.134} = .6921$$
$$R^2_{3.124} = .7495$$
$$R^2_{4.123} = .5947$$

$R^2_{3.124}$ is the largest so the fourth component will be based on the regression equation used to predict x_3:

$$y_4 = x_3 - b_{31.24}x_1 - b_{32.14}x_2 - b_{34.12}x_4$$
$$= x_3 - .2451x_1 - .2491x_2 - .3658x_4$$

To determine the next identified variable, one must obtain the next hierarchy of multiple correlations. These are $R_{1.24}^2 = .5395$, $R_{2.14}^2 = .6466$, and $R_{4.12}^2 = .4570$ and are used to determine the minimum of

$$(1 + b_{31.24})^2(1 - R_{1.24}^2) = .7139$$
$$(1 + b_{32.14})^2(1 - R_{2.14}^2) = .5514$$
$$(1 + b_{34.12})^2(1 - R_{4.12}^2) = 1.0129$$

This would indicate that x_2 is the next one to go and this produces the component

$$y_3 = x_2 - b_{21.4}x_1 - b_{24.1}x_4$$
$$= x_2 - .6144x_1 - .4726x_4$$

Now there are only two variables left, x_1 and x_4 for which $r^2 = .2872$. Similarly to the process for the third component, one now needs to compare

$$(1 + b_{21.4})^2(1 - r^2) = 1.8578$$
$$(1 + b_{24.1})^2(1 - r^2) = 1.5457$$

This implies that

$$y_2 = x_4 - b_{4.1}x_1$$
$$= x_4 - .5262x_1$$

Now we are down to x_1 and the last component is simply

$$y_1 = x_1$$

If the variables in the covariance matrix had been reordered 1, 4, 2, 3 and Rao's algorithm employed, the results would be identical to these regression components. This backward elimination scheme implies that x_1 is most important, followed by x_4, x_2, and x_3.

Ottestad illustrated his procedure with a number of examples, including the Seventh-Grade example introduced in Section 3.3.4. In that example, he stated that the first two regression components accounted for more than the first two pc's but he used a different criterion, the variance of the sum of the variables, rather than the sum of the variances as is customary.

There are a number of other triangular techniques but not all of them are as useful. The *Cholesky* decomposition, available in many computer packages, is of the form $S = T'T$, where T is a triangular matrix. However, this algorithm does not produce uncorrelated components. For the Ballistic Missile data, this produced components more highly correlated than were the original variables.

18.4 ARBITRARY COMPONENTS

Another alternative is to make up one's own components based on subjective criteria. There is nothing wrong with this, although one should investigate the properties of these components, particularly with regard to the extent to which they are correlated and the extent to which they account for variability in the original variables.

A more systematic way of obtaining these components is the *subjective principal component* scheme of Korhohen (1984). This is an interactive procedure in which the user attempts to find "components" that best correlate with those original variables thought to be most important—a sort of intangible weighted PCA. This procedure combines judgments and algorithms designed to provide feasible choices for these judgments. A graphical microcomputer program designed for the end-user (or at least in collaboration with that individual) was produced by Korhohen and Lasko (1986). As obtained by Korhohen, these subjective pc's are uncorrelated and, if one obtains all p of them, all of the variability will have been accounted for. However, the first subjective pc will not necessarily account for more than some of the others; in his example it did, but the fourth spc (out of five) accounted for almost as much and more than either the second or third.

Another proposal, due to Kaiser (1967), is to obtain "components" by a linear transformation of the original variables of the form $Z = XA$. Here A is chosen such that $Tr(A)$ is maximized so that each column of X is paired with a column of Z, the sums of correlations over all p pairs being as large as possible. Additional strategies for solutions were given by Price and Nicewander (1977). Yet another possibility is to reduce the covariance or correlation matrix by any known linear combinations that are relevant and perform a PCA on the residual matrix resulting from that operation.

18.5 SUBSETS OF VARIABLES

The most simple and popular alternative to PCA is simply to delete some of the variables from the set. The strategy, here, is to delete as many variables as possible and still have the variability of these deleted variables reasonably well accounted for by those that remain. This was discussed in Section 14.4 in connection with the use of PCA to perform this.

Recall from Section 14.4.2 that Jolliffe (1972, 1973) had a number of alternative variable-selection procedures, some of which used PCA and were discussed in detail there. Some of these turned out to be losers but will be included here since, on the surface, they seemed rational and hence someone might be tempted to use one of them. The entire list, all of them operating on correlation matrices, is as follows:

A. *Correlation Methods*

    A_1: Retain a set of p variables that maximize the minimum correlation between these and any rejected $k - p$ variables (Beale et al., 1967).

    A_2: Stepwise version of A_1. Described in Section 14.4.2.

B. *PCA Methods.* (All but B_3 are discussed in Section 14.4.2)

    B_2: Backward elimination.

    B_1: An iterative version of B_2 in which PCA is repeatedly applied to the remaining variables (Beale et al., 1967).

    B_4: The reverse of B_2.

    B_3: Eliminates the $p - k$ variables, x_i, for which

$$\sum_{j=k+1}^{p} l_j u_{ji}^2$$

    is a minimum.

C. *Hierarchical Cluster Methods.* Both methods used *inner clustering* in which the selected variable is one of the original members of the group before the last one is included and *outer clustering* in which the last variable to join a group is the one selected.

    C_1: Single Linkage.

    C_2: Average Linkage.

In making some comparisons on both real and artificial data, Jolliffe did not use A_1 or B_1 because they were too time consuming, nor B_3 because preliminary work demonstrated that it was clearly inferior. That left A_2, B_2, B_4, C_1, and C_2. For the clustering methods, outer clustering did not work out well. For the artificial data, A_2, C_1, and C_2 using inner clustering appeared to be most consistent with C_2 better than C_1. B_4 retained best sets more than these methods but also retained some moderate or bad ones. B_2 was in the middle. In other words, the PCA methods had a smaller Type II error and a larger Type I error. For the real data, the results were less conclusive.

Another variate selection method using PCA, the method of *principal variables* was described in Section 14.4.3. Both this method and the B-series of Jolliffe are competitors of PCA in the sense that the final variables used are not pc's even though PCA is used to determine them. By the same token, the rotation methods given in Chapter 8, in addition to clustering variables, may provide some guide to variable deletion.

If the variables are all in the same units, a fairly straightforward method not involving PCA is to select as the first variable in the retained set that one having the largest variance. All of the remaining variables are regressed against this.

The variable having the largest residual variance is then placed in the retained set and the process is repeated. This continues until the residual variances of the remaining variables are all less than some specified amount, and at this point the remaining variables are deleted.

18.6 ANDREWS' FUNCTION PLOTS

Like most of the techniques discussed in this chapter, *Andrews' function plots* (Andrews, 1972) also operate with a linear combination of the original variables. However, there is only a single transformation and, instead of transforming the original variables into a single new variable, it transforms them into a continuous function.

The transformation can be any one of a set of orthogonal functions of which the two most employed are:

$$
\begin{aligned}
f_1(\theta) = x_1/\sqrt{2} + x_2 \sin(\theta) + x_3 \cos(\theta) + x_4 \sin(2\theta) \\
+ x_5 \cos(2\theta) + x_6 \sin(3\theta) + \cdots
\end{aligned}
\tag{18.6.1}
$$

when p is odd, and

$$
f_2(\theta) = x_1 \sin(\theta) + x_2 \cos(\theta) + x_3 \sin(2\theta) + \cdots
\tag{18.6.2}
$$

when p is even, these two functions differing only in the leading term. θ is defined over the range $-\pi < \theta < \pi$. If there are p variables, the first p terms of one of these equations would be used. It is common practice to use the variables as deviations from their means so that the Andrews' function will have a mean of zero.

The Andrews' function procedure produces, for each data vector, a continuous curve over the range $-\pi < \theta < \pi$. Andrews proposed this function as a data-analysis tool, plotting all of the curves for a set of data on the same plot, and used this to perform a visual cluster analysis of some anthropological data.

Kulkarni and Paranjape (1984) carried this procedure one step farther by constructing limits for $\mathbf{f}(\theta)$ that are a function of the covariance matrix:

$$
f(\theta, \bar{\mathbf{x}}) \pm \sqrt{T_{p,n}^2 [\mathbf{f}(\theta)]' \mathbf{S}[\mathbf{f}(\theta)]}
\tag{18.6.3}
$$

where $[f(\theta, \bar{\mathbf{x}})]$ denotes an Andrews' function with x_i being replaced by $\bar{x}_i$ and is also a continuous function of θ.

This technique is not really designed for the two-variable case as a simple scatter plot would do a better job, but to see how this technique works, consider the Chemical problem from Chapter 1. Since $p = 2$, we shall use (18.6.2) for this example, viz.,

$$
f_2(\theta, \bar{\mathbf{x}}) = (x_1 - \bar{x}_1) \sin(\theta) + (x_2 - \bar{x}_2) \cos(\theta)
$$

For the two-variable case, $f_2(\theta)$ is symmetric about $\theta = 0$ so we need plot this function only over the range $0 < \theta < \pi$. Because the reference line for this plot is 0 rather than $\bar{x}_1 \sin(t) + \bar{x}_2 \cos(t)$, the 95% limits for this function become

$$\pm \sqrt{T^2_{2,15,05}[\sin(\theta) \ \cos(\theta)]\begin{bmatrix} .7986 & .6793 \\ .6793 & .7343 \end{bmatrix}\begin{bmatrix} \sin(\theta) \\ \cos(\theta) \end{bmatrix}}$$

The situation for two variables is a very special case because $x_1 - \bar{x}_1$ is displayed at $\theta = \pi/2$, $x_2 - \bar{x}_2$ at $\theta = 0$, their sum divided by $\sqrt{2}$ at $\theta = \pi/4$, and their difference divided by $\sqrt{2}$ at $\theta = 3\pi/4$. Figure 18.1 shows the first seven observations from Table 1.1, all of them being within limits. Figure 18.2 shows the four aberrant observations A, B, C, and D. A is just barely outside of the limits between $0 < \theta < \pi/4$ as it was barely outside the control ellipse in Figure 1.6. B is farther out in the same region but in the opposite direction. C, which represented a mismatch between the methods is out of control at $\theta = 3\pi/4$, and D is out of control over a fairly long range, indicating its dual problem. If more than two variables are being considered, the plotted function and its limits become more irregular and one must use the entire range $-\pi < \theta < \pi$.

It is probably not apparent from this example that the Andrews' plots have any advantage over other methods, partly because the example was for $p = 2$ and also because this example has been used so often that the reader is quite

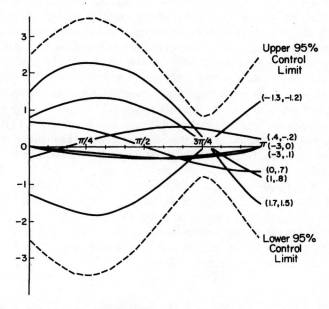

FIGURE 18.1. Chemical Example: Andrews plots; observations 1–7. Reproduced from Jackson (1985) with permission of Marcel Dekker.

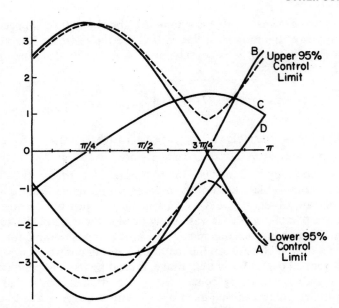

FIGURE 18.2. Chemical Example: Andrews plot; observations A, B, C, and D. Reproduced from Jackson (1985) with permission of Marcel Dekker.

familiar with it and no surprises remain. The advantage of the Andrews' plot is in its relative simplicity, since there is only a single transformed variable with which to work. For a larger number of variables, the advantage of this simplicity is more apparent. However, the functions will be more difficult to interpret and one should not expect to find the easy interpretations found in the above example.

Andrews also recommended using this technique for the pc's themselves. If less than a full set of pc's were employed, these results might appear different from those using the original data because this would represent only variability associated the principal component model itself. Equations (18.6.1) or (18.6.2) would be the same except that they would, presumably, have fewer terms and would be in terms of the pc's rather than the original variables. The covariance matrix in (18.6.3) would have **S** replaced by a unit matrix if the pc's were produced by **W**-vectors or by **L** if they were produced by **U**-vectors.

Conversely, one could also obtain Andrews' plots for the residuals after k pc's had been obtained. In this case it is the residual variability that is being characterized and the plot might indicate clusters of residual variability behavior. The functions would now be in terms of $\mathbf{x} - \hat{\mathbf{x}}$ and in (18.6.3) **S** would be replaced by the residual covariance matrix.

Conclusion

One of the goals stated in the Preface was to produce a book that could be used by practitioners of multivariate data analysis. Now that we are finished, it is hoped that you are comfortable with PCA, have been motivated to investigate some of the special techniques described in the book, and appreciate the difference between PCA and factor analysis.

Principal component analysis is one of the most widely used tools for working with multivariate data. One of the reasons for its popularity, of late, is the availability of computer software to carry out some of the operations, particularly the backbreaking job of obtaining characteristic roots and vectors. However, most of this software is limited in what it can do. There are lot of inferential procedures and other "whistles and bells" associated with PCA that the statistical packages do not address. As stated earlier, most of these procedures are fairly easy to program and can be performed as "add-ons" to some of the statistical packages. One of the reasons for the many numerical examples in this book is to furnish readers material with which to check out both existing statistical packages and any other procedures that they might wish to use. As my colleagues and I investigated many of the PCA and factor analysis techniques, we were continually running into problems of notation and nomenclature from various parts of the far-flung literature on the subject. It is hoped that this book may serve to make things easier for others. Do not be afraid to consult journals in fields of application other than your own. They may use some strange words to describe what they are doing but, apart from that, they are solving the same sort of problems you are and the more we talk to each other, the more we will all learn.

Always try out new packages or procedures on some data sets with which you feel comfortable. All may not be as it seems. Use more than one set. They may furnish more experience with interpretation. If, in your experience with PCA, you discover something that surprises you, do not assume that it is a property of PCA known to everyone but you. If no one else has written about it, sit down with pencil and paper (*keyboard* and *screen* would probably be more appropriate) and share it with the rest of us. Similarly, although it may appear that the pioneers in the field have skimmed off all of the cream, there are still plenty of development opportunities to make PCA an even more useful technique. The increase in articles during the past few years bears this out. Come aboard.

APPENDIX A

Matrix Properties

A.1 INTRODUCTION

This appendix is concerned with the basic definitions and operations associated with matrix algebra. The actual operations involving the use of determinants and inverses are not germane to this book and will not be included. Those wishing more information than is to be found in Appendices A, B, and C should consult one of the many texts available dealing with matrix algebra.

A.2 DEFINITIONS

A.2.1 Definitions of Matrices

A *matrix* is defined as an orderly array of numbers. Examples of matrices are the following:

$$\begin{bmatrix} 2 & 3 \\ 0 & 4 \end{bmatrix} \quad \begin{bmatrix} 1 & 2 & 7 \\ -1 & 6 & 0 \end{bmatrix} \quad \begin{bmatrix} 1 \\ 2 \\ 4 \\ 3 \end{bmatrix}$$

In particular, the $r \times c$ matrix $\mathbf{A}$ denotes an array of numbers consisting of r rows and c columns, viz.,

$$\mathbf{A} = \begin{bmatrix} a_{11} & a_{12} & \cdots & a_{1c} \\ a_{21} & a_{22} & \cdots & a_{2c} \\ \vdots & \vdots & & \vdots \\ a_{r1} & a_{r2} & \cdots & a_{rc} \end{bmatrix}$$

If $r = c$, the numbers of rows and columns are the same and the matrix is said to be a *square matrix*.

A.2.2 Symmetric Matrices

If, for all of the nondiagonal elements of a square matrix, the relation

$$a_{ij} = a_{ji}$$

holds, the matrix is said to be *symmetric*. For instance,

$$\mathbf{A} = \begin{bmatrix} 3 & 4 & 2 \\ 4 & 5 & -1 \\ 2 & -1 & 6 \end{bmatrix}$$

is a symmetric matrix. Symmetric matrices are often *half-stated*, in which the area above or below the diagonal is left blank.

A *skew-symmetric* matrix is defined by the relations

$$a_{ii} = 0 \qquad a_{ij} = -a_{ji} \qquad i \neq j$$

An example of a skew-symmetric matrix is

$$\mathbf{A} = \begin{bmatrix} 0 & 1 & -3 \\ -1 & 0 & 2 \\ 3 & -2 & 0 \end{bmatrix}$$

A.2.3 Matrix Transpose

If, for any matrix $\mathbf{A}$, a new matrix $\mathbf{B}$ is formed by interchanging the rows and columns (i.e., $a_{ij} = b_{ji}$), the resultant matrix is said to be the *transpose* of the original matrix and is denoted by $\mathbf{A}'$. For instance, if

$$\mathbf{A} = \begin{bmatrix} 1 & 4 & 7 \\ 3 & 0 & -1 \end{bmatrix} \qquad \mathbf{B} = \begin{bmatrix} 1 & 3 \\ 4 & 0 \\ 7 & -1 \end{bmatrix} = \mathbf{A}'$$

For a symmetric matrix, $\mathbf{A} = \mathbf{A}'$.

A.2.4 Diagonal Matrices

A square matrix whose only nonzero elements are on the diagonal is called a *diagonal* matrix, viz.,

$$\mathbf{D} = \begin{bmatrix} 3 & 0 & 0 \\ 0 & 2 & 0 \\ 0 & 0 & 4 \end{bmatrix}$$

In particular, a square matrix that has ones on the diagonal and zeros elsewhere is denoted by **I** and is called a *unit* or *identity* matrix, viz.,

$$\mathbf{I} = \begin{bmatrix} 1 & 0 & 0 \\ 0 & 1 & 0 \\ 0 & 0 & 1 \end{bmatrix}$$

A.2.5 Vectors

A matrix that has only a single row is called a *row vector* and a matrix that has only a single column is called a *column vector*. The transpose of a row vector is a column vector and vice versa. A matrix with only a single row and column is a *scalar*.

A.2.6 Orthonormal and Orthogonal Matrices

An orthonormal matrix is a square matrix with the following properties:

1. $|\mathbf{A}| = \pm 1$, where $|\mathbf{A}|$ is the determinant of **A**, to be defined below in Section A.3.1.
2. $\sum_{i=1}^{p} a_{ij}^2 = \sum_{j=1}^{p} a_{ij}^2 = 1$ for all i, j. The sum of squares of any row or column is equal to unity.
3. $\sum_{i=1}^{p} a_{ij} a_{ik} = 0$ for all $j \neq k$. The sum of crossproducts of any two columns is equal to zero and implies that the coordinate axes, which these two columns represent, intersect at an angle of $90°$.

This implies that $\mathbf{AA'} = \mathbf{I}$. if **A** is orthonormal, $\mathbf{A}^{-1} = \mathbf{A'}$ where $\mathbf{A}^{-1}$ is the inverse of **A**, to be defined in Section A.3.4. A matrix that satisfies Condition 3 but not Conditions 1 and 2 is said to be *orthogonal*. The matrix of characteristic vectors, **U**, in which the vectors are scaled to unity, is orthonormal. The other scalings of characteristic vectors, **V** and **W**, are orthogonal.

A.2.7   Triangular Matrices

A matrix that has only zero elements below (or above) the diagonal is called a *triangular* matrix, viz.,

$$
\mathbf{T} = \begin{bmatrix} 4 & 3 & -1 & 7 \\ 0 & 2 & 4 & -2 \\ 0 & 0 & 1 & 6 \\ 0 & 0 & 0 & 3 \end{bmatrix}
$$

A.2.8   Partitioned Matrices

Sometimes a matrix is divided into two or more subsets, each being treated somewhat as a matrix by itself. This procedure is known as *partitioning*. For instance,

$$
\begin{bmatrix} a_{11} & a_{12} & a_{13} & a_{14} & a_{15} \\ a_{21} & a_{22} & a_{23} & a_{24} & a_{25} \\ a_{31} & a_{32} & a_{33} & a_{34} & a_{35} \\ a_{41} & a_{42} & a_{43} & a_{44} & a_{45} \end{bmatrix} = \begin{bmatrix} \mathbf{A}_{11} & \mathbf{A}_{12} \\ \mathbf{A}_{21} & \mathbf{A}_{22} \end{bmatrix}
$$

where

$$
\mathbf{A}_{11} = \begin{bmatrix} a_{11} & a_{12} & a_{13} \\ a_{21} & a_{22} & a_{23} \end{bmatrix} \quad \text{etc.}
$$

No operations are involved here; this is merely a notation of convenience.

A.2.9   Hermitian Matrices

A *Hermitian* matrix is a special matrix made up of complex numbers such that the diagonal elements are real and the pairs of off-diagonal elements are complex conjugates of each other. An example of a Hermitian matrix is

$$
\mathbf{H} = \begin{bmatrix} 2 & 1+i & 2-2i \\ 1-i & 4 & 3+2i \\ 2+2i & 3-2i & 5 \end{bmatrix}
$$

where $i = \sqrt{-1}$. Another concept of a Hermitian matrix is that it is made up of the sum of a real part that is symmetric and an imaginary part that is skew-symmetric, viz.,

$$\mathbf{H} = \begin{bmatrix} 2 & 1 & 2 \\ 1 & 4 & 3 \\ 2 & 3 & 5 \end{bmatrix} + \begin{bmatrix} 0 & 1 & -2 \\ -1 & 0 & 2 \\ 2 & -2 & 0 \end{bmatrix} i$$

A.3 OPERATIONS WITH MATRICES

A.3.1 Determinants

The *determinant*, $|\mathbf{A}|$, of an $m \times m$ square matrix $\mathbf{A}$, is a single value associated with square matrices and in many cases is related to a measure of volume. The actual operation for obtaining this quantity is rather involved and will not be given here. However for the cases $m = 2$ and $m = 3$ only, a special technique known as the "criss-cross" method may be employed.

$m = 2$: If

$$\mathbf{A} = \begin{bmatrix} a_{11} & a_{12} \\ a_{21} & a_{22} \end{bmatrix}$$

then

$$|\mathbf{A}| = a_{11}a_{22} - a_{21}a_{12}$$

$m = 3$: If

$$\mathbf{A} = \begin{bmatrix} a_{11} & a_{12} & a_{13} \\ a_{21} & a_{22} & a_{23} \\ a_{31} & a_{32} & a_{33} \end{bmatrix}$$

then

$$|\mathbf{A}| = a_{11}a_{22}a_{33} + a_{12}a_{23}a_{31} + a_{13}a_{21}a_{32}$$
$$- [a_{31}a_{22}a_{13} + a_{21}a_{12}a_{33} + a_{11}a_{32}a_{23}]$$

For example, if

$$A = \begin{bmatrix} 3 & 0 & 1 \\ -1 & 2 & 4 \\ 5 & 1 & 1 \end{bmatrix}$$

then

$$|A| = (3)(2)(1) + (-1)(1)(1) + (5)(0)(4)$$
$$- [(5)(2)(1) + (-1)(0)(1) + (3)(1)(4)]$$
$$= [6 - 1 + 0] - [10 + 0 + 12] = -17$$

If any two rows or two columns of a square matrix are dependent, the determinant of that matrix is zero.

The *rank* of a matrix is equal to the size of the largest nonzero determinant that can be found in that matrix. If the $m \times m$ (mth order, as it is usually called) determinant of an $m \times m$ matrix is not equal to zero, the matrix is said to be *nonsingular*.

A.3.2 Addition and Subtraction

Two matrices, **A** and **B**, can be added together term by term *provided* both are of the same size, that is, provided both **A** and **B** have the same number of rows and the same number of columns:

$$A + B = \begin{bmatrix} a_{11} + b_{11} & a_{12} + b_{12} & \cdots & a_{1n} + b_{1n} \\ a_{21} + b_{21} & a_{22} + b_{22} & \cdots & a_{2n} + b_{2n} \\ \vdots & \vdots & & \vdots \\ a_{m1} + b_{m1} & a_{m2} + b_{m2} & \cdots & a_{mn} + b_{mn} \end{bmatrix}$$

If

$$A = \begin{bmatrix} 3 & 2 & 1 \\ 4 & 0 & -1 \end{bmatrix} \quad \text{and} \quad B = \begin{bmatrix} -1 & 4 & 0 \\ 3 & 10 & 5 \end{bmatrix}$$

then

$$A + B = \begin{bmatrix} 2 & 6 & 1 \\ 7 & 10 & 4 \end{bmatrix}$$

One matrix can be subtracted from another in the same manner:

$$A - B = \begin{bmatrix} 4 & -2 & 1 \\ 1 & -10 & -6 \end{bmatrix}$$

A.3.3 Multiplication

Two matrices A and B can be multiplied together to form the product AB provided the *number of columns* of A is equal to the *number of rows* of B. Suppose A is an $m \times n$ matrix and B is $n \times p$, then the product AB would be $m \times p$. This procedure multiplies rows of A by columns of B, term by term, and sums:

$$
AB = \begin{bmatrix}
\sum_{j=1}^{n} a_{1j}b_{j1} & \sum_{j=1}^{n} a_{1j}b_{j2} & \cdots & \sum_{j=1}^{n} a_{1j}b_{jp} \\
\sum_{j=1}^{n} a_{2j}b_{j1} & \sum_{j=1}^{n} a_{2j}b_{j2} & \cdots & \sum_{j=1}^{n} a_{2j}b_{jp} \\
\vdots & \vdots & & \vdots \\
\sum_{j=1}^{n} a_{mj}b_{j1} & \sum_{j=1}^{n} a_{mj}b_{j2} & \cdots & \sum_{j=1}^{n} a_{mj}b_{jp}
\end{bmatrix}
$$

If

$$
A = \begin{bmatrix} 4 & 1 & 3 \\ 1 & -1 & 2 \\ 3 & 0 & 1 \end{bmatrix} \quad \text{and} \quad B = \begin{bmatrix} 3 & 1 \\ -1 & 1 \\ 0 & 2 \end{bmatrix}
$$

then

$$
AB = \begin{bmatrix} (4)(3) + (1)(-1) + (3)(0) = 11 & 11 \\ 4 & 4 \\ 9 & 5 \end{bmatrix}
$$

Note, in this case, that the product BA cannot be formed since the number of columns of B does not equal the number of rows of A. Even in the cases where A and B are square matrices of the same size, so that the products can be obtained in either order, in general, $AB \neq BA$. For instance, if

$$
A = \begin{bmatrix} 1 & 2 \\ -1 & 0 \end{bmatrix} \quad \text{and} \quad B = \begin{bmatrix} 3 & 1 \\ 1 & 2 \end{bmatrix}
$$

$$
AB = \begin{bmatrix} 5 & 5 \\ -3 & -1 \end{bmatrix} \quad \text{while} \quad BA = \begin{bmatrix} 2 & 6 \\ -1 & 2 \end{bmatrix}
$$

Any matrix when multiplied by the identity matrix I will be unchanged: $AI = A$.

It is important to note that the product of two *symmetric* matrices, will, in general, be asymmetric.

A.3.4 Inversion

There is no division operation, as such, in matrix algebra. Instead of dividing a matrix A by a matrix B, matrix A is multiplied by the *inverse*, B^{-1}, of the matrix B. However, B must be a square nonsingular matrix. B^{-1} is defined as that matrix which when multiplied by the matrix B will yield the identity matrix I (i.e., $BB^{-1} = I = B^{-1}B$). B^{-1} is unique. If

$$A = \begin{bmatrix} 4 & 2 \\ 1 & 3 \end{bmatrix}$$

then

$$A^{-1} = \begin{bmatrix} .3 & -.2 \\ -.1 & .4 \end{bmatrix} \quad \text{and} \quad AA^{-1} = \begin{bmatrix} 1 & 0 \\ 0 & 1 \end{bmatrix}$$

The actual inversion operation is much more complicated than the other operations discussed in this appendix. Although this operation is used throughout the book, knowledge of the mechanics of the operation is not necessary to understand the concepts developed. Nearly any book on matrix algebra or numerical analysis should have adequate coverage of this topic and hence it will not be given here. All of the principal statistical computer packages will have an inversion subroutine.

We should not leave this topic, however, without mentioning some of the problems that can arise in obtaining an inverse of a matrix. It is a rather complex operation. For large matrices in particular, it is essential that rounding errors are controlled. Moreover, some matrices may be singular and hence not have an inverse. More likely, in statistical problems, the matrix may be nonsingular but just barely so, in which case the inverse may not be obtained with the desired precision. In particular, covariance or correlation matrices representing variables that are highly correlated or exhibit some strong linear relationship will produce inversion problems and, as mentioned in the body of this book, these situations are ones for which PCA methods may be most appropriate. However, as shown by equation (1.6.11), $S^{-1} = WW'$, so an inverse may be obtained directly from the characteristic vectors if they are available. If the matrix is ill-conditioned, an examination of the characteristic roots and vectors may be useful in ascertaining the nature of the problem.

There are also *generalized inverses* of rectangular matrices R denoted by R^{-}. These inverses may be obtained using singular value decomposition as described in Section 10.3.

A.3.5 Direct Product Matrices

A *direct product* or *Kronecker* matrix of the form $C = A \otimes B$ implies that each element of A is multiplied by the matrix B, viz.,

$$C = \begin{bmatrix} a_{11}B & a_{12}B & \cdots \\ a_{21}B & a_{22}B & \cdots \\ \cdot & \cdot & \cdots \end{bmatrix}$$

If a matrix is multiplied by a constant, every element of that matrix is multiplied by that constant and in this case it means that the element a_{11} will be replaced by the product of every element in B by a_{11}. If A is $(m \times n)$ and B is $(p \times q)$, then C will be $(mp \times nq)$. Let

$$A = \begin{bmatrix} 4 \\ 2 \end{bmatrix} \qquad B = \begin{bmatrix} 3 & 1 \\ -1 & 0 \\ 2 & 1 \end{bmatrix}$$

Then C will be a 6×2 matrix

$$C = \begin{bmatrix} 12 & 4 \\ -4 & 0 \\ 8 & 4 \\ 6 & 2 \\ -2 & 0 \\ 4 & 2 \end{bmatrix}$$

These matrices are employed in Sections 10.9, 10.10, and 17.9.6. For some special operations with direct product matrices used in statistical operations, see Neudecker and Wansbeek (1983).

Matrix Algebra Associated with Principal Component Analysis

This appendix is an extension of the previous one and will contain the properties associated with characteristic roots and vectors. Much has been written about this topic and for a more detailed description of these properties, along with the theorems and proofs that underlie them, the reader may refer to any number of books on matrix algebra and/or multivariate analysis. Rao (1979, 1980) contains a number of theorems directly related to PCA as well as procrustes rotation and principal component regression. The intent here is merely to list a few definitions and theorems without proof to serve as a ready reference for this book. With the exception of the material on singular value decomposition, this appendix will be restricted to the case where the matrix on which these operations are to be carried out must be symmetric.

It will be assumed that the characteristic roots must be *distinct* and *real*. This is a reasonable restriction in that it is highly unlikely that a PCA of real data would produce multiple roots and that any dispersion or correlation matrix that one has would, at worst, be positive-semidefinite. One exception to these conditions could occur in the analysis of complex variable data such as encountered in cross-spectral analysis. Another exception might deal with the multivariate analysis of variance, which could produce components of variance matrices having negative or complex roots; in this case, the real problem is not the nature of these matrices but rather the estimation procedures that produced them in the first place.

As much as possible, all of the definitions and theorems in this appendix will be in terms either of the sample covariance matrix S or the original data X in terms of deviations from the variable means. However, the relationships for S will hold for any symmetric matrix, be it the population covariance matrix, the correlation matrix or any arbitrary symmetric matrix. For instance, if one wishes to state properties about the population covariance matrix, S would be replaced with Σ, L with Λ, l with λ, and so on.

Theorem B1.   Let **L** be a $p \times p$ diagonal matrix consisting of the characteristic roots of **S** and let **U** be a $p \times p$ matrix whose columns are the p characteristic vectors of **S**. Then,

$$U'SU = L \tag{B.1}$$

and

$$ULU' = S \tag{B.2}$$

The matrix **U** will be *orthonormal*. This is the decomposition of the matrix **S** to *canonical* form.

Definition B1.   If (B.1) holds, then S is said to be *diagonalizable, normalizable* or *nondefective*.   □

Theorem B2.   Let **x** have covariance matrix **S** and let $z = U'x$ be the *principal components* of **x**. Then the covariance matrix of the principal components, **z**, is

$$U'SU = L \tag{B.3}$$

Theorem B3.   If $u_i \neq 0$ is a p-dimensional vector, it will be a characteristic vector of **S** corresponding to characteristic root l_i if

$$Su_i = l_i u_i \tag{B.4}$$

In addition to being the definition of a characteristic vector, it is also the basis for the *power* method of calculating characteristic vectors, which will be discussed in Appendix C.

Definition B2.   The *characteristic equation* for the matrix **S** is of the form:

$$|S - lI| = 0 \tag{B.5}$$

This becomes a pth-order polynomial in l called the *characteristic polynomial* and will have solutions $l_1, l_2, \ldots, l_p$.   □

Theorem B4.   For a given characteristic root $l_i > 0$, its corresponding characteristic vector u_i may be obtained from the solution of the set of homogeneous linear equations

$$[S - l_i I]u_i = 0 \tag{B.6}$$

REMARK.   If $p - k$ characteristic roots are equal to zero, their corresponding characteristic vectors are an arbitrary set of vectors that are orthogonal to each other and the first k vectors.

Definition B.3. If two $p \times p$ matrices have the same characteristic polynomial, they are said to be *similar*. $\qquad \square$

Theorem B5. If S is positive definite, all of the characteristic roots will be positive. If S is positive-semidefinite, at least one of the characteristic roots will be positive and the remaining roots will be either positive or zero.

Theorem B6. The characteristic roots of a *diagonal* matrix are its diagonal elements.

Theorem B7. The characteristic roots of a *triangular* matrix are its diagonal elements.

Theorem B8. If l_i is the ith characteristic root of S with characteristic vector $\mathbf{u}_i$ and c is a constant, then:

1. cl_i is a characteristic root of cS with characteristic vector $\mathbf{u}_i$.
2. $l_i - c$ is a characteristic root of $S - c\mathbf{I}$ with characteristic vector $\mathbf{u}_i$.
3. l_i^c is a characteristic root of S^c with characteristic vector $\mathbf{u}_i$.
4. l_i^{-1} is a characteristic root of S^{-1} with characteristic vector $\mathbf{u}_i$. This is a special case of Part 3 with $c = -1$.

Definition B4. The *spectral radius* of S is the size of its largest characteristic root. $\qquad \square$ ′

Theorem B9. $$S^c = UL^cU' \qquad (B.7)$$

This is another extension of Theorem B8, Part 3. This relationship is useful in obtaining powers of symmetric matrices, in particular, the square root of a symmetric matrix.

Definition B5. The quantity

$$\mathbf{u}'S\mathbf{u}/\mathbf{u}'\mathbf{u} = l \qquad (B.8)$$

is called the *Rayleigh quotient*. (So far, $\mathbf{u}$ has been scaled such that $\mathbf{u}'\mathbf{u} = 1$. However, the Rayleigh quotient will hold for any scaling of the characteristic vector. In particular, we shall shortly introduce two of them, $\mathbf{v}$ and $\mathbf{w}$.) $\qquad \square$

Theorem B10. For any arbitrary vector $\mathbf{t} \neq 0$,

$$l_1 \geqslant \mathbf{t}'S\mathbf{t}/\mathbf{t}'\mathbf{t} \geqslant l_p \qquad (B.9)$$

Theorem B11. Let $L^{1/2}$ be a diagonal matrix with diagonal elements $\sqrt{l_1}, \ldots, \sqrt{l_p}$ and let $V = L^{1/2}U$ and $W = L^{-1/2}U$. Then $z^* = V'x$ will have covariance matrix L^2 and $y = W'x$ will have covariance matrix I.

Theorem B12. $V'V = L$.

Theorem B13. $VV' = S$.

The following theorems deal with the *singular value decomposition* of an $n \times p$ matrix X and, for this discussion, we shall assume that $n > p$, that is, there are more observations than variables, although most of these results will follow anyway. Assume that $X'X$ has rank p. Let the characteristic roots of $X'X$ be denoted by $l_1, \ldots, l_p$ with corresponding vectors $u_1, \ldots, u_p$ as usual.

Theorem B14. The characteristic roots of the $n \times n$ matrix XX' are $l_1, \ldots, l_p$ with the remaining $n - p$ roots all equal to zero.

Theorem B15. If the rank of $X'X$ is k, then the rank of XX' will also be equal to k.

Theorem B16. The singular value decomposition of the matrix X is

$$X = U^*L^{1/2}U' \tag{B.10}$$

U is the same $p \times p$ matrix of characteristic vectors as obtained from $X'X$. U^* is the $n \times p$ matrix of the first p characteristic vectors obtained from XX' (all the ones corresponding to nonzero characteristic roots). The characteristic roots are the square roots of those obtained from $X'X$ and XX'.

Theorem B17. $$U^{*'}XU = L^{1/2}. \tag{B.11}$$

APPENDIX C

Computational Methods

C.1 INTRODUCTION

One of the reasons that principal components has only recently enjoyed much popularity is the computational requirements of the method. Operations that now require a few seconds on a microcomputer and even less on a mainframe used to require a week or more on a desk calculator. Now that the labor is gone out of principal components, its use has become widespread. There are a number of methods for obtaining characteristic roots and vectors. In this appendix, we shall describe only two, neither of which is very efficient, but which are easy to understand and use. These are the solution of the characteristic equation and the *power* method. Most computer packages now employ more powerful methods such as the Jacobi method and the QR algorithm. Because these methods are much more involved, only a brief description of them will be given here. There are a number of suitable references for those wishing more detailed information. Among those which may be useful are Blum (1972), Ralston and Rabinowitz (1978), Stewart (1973), and Nash (1979). The book by Nash includes BASIC program code for many of the algorithms discussed in this appendix.

C.2 SOLUTION OF THE CHARACTERISTIC EQUATION

To illustrate these techniques we shall use the two-variable Chemical example. As in Appendix B, we shall use symbols dealing with sample estimates, the sample covariance matrix S, and the sample characteristic roots l_i.

The starting place for these methods is the sample covariance matrix, correlation matrix, or other dispersion matrix. For the Chemical example

$$\mathbf{S} = \begin{bmatrix} .7986 & .6793 \\ .6793 & .7343 \end{bmatrix}$$

The first method, solution of the characteristic equation, begins with equation (B.5) to obtain the characteristic roots. For this example

$$|S - lI| = \begin{vmatrix} .7986 - l & .6793 \\ .6793 & .7343 - l \end{vmatrix} = 0$$

This becomes a quadratic equation in l:

$$l^2 - 1.5329l + .12496 = 0$$

with the solution $l_1 = 1.4465$ and $l_2 = .0864$.

To obtain the corresponding characteristic vectors, we employ equation (B.6) with t_i replacing u_i, because we are working with a set of *homogeneous* linear equations. There are an infinite number of solutions so that it takes two steps to get the final vectors. Equation (B.6) is employed once for each characteristic root. For the first root, $l_1 = 1.4465$,

$$[S - l_i I] = \begin{bmatrix} .7986 - 1.4465 & .6793 \\ .6793 & .7343 - 1.4465 \end{bmatrix} \begin{bmatrix} t_{11} \\ t_{21} \end{bmatrix} = \begin{bmatrix} 0 \\ 0 \end{bmatrix}$$

Since these are two homogeneous linear equations in two unknowns, we can arbitrarily set $t_{11} = 1$ and work with just one equation:

$$-.6479 + .6793t_{21} = 0$$

From this, $t_{21} = .9538$. This vector, t_1, must be normalized to length 1.0, viz.,

$$u_1 = t_1/\sqrt{(t_1't_1)} = \begin{bmatrix} 1/\sqrt{1.9097} \\ .9538/\sqrt{1.9097} \end{bmatrix} = \begin{bmatrix} .7236 \\ .6902 \end{bmatrix}$$

Similarly, using $l_2 = .0864$ we obtain:

$$u_2 = \begin{bmatrix} -.6902 \\ .7236 \end{bmatrix}$$

Although this method is simple to follow through, it is very limited for more than three variables because of the difficulty of solving the determinental equation (B.5).

C.3   THE POWER METHOD

Over the years, the workhorse of principal components has been the *power* method, which comes from the equation associated with Theorem B3, in the present notation:

$$St = lt$$

The first approximation to t_1 is commonly taken to be a vector of 1's so that the first approximation to St_1 is the column sums of S. This quantity is divided through by its largest element to obtain the next approximation to t_1, and so on until the values of t_1 converge. Unlike the direct solution of the characteristic equation, the power method is applicable for any size of matrix. The power method is sequential in that one obtains u_1, then u_2, and so on, so that if one only needs the first k vectors, the procedure may be terminated at that point. However, the last $p - k$ roots are not available for tests of significance and procedures for residuals.

For this example, the successive iterations are:

Iteration	t_1	St_1	
1	$\begin{bmatrix} 1 \\ 1 \end{bmatrix}$	$\begin{bmatrix} 1.4779 \\ 1.4136 \end{bmatrix}$	Divide by 1.4779
2	$\begin{bmatrix} 1 \\ .9565 \end{bmatrix}$	$\begin{bmatrix} 1.4484 \\ 1.3817 \end{bmatrix}$	Divide by 1.4484
3	$\begin{bmatrix} 1 \\ .9539 \end{bmatrix}$	$\begin{bmatrix} 1.4466 \\ 1.3797 \end{bmatrix}$	and so on
4	$\begin{bmatrix} 1 \\ .9538 \end{bmatrix}$	$\begin{bmatrix} 1.4465 \\ 1.3797 \end{bmatrix}$	
5	$\begin{bmatrix} 1 \\ .9538 \end{bmatrix}$		

The process has converged in five iterations with the same results as in the first method. The last divisor, 1.4465, is l_1, the first characteristic root, and the first vector u_1 is found as in the first method.

To obtain the second vector, we must now reduce S by the amount explained by the first principal component. This is equal to $v_1 v_1'$, where

$$v_1 = \sqrt{l_1}\, u_1 = \sqrt{1.4465} \begin{bmatrix} .7236 \\ .6902 \end{bmatrix} = \begin{bmatrix} .8703 \\ .8301 \end{bmatrix}$$

The residual (sometimes called *deflated*) covariance matrix is $S - v_1 v_1'$, which is

$$\begin{bmatrix} .7986 & .6793 \\ .6793 & .7343 \end{bmatrix} - \begin{bmatrix} .7574 & .7224 \\ .7224 & .6891 \end{bmatrix} = \begin{bmatrix} .0412 & -.0431 \\ -.0431 & .0452 \end{bmatrix}$$

(The determinant of this residual matrix is zero because the rank has been reduced by one.) To obtain the second characteristic root and vector, one would operate on $S - v_1 v_1'$ in the same manner as the operations on S to obtain u_1. The quantity $v_2 v_2'$ will be the amount explained by the second principal component and the matrix

$$S - v_1 v_1' - v_2 v_2'$$

should be all zeros because there is nothing left to explain. In larger matrices, this would not generally be the case and one could then obtain the third vector if desired, and so on.

The occasion may arise where the smallest root and vector associated with a matrix are desired. In this case, one may employ Theorem B8 (Part 4), obtain the inverse of the matrix, and obtain the first root and vector from it.

The author recalls, vividly, grinding out these computations on a desk calculator in the late 1940s and early 1950s. The Color Film example, for instance, probably took a week or so and we were always scanning the literature for "new tricks" that would put some light at the end of the tunnel. These procedures, some due to Hotelling himself (Hotelling, 1936a, 1943a,b, 1949) were quite ingenious but are now forgotten relics of another age, of interest only to mathematical archaeologists. For instance, the rate of convergence of the approximations to u_i is a function of the ratio of l_i/l_{i+1}; the closer this is to unity, the slower the convergence. One way of speeding this up is to make use of Theorem B8 (Part 3) and Theorem B9. The characteristic roots of S^c are roots $l_1^c, l_2^c, \ldots, l_p^c$. For $c > 1$, this powering of S separates the roots while keeping the vectors the same. Squaring S cuts the number of iterations in half, the fourth power reduces the number by 75%, and so on. (One of the saving graces of the power method in the days of desk calculators was that if one made a clerical or computing error in the procedure, the process would still converge, albeit more slowly!)

The power method may also be used for singular value decomposition, which involves working with two sets of vectors instead of one. One set will be associated with U and the other with U^*. The procedures generally obtain an estimate for u_i, from that obtain an estimate for u_i^*, from that get a second estimate for u_i, and so on. This is sometimes referred to as "ping-ponging" and is used in the NIPALS procedure (Wold, 1966a,b), alternating least squares, and the partial least-squares alternative to principal components regression discussed in Section 12.5.

C.4 HIGHER-LEVEL TECHNIQUES

Most computer packages have replaced the power method with more efficient procedures. This efficiency is paid for by the complexity of the methods, many of which appear to the user to be little more than "black boxes." However, given their efficiency and accuracy, the average user is probably happy for the exchange.

The first of the higher-level procedures was the *Jacobi* method. This procedure starts with the relationship $U'SU = L$ and reduces S to L by a series of orthogonal matrices $J_1, J_2, \ldots$ such that

$$\lim_{k \to \infty} J_1 J_2 \ldots J_q = U \qquad (C.4.1)$$

The Jacobi method obtains all of the characteristic roots and vectors in one operation and was the method used to obtain most of the results in this book. It is very compact (Nash, 1979, p. 106) and requires little storage.

More efficient for larger matrices (Nash feels that the Jacobi method loses its competitive edge for $p \geqslant 10$) is the *Givens* method. This procedure first reduces S to a tridiagonal matrix, one which has a main diagonal in addition to another diagonal above and below it. From there, a fairly simple iterative procedure will obtain the characteristic roots and in turn the corresponding vectors. (A popular one is the QR algorithm.) A modification of this, called the *Householder* method, has a more efficient method of obtaining the tridiagonal matrix.

Similar procedures exist for asymmetric matrices and for SVD of data matrices. More detail on these higher-level procedures may be found in the references listed at the beginning of this appendix or any similar text dealing with numerical analysis procedures for matrix operations. Currently, there is considerable interest in these techniques particularly in such journals as the *SIAM Journal of Scientific and Statistical Computing*.

C.5 COMPUTER PACKAGES

I had originally intended to make a comparison of the major statistical computer packages with regard to PCA and, to some extent, factor analysis as has been done earlier (see for example, Jackson and Hearne, 1979b, Schiffman et al., 1982, or MacCallum, 1983), but decided that discretion might be the better part of valor. Between the continual updating of packages already on the market and the proliferation of new microcomputer statistical packages (Dallal, 1988, cited the existence of over 200 such packages for IBM-PC compatibles alone), whatever was current when this manuscript went to the printer might well be invalid by the time the published volume became available. Instead, this section will contain a checklist of things to look for in these packages. There are enough examples in this book that most options of these packages can be checked out.

Some things to investigate when choosing a package are the following:

1. *Is there a separate set of procedures for PCA or is it buried somewhere in factor analysis?* If the latter is the case, one may have to be diligent. One

of the early mainframe packages had, as the default for their factor analysis procedure, the Little Jiffy without any options, which meant that you got a PCA of the correlation matrix followed by a Varimax rotation—take it or leave it.

2. *Data Entry.* Does the package allow the option of starting with the original data or with a dispersion matrix? Some will not accept the latter.

3. *Dispersion Matrix.* Almost every package will perform a PCA with a correlation matrix but not all of them will work with a covariance matrix and even fewer will permit a product matrix. There is at least one case where a package accepted a covariance matrix but automatically converted it to a correlation matrix before performing the PCA.

4. *Scaling Vectors.* You will not likely get all three scalings of vectors, **U, V** and **W**, used in this book. At most, a package may have two scalings but there is no general custom about this, nor the names assigned to them (see Appendix D). Again, try an example to find out what you have.

5. *Pc Calculations.* There is no uniformity about this either. Some packages produce z-scores (equation 1.4.1), others produce y-scores (equation 1.6.9) and a few may use equation (1.6.6). The most likely term applied to these will be *pc scores*—if they are not still calling them factor scores (see Appendix D). Some packages do not compute any of them.

6. *Residual Analysis.* Very few packages have any tests for residuals let alone any display of the residuals themselves. A number do not even compute them at all.

7. *Stopping Rules.* The favorite ones are percentage of the trace accounted for or, for correlation matrices, using all the pc's whose characteristic roots are greater than 1. A few also have SCREE charts. Very few have significance tests.

This may seem a rather pessimistic view of these packages, but the fact is that a lot of rather useful techniques are not generally available and some of the documentation leaves a bit to be desired. On the other hand, with many of these packages, it is possible to store results of the analysis in the form of data sets for use in subsequent analysis utilizing some of these special techniques. Most of the procedures discussed in this book are not complicated and can easily be programmed. With the exception of some of the multidimensional scaling and factor analysis examples, nearly all of the calculations in this book were performed on a 128K computer.

A Directory of Symbols and Definitions for PCA

D.1 SYMBOLS

One difficulty in reading the literature on principal components is that there is no semblance of uniformity of notation, a situation shared by most other topics in multivariate analysis as well. Three normalizations of characteristic vectors have been introduced in this book but few of the references use two of them, let alone three. A similar condition exists for the pc's themselves. Table D.1 contains a list of symbols used by a number of key references for a few of the quantities widely used in this book. If a particular space is blank, it means that that particular quantity was not discussed in that publication. Some of the references have obtained maximum likelihood estimates by using $[(n - 1)/n]S$ instead of S and while this will affect the characteristic root and the V- and W-vectors by that factor or its square root, that distinction is not made here. Insofar as possible, all of these symbols refer to *sample* estimates although in some cases it is not clearly stated that this is the case. The use of λ to denote characteristic roots in matrix algebra is so widespread that it has carried over into principal components, even to the point of being used by many writers to denote a sample estimate. Where both a scalar and a matrix are given for the characteristic root, it means that a particular letter is used for the roots themselves but the corresponding capital letter is not used for the matrix of the roots.

From this table, it is apparent that when one is *writing* about principal components, the most commonly used vectors are the U-vectors and their corresponding pc, z. This choice is quite understandable since the U-vectors are orthonormal and produce much neater matrix equations in presenting the properties of PCA, including the process of using pc's to predict the original data. For actual operations, however, other normalizations appear to have more utility. At this writing, of the three mainframe statistical packages—BMDP, SAS, and SPSS—only SAS lists the U-vectors at all while all three packages

Table D.1. Symbols Used for PCA in Selected References

Reference	Characteristic root	Characteristic Vectors			Principal Components	
		U	V	W		
This book:	l				z	y
Affi and Clark (1984)	Var C	**A**			C	
Anderson (1984a)	l	**B**			u	
Chatfield and Collins (1980)	λ	**A**				
Cooley and Lohnes (1971)	$\lambda, \mathbf{L}$	**V**	**S**	**B**	y	f
Dillon and Goldstein (1984)	l	**A**			y	
Dunteman (1989)	λ	**A**			y	
Everitt and Dunn (1983)	λ	**A**	**A***		y	
Flury (1988)	l	**B**			U	
Gnanadesikan (1977)						
Green (1978)	c	**A**	**F**	$\mathbf{UD}^{-1/2}$	z	z_s
Harmon (1976)	d	**U**	**A**	$\mathbf{A}\boldsymbol{\Lambda}^{-1}$	z	f
Harris (1975)	$\lambda, \mathbf{L}$	**B**			pc	
Hotelling (1933)	k		a_{ij}	A_{ij}		
Johnson and Wichern (1982)	λ	**E**			Y	γ
Jolliffe (1986)	l	**A**		$\tilde{\mathbf{A}}$	z	
Karson (1982)	l	**A**			y	z^*

Table D.1. *Continued*

Reference	Characteristic root	Characteristic Vectors — V	Characteristic Vectors — W	Principal Components	
This book:	l	$\mathbf{U}$	$\mathbf{W}$	z	y
Kendall (1957)	λ	$\mathbf{L}$		ζ	
Krzanowski (1988)	l	$\mathbf{A}$		y	
Kshirsager (1972)	d	$\mathbf{C}$		u	
Lawley and Maxwell (1971)	λ	$\mathbf{U}$, $\mathbf{W}$	$\mathbf{U\Lambda}^{-1/2}$	y	z
Lebart *et al.* (1984)	λ	$\mathbf{U}$, $\mathbf{U\Lambda}^{1/2}$	$\mathbf{U\Lambda}^{-1/2}$		v
Lewi (1982)	VXF	UCF		XRF	
Marascuilo and Levin (1983)	λ	$\mathbf{A}$		$\mathbf{Y}$	
Mardia *et al.* (1979)	l	$\mathbf{G}$		y	
Maxwell (1977) (1978)	λ	$\mathbf{U}$, $\mathbf{W}$	$\mathbf{W}$	y	z
Morrison (1976)	l	$\mathbf{A}$		y	
Press (1972)	λ	Γ, α_{ij}		z	
Rao (1964)	λ	$\mathbf{P}$, $\mathbf{P\Lambda}^{1/2}$	$\mathbf{P\Lambda}^{-1/2}$	y	
Seber (1984)	λ	$\mathbf{T}$		y	
Srivastava and Carter (1983)	l	$\mathbf{C}$		Y (covariance matrix) Z (correlation matrix)	
Taylor (1977)	λ	$\mathbf{C}$	$\mathbf{L}$	f (corresponding to our $\sqrt{lz}$)	
Timm (1975)	$\boldsymbol{\Lambda}$	$\mathbf{P}$	$\mathbf{W}$	x	

list *both* **V** and **W**. Although the literature discusses primarily *z*-scores, these packages use *y*-scores, as have most of the examples in this book.

D.2 DEFINITIONS

This section will present some of the definitions commonly used for some of the quantities listed in Table D.1. Fortunately, there is a little more uniformity in the definitions than the symbols, but only a little.

There is almost complete agreement that *l* is the *characteristic root* or *eigenvalue*. A few references will use *latent root, principal value* or *singular value*. If the original variables are complex, the roots will be complex also and the largest one will be called the *spectral radius*, that which will contain all of the rest in the complex domain.

Similarly, the U-vectors are almost universally known as either *characteristic vectors* or *eigenvectors*. Occasionally, the term *latent vector* may be used.

Many of the definitions for the other quantities reflect the confusion in earlier years between PCA and factor analysis and the fact that some packages use the same output format for both. The V-vectors are often referred to in the literature as *factor loadings*. (Green, 1978, defined them as *component* loadings but Massey, 1965, used this term for the **W**-matrix.) For the computer packages, SAS defines them as the *factor pattern* while BMDP calls them *factor loadings* (*pattern*).

Since there is very little use made of **W**-vectors in the literature, there does not appear to be much in the line of definitions for them. However, since the computer packages do use them, they also define them. SAS calls them *standardized scoring coefficients* and BMDP calls them *factor score coefficients*.

Finally, the principal components themselves. The influence of factor analysis is seen in the *z*-scores because they are generally referred to as scores of some kind or other, most commonly *principal component scores* or *scores on the principal components*. Again, many computer packages do not use this quantity, preferring to use *y*-scores instead. SAS calls *these* principal component scores while BMDP refers to them as *estimated factor scores*. Green called them *standardized component scores*.

So there it is. It is hoped that this little appendix will enable the reader to feel a little more comfortable with the literature and use some of the available software with more assurance than might have been the case otherwise.

APPENDIX E

Some Classic Examples

E.1 INTRODUCTION

This appendix will include a brief description of a few examples, not already cited, that have enjoyed a great deal of exposure in the role of illustrating the method of principal components and may be useful in exploring some of the techniques described in this book. Each of the examples listed here will include a brief description of the example and indicate where the data may be found. Some of these examples include the raw data while others begin with either the covariance or correlation matrix. A few other classic examples have been used in this book and will be listed in Appendix F. Other data sets have been cited in the body of the text.

E.2 EXAMPLES FOR WHICH THE ORIGINAL DATA ARE AVAILABLE

E.2.1 The "Open-Book–Closed-Book" Examination Data

The data for this example may be found in Mardia et al. (1979) and consist of examination grades for $n = 100$ students for $p = 5$ subjects. Two of these examinations, Mechanics and Vectors, were closed-book. The other three, Algebra, Analysis, and Statistics were open-book. The first pc, accounting for 62% of the trace of the covariance matrix, represented overall test results, that is, all coefficients positive and nearly equal. The second pc (18%) was the difference between the open-book and closed-book exams, although it could be argued that the nature of the courses (engineering vs. mathematical) were confounded with this. It would be interesting to know how these data were chosen, because the pc-scores of the first component were in descending order for the data as presented.

E.2.2 Crime Data

There have been a number of examples where crime statistics as a function of time have been analyzed. This particular data set may be found in Ahamad (1967) and consists of frequencies of occurrence of $p = 18$ classes of crime for $n = 14$ years, 1950–1963, in England and Wales. This data set is interesting in a number of ways, one of which is that $p > n$ so that the last five characteristic roots are zero. This makes it a good example with which to check out computer programs to see how well they deal with this. (Anyone working with this example should start with the original data, not the published correlation matrix, which has some errors in it.) Although the units for the variables are the same, the variances are not homogeneous by any means and one would undoubtedly elect to use the correlation matrix.

The other thing interesting about these data is the fact that the correlation matrix has a few minus signs in it so that first pc (72%) did not represent the overall level of crime; 13 of the 18 coefficients were about equal but two of the others, assault and homosexual offense, were negative, and another, homicide, was almost zero. Assault and wounding were negatively correlated, possibly because of differences in reporting over that time period. In fact, the second pc (15%) was related to reporting differences such as assault–wounding and homosexuality as an example of increased nonreporting. This may be illustrated with a two-dimensional plot of the coefficients of the first two vectors. Ahamad referred to this second pc as a measure of the quality of the data. The third pc (15%) was essentially homicide, occurrences of that crime being relatively uncorrelated with the others. Ahamad extended this example to take age into account although those data were not included. There was some controversy at the time over the appropriateness of PCA for this example (Walker, 1967; Ahamad, 1968).

E.2.3 Spectrophotometric Data

This example has been widely cited in the field of analytical chemistry (Wernimont, 1967). The data consist of $n = 18$ spectrophotometric curves with readings obtained at $p = 20$ different wavelengths over the range 230–400 millimicrons. These 18 curves were in the nature of a factorial experiment. Solutions containing three different concentrations of potassium dichromate were measured on three different spectrophotometers on two different days. This was actually part of a larger experiment involving nine instruments. The first pc represented changes in concentration and the second pc, primarily, difference among instruments. Wernimont concluded that if the instruments were calibrated properly, a single component system should result. For this particular example, it appeared that PCA was a more simple way in which to evaluate the instruments than by an actual physical examination.

E.2.4 The Periodic Table

A completely different type of example may be found in Lewi (1982). The data consist of $p = 14$ physical properties of the first $n = 54$ elements in the periodic table. Lewi studied various subsets of these variables and in his main reported example used only seven of them. (Some of the variables were deleted in this example because they were unknown to Mendeléeff.) The first pc (66%) was more or less related to the periods in the table and the second pc (27%) to groups of elements (i.e., the metallic elements were negative, the inert gases were the highest positive). Hydrogen and helium were outliers and had been removed from the analysis.

E.2.5 The Longley Data

The "Longley" data consist of data obtained from the Bureau of Labor Statistics for the years 1947–1962. These figures, for the most part, deal with employment in various categories and contain very large numbers. Longley (1967) used this example to show what sort of computing problems can come up using data of this sort in multiple regression. (The covariance matrix of the six predictors had a determinant equal to 5.7×10^{25}, while the determinant of the correlation matrix was 1.6×10^{-8}.) While initially chosen for regression examples, it has subsequently been used for other multivariate techniques and could be used as a check on the precision of PCA or factor analysis programs.

E.3 COVARIANCE OR CORRELATION MATRICES ONLY

E.3.1 Pitprop Data

An example that has been cited a number of times is the Pitprop data of Jeffers (1967). Pitprops are the timbers used to shore up passageways in mines and these data were taken from a study to determine whether Corsican pine grown locally was sufficiently strong for the purpose. Physical measurements ($p = 13$) relating to size, density, rings, knots, and so on, were obtained on $n = 180$ pitprops. This example was a bit "sandier" than those in the previous section, the first five pc's only accounting for about 80% of the trace of the correlation matrix. The average root rule would have retained only the first four pc's (74%); the first pc accounted for only 32%. Nevertheless, this is typical of some PCA studies and the first few pc's do have reasonable interpretation.

   This example was used by Jolliffe (1986, p. 147) in comparing a number of variate-selection strategies.

E.3.2 Alate Adelges

This is another example introduced by Jeffers (1967) and is typical of many similar biological studies. Adelges are winged aphids and the sample consisted of $n = 40$ of them on which $p = 19$ physical measurements were made. This time the first pc accounted for 73% of the trace of the correlation matrix and the first four pc's accounted for 92%. The first pc represented overall size; the next three pc's were generally related to at most three variables per component. A plot of the first pc versus the second indicated a strong nonlinear relationship between them, which he attributed to the fact that these 40 aphids could be divided into four clusters.

APPENDIX F

Data Sets Used In This Book

F.1 INTRODUCTION

This appendix contains a catalog of all the numerical examples used in this book. The description will include the nature, form, and location of the original data; a list of the applications for which these data were used; and the location of each application.

F.2 CHEMICAL EXAMPLE

Fictitious data consisting of $n = 15$ observations on each of $p = 2$ test procedures.

Data: Table 1.1: Additional data: Section 1.7.1
Covariance matrix: Section 1.2
Roots: Section 1.3
    Confidence limits: Section 4.2.1
    Test for sphericity: Section 15.2
Vectors, U: Section 1.3
          Standard errors: Section 4.2.1
          Tests for vectors: Section 4.6.2
          Confidence limits: Section 4.6.3
          V: Section 1.6
          W: Section 1.6
          Inference for orthogonal regression line: Section 15.2
          Simplified: Section 7.3
z-scores: Table 1.2
y-scores: Table 1.2

T^2-control chart: Figure 1.5, Table 1.2
Control ellipse: Section 15.1, Table 15.1, Figure 1.6
Rotation: Section 8.3
Influence functions: Table 16.3
Singular value decomposition: Section 10.3
JK biplot: Figure 10.2
Correlation matrix: Section 3.3.2
   Vectors
      U: Section 3.3.2
      V: Section 3.3.2
      V*: Section 3.3.2
      Reduced major axis: Section 15.4
      W: Section 3.3.2
      W*: Section 3.3.2
Exponentially weighted data: Section 3.6
   Covariance matrix: Section 3.6
   Roots: Section 3.6
   Vectors, U: Section 3.6
Andrews' function: Section 18.6, Figures 18.1 and 18.2
Augmented covariance matrix, $x_3 = x_1 + x_2$: Section 2.4
Characteristic equation solution: Appendix C.2
Power method: Appendix C.3

F.3 GROUPED CHEMICAL EXAMPLE

This example is similar to the Chemical example except that the data are now arranged in three groups of four observations each, $p = 2$ as before.

Data: Table 13.1
Vectors
   U: Section 13.4
   W: Section 13.4
y-scores: Table 13.4
ANOVA: Table 13.2
MANOVA: Section 13.3, Table 13.3
pc-ANOVA: Table 13.5
Dempster's stepdown pc-ANOVA: Section 13.4
g-Group procedures: Sections 16.6.2 and 16.6.3
Discriminant analysis: Section 14.5.2, Table 14.4

F.4 BALLISTIC MISSILE EXAMPLE

This set of real data consists of $p = 4$ test procedure results on each of $n = 40$ ballistic missiles. Original data are not included.

Covariance matrix: Table 2.1
  Gleason-Staelin statistic: Section 2.8.2
  Index: Section 2.8.2
Roots: Table 2.2
  Bounds on roots: Sections 4.2.3 and 4.5.2
  Bias of estimates: Table 4.1
  Standard errors: Table 4.1
  Test for maximum root: Section 4.5.2
  Test for equality of roots: Section 2.6, Table 4.2
  Test that last roots are equal to a constant: Section 4.4.2
  Joint confidence limits: Section 4.5.3
Vectors
  U: Table 2.2
  Correlation matrix of coefficients: Table 7.3
  Test for vectors: Section 4.6.4
  Effect of rounding: Section 16.4.4
  V: Table 2.3
  W: Table 2.3
  Sensitivity to change of root: Section 16.4.2
Q-test: Section 2.7.2
Hawkins statistics: Sections 2.7.3 and 2.7.4
Broken stick: Section 2.8.8
Velicer stopping rule: Section 2.8.9
Principal variables: Section 14.4.3, Table 14.3
Group data computations: Table 6.1
  T^2 tests: Section 6.3
  Q-tests: Section 6.4.1
  Hawkins statistics: Section 6.4.2
Sweep-out solution: Section 18.3
Regression components: Section 18.3
Correlation matrix
  W-vectors: Section 3.3.3

F.5 BLACK-AND-WHITE FILM EXAMPLE

This set of real data consists of densities at $p = 14$ exposure levels of $n = 232$ film samples. Data are not included.

Covariance matrix: Table 2.4
 Gleason–Staelin statistic: Section 2.9.2
Roots: Table 2.5
Vectors, **V**: Table 2.5
 Correlation matrix of coefficients: Table 7.4
Test for roots: Table 2.6
Comparison of seven stopping rules: Section 2.9.3
Correlation matrix: Table 7.5
Variate selection: Section 14.4.2, Table 14.2
Growth curve smoothing: Section 16.7.2, Table 16.4
Comparison with eigenfunctions: Table 16.5

F.6   COLOR FILM EXAMPLE

This set of real data consists of color density measurements (three colors, three density levels: $p = 9$) on $n = 139$ film samples, grouped by weeks with an unequal number of observations per week. Data are not included. Covariance matrix is based on within-week variability.

Covariance matrix: Table 6.2
Roots: Table 6.4
Vectors, **W**: Table 6.4
 Simplified: Table 7.1
 Tests for simplified vectors: Section 7.4
 Correlation of actual and simplified: Table 7.2
y-Score control charts: Figure 6.4
T^2-tests: Figure 6.5
Residual tests: Tables 6.5 and 6.6
Correlation matrix: Table 6.3

F.7   COLOR·PRINT EXAMPLE

Real data consisting of judgments of $n = 18$ respondents, each of whom ranked $p(p - 1)/2$ pairs of color prints on the basis of overall similarity. $p = 6$.

Data: Table 11.1
Dissimilarity matrix: Table 11.2
Scalar products matrix: Table 11.3
Vectors, **V**: Table 11.4
Nonmetric solution: Tables 11.5 and 11.6, Figure 11.6

Similarity plot: Figure 11.5
INDSCAL plot: Figure 11.9
INDSCAL saliences: Table 11.8, Figure 11.10

F.8 SEVENTH-GRADE TESTS

Real data consisting of results of $p = 4$ tests given to $n = 140$ students. Data are not included. Source: Hotelling (1933).

Correlation matrix: Table 3.1
Roots: Table 3.2
    Covariance matrix of roots: Table 4.3
    Test for equality of roots: Table 4.4
Vectors
    U: Table 3.2
    V: Table 3.2
Image analysis: Section 18.2

F.9 ABSORBENCE CURVES

Fictitious data consisting of absorbences of $n = 10$ chemical samples measured at $p = 7$ wavelengths.

Data: Table 3.3
Product matrix: Table 3.4
Roots: Section 3.4
Vectors
    U: Section 3.4
    V: Section 3.4
    W: Section 3.4

F.10 COMPLEX VARIABLES EXAMPLE

Fictitious example. $p = 2$. No data.
Hermitian matrix: Section 3.7
Roots: Section 3.7
Vectors
    U: Section 3.7
    V: Section 3.7

F.11　AUDIOMETRIC EXAMPLE

Real data consisting of audiometric measurements (four frequencies, both ears; $p = 8$) for $n = 100$ males, aged 39.

Data: Table 5.1

Covariance matrix: Table 5.2

Correlation matrix: Table 5.3
　　Gleason–Staelin statistic: Section 5.3
　　Index: Section 5.3

Residual correlation matrix: Tables 5.8 and 17.2

Roots: Table 5.4
　　Test for roots: Table 5.4
　　Covariance matrix of first four roots: Table 5.5
　　SCREE plot: Figure 5.1
　　Proportion of each variable explained by each principal component: Table 5.7
　　Cross-validation: Section 16.3.2, Table 16.2

Vectors
　　U: Table 5.6, Table 17.1
Standard errors: Table 5.6
Correlation matrix of coefficients: Table 5.9
　　V: Table 5.10, Table 8.3
　　V*: Table 5.11
　　W: Table 5.10
　　W*: Table 5.11

y-scores: Table 5.13
　　Two-way plots: Figure 5.3

T^2-test: Table 5.13

Q-test: Table 5.13

Selected residuals: Table 5.14

Plot of residuals vs. data, x_8: Figure 5.2

Varimax rotation: Tables 8.4 and 8.5

Quartimax rotation: Tables 8.4 and 8.5

Missing data estimation: Section 14.1.6

Factor analysis
　　Principal factor solution: Table 17.1
Residual matrix: Table 17.2
　　Other solutions: Section 17.6.3
　　Confirmatory factor analysis: Section 17.8.2, Table 17.8

F.12   AUDIOMETRIC CASE HISTORY

An extension of the audiometric example to a case study involving a total sample size of 10,358 males, subgrouped by age. Data are not included. All operations are in Chapter 9 except where noted.

Procrustes fit to Audiometric example: Tables 8.6 and 8.7
Confirmatory factor analysis: Section 17.8.2, Table 17.8

F.13   ROTATION DEMONSTRATION

Fictitious example. No data.
Rotation: Section 8.3, Figure 8.1

F.14   PHYSICAL MEASUREMENTS

Real data consisting of $p = 8$ physical measurements on $n = 305$ girls. Data are not included. Source: Harmon (1976) with permission of the University of Chicago Press, Chicago, IL.

Correlation matrix: Table 8.1
Roots: Table 8.2
Vectors
V: Table 8.2, Table 17.5
Residuals: Table 17.6
Varimax rotation: Section 8.4.2, Figure 8.2
HK-rotation: Section 8.4.3, Figure 8.3
Image analysis: Table 17.5
Factor analysis
Maximum likelihood: Table 17.5
Principal factors: Table 17.5
ULS: Table 17.5
α-Analysis: Table 17.5
Residuals: Table 17.6
Factor score coefficients: Table 17.7

F.15   RECTANGULAR DATA MATRIX

Fictitious $p = 3$ by $n = 4$ data matrix.

Data: Section 10.2
Covariance matrix (variables): Section 10.2
Roots and U, V, and W-vectors: Section 10.2

Covariance matrix (cases): Section 10.2
   Roots and U, V, and W-vectors: Section 10.2
Singular value decomposition: Section 10.3
Generalized inverse: Section 10.3
SQ biplot: Section 10.5, Figure 10.1
GH biplot: Section 10.5
JK biplot: Section 10.5
Distance calculations: Section 11.2

F.16   HORSESHOE EXAMPLE

Fictitious example of interpoint distance ranks among $p = 6$ stimuli.

   Data: Table 11.7
   Similarity plot: Figure 11.8

F.17   PRESIDENTIAL HOPEFULS

Real data consisting of graphic ratings of $p = 6$ candidates by $n = 18$ respondents. The same respondents also performed ranked pair similarity judgments for the same candidates but these data are not included. A different group of respondents rated a different set of candidates on a number of criteria. The data are not included.

   Data: Table 10.1
   Roots: Table 10.2
   U-vectors: Table 10.2
   z-scores: Table 10.3
   MDPREF-plot, 2D: Figure 10.3
      Correlation of predicted and actual ratings: Table 10.4
   MDPREF-plot, 3D: Figure 10.4
   Point-point plot: Figure 10.6
   INDSCAL plot: Figure 11.11
   INDSCAL saliences: Figure 11.12
   PREFMAP plot: Figure 11.14
   PROFIT plot: Figure 11.15 (Different data set)
   Double-centered matrix: Table 13.7
      ANOVA: Table 13.6
   Vectors
      U: Table 13.8

z-scores: Table 13.9
MDPREF plot: Figure 13.1
Inherent variability: Section 13.7.6
Simultaneous inference: Section 13.7.6, Table 13.10

F.18   CONTINGENCY TABLE DEMO:BRAND vs. SEX

Fictitious data in the form of a 2 × 3 contingency table of counts.

Data: Table 10.5

F.19   CONTINGENCY TABLE DEMO:BRAND vs. AGE

Fictitious data in the form of a 3 × 4 contingency table of counts.

Data: Table 10.6
Single value decomposition: Section 10.8.2
Correspondence analysis plot: Figure 10.7
Burt matrix: Table 10.11
U-vectors: Table 10.13
MCA computations: Sections 10.8.2 and 10.8.5

F.20   THREE-WAY CONTINGENCY TABLE

Fictitious data in the form of a 2 × 3 × 4 contingency table of counts.

Data: Table 10.14
Burt matrix: Table 10.15
U-vectors: Table 10.16
MCA computations: Table 10.17
MCA plot: Figure 10.9

F.21   OCCURRENCE OF PERSONAL ASSAULT

Real data consisting of number of assaults by years and seasons in the form of
a 10 × 4 contingency table. Source: Leffingwell (1892)

Data: Table 10.7
H-matrix: Table 10.9

Singular value decomposition: Section 10.8.3
Correspondence analysis plot: Figure 10.8

F.22 LINNERUD DATA

Real data consisting of three physical measurements and three physical tests on each of $n = 20$ men. Source: Young and Sarle (1983) with permission of SAS Institute, Inc., Cary NC.

Data: Tables 12.1 and 12.2
Correlation matrix: Table 12.3
Vectors
    U: Section 12.3.1
OLS solution: Section 12.2.2, Table 12.4
PCR solutions: Sections 12.3.1 and 12.3.2, Tables 12.5–12.7
Latent root regression: Section 12.3.4
PLSR solution: Section 12.5.4
    PLSR stopping rules: Section 12.5.5
Canonical correlation: Section 12.6.2, Figure 12.1
    RV-coefficient: Section 12.6.2
Redundancy: Section 12.6.3, Table 12.8
Maximum redundancy: Section 12.6.4

F.23 BIVARIATE NONNORMAL DISTRIBUTION

Fictitious relative frequency data.

Data: Table 14.1

F.24 CIRCLE DATA

Two-way array of points representing a circle.

Data: Table 16.1
Covariance matrix: Section 16.2
Vectors
    U: Section 16.2

F.25   U.S. BUDGET

Data consisting of breakdown of U.S. Budget by years for the period 1967–1986.

Data: Tables 16.6–16.8
Covariance matrix: Table 16.9
Vectors
    U: Table 16.10

APPENDIX G

Tables

This appendix contains most of the tables that the reader will need in conjunction with this book. These include tables of the normal, t, chi-square, and F-distributions; the Lawley–Hotelling statistic; and the distribution of the largest and smallest characteristic roots of a covariance matrix. The reader may find tables for some of the more specialized applications dealing with characteristic roots and vectors from references given in Chapter 4.

APPENDIX G.1

Table of the Normal Distribution

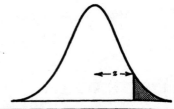

z	0.00	0.01	0.02	0.03	0.04	0.05	0.06	0.07	0.08	0.09
0.0	0.5000	0.4960	0.4920	0.4880	0.4840	0.4801	0.4761	0.4721	0.4681	0.4641
0.1	0.4602	0.4562	0.4522	0.4483	0.4443	0.4404	0.4364	0.4325	0.4286	0.4247
0.2	0.4207	0.4168	0.4129	0.4090	0.4052	0.4013	0.3974	0.3936	0.3897	0.3859
0.3	0.3821	0.3783	0.3745	0.3707	0.3669	0.3632	0.3594	0.3557	0.3520	0.3483
0.4	0.3446	0.3409	0.3372	0.3336	0.3300	0.3264	0.3228	0.3192	0.3156	0.3121
0.5	0.3085	0.3050	0.3015	0.2981	0.2946	0.2912	0.2877	0.2843	0.2810	0.2776
0.6	0.2743	0.2709	0.2676	0.2643	0.2611	0.2578	0.2546	0.2514	0.2483	0.2451
0.7	0.2420	0.2389	0.2358	0.2327	0.2296	0.2266	0.2236	0.2206	0.2177	0.2148
0.8	0.2119	0.2090	0.2061	0.2033	0.2005	0.1977	0.1949	0.1922	0.1894	0.1867
0.9	0.1841	0.1814	0.1788	0.1762	0.1736	0.1711	0.1685	0.1660	0.1635	0.1611
1.0	0.1587	0.1562	0.1539	0.1515	0.1492	0.1469	0.1446	0.1423	0.1401	0.1379
1.1	0.1357	0.1335	0.1314	0.1292	0.1271	0.1251	0.1230	0.1210	0.1190	0.1170
1.2	0.1151	0.1131	0.1112	0.1093	0.1075	0.1056	0.1038	0.1020	0.1003	0.0985
1.3	0.0968	0.0951	0.0934	0.0918	0.0901	0.0885	0.0869	0.0853	0.0838	0.0823
1.4	0.0808	0.0793	0.0778	0.0764	0.0749	0.0735	0.0721	0.0708	0.0694	0.0681
1.5	0.0668	0.0655	0.0643	0.0630	0.0618	0.0606	0.0594	0.0582	0.0571	0.0559
1.6	0.0548	0.0537	0.0526	0.0516	0.0505	0.0495	0.0485	0.0475	0.0465	0.0455
1.7	0.0446	0.0436	0.0427	0.0418	0.0409	0.0401	0.0392	0.0384	0.0375	0.0367
1.8	0.0359	0.0351	0.0344	0.0336	0.0329	0.0322	0.0314	0.0307	0.0301	0.0294
1.9	0.0287	0.0281	0.0274	0.0268	0.0262	0.0256	0.0250	0.0244	0.0239	0.0233
2.0	0.0228	0.0222	0.0217	0.0212	0.0207	0.0202	0.0197	0.0192	0.0188	0.0183
2.1	0.0179	0.0174	0.0170	0.0166	0.0162	0.0158	0.0154	0.0150	0.0146	0.0143
2.2	0.0139	0.0136	0.0132	0.0129	0.0125	0.0122	0.0119	0.0116	0.0113	0.0110
2.3	0.0107	0.0104	0.0102	0.0099	0.0096	0.0094	0.0091	0.0089	0.0087	0.0084
2.4	0.0082	0.0080	0.0078	0.0075	0.0073	0.0071	0.0069	0.0068	0.0066	0.0064
2.5	0.0062	0.0060	0.0059	0.0057	0.0055	0.0054	0.0052	0.0051	0.0049	0.0048
2.6	0.0047	0.0045	0.0044	0.0043	0.0041	0.0040	0.0039	0.0038	0.0037	0.0036
2.7	0.0035	0.0034	0.0033	0.0032	0.0031	0.0030	0.0029	0.0028	0.0027	0.0026
2.8	0.0026	0.0025	0.0024	0.0023	0.0023	0.0022	0.0021	0.0021	0.0020	0.0019
2.9	0.0019	0.0018	0.0018	0.0017	0.0016	0.0016	0.0015	0.0015	0.0014	0.0014
3.0	0.0013	0.0013	0.0013	0.0012	0.0012	0.0011	0.0011	0.0011	0.0010	0.0010
3.1	0.0010	0.0009	0.0009	0.0009	0.0008	0.0008	0.0008	0.0008	0.0007	0.0007
3.2	0.0007	0.0007	0.0006	0.0006	0.0006	0.0006	0.0006	0.0005	0.0005	0.0005
3.3	0.0005	0.0005	0.0005	0.0004	0.0004	0.0004	0.0004	0.0004	0.0004	0.0003
3.4	0.0003	0.0003	0.0003	0.0003	0.0003	0.0003	0.0003	0.0003	0.0003	0.0002
3.5	0.0002	0.0002	0.0002	0.0002	0.0002	0.0002	0.0002	0.0002	0.0002	0.0002
3.6	0.0002	0.0002	0.0001	0.0001	0.0001	0.0001	0.0001	0.0001	0.0001	0.0001
3.7	0.0001	0.0001	0.0001	0.0001	0.0001	0.0001	0.0001	0.0001	0.0001	0.0001
3.8	0.0001	0.0001	0.0001	0.0001	0.0001	0.0001	0.0001	0.0001	0.0001	0.0001
3.9	0.0000	0.0000	0.0000	0.0000	0.0000	0.0000	0.0000	0.0000	0.0000	0.0000

Reproduced with permission of John Wiley and Sons, Inc. from Box, Hunter, and Hunter (1978).

APPENDIX G.2

Table of the t-Distribution

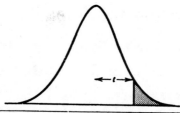

ν	| tail area probability									
	0.4	**0.25**	**0.1**	**0.05**	**0.025**	**0.01**	**0.005**	**0.0025**	**0.001**	**0.0005**
1	0.325	1.000	3.078	6.314	12.706	31.821	63.657	127.32	318.31	636.62
2	0.289	0.816	1.886	2.920	4.303	6.965	9.925	14.089	22.326	31.598
3	0.277	0.765	1.638	2.353	3.182	4.541	5.841	7.453	10.213	12.924
4	0.271	0.741	1.533	2.132	2.776	3.747	4.604	5.598	7.173	8.610
5	0.267	0.727	1.476	2.015	2.571	3.365	4.032	4.773	5.893	6.869
6	0.265	0.718	1.440	1.943	2.447	3.143	3.707	4.317	5.208	5.959
7	0.263	0.711	1.415	1.895	2.365	2.998	3.499	4.029	4.785	5.408
8	0.262	0.706	1.397	1.860	2.306	2.896	3.355	3.833	4.501	5.041
9	0.261	0.703	1.383	1.833	2.262	2.821	3.250	3.690	4.297	4.781
10	0.260	0.700	1.372	1.812	2.228	2.764	3.169	3.581	4.144	4.587
11	0.260	0.697	1.363	1.796	2.201	2.718	3.106	3.497	4.025	4.437
12	0.259	0.695	1.356	1.782	2.179	2.681	3.055	3.428	3.930	4.318
13	0.259	0.694	1.350	1.771	2.160	2.650	3.012	3.372	3.852	4.221
14	0.258	0.692	1.345	1.761	2.145	2.624	2.977	3.326	3.787	4.140
15	0.258	0.691	1.341	1.753	2.131	2.602	2.947	3.286	3.733	4.073
16	0.258	0.690	1.337	1.746	2.120	2.583	2.921	3.252	3.686	4.015
17	0.257	0.689	1.333	1.740	2.110	2.567	2.898	3.222	3.646	3.965
18	0.257	0.688	1.330	1.734	2.101	2.552	2.878	3.197	3.610	3.922
19	0.257	0.688	1.328	1.729	2.093	2.539	2.861	3.174	3.579	3.883
20	0.257	0.687	1.325	1.725	2.086	2.528	2.845	3.153	3.552	3.850
21	0.257	0.686	1.323	1.721	2.080	2.518	2.831	3.135	3.527	3.819
22	0.256	0.686	1.321	1.717	2.074	2.508	2.819	3.119	3.505	3.792
23	0.256	0.685	1.319	1.714	2.069	2.500	2.807	3.104	3.485	3.767
24	0.256	0.685	1.318	1.711	2.064	2.492	2.797	3.091	3.467	3.745
25	0.256	0.684	1.316	1.708	2.060	2.485	2.787	3.078	3.450	3.725
26	0.256	0.684	1.315	1.706	2.056	2.479	2.779	3.067	3.435	3.707
27	0.256	0.684	1.314	1.703	2.052	2.473	2.771	3.057	3.421	3.690
28	0.256	0.683	1.313	1.701	2.048	2.467	2.763	3.047	3.408	3.674
29	0.256	0.683	1.311	1.699	2.045	2.462	2.756	3.038	3.396	3.659
30	0.256	0.683	1.310	1.697	2.042	2.457	2.750	3.030	3.385	3.646
40	0.255	0.681	1.303	1.684	2.021	2.423	2.704	2.971	3.307	3.551
60	0.254	0.679	1.296	1.671	2.000	2.390	2.660	2.915	3.232	3.460
120	0.254	0.677	1.289	1.658	1.980	2.358	2.617	2.860	3.160	3.373
∞	0.253	0.674	1.282	1.645	1.960	2.326	2.576	2.807	3.090	3.291

APPENDIX G.3

Table of the Chi-Square Distribution

Example
$\Pr (\chi^2 > 23.8277) = 0.25$
$\Pr (\chi^2 > 31.4104) = 0.05$ for df = 20
$\Pr (\chi^2 > 37.5662) = 0.01$

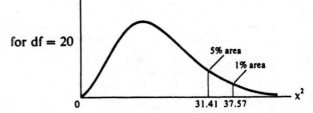

df \ Pr	0.250	0.100	0.050	0.025	0.010	0.005	0.001
1	1.32330	2.70554	3.84146	5.02389	6.63490	7.87944	10.828
2	2.77259	4.60517	5.99146	7.37776	9.21034	10.5966	13.816
3	4.10834	6.25139	7.81473	9.34840	11.3449	12.8382	16.266
4	5.38527	7.77944	9.48773	11.1433	13.2767	14.8603	18.467
5	6.62568	9.23636	11.0705	12.8325	15.0863	16.7496	20.515
6	7.84080	10.6446	12.5916	14.4494	16.8119	18.5476	22.458
7	9.03715	12.0170	14.0671	16.0128	18.4753	20.2777	24.322
8	10.2189	13.3616	15.5073	17.5345	20.0902	21.9550	26.125
9	11.3888	14.6837	16.9190	19.0228	21.6660	23.5894	27.877
10	12.5489	15.9872	18.3070	20.4832	23.2093	25.1882	29.588
11	13.7007	17.2750	19.6751	21.9200	24.7250	26.7568	31.264
12	14.8454	18.5493	21.0261	23.3367	26.2170	28.2995	32.909
13	15.9839	19.8119	22.3620	24.7356	27.6882	29.8195	34.528
14	17.1169	21.0641	23.6848	26.1189	29.1412	31.3194	36.123
15	18.2451	22.3071	24.9958	27.4884	30.5779	32.8013	37.697
16	19.3689	23.5418	26.2962	28.8454	31.9999	34.2672	39.252
17	20.4887	24.7690	27.5871	30.1910	33.4087	35.7185	40.790
18	21.6049	25.9894	28.8693	31.5264	34.8053	37.1565	42.312
19	22.7178	27.2036	30.1435	32.8523	36.1909	38.5823	43.820

† For df greater than 100, the expression

$$\sqrt{2\chi^2} - \sqrt{(2k - 1)} = Z$$

follows the standardized normal distribution, where k represents the degrees of freedom.

Pr df	0.250	0.100	0.050	0.025	0.010	0.005	0.001
20	23.8277	28.4120	31.4104	34.1696	37.5662	39.9968	45.315
21	24.9348	29.6151	32.6706	35.4789	38.9322	41.4011	46.797
22	26.0393	30.8133	33.9244	36.7807	40.2894	42.7957	48.268
23	27.1413	32.0069	35.1725	38.0756	41.6384	44.1813	49.728
24	28.2412	33.1962	36.4150	39.3641	42.9798	45.5585	51.179
25	29.3389	34.3816	37.6525	40.6465	44.3141	46.9279	52.618
26	30.4346	35.5632	38.8851	41.9232	45.6417	48.2899	54.052
27	31.5284	36.7412	40.1133	43.1945	46.9629	49.6449	55.476
28	32.6205	37.9159	41.3371	44.4608	48.2782	50.9934	56.892
29	33.7109	39.0875	42.5570	45.7223	49.5879	52.3356	58.301
30	34.7997	40.2560	43.7730	46.9792	50.8922	53.6720	59.703
40	45.6160	51.8051	55.7585	59.3417	63.6907	66.7660	73.402
50	56.3336	63.1671	67.5048	71.4202	76.1539	79.4900	86.661
60	66.9815	74.3970	79.0819	83.2977	88.3794	91.9517	99.607
70	77.5767	85.5270	90.5312	95.0232	100.425	104.215	112.317
80	88.1303	96.5782	101.879	106.629	112.329	116.321	124.839
90	98.6499	107.565	113.145	118.136	124.116	128.299	137.208
100	109.141	118.498	124.342	129.561	135.807	140.169	149.449
Z†	+ 0.6745	+ 1.2816	+ 1.6449	+ 1.9600	+ 2.3263	+ 2.5758	+ 3.0902

Reproduced with permission of the Biometrika Trustees from Table 8, Pearson and Hartley (1966).

APPENDIX G.4

Table of the F-Distribution: upper 25% points

v_2 \ v_1	1	2	3	4	5	6	7	8	9	10	12	15	20	24	30	40	60	120	∞
1	5.83	7.50	8.20	8.58	8.82	8.98	9.10	9.19	9.26	9.32	9.41	9.49	9.58	9.63	9.67	9.71	9.76	9.80	9.85
2	2.57	3.00	3.15	3.23	3.28	3.31	3.34	3.35	3.37	3.38	3.39	3.41	3.43	3.43	3.44	3.45	3.46	3.47	3.48
3	2.02	2.28	2.36	2.39	2.41	2.42	2.43	2.44	2.44	2.44	2.45	2.46	2.46	2.46	2.47	2.47	2.47	2.47	2.47
4	1.81	2.00	2.05	2.06	2.07	2.08	2.08	2.08	2.08	2.08	2.08	2.08	2.08	2.08	2.08	2.08	2.08	2.08	2.08
5	1.69	1.85	1.88	1.89	1.89	1.89	1.89	1.89	1.89	1.89	1.89	1.89	1.88	1.88	1.88	1.88	1.87	1.87	1.87
6	1.62	1.76	1.78	1.79	1.79	1.78	1.78	1.78	1.77	1.77	1.77	1.76	1.76	1.75	1.75	1.75	1.74	1.74	1.74
7	1.57	1.70	1.72	1.72	1.71	1.71	1.70	1.70	1.69	1.69	1.68	1.68	1.67	1.67	1.66	1.66	1.65	1.65	1.65
8	1.54	1.66	1.67	1.66	1.66	1.65	1.64	1.64	1.63	1.63	1.62	1.62	1.61	1.60	1.60	1.59	1.59	1.58	1.58
9	1.51	1.62	1.63	1.63	1.62	1.61	1.60	1.60	1.59	1.59	1.58	1.57	1.56	1.56	1.55	1.54	1.54	1.53	1.53
10	1.49	1.60	1.60	1.59	1.59	1.58	1.57	1.56	1.56	1.55	1.54	1.53	1.52	1.52	1.51	1.51	1.50	1.49	1.48
11	1.47	1.58	1.58	1.57	1.56	1.55	1.54	1.53	1.53	1.52	1.51	1.50	1.49	1.49	1.48	1.47	1.47	1.46	1.45
12	1.46	1.56	1.56	1.55	1.54	1.53	1.52	1.51	1.51	1.50	1.49	1.48	1.47	1.46	1.45	1.45	1.44	1.43	1.42
13	1.45	1.55	1.55	1.53	1.52	1.51	1.50	1.49	1.49	1.48	1.47	1.46	1.45	1.44	1.43	1.42	1.42	1.41	1.40
14	1.44	1.53	1.53	1.52	1.51	1.50	1.49	1.48	1.47	1.46	1.45	1.44	1.43	1.42	1.41	1.41	1.40	1.39	1.38
15	1.43	1.52	1.52	1.51	1.49	1.48	1.47	1.46	1.46	1.45	1.44	1.43	1.41	1.41	1.40	1.39	1.38	1.37	1.36
16	1.42	1.51	1.51	1.50	1.48	1.47	1.46	1.45	1.44	1.44	1.43	1.41	1.40	1.39	1.38	1.37	1.36	1.35	1.34
17	1.42	1.51	1.50	1.49	1.47	1.46	1.45	1.44	1.43	1.43	1.41	1.40	1.39	1.38	1.37	1.36	1.35	1.34	1.33
18	1.41	1.50	1.49	1.48	1.46	1.45	1.44	1.43	1.42	1.42	1.40	1.39	1.38	1.37	1.36	1.35	1.34	1.33	1.32
19	1.41	1.49	1.49	1.47	1.46	1.44	1.43	1.42	1.41	1.41	1.40	1.38	1.37	1.36	1.35	1.34	1.33	1.32	1.30
20	1.40	1.49	1.48	1.47	1.45	1.44	1.43	1.42	1.41	1.40	1.39	1.37	1.36	1.35	1.34	1.33	1.32	1.31	1.29
21	1.40	1.48	1.48	1.46	1.44	1.43	1.42	1.41	1.40	1.39	1.38	1.37	1.35	1.34	1.33	1.32	1.31	1.30	1.28
22	1.40	1.48	1.47	1.45	1.44	1.42	1.41	1.40	1.39	1.39	1.37	1.36	1.34	1.33	1.32	1.31	1.30	1.29	1.28
23	1.39	1.47	1.47	1.45	1.43	1.42	1.41	1.40	1.39	1.38	1.37	1.35	1.34	1.33	1.32	1.31	1.30	1.28	1.27
24	1.39	1.47	1.46	1.44	1.43	1.41	1.40	1.39	1.38	1.38	1.36	1.35	1.33	1.32	1.31	1.30	1.29	1.28	1.26
25	1.39	1.47	1.46	1.44	1.42	1.41	1.40	1.39	1.38	1.37	1.36	1.34	1.33	1.32	1.31	1.29	1.28	1.27	1.25
26	1.38	1.46	1.45	1.44	1.42	1.41	1.39	1.38	1.37	1.37	1.35	1.34	1.32	1.31	1.30	1.29	1.28	1.26	1.25
27	1.38	1.46	1.45	1.43	1.42	1.40	1.39	1.38	1.37	1.36	1.35	1.33	1.32	1.31	1.30	1.28	1.27	1.26	1.24
28	1.38	1.46	1.45	1.43	1.41	1.40	1.39	1.38	1.37	1.36	1.34	1.33	1.32	1.31	1.29	1.28	1.27	1.25	1.24
29	1.38	1.45	1.45	1.43	1.41	1.40	1.38	1.37	1.36	1.35	1.34	1.32	1.31	1.30	1.29	1.27	1.26	1.25	1.23
30	1.38	1.45	1.44	1.42	1.41	1.39	1.38	1.37	1.36	1.35	1.34	1.32	1.30	1.29	1.28	1.27	1.26	1.24	1.23
40	1.36	1.44	1.42	1.40	1.39	1.37	1.36	1.35	1.34	1.33	1.31	1.30	1.28	1.26	1.25	1.24	1.22	1.21	1.19
60	1.35	1.42	1.41	1.38	1.37	1.35	1.33	1.32	1.31	1.30	1.29	1.27	1.25	1.24	1.22	1.21	1.19	1.17	1.15
120	1.34	1.40	1.39	1.37	1.35	1.33	1.31	1.30	1.29	1.28	1.26	1.24	1.22	1.21	1.19	1.18	1.16	1.13	1.10
∞	1.32	1.39	1.37	1.35	1.33	1.31	1.29	1.28	1.27	1.25	1.24	1.22	1.19	1.18	1.16	1.14	1.12	1.08	1.00

Table of the F-Distribution: upper 10% points

v_2 \ v_1	1	2	3	4	5	6	7	8	9	10	12	15	20	24	30	40	60	120	∞
1	39.86	49.50	53.59	55.83	57.24	58.20	58.91	59.44	59.86	60.19	60.71	61.22	61.74	62.00	62.26	62.53	62.79	63.06	63.33
2	8.53	9.00	9.16	9.24	9.29	9.33	9.35	9.37	9.38	9.39	9.41	9.42	9.44	9.45	9.46	9.47	9.47	9.48	9.49
3	5.54	5.46	5.39	5.34	5.31	5.28	5.27	5.25	5.24	5.23	5.22	5.20	5.18	5.18	5.17	5.16	5.15	5.14	5.13
4	4.54	4.32	4.19	4.11	4.05	4.01	3.98	3.95	3.94	3.92	3.90	3.87	3.84	3.83	3.82	3.80	3.79	3.78	3.76
5	4.06	3.78	3.62	3.52	3.45	3.40	3.37	3.34	3.32	3.30	3.27	3.24	3.21	3.19	3.17	3.16	3.14	3.12	3.10
6	3.78	3.46	3.29	3.18	3.11	3.05	3.01	2.98	2.96	2.94	2.90	2.87	2.84	2.82	2.80	2.78	2.76	2.74	2.72
7	3.59	3.26	3.07	2.96	2.88	2.83	2.78	2.75	2.72	2.70	2.67	2.63	2.59	2.58	2.56	2.54	2.51	2.49	2.47
8	3.46	3.11	2.92	2.81	2.73	2.67	2.62	2.59	2.56	2.54	2.50	2.46	2.42	2.40	2.38	2.36	2.34	2.32	2.29
9	3.36	3.01	2.81	2.69	2.61	2.55	2.51	2.47	2.44	2.42	2.38	2.34	2.30	2.28	2.25	2.23	2.21	2.18	2.16
10	3.29	2.92	2.73	2.61	2.52	2.46	2.41	2.38	2.35	2.32	2.28	2.24	2.20	2.18	2.16	2.13	2.11	2.08	2.06
11	3.23	2.86	2.66	2.54	2.45	2.39	2.34	2.30	2.27	2.25	2.21	2.17	2.12	2.10	2.08	2.05	2.03	2.00	1.97
12	3.18	2.81	2.61	2.48	2.39	2.33	2.28	2.24	2.21	2.19	2.15	2.10	2.06	2.04	2.01	1.99	1.96	1.93	1.90
13	3.14	2.76	2.56	2.43	2.35	2.28	2.23	2.20	2.16	2.14	2.10	2.05	2.01	1.98	1.96	1.93	1.90	1.88	1.85
14	3.10	2.73	2.52	2.39	2.31	2.24	2.19	2.15	2.12	2.10	2.05	2.01	1.96	1.94	1.91	1.89	1.86	1.83	1.80
15	3.07	2.70	2.49	2.36	2.27	2.21	2.16	2.12	2.09	2.06	2.02	1.97	1.92	1.90	1.87	1.85	1.82	1.79	1.76
16	3.05	2.67	2.46	2.33	2.24	2.18	2.13	2.09	2.06	2.03	1.99	1.94	1.89	1.87	1.84	1.81	1.78	1.75	1.72
17	3.03	2.64	2.44	2.31	2.22	2.15	2.10	2.06	2.03	2.00	1.96	1.91	1.86	1.84	1.81	1.78	1.75	1.72	1.69
18	3.01	2.62	2.42	2.29	2.20	2.13	2.08	2.04	2.00	1.98	1.93	1.89	1.84	1.81	1.78	1.75	1.72	1.69	1.66
19	2.99	2.61	2.40	2.27	2.18	2.11	2.06	2.02	1.98	1.96	1.91	1.86	1.81	1.79	1.76	1.73	1.70	1.67	1.63
20	2.97	2.59	2.38	2.25	2.16	2.09	2.04	2.00	1.96	1.94	1.89	1.84	1.79	1.77	1.74	1.71	1.68	1.64	1.61
21	2.96	2.57	2.36	2.23	2.14	2.08	2.02	1.98	1.95	1.92	1.87	1.83	1.78	1.75	1.72	1.69	1.66	1.62	1.59
22	2.95	2.56	2.35	2.22	2.13	2.06	2.01	1.97	1.93	1.90	1.86	1.81	1.76	1.73	1.70	1.67	1.64	1.60	1.57
23	2.94	2.55	2.34	2.21	2.11	2.05	1.99	1.95	1.92	1.89	1.84	1.80	1.74	1.72	1.69	1.66	1.62	1.59	1.55
24	2.93	2.54	2.33	2.19	2.10	2.04	1.98	1.94	1.91	1.88	1.83	1.78	1.73	1.70	1.67	1.64	1.61	1.57	1.53
25	2.92	2.53	2.32	2.18	2.09	2.02	1.97	1.93	1.89	1.87	1.82	1.77	1.72	1.69	1.66	1.63	1.59	1.56	1.52
26	2.91	2.52	2.31	2.17	2.08	2.01	1.96	1.92	1.88	1.86	1.81	1.76	1.71	1.68	1.65	1.61	1.58	1.54	1.50
27	2.90	2.51	2.30	2.17	2.07	2.00	1.95	1.91	1.87	1.85	1.80	1.75	1.70	1.67	1.64	1.60	1.57	1.53	1.49
28	2.89	2.50	2.29	2.16	2.06	2.00	1.94	1.90	1.87	1.84	1.79	1.74	1.69	1.66	1.63	1.59	1.56	1.52	1.48
29	2.89	2.50	2.28	2.15	2.06	1.99	1.93	1.89	1.86	1.83	1.78	1.73	1.68	1.65	1.62	1.58	1.55	1.51	1.47
30	2.88	2.49	2.28	2.14	2.05	1.98	1.93	1.88	1.85	1.82	1.77	1.72	1.67	1.64	1.61	1.57	1.54	1.50	1.46
40	2.84	2.44	2.23	2.09	2.00	1.93	1.87	1.83	1.79	1.76	1.71	1.66	1.61	1.57	1.54	1.51	1.47	1.42	1.38
60	2.79	2.39	2.18	2.04	1.95	1.87	1.82	1.77	1.74	1.71	1.66	1.60	1.54	1.51	1.48	1.44	1.40	1.35	1.29
120	2.75	2.35	2.13	1.99	1.90	1.82	1.77	1.72	1.68	1.65	1.60	1.55	1.48	1.45	1.41	1.37	1.32	1.26	1.19
∞	2.71	2.30	2.08	1.94	1.85	1.77	1.72	1.67	1.63	1.60	1.55	1.49	1.42	1.38	1.34	1.30	1.24	1.17	1.00

Table of the *F*-Distribution: upper 5% points

v_2 \ v_1	1	2	3	4	5	6	7	8	9	10	12	15	20	24	30	40	60	120	∞
1	161.4	199.5	215.7	224.6	230.2	234.0	236.8	238.9	240.5	241.9	243.9	245.9	248.0	249.1	250.1	251.1	252.2	253.3	254.3
2	18.51	19.00	19.16	19.25	19.30	19.33	19.35	19.37	19.38	19.40	19.41	19.43	19.45	19.45	19.46	19.47	19.48	19.49	19.50
3	10.13	9.55	9.28	9.12	9.01	8.94	8.89	8.85	8.81	8.79	8.74	8.70	8.66	8.64	8.62	8.59	8.57	8.55	8.53
4	7.71	6.94	6.59	6.39	6.26	6.16	6.09	6.04	6.00	5.96	5.91	5.86	5.80	5.77	5.75	5.72	5.69	5.66	5.63
5	6.61	5.79	5.41	5.19	5.05	4.95	4.88	4.82	4.77	4.74	4.68	4.62	4.56	4.53	4.50	4.46	4.43	4.40	4.36
6	5.99	5.14	4.76	4.53	4.39	4.28	4.21	4.15	4.10	4.06	4.00	3.94	3.87	3.84	3.81	3.77	3.74	3.70	3.67
7	5.59	4.74	4.35	4.12	3.97	3.87	3.79	3.73	3.68	3.64	3.57	3.51	3.44	3.41	3.38	3.34	3.30	3.27	3.23
8	5.32	4.46	4.07	3.84	3.69	3.58	3.50	3.44	3.39	3.35	3.28	3.22	3.15	3.12	3.08	3.04	3.01	2.97	2.93
9	5.12	4.26	3.86	3.63	3.48	3.37	3.29	3.23	3.18	3.14	3.07	3.01	2.94	2.90	2.86	2.83	2.79	2.75	2.71
10	4.96	4.10	3.71	3.48	3.33	3.22	3.14	3.07	3.02	2.98	2.91	2.85	2.77	2.74	2.70	2.66	2.62	2.58	2.54
11	4.84	3.98	3.59	3.36	3.20	3.09	3.01	2.95	2.90	2.85	2.79	2.72	2.65	2.61	2.57	2.53	2.49	2.45	2.40
12	4.75	3.89	3.49	3.26	3.11	3.00	2.91	2.85	2.80	2.75	2.69	2.62	2.54	2.51	2.47	2.43	2.38	2.34	2.30
13	4.67	3.81	3.41	3.18	3.03	2.92	2.83	2.77	2.71	2.67	2.60	2.53	2.46	2.42	2.38	2.34	2.30	2.25	2.21
14	4.60	3.74	3.34	3.11	2.96	2.85	2.76	2.70	2.65	2.60	2.53	2.46	2.39	2.35	2.31	2.27	2.22	2.18	2.13
15	4.54	3.68	3.29	3.06	2.90	2.79	2.71	2.64	2.59	2.54	2.48	2.40	2.33	2.29	2.25	2.20	2.16	2.11	2.07
16	4.49	3.63	3.24	3.01	2.85	2.74	2.66	2.59	2.54	2.49	2.42	2.35	2.28	2.24	2.19	2.15	2.11	2.06	2.01
17	4.45	3.59	3.20	2.96	2.81	2.70	2.61	2.55	2.49	2.45	2.38	2.31	2.23	2.19	2.15	2.10	2.06	2.01	1.96
18	4.41	3.55	3.16	2.93	2.77	2.66	2.58	2.51	2.46	2.41	2.34	2.27	2.19	2.15	2.11	2.06	2.02	1.97	1.92
19	4.38	3.52	3.13	2.90	2.74	2.63	2.54	2.48	2.42	2.38	2.31	2.23	2.16	2.11	2.07	2.03	1.98	1.93	1.88
20	4.35	3.49	3.10	2.87	2.71	2.60	2.51	2.45	2.39	2.35	2.28	2.20	2.12	2.08	2.04	1.99	1.95	1.90	1.84
21	4.32	3.47	3.07	2.84	2.68	2.57	2.49	2.42	2.37	2.32	2.25	2.18	2.10	2.05	2.01	1.96	1.92	1.87	1.81
22	4.30	3.44	3.05	2.82	2.66	2.55	2.46	2.40	2.34	2.30	2.23	2.15	2.07	2.03	1.98	1.94	1.89	1.84	1.78
23	4.28	3.42	3.03	2.80	2.64	2.53	2.44	2.37	2.32	2.27	2.20	2.13	2.05	2.01	1.96	1.91	1.86	1.81	1.76
24	4.26	3.40	3.01	2.78	2.62	2.51	2.42	2.36	2.30	2.25	2.18	2.11	2.03	1.98	1.94	1.89	1.84	1.79	1.73
25	4.24	3.39	2.99	2.76	2.60	2.49	2.40	2.34	2.28	2.24	2.16	2.09	2.01	1.96	1.92	1.87	1.82	1.77	1.71
26	4.23	3.37	2.98	2.74	2.59	2.47	2.39	2.32	2.27	2.22	2.15	2.07	1.99	1.95	1.90	1.85	1.80	1.75	1.69
27	4.21	3.35	2.96	2.73	2.57	2.46	2.37	2.31	2.25	2.20	2.13	2.06	1.97	1.93	1.88	1.84	1.79	1.73	1.67
28	4.20	3.34	2.95	2.71	2.56	2.45	2.36	2.29	2.24	2.19	2.12	2.04	1.96	1.91	1.87	1.82	1.77	1.71	1.65
29	4.18	3.33	2.93	2.70	2.55	2.43	2.35	2.28	2.22	2.18	2.10	2.03	1.94	1.90	1.85	1.81	1.75	1.70	1.64
30	4.17	3.32	2.92	2.69	2.53	2.42	2.33	2.27	2.21	2.16	2.09	2.01	1.93	1.89	1.84	1.79	1.74	1.68	1.62
40	4.08	3.23	2.84	2.61	2.45	2.34	2.25	2.18	2.12	2.08	2.00	1.92	1.84	1.79	1.74	1.69	1.64	1.58	1.51
60	4.00	3.15	2.76	2.53	2.37	2.25	2.17	2.10	2.04	1.99	1.92	1.84	1.75	1.70	1.65	1.59	1.53	1.47	1.39
120	3.92	3.07	2.68	2.45	2.29	2.17	2.09	2.02	1.96	1.91	1.83	1.75	1.66	1.61	1.55	1.50	1.43	1.35	1.25
∞	3.84	3.00	2.60	2.37	2.21	2.10	2.01	1.94	1.88	1.83	1.75	1.67	1.57	1.52	1.46	1.39	1.32	1.22	1.00

Table of the *F*-Distribution: upper 1% points

v_2 \ v_1	1	2	3	4	5	6	7	8	9	10	12	15	20	24	30	40	60	120	∞
1	4052	4999.50	5403	5625	5764	5859	5928	5982	6022	6056	6106	6157	6209	6235	6261	6287	6313	6339	6366
2	98.50	99.00	99.17	99.25	99.30	99.33	99.36	99.37	99.39	99.40	99.42	99.43	99.45	99.46	99.47	99.47	99.48	99.49	99.50
3	34.12	30.82	29.46	28.71	28.24	27.91	27.67	27.49	27.35	27.23	27.05	26.87	26.69	26.60	26.50	26.41	26.32	26.22	26.13
4	21.20	18.00	16.69	15.98	15.52	15.21	14.98	14.80	14.66	14.55	14.37	14.20	14.02	13.93	13.84	13.75	13.65	13.56	13.46
5	16.26	13.27	12.06	11.39	10.97	10.67	10.46	10.29	10.16	10.05	9.89	9.72	9.55	9.47	9.38	9.29	9.20	9.11	9.02
6	13.75	10.92	9.78	9.15	8.75	8.47	8.26	8.10	7.98	7.87	7.72	7.56	7.40	7.31	7.23	7.14	7.06	6.97	6.88
7	12.25	9.55	8.45	7.85	7.46	7.19	6.99	6.84	6.72	6.62	6.47	6.31	6.16	6.07	5.99	5.91	5.82	5.74	5.65
8	11.26	8.65	7.59	7.01	6.63	6.37	6.18	6.03	5.91	5.81	5.67	5.52	5.36	5.28	5.20	5.12	5.03	4.95	4.86
9	10.56	8.02	6.99	6.42	6.06	5.80	5.61	5.47	5.35	5.26	5.11	4.96	4.81	4.73	4.65	4.57	4.48	4.40	4.31
10	10.04	7.56	6.55	5.99	5.64	5.39	5.20	5.06	4.94	4.85	4.71	4.56	4.41	4.33	4.25	4.17	4.08	4.00	3.91
11	9.65	7.21	6.22	5.67	5.32	5.07	4.89	4.74	4.63	4.54	4.40	4.25	4.10	4.02	3.94	3.86	3.78	3.69	3.60
12	9.33	6.93	5.95	5.41	5.06	4.82	4.64	4.50	4.39	4.30	4.16	4.01	3.86	3.78	3.70	3.62	3.54	3.45	3.36
13	9.07	6.70	5.74	5.21	4.86	4.62	4.44	4.30	4.19	4.10	3.96	3.82	3.66	3.59	3.51	3.43	3.34	3.25	3.17
14	8.86	6.51	5.56	5.04	4.69	4.46	4.28	4.14	4.03	3.94	3.80	3.66	3.51	3.43	3.35	3.27	3.18	3.09	3.00
15	8.68	6.36	5.42	4.89	4.56	4.32	4.14	4.00	3.89	3.80	3.67	3.52	3.37	3.29	3.21	3.13	3.05	2.96	2.87
16	8.53	6.23	5.29	4.77	4.44	4.20	4.03	3.89	3.78	3.69	3.55	3.41	3.26	3.18	3.10	3.02	2.93	2.84	2.75
17	8.40	6.11	5.18	4.67	4.34	4.10	3.93	3.79	3.68	3.59	3.46	3.31	3.16	3.08	3.00	2.92	2.83	2.75	2.65
18	8.29	6.01	5.09	4.58	4.25	4.01	3.84	3.71	3.60	3.51	3.37	3.23	3.08	3.00	2.92	2.84	2.75	2.66	2.57
19	8.18	5.93	5.01	4.50	4.17	3.94	3.77	3.63	3.52	3.43	3.30	3.15	3.00	2.92	2.84	2.76	2.67	2.58	2.49
20	8.10	5.85	4.94	4.43	4.10	3.87	3.70	3.56	3.46	3.37	3.23	3.09	2.94	2.86	2.78	2.69	2.61	2.52	2.42
21	8.02	5.78	4.87	4.37	4.04	3.81	3.64	3.51	3.40	3.31	3.17	3.03	2.88	2.80	2.72	2.64	2.55	2.46	2.36
22	7.95	5.72	4.82	4.31	3.99	3.76	3.59	3.45	3.35	3.26	3.12	2.98	2.83	2.75	2.67	2.58	2.50	2.40	2.31
23	7.88	5.66	4.76	4.26	3.94	3.71	3.54	3.41	3.30	3.21	3.07	2.93	2.78	2.70	2.62	2.54	2.45	2.35	2.26
24	7.82	5.61	4.72	4.22	3.90	3.67	3.50	3.36	3.26	3.17	3.03	2.89	2.74	2.66	2.58	2.49	2.40	2.31	2.21
25	7.77	5.57	4.68	4.18	3.85	3.63	3.46	3.32	3.22	3.13	2.99	2.85	2.70	2.62	2.54	2.45	2.36	2.27	2.17
26	7.72	5.53	4.64	4.14	3.82	3.59	3.42	3.29	3.18	3.09	2.96	2.81	2.66	2.58	2.50	2.42	2.33	2.23	2.13
27	7.68	5.49	4.60	4.11	3.78	3.56	3.39	3.26	3.15	3.06	2.93	2.78	2.63	2.55	2.47	2.38	2.29	2.20	2.10
28	7.64	5.45	4.57	4.07	3.75	3.53	3.36	3.23	3.12	3.03	2.90	2.75	2.60	2.52	2.44	2.35	2.26	2.17	2.06
29	7.60	5.42	4.54	4.04	3.73	3.50	3.33	3.20	3.09	3.00	2.87	2.73	2.57	2.49	2.41	2.33	2.23	2.14	2.03
30	7.56	5.39	4.51	4.02	3.70	3.47	3.30	3.17	3.07	2.98	2.84	2.70	2.55	2.47	2.39	2.30	2.21	2.11	2.01
40	7.31	5.18	4.31	3.83	3.51	3.29	3.12	2.99	2.89	2.80	2.66	2.52	2.37	2.29	2.20	2.11	2.02	1.92	1.80
60	7.08	4.98	4.13	3.65	3.34	3.12	2.95	2.82	2.72	2.63	2.50	2.35	2.20	2.12	2.03	1.94	1.84	1.73	1.60
120	6.85	4.79	3.95	3.48	3.17	2.96	2.79	2.66	2.56	2.47	2.34	2.19	2.03	1.95	1.86	1.76	1.66	1.53	1.38
∞	6.63	4.61	3.78	3.32	3.02	2.80	2.64	2.51	2.41	2.32	2.18	2.04	1.88	1.79	1.70	1.59	1.47	1.32	1.00

Table of the F-Distribution: upper 0.1% points

$\nu_2 \backslash \nu_1$	1	2	3	4	5	6	7	8	9	10	12	15	20	24	30	40	60	120	∞
1	4053*	5000*	5404*	5625*	5764*	5859*	5929*	5981*	6023*	6056*	6107*	6158*	6209*	6235*	6261*	6287*	6313*	6340*	6366*
2	998.5	999.0	999.2	999.2	999.3	999.3	999.4	999.4	999.4	999.4	999.4	999.4	999.4	999.5	999.5	999.5	999.5	999.5	999.5
3	167.0	148.5	141.1	137.1	134.6	132.8	131.6	130.6	129.9	129.2	128.3	127.4	126.4	125.9	125.4	125.0	124.5	124.0	123.5
4	74.14	61.25	56.18	53.44	51.71	50.53	49.66	49.00	48.47	48.05	47.41	46.76	46.10	45.77	45.43	45.09	44.75	44.40	44.05
5	47.18	37.12	33.20	31.09	29.75	28.84	28.16	27.64	27.24	26.92	26.42	25.91	25.39	25.14	24.87	24.60	24.33	24.06	23.79
6	35.51	27.00	23.70	21.92	20.81	20.03	19.46	19.03	18.69	18.41	17.99	17.56	17.12	16.89	16.67	16.44	16.21	15.99	15.75
7	29.25	21.69	18.77	17.19	16.21	15.52	15.02	14.63	14.33	14.08	13.71	13.32	12.93	12.73	12.53	12.33	12.12	11.91	11.70
8	25.42	18.49	15.83	14.39	13.49	12.86	12.40	12.04	11.77	11.54	11.19	10.84	10.48	10.30	10.11	9.92	9.73	9.53	9.33
9	22.86	16.39	13.90	12.56	11.71	11.13	10.70	10.37	10.11	9.89	9.57	9.24	8.90	8.72	8.55	8.37	8.19	8.00	7.81
10	21.04	14.91	12.55	11.28	10.48	9.92	9.52	9.20	8.96	8.75	8.45	8.13	7.80	7.64	7.47	7.30	7.12	6.94	6.76
11	19.69	13.81	11.56	10.35	9.58	9.05	8.66	8.35	8.12	7.92	7.63	7.32	7.01	6.85	6.68	6.52	6.35	6.17	6.00
12	18.64	12.97	10.80	9.63	8.89	8.38	8.00	7.71	7.48	7.29	7.00	6.71	6.40	6.25	6.09	5.93	5.76	5.59	5.42
13	17.81	12.31	10.21	9.07	8.35	7.86	7.49	7.21	6.98	6.80	6.52	6.23	5.93	5.78	5.63	5.47	5.30	5.14	4.97
14	17.14	11.78	9.73	8.62	7.92	7.43	7.08	6.80	6.58	6.40	6.13	5.85	5.56	5.41	5.25	5.10	4.94	4.77	4.60
15	16.59	11.34	9.34	8.25	7.57	7.09	6.74	6.47	6.26	6.08	5.81	5.54	5.25	5.10	4.95	4.80	4.64	4.47	4.31
16	16.12	10.97	9.00	7.94	7.27	6.81	6.46	6.19	5.98	5.81	5.55	5.27	4.99	4.85	4.70	4.54	4.39	4.23	4.06
17	15.72	10.66	8.73	7.68	7.02	6.56	6.22	5.96	5.75	5.58	5.32	5.05	4.78	4.63	4.48	4.33	4.18	4.02	3.85
18	15.38	10.39	8.49	7.46	6.81	6.35	6.02	5.76	5.56	5.39	5.13	4.87	4.59	4.45	4.30	4.15	4.00	3.84	3.67
19	15.08	10.16	8.28	7.26	6.62	6.18	5.85	5.59	5.39	5.22	4.97	4.70	4.43	4.29	4.14	3.99	3.84	3.68	3.51
20	14.82	9.95	8.10	7.10	6.46	6.02	5.69	5.44	5.24	5.08	4.82	4.56	4.29	4.15	4.00	3.86	3.70	3.54	3.38
21	14.59	9.77	7.94	6.95	6.32	5.88	5.56	5.31	5.11	4.95	4.70	4.44	4.17	4.03	3.88	3.74	3.58	3.42	3.26
22	14.38	9.61	7.80	6.81	6.19	5.76	5.44	5.19	4.99	4.83	4.58	4.33	4.06	3.92	3.78	3.63	3.48	3.32	3.15
23	14.19	9.47	7.67	6.69	6.08	5.65	5.33	5.09	4.89	4.73	4.48	4.23	3.96	3.82	3.68	3.53	3.38	3.22	3.05
24	14.03	9.34	7.55	6.59	5.98	5.55	5.23	4.99	4.80	4.64	4.39	4.14	3.87	3.74	3.59	3.45	3.29	3.14	2.97
25	13.88	9.22	7.45	6.49	5.88	5.46	5.15	4.91	4.71	4.56	4.31	4.06	3.79	3.66	3.52	3.37	3.22	3.06	2.89
26	13.74	9.12	7.36	6.41	5.80	5.38	5.07	4.83	4.64	4.48	4.24	3.99	3.72	3.59	3.44	3.30	3.15	2.99	2.82
27	13.61	9.02	7.27	6.33	5.73	5.31	5.00	4.76	4.57	4.41	4.17	3.92	3.66	3.52	3.38	3.23	3.08	2.92	2.75
28	13.50	8.93	7.19	6.25	5.66	5.24	4.93	4.69	4.50	4.35	4.11	3.86	3.60	3.46	3.32	3.18	3.02	2.86	2.69
29	13.39	8.85	7.12	6.19	5.59	5.18	4.87	4.64	4.45	4.29	4.05	3.80	3.54	3.41	3.27	3.12	2.97	2.81	2.64
30	13.29	8.77	7.05	6.12	5.53	5.12	4.82	4.58	4.39	4.24	4.00	3.75	3.49	3.36	3.22	3.07	2.92	2.76	2.59
40	12.61	8.25	6.60	5.70	5.13	4.73	4.44	4.21	4.02	3.87	3.64	3.40	3.15	3.01	2.87	2.73	2.57	2.41	2.23
60	11.97	7.76	6.17	5.31	4.76	4.37	4.09	3.87	3.69	3.54	3.31	3.08	2.83	2.69	2.55	2.41	2.25	2.08	1.89
120	11.38	7.32	5.79	4.95	4.42	4.04	3.77	3.55	3.38	3.24	3.02	2.78	2.53	2.40	2.26	2.11	1.95	1.76	1.54
∞	10.83	6.91	5.42	4.62	4.10	3.74	3.47	3.27	3.10	2.96	2.74	2.51	2.27	2.13	1.99	1.84	1.66	1.45	1.00

* Multiply these entries by 100.

Reproduced with permission of the Biometrika Trustees from Merrington and Thompson (1943).

APPENDIX G.5

Table of the Lawley–Hotelling Trace Statistic

m_E \ m_H	2	3	4	5	6	8	d = 2 10	12	15	20	25	40	60
5% 2	9.8591†	10.659†	11.098†	11.373†	11.562†	11.804†	11.952†	12.052†	12.153†	12.254†	12.316†	12.409†	12.461†
3	58.428	58.915	59.161	59.308	59.407	59.531	59.606	59.655	59.705	59.755	59.785	59.830	59.855
4	23.999	23.312	22.918	22.663	22.484	22.250	22.104	22.003	21.901	21.797	21.733	21.636	21.582
5	15.639	14.864	14.422	14.135	13.934	13.670	13.504	13.391	13.275	13.156	13.083	12.972	12.909
6	12.175	11.411	10.975	10.691	10.491	10.228	10.063	9.9489	9.8320	9.7118	9.6381	9.5251	9.4610
7	10.334	9.5937	9.1694	8.8927	8.6975	8.4396	8.2765	8.1639	8.0480	7.9285	7.8549	7.7417	7.6773
8	9.2069	8.4881	8.0752	7.8054	7.6145	7.3614	7.2008	7.0896	6.9748	6.8560	6.7826	6.6694	6.6048
10	7.9095	7.2243	6.8294	6.5702	6.3860	6.1405	5.9837	5.8745	5.7612	5.6433	5.5701	5.4564	5.3910
12	7.1902	6.5284	6.1461	5.8942	5.7147	5.4744	5.3200	5.2122	5.0997	4.9820	4.9085	4.7938	4.7274
14	6.7350	6.0902	5.7168	5.4703	5.2941	5.0574	4.9048	4.7977	4.6856	4.5678	4.4939	4.3780	4.3105
16	6.4217	5.7895	5.4230	5.1804	5.0067	4.7727	4.6213	4.5147	4.4028	4.2846	4.2102	4.0930	4.0243
18	6.1932	5.5708	5.2095	4.9700	4.7982	4.5663	4.4157	4.3094	4.1976	4.0791	4.0042	3.8855	3.8158
20	6.0192	5.4046	5.0475	4.8105	4.6402	4.4099	4.2600	4.1539	4.0420	3.9231	3.8477	3.7278	3.6569
25	5.7244	5.1237	4.7741	4.5415	4.3740	4.1465	3.9977	3.8919	3.7798	3.6598	3.5832	3.4605	3.3868
30	5.5401	4.9487	4.6040	4.3743	4.2086	3.9829	3.8347	3.7291	3.6166	3.4957	3.4181	3.2926	3.2168
35	5.4140	4.8291	4.4880	4.2604	4.0959	3.8715	3.7237	3.6181	3.5054	3.3836	3.3051	3.1774	3.1000
40	5.3224	4.7424	4.4039	4.1778	4.0143	3.7908	3.6433	3.5377	3.4247	3.3022	3.2230	3.0933	3.0140
50	5.1981	4.6249	4.2900	4.0661	3.9039	3.6817	3.5346	3.4289	3.3154	3.1919	3.1115	2.9787	2.8965
60	5.1178	4.5490	4.2166	3.9941	3.8328	3.6114	3.4646	3.3588	3.2450	3.1206	3.0392	2.9041	2.8196
70	5.0616	4.4960	4.1653	3.9439	3.7831	3.5624	3.4157	3.3099	3.1957	3.0706	2.9886	2.8516	2.7652
80	5.0200	4.4569	4.1275	3.9068	3.7465	3.5262	3.3796	3.2737	3.1594	3.0338	2.9512	2.8126	2.7247
100	4.9628	4.4030	4.0754	3.8557	3.6961	3.4764	3.3300	3.2240	3.1093	2.9829	2.8994	2.7586	2.6683
200	4.8514	4.2982	3.9742	3.7567	3.5983	3.3798	3.2336	3.1275	3.0120	2.8838	2.7984	2.6520	2.5559
∞	4.7442	4.1973	3.8769	3.6614	3.5044	3.2870	3.1410	3.0346	2.9182	2.7879	2.7002	2.5470	2.4428
1% 2	2.4673*	2.6666*	2.7758*	2.8444*	2.8914*	2.9517*	2.9886*	3.0135*	3.0387*	3.0641*	3.0796*	3.1025*	3.1154*
3	2.9849†	2.9898†	2.9923†	2.9938†	2.9948†	2.9961†	2.9968†	2.9973†	2.9978†	2.9983†	2.9985†	2.9991†	2.9993†
4	74.275	71.026	69.244	68.116	67.337	66.332	65.712	65.290	64.862	64.427	64.163	63.763	63.538
5	38.295	35.567	34.070	33.121	32.465	31.615	31.088	30.729	30.364	29.993	29.766	29.422	29.228
6	26.118	23.794	22.517	21.706	21.143	20.413	19.958	19.648	19.332	19.009	18.812	18.511	18.341
7	20.388	18.326	17.191	16.469	15.967	15.313	14.905	14.626	14.341	14.049	13.870	13.596	13.442
8	17.152	15.268	14.229	13.567	13.106	12.504	12.127	11.868	11.603	11.331	11.165	10.909	10.764
10	13.701	12.038	11.120	10.531	10.121	9.5819	9.2431	9.0096	8.7694	8.5215	8.3688	8.1334	7.9990
12	11.920	10.388	9.5405	8.9961	8.6148	8.1132	7.7962	7.5770	7.3505	7.1159	6.9707	6.7457	6.6166
14	10.844	9.3990	8.5974	8.0816	7.7196	7.2419	6.9389	6.7287	6.5109	6.2843	6.1435	5.9244	5.7980
16	10.128	8.7432	7.9743	7.4786	7.1301	6.6691	6.3758	6.1718	5.9598	5.7385	5.6002	5.3847	5.2596
18	9.6174	8.2781	7.5334	7.0527	6.7142	6.2655	5.9793	5.7797	5.5718	5.3540	5.2178	5.0040	4.8794
20	9.2360	7.9316	7.2056	6.7365	6.4057	5.9664	5.6855	5.4893	5.2844	5.0692	4.9342	4.7214	4.5970
25	8.6044	7.3601	6.6663	6.2169	5.8993	5.4760	5.2042	5.0134	4.8135	4.6021	4.4686	4.2577	4.1314
30	8.2188	7.0127	6.3394	5.9026	5.5933	5.1801	4.9138	4.7264	4.5293	4.3199	4.1870	3.9745	3.8479
35	7.9592	6.7796	6.1205	5.6923	5.3888	4.9825	4.7200	4.5348	4.3395	4.1312	3.9985	3.7850	3.6581
40	7.7727	6.6125	5.9638	5.5420	5.2426	4.8413	4.5816	4.3980	4.2038	3.9962	3.8635	3.6488	3.5191
50	7.5228	6.3891	5.7545	5.3414	5.0477	4.6533	4.3972	4.2156	4.0230	3.8161	3.6830	3.4661	3.3335
60	7.3630	6.2465	5.6212	5.2138	4.9238	4.5338	4.2800	4.0998	3.9081	3.7014	3.5676	3.3490	3.2139
70	7.2520	6.1478	5.5290	5.1255	4.8382	4.4512	4.1991	4.0197	3.8285	3.6219	3.4880	3.2675	3.1303
80	7.1705	6.0753	5.4613	5.0608	4.7754	4.3906	4.1398	3.9610	3.7703	3.5636	3.4294	3.2074	3.0685
100	7.0588	5.9761	5.3688	4.9723	4.6896	4.3081	4.0587	3.8808	3.6906	3.4839	3.3491	3.1247	2.9831
200	6.8435	5.7852	5.1910	4.8025	4.5251	4.1497	3.9034	3.7271	3.5377	3.3304	3.1941	2.9640	2.8152
∞	6.6385	5.6040	5.0226	4.6419	4.3695	4.0000	3.7566	3.5817	3.3928	3.1845	3.0462	2.8062	2.6492

† Multiply entry by 100.   * Multiply entry by 10^4.

d = 3

5%

m_E \ m_H	3	4	5	6	8	10	12	16	20	25	40	60
3	25·930*	26·996*	27·665*	28·125*	28·712*	29·073*	29·316*	29·561*	29·809*	29·959*	30·19*	30·31*
4	1·1880*	1·1929*	1·1959*	1·1978*	1·2005*	1·2018*	1·2028*	1·2038*	1·2048*	1·2054*	1·2063*	1·2065*
5	42·474	41·764	·305	40·983	40·563	40·300	40·120	39·937	39·750	39·635	39·463	39·366
6	25·456	24·715	24·235	23·899	23·68	23·183	22·992	22·799	22·600	22·479	22·294	22·190
7	18·762	18·056	17·605	17·288	16·870	16·608	16·437	16·241	16·051	15·934	15·755	15·453
8	15·308	14·657	14·233	13·934	13·540	13·290	13·118	12·941	12·758	12·646	12·473	12·375
10	11·898	11·306	10·921	10·649	10·287	10·057	9·8974	9·7320	9·5603	9·4541	9·2897	9·1955
12	10·229	9·6825	9·3234	9·0680	8·7271	8·5088	8·3566	8·1983	8·0330	7·9301	7·7700	7·6777
14	9·2550	8·7356	8·3935	8·1495	7·8225	7·6122	7·4649	7·3110	7·1497	7·0488	6·8908	6·7991
16	8·6180	8·1183	7·7884	7·5528	7·2355	7·0307	6·8868	6·7360	6·5772	6·4774	6·3204	6·2287
18	8·1701	7·6851	7·3644	7·1347	6·8251	6·6244	6·4830	6·3343	6·1771	6·0780	5·9212	5·8292
20	7·8384	7·3649	7·0513	6·8263	6·5224	6·3249	6·1853	6·0383	5·8822	5·7834	5·6266	5·5341
25	7·2943	6·8407	6·5394	6·3227	6·0287	5·8365	5·7001	5·5555	5·4010	5·3025	5·1446	5·0503
30	6·9054	6·5345	6·2311	6·0196	5·7319	5·5431	5·4085	5·2654	5·1116	5·0129	4·8535	4·7576
35	6·7453	6·3132	6·0253	5·8175	5·5341	5·3476	5·2143	5·0720	4·9185	4·8175	4·6586	4·5608
40	6·5877	6·1621	5·8783	5·6732	5·3929	5·2081	5·0757	4·9340	4·7806	4·6813	4·5189	4·4185
50	6·3773	5·9606	5·6823	5·4809	5·2050	5·0234	4·8911	4·7502	4·5967	4·4968	4·3319	4·2297
60	6·2433	5·8334	5·5677	5·3587	5·0856	4·9044	4·7739	4·6334	4·4798	4·3793	4·2123	4·1078
70	6·1604	5·7435	5·4715	5·2743	5·0031	4·8229	4·6929	4·5526	4·3988	4·2979	4·1292	4·0227
80	6·0823	5·6786	5·4094	5·2123	4·9426	4·7632	4·6336	4·4935	4·3395	4·2381	4·0680	3·9600
100	5·9981	5·6896	5·3230	5·1276	4·8601	4·6817	4·5525	4·4126	4·2583	4·1563	3·9840	3·8734
200	5·8099	5·5186	5·1563	4·9653	4·7017	4·5255	4·3970	4·2574	4·1023	3·9988	3·8212	3·7042
∞	5·6397	5·2565	4·9992	4·8116	4·5519	4·3773	4·2499	4·1104	3·9541	3·8487	3·6642	3·5384

1%

m_E \ m_H	3	4	5	6	8	10	12	16	20	25	40	60
3	6·4845†	6·7500†	6·9169†	7·0313†	7·1778†	7·2676†	7·3281†	7·3911†	7·4511†	7·4883†		—
4	5·9486*	5·9946*	5·9976*	5·9996*	6·0021*	6·0035*	6·0046*	6·0056*	6·0067*	6·0071*	6·008*	6·008*
5	1·2738*	1·2420*	1·3219*	1·2090*	1·1901*	1·1790*	1·1715*	1·1638*	1·1561*	1·1514*	1·144*	1·141*
6	59·507	57·032	55·462	54·277	52·973	52·102	51·509	50·906	50·292	49·918	49·349	49·04
7	37·994	35·993	34·721	33·840	32·695	31·984	31·498	31·002	30·496	30·188	29·718	29·452
8	29·737	26·699	25·611	24·755	23·771	23·157	22·737	22·308	21·868	21·599	21·188	20·955
10	19·737	17·471	16·455	13·448?	12·737	12·288	11·978	11·659	11·328	11·124	10·809	10·658
12	15·973	14·765	13·990	12·737	12·288	11·978	11·659					
14	13·905	12·803	12·096	11·599	10·945	10·530	10·243	9·9462	9·6377	9·4463	9·1490	8·9780
16	12·610	11·681	10·918	10·453	9·8359	9·4444	9·1724	8·8900	8·5955	8·4121	8·1260	7·9605
18	11·729	10·761	10·120	9·6756	9·0670	8·7117	8·4503	8·1782	7·8924	7·7154	7·4365	7·2743
20	11·091	10·162	9·5452	9·1173	8·5492	8·1861	7·9325	7·6679	7·3901	7·2159	6·9419	6·7818
25	10·075	9·2005	8·6339	8·2333	7·6992	7·3560	7·1162	6·8627	6·6868	6·4273	6·1598	6·0019
30	9·4785	8·6441	8·1022	7·7183	7·2050	6·8739	6·6407	6·3953	6·1346	5·9690	5·7002	5·6464
35	9·0874	8·2798	7·7548	7·3822	6·8819	6·5598	6·3317	6·0909	5·8339	5·6700	5·4013	5·2478
40	8·8113	8·0233	7·5106	7·1460	6·6564	6·3392	6·1147	5·8771	5·6227	5·4598	5·1942	5·0367
50	8·4479	7·6861	7·1894	6·8358	6·3599	6·0503	5·8305	5·5970	5·3457	5·1838	4·9146	4·7578
60	8·2195	7·4745	6·9882	6·6416	6·1744	5·8696	5·6528	5·4228	5·1722	5·0108	4·7455	4·5816
70	8·0627	7·3395	6·8504	6·5087	6·0474	5·7460	5·5312	5·3019	5·0535	4·8922	4·6258	4·4598
80	7·9485	7·2239	6·7602	6·4120	5·9651	5·6628	5·4427	5·2147	4·9670	4·8058	4·5383	4·3706
100	7·7932	7·0805	6·6141	6·2909	5·8300	5·5344	5·3220	5·0945	4·8497	4·6883	4·4190	4·2484
200	7·4980	6·8083	6·3661	6·0333	5·5930	5·3037	5·0961	4·8725	4·6270	4·4650	4·1904	4·0124
∞	7·2220	6·5542	6·1164	5·8009	5·3725	5·0982	4·8849	4·6638	4·4190	4·2557	3·9738	3·7643

* Multiply entry by 100. † Multiply entry by 10^4.

486

5%

$m_E^{m_H}$	4	5	6	8	10	12	15	20	25	40	60
4	49·364*	51·204*	52·054*	53·142*	53·808*	54·258*	54·71*	55·17*	55·46*	—	—
5	1·9964*	2·0013*	2·0046*	2·0087*	2·0112*	2·0128*	2·0145*	2·0161*	2·0171*	2·019*	—
6	65·715	64·999	64·497	63·841	63·432	63·151	62·866	62·573	62·396	62·13	—
7	37·343	36·629	36·129	35·474	35·064	34·782	34·495	34·200	34·019	33·75	—
8	26·516	25·868	25·413	24·814	24·437	24·178	23·912	23·639	23·471	23·214	23·072
10	17·875	17·326	16·938	16·424	16·098	15·872	15·640	15·399	15·250	15·021	14·891
12	14·338	13·848	13·500	13·037	12·741	12·535	12·321	12·099	11·961	11·747	11·624
14	12·455	12·002	11·680	11·248	10·972	10·778	10·577	10·366	10·234	10·029	9·9103
16	11·295	10·868	10·563	10·164	9·8904	9·7054	9·5119	9·3085	9·1810	8·9808	8·8444
18	10·512	10·104	9·8121	9·4190	9·1647	8·9857	8·7978	8·5996	8·4748	8·2778	8·1626
20	9·9500	9·5560	9·2736	8·8926	8·6453	8·4708	8·2571	8·0926	7·9696	7·7748	7·6401
25	9·0585	8·6884	8·4223	8·0616	7·8261	7·6590	7·4821	7·2933	7·1730	6·9805	6·8669
30	8·5377	8·1825	7·9265	7·5784	7·3502	7·1876	7·0147	6·8291	6·7101	6·5181	6·4026
35	8·1968	7·8517	7·6026	7·2631	7·0397	6·8801	6·7099	6·5262	6·4079	6·2156	6·0989
40	7·9566	7·6188	7·3746	7·0413	6·8214	6·6640	6·4955	6·3131	6·1952	6·0023	5·8844
50	7·6404	7·3125	7·0751	6·7501	6·5350	6·3804	6·2143	6·0334	5·9157	5·7214	5·6011
60	7·4417	7·1202	6·8872	6·5676	6·3555	6·2027	6·0381	5·8581	5·7403	5·5446	5·4222
70	7·3054	6·9884	6·7584	6·4428	6·2325	6·0809	5·9173	5·7378	5·6200	5·4230	5·2987
80	7·2061	6·8924	6·6646	6·3515	6·1430	5·9924	5·8294	5·6503	5·5323	5·3343	5·2084
100	7·0711	6·7619	6·5372	6·2279	6·0215	5·8721	5·7101	5·5313	5·4131	5·2133	5·0849
200	6·8143	6·5139	6·2952	5·9933	5·7910	5·6439	5·4836	5·3053	5·1863	4·9819	4·8471
∞	6·5741	6·2821	6·0692	5·7743	5·5758	5·4309	5·2721	5·0940	4·9737	4·7629	4·6190

1%

$m_E^{m_H}$	4	5	6	8	10	12	15	20	25	40	60
4	12·491†	12·800†	13·012†	13·283†	13·449†	13·561†	13·67†	13·79†	13·87†	—	—
5	9·9992*	10·004*	10·008*	10·012*	10·014*	10·016*	10·018*	10·02*	10·02*	—	—
6	1·9377*	1·9064*	1·8848*	1·8570*	1·8398*	1·8281*	1·8162*	1·8041*	1·7969*	—	—
7	85·053	82·731	81·125	79·047	77·759	76·882	75·989	75·082	74·622	—	—
8	51·991	50·178	48·921	47·290	46·276	45·583	44·877	44·156	43·715	43·04	22·95
10	29·789	28·478	27·566	26·376	25·632	25·121	24·597	24·060	23·731	23·224	16·281
12	21·965	20·889	20·138	19·164	18·634	18·108	17·668	17·215	16·936	16·505	
14	18·142	17·199	16·639	15·670	15·121	14·742	14·349	13·943	13·691	13·301	13·077
16	15·916	15·069	14·457	13·662	13·157	12·807	12·444	12·066	11·831	11·466	11·255
18	14·473	13·674	13·112	12·358	11·894	11·564	11·221	10·863	10·639	10·289	10·086
20	13·466	12·710	12·177	11·470	11·018	10·703	10·374	10·030	9·8138	9·4748	9·2771
25	11·924	11·237	10·761	10·103	9·6871	9·3951	9·0890	8·7658	8·5618	8·2383	8·0476
30	11·055	10·409	9·9509	9·3382	8·9430	8·6646	8·3715	8·0602	7·8626	7·5468	7·3586
35	10·499	9·8801	9·4405	8·8611	8·4695	8·2000	7·9163	7·6115	7·4177	7·1059	6·9186
40	10·114	9·5138	9·0872	8·5142	8·1424	7·8791	7·6002	7·3015	7·1102	6·8006	6·6132
50	9·6141	9·0400	8·6308	8·0795	7·7204	7·4652	7·1938	6·9015	6·7131	6·4054	6·2168
60	9·3053	8·7472	8·3490	7·8114	7·4603	7·2101	6·9434	6·6548	6·4679	6·1606	5·9704
70	9·0954	8·5485	8·1578	7·6297	7·2840	7·0373	6·7736	6·4875	6·3015	5·9940	5·8022
80	8·9437	8·4048	8·0197	7·4984	7·1567	6·9124	6·6510	6·3666	6·1812	5·8733	5·6799
100	8·7388	8·2111	7·8334	7·3214	6·9851	6·7443	6·4858	6·2036	6·0189	5·7100	5·5139
200	8·3542	7·8476	7·4842	6·9900	6·6639	6·4293	6·1763	5·8979	5·7138	5·4012	5·1976
∞	8·0000	7·5132	7·1633	6·6857	6·3691	6·1402	5·8920	5·6164	5·4323	5·1133	4·8981

* Multiply entry by 100. † Multiply entry by 10⁴.

$d = 5$

$m_E \backslash m_H$	5	6	8	10	12	15	20	25	40	60
5%										
5	81.991 +	83.352 +	85.093 +	86.160 +	86.88 +	—	—	—	—	—
6	3.0093+	3.042+	3.0204+	3.0241+	3.0266+	3.0291+	3.032 +	—	—	—
7	93.762	93.042	92.102	91.515	91.113	90.705	90.29	90.04	—	—
8	51.339	50.646	49.739	49.170	48.780	48.382	47.973	47.723	47.35	—
10	27.667	27.115	26.387	25.927	25.610	25.284	24.947	24.740	24.422	—
12	20.169	19.701	19.079	18.683	18.409	18.124	17.830	17.647	17.365	17.20
14	16.643	16.224	15.666	15.309	15.059	14.800	14.530	14.361	14.100	13.95
16	14.624	14.239	13.722	13.389	13.157	12.914	12.659	12.499	12.250	12.105
18	13.326	12.963	12.476	12.161	11.939	11.708	11.463	11.310	11.068	10.928
20	12.424	12.078	11.612	11.310	11.097	10.874	10.637	10.488	10.252	10.113
25	11.046	10.728	10.297	10.016	9.8168	9.6061	9.3814	9.2386	9.0102	8.8745
30	10.270	9.9689	9.5592	9.2907	9.0995	8.8964	8.6785	8.5389	8.3141	8.1790
35	9.7739	9.4836	9.0879	8.8277	8.6419	8.4437	8.2301	8.0926	7.8693	7.7339
40	9.4292	9.1469	8.7613	8.5070	8.3250	8.1303	7.9195	7.7833	7.5607	7.4247
50	8.9825	8.7107	8.3385	8.0921	7.9150	7.7248	7.5177	7.3829	7.1605	7.0229
60	8.7057	8.4406	8.0769	7.8355	7.6615	7.4741	7.2692	7.1351	6.9124	6.7730
70	8.5174	8.2570	7.8991	7.6612	7.4894	7.3039	7.1004	6.9667	6.7434	6.6024
80	8.3811	8.1241	7.7705	7.5351	7.3648	7.1807	6.9782	6.8448	6.6208	6.4785
100	8.1969	7.9446	7.5969	7.3649	7.1968	7.0145	6.8133	6.6801	6.4550	6.3103
200	7.8505	7.6070	7.2706	7.0451	6.8811	6.7023	6.5032	6.3702	6.1416	5.9908
∞	7.5305	7.2955	6.9698	6.7505	6.5902	6.4144	6.2171	6.0838	5.8499	5.6699
1%										
5	20.495 *	20.834 *	21.267 *	21.53 *	—	—	—	—	—	—
6	15.014 +	15.019 +	15.025 +	15.029 +	15.033 *	15.03 +	15.06 +	—	—	—
7	2.7354+	2.7045+	2.6646+	2.6400+	2.6232+	2.6064+	2.590 +	2.579 +	—	—
8	1.1498+	1.1276+	1.0989+	1.0811+	1.0689+	1.0567+	1.0440+	1.0564+	—	—
10	48.048	46.670	44.877	43.758	42.992	42.210	41.408	40.921	—	—
12	31.108	30.065	28.701	27.846	27.257	26.653	26.031	25.648	25.06	24.71
14	24.016	23.145	22.001	21.279	20.781	20.268	19.736	19.408	18.90	18.61
16	20.240	19.472	18.459	17.817	17.373	16.913	16.435	16.138	15.678	15.412
18	17.929	17.228	16.302	15.713	15.304	14.878	14.435	14.159	13.727	13.478
20	16.380	15.727	14.862	14.310	13.925	13.525	13.105	12.843	12.431	12.192
25	14.107	13.529	12.759	12.265	11.918	11.555	11.172	10.930	10.547	10.322
30	12.880	12.345	11.629	11.167	10.842	10.500	10.136	9.9059	9.5378	9.3188
35	12.115	11.607	10.926	10.486	10.174	9.8453	9.4944	9.2706	8.9106	8.6946
40	11.593	11.105	10.448	10.022	9.7204	9.4006	9.0581	8.8387	8.4638	8.2691
50	10.928	10.465	9.8408	9.4336	9.1141	8.8361	8.5041	8.2901	7.9404	7.7261
60	10.523	10.076	9.4712	9.0758	8.7938	8.4930	8.1674	7.9563	7.6090	7.3940
70	10.251	9.8142	9.2229	8.8354	8.5586	8.2626	7.9411	7.7319	7.3858	7.1697
80	10.055	9.6261	9.0446	8.6629	8.3899	8.0973	7.7787	7.5708	7.2251	7.0078
100	9.7929	9.3742	8.8058	8.4319	8.1638	7.8758	7.5611	7.3547	7.0093	6.7897
200	9.3055	8.9065	8.3629	8.0036	7.7448	7.4652	7.1572	6.9532	6.6062	6.3798
∞	8.8628	8.4820	7.9613	7.6154	7.3650	7.0929	6.7903	6.5878	6.2361	5.9984

+ Multiply entry by 100. * Multiply entry by 10^4.

$d = 6$

m_g \ m_H	6	8	10	12	15	20	25	30	35
5% 10	45.722	44.677	44.019	43.567	43.103	42.626	42.334	42.136	41.993
12	28.959	28.121	27.590	27.223	26.843	26.451	26.209	26.044	25.925
14	22.321	21.600	21.141	20.821	20.489	20.144	19.929	19.783	19.677
16	18.858	18.210	17.795	17.505	17.202	16.886	16.688	16.553	16.455
18	16.755	16.157	15.772	15.501	15.218	14.921	14.735	14.607	14.513
20	15.351	14.788	14.424	14.168	13.899	13.615	13.436	13.313	13.223
25	13.293	12.786	12.456	12.222	11.975	11.711	11.544	11.428	11.343
30	12.180	11.705	11.395	11.173	10.939	10.687	10.526	10.414	10.331
35	11.484	11.031	10.733	10.520	10.293	10.049	9.8921	9.7820	9.7003
40	11.009	10.571	10.282	10.075	9.8535	9.6142	9.4596	9.3508	9.2699
50	10.402	9.9832	9.7060	9.5067	9.2927	9.0598	8.9082	8.8009	8.7207
60	10.031	9.6246	9.3547	9.1602	8.9507	8.7215	8.5717	8.4651	8.3851
70	9.7813	9.3830	9.1182	8.9269	8.7204	8.4938	8.3450	8.2388	8.1589
80	9.6014	9.2093	8.9480	8.7591	8.5548	8.3300	8.1819	8.0759	7.9959
100	9.3598	8.9760	8.7197	8.5340	8.3326	8.1102	7.9629	7.8572	7.7771
200	8.9099	8.5419	8.2950	8.1153	7.9193	7.7011	7.5552	7.4494	7.3685
500	8.6594	8.3002	8.0587	7.8823	7.6894	7.4734	7.3280	7.2219	7.1403
1000	8.5788	8.2226	7.9827	7.8075	7.6155	7.4002	7.2550	7.1487	7.0668
∞	8.4997	8.1463	7.9082	7.7340	7.5430	7.3284	7.1832	7.0768	6.9945
1% 10	86.397	83.565	81.804	80.602	79.376	78.124	77.360	76.845	76.474
12	46.027	44.103	42.899	42.073	41.227	40.359	39.826	39.466	39.206
14	32.433	30.918	29.966	29.309	28.634	27.936	27.507	27.215	27.004
16	25.977	24.689	23.875	23.311	22.729	22.126	21.753	21.498	21.314
18	22.292	21.146	20.418	19.913	19.389	18.844	18.505	18.273	18.105
20	19.935	18.886	18.217	17.752	17.267	16.761	16.445	16.229	16.071
25	16.642	15.737	15.156	14.749	14.324	13.875	13.592	13.397	13.254
30	14.944	14.118	13.586	13.211	12.816	12.398	12.133	11.949	11.814
35	13.913	13.138	12.635	12.281	11.906	11.506	11.252	11.074	10.943
40	13.223	12.482	12.000	11.659	11.298	10.911	10.663	10.490	10.361
50	12.358	11.661	11.206	10.882	10.538	10.167	9.9271	9.7587	9.6333
60	11.839	11.169	10.730	10.417	10.083	9.7206	9.4860	9.3202	9.1963
70	11.493	10.841	10.413	10.107	9.7795	9.4238	9.1922	9.0281	8.9050
80	11.246	10.607	10.187	9.8859	9.5634	9.2121	8.9826	8.8195	8.6968
100	10.917	10.295	9.8857	9.5918	9.2758	8.9301	8.7033	8.5414	8.4193
200	10.312	9.7231	9.3330	9.0517	8.7476	8.4121	8.1897	8.0295	7.9075
500	9.9799	9.4089	9.0296	8.7553	8.4576	8.1275	7.9072	7.7473	7.6249
1000	9.8738	9.3085	8.9328	8.6607	8.3651	8.0366	7.8168	7.6570	7.5344
∞	9.7699	9.2103	8.8379	8.5680	8.2744	7.9475	7.7283	7.5685	7.4456

$d = 7$

m_r \ m_H	8	10	12	15	20	25	30	35
5% 10	85.040	84.082	83.426	82.755	82.068	81.648	81.364	81.159
12	42.850	42.126	41.627	41.113	40.583	40.257	40.037	39.877
14	29.968	29.373	28.961	28.534	28.091	27.817	27.631	27.495
16	24.038	23.519	23.158	22.781	22.389	22.145	21.978	21.857
18	20.692	20.222	19.893	19.549	19.189	18.964	18.809	18.696
20	18.561	18.125	17.819	17.498	17.159	16.947	16.800	16.694
25	15.587	15.202	14.930	14.642	14.337	14.143	14.009	13.911
30	14.049	13.693	13.440	13.172	12.884	12.701	12.573	12.478
35	13.113	12.776	12.535	12.278	12.002	11.825	11.700	11.608
40	12.485	12.160	11.927	11.679	11.411	11.237	11.115	11.025
50	11.695	11.386	11.165	10.927	10.668	10.500	10.381	10.292
60	11.219	10.921	10.706	10.475	10.221	10.056	9.9383	9.8500
70	10.901	10.610	10.400	10.173	9.9233	9.7596	9.6429	9.5550
80	10.674	10.388	10.181	9.9567	9.7102	9.5478	9.4317	9.3440
100	10.371	10.091	9.8886	9.6688	9.4259	9.2652	9.1498	9.0622
200	9.8118	9.5448	9.3504	9.1384	8.9021	8.7441	8.6295	8.5419
500	9.5037	9.2438	9.0539	8.8462	8.6134	8.4568	8.3424	8.2543
1000	9.4051	9.1475	8.9591	8.7527	8.5211	8.3648	8.2504	8.1621
∞	9.3085	9.0531	8.8662	8.6612	8.4306	8.2747	8.1603	8.0718
1% 10	185.93	182.94	180.90	178.83	176.73	175.44	174.57	173.92
12	71.731	69.978	68.779	67.552	66.296	65.528	65.010	64.636
14	44.255	42.978	42.099	41.197	40.269	39.698	39.311	39.032
16	33.097	32.057	31.339	30.599	29.834	29.361	29.039	28.806
18	27.273	26.374	25.750	25.105	24.435	24.019	23.735	23.529
20	23.757	22.949	22.388	21.804	21.195	20.816	20.556	20.367
25	19.117	18.440	17.965	17.469	16.947	16.619	16.392	16.227
30	16.848	16.239	15.810	15.360	14.882	14.580	14.370	14.216
35	15.512	14.945	14.544	14.121	13.670	13.383	13.183	13.036
40	14.634	14.095	13.713	13.309	12.876	12.599	12.405	12.262
50	13.553	13.049	12.691	12.310	11.899	11.634	11.448	11.309
60	12.914	12.432	12.088	11.720	11.323	11.065	10.882	10.746
70	12.492	12.024	11.690	11.332	10.942	10.689	10.509	10.374
80	12.193	11.736	11.408	11.056	10.673	10.422	10.244	10.110
100	11.797	11.353	11.034	10.691	10.316	10.070	9.8935	9.7607
200	11.077	10.658	10.356	10.028	9.6670	9.4273	9.2545	9.1228
500	10.685	10.280	9.9866	9.6679	9.3140	9.0776	8.9060	8.7744
1000	10.561	10.160	9.8693	9.5533	9.2017	8.9663	8.7950	8.6634
∞	10.439	10.043	9.7547	9.4413	9.0920	8.8575	8.6865	8.5548

$d = 8$

m_E \ m_H		8	10	12	15	20	25	30	35
5%	14	42.516	41.737	41.198	40.641	40.066	39.711	39.470	39.296
	16	31.894	31.242	30.788	30.318	29.829	29.525	29.318	29.167
	18	26.421	25.847	25.446	25.028	24.591	24.319	24.132	23.996
	20	23.127	22.605	22.239	21.856	21.454	21.201	21.028	20.902
	25	18.770	18.324	18.009	17.677	17.325	17.102	16.947	16.834
	30	16.626	16.221	15.934	15.629	15.303	15.095	14.950	14.843
	35	15.356	14.977	14.707	14.418	14.109	13.910	13.771	13.668
	40	14.518	14.156	13.898	13.621	13.322	13.129	12.994	12.893
	50	13.482	13.142	12.898	12.636	12.351	12.165	12.034	11.936
	60	12.866	12.540	12.305	12.051	11.774	11.593	11.465	11.368
	70	12.459	12.142	11.912	11.665	11.393	11.215	11.088	10.992
	80	12.169	11.858	11.634	11.390	11.122	10.946	10.820	10.725
	100	11.785	11.483	11.264	11.026	10.763	10.590	10.465	10.370
	200	11.084	10.798	10.589	10.362	10.108	9.9389	9.8159	9.7218
	500	10.701	10.423	10.221	9.9993	9.7509	9.5836	9.4614	9.3673
	1000	10.579	10.304	10.104	9.8840	9.6371	9.4704	9.3484	9.2543
	∞	10.459	10.188	9.9892	9.7712	9.5258	9.3598	9.2379	9.1437
1%	14	65.793	64.035	62.828	61.592	60.323	59.545	59.019	58.639
	16	44.977	43.633	42.707	41.754	40.771	40.164	39.753	39.456
	18	35.265	34.146	33.373	32.573	31.745	31.232	30.882	30.629
	20	29.786	28.808	28.129	27.425	26.691	26.235	25.924	25.697
	25	23.001	22.212	21.661	21.085	20.480	20.100	19.838	19.647
	30	19.867	19.173	18.686	18.173	17.631	17.288	17.051	16.876
	35	18.077	17.440	16.991	16.516	16.011	15.690	15.466	15.301
	40	16.924	16.324	15.900	15.451	14.970	14.662	14.447	14.288
	50	15.528	14.975	14.582	14.163	13.711	13.420	13.216	13.063
	60	14.715	14.190	13.815	13.414	12.980	12.698	12.499	12.351
	70	14.184	13.677	13.313	12.925	12.502	12.226	12.031	11.885
	80	13.810	13.315	12.960	12.580	12.165	11.894	11.701	11.556
	100	13.317	12.839	12.496	12.127	11.722	11.457	11.267	11.124
	200	12.429	11.983	11.660	11.311	10.925	10.669	10.484	10.343
	500	11.951	11.521	11.210	10.871	10.495	10.244	10.061	9.9208
	1000	11.800	11.375	11.067	10.732	10.359	10.109	9.9270	9.7870
	∞	11.652	11.233	10.928	10.597	10.227	9.9778	9.7963	9.6564

$d=9$

m_g \ m_H	10	12	15	20	25	30	35
5% 14	61.915	61.196	60.456	59.694	59.224	58.907	58.68
16	42.157	41.583	40.988	40.372	39.990	39.730	39.542
18	33.171	32.680	32.169	31.637	31.305	31.079	30.914
20	28.140	27.702	27.245	26.766	26.466	26.260	26.110
25	21.911	21.548	21.165	20.759	20.503	20.326	20.196
30	19.020	18.695	18.350	17.982	17.747	17.584	17.464
35	17.361	17.059	16.737	16.391	16.170	16.015	15.900
40	16.287	16.000	15.694	15.363	15.150	15.000	14.889
50	14.982	14.714	14.427	14.115	13.912	13.768	13.661
60	14.218	13.962	13.687	13.385	13.188	13.049	12.944
70	13.717	13.469	13.201	12.907	12.714	12.577	12.473
80	13.364	13.121	12.859	12.570	12.380	12.243	12.141
100	12.898	12.663	12.407	12.125	11.938	11.804	11.703
200	12.055	11.833	11.591	11.321	11.141	11.010	10.910
500	11.600	11.385	11.150	10.887	10.710	10.580	10.480
1000	11.455	11.243	11.011	10.749	10.573	10.444	10.344
∞	11.315	11.105	10.874	10.615	10.440	10.311	10.211
1% 14	101.87	100.15	98.387	96.583	95.478	94.74	94.2
16	60.990	59.770	58.518	57.229	56.437	55.90	55.51
18	44.668	43.697	42.697	41.662	41.022	40.587	40.272
20	36.244	35.419	34.564	33.676	33.126	32.750	32.477
25	26.601	25.960	25.292	24.591	24.152	23.850	23.629
30	22.443	21.890	21.310	20.696	20.308	20.040	19.843
35	20.153	19.650	19.120	18.556	18.198	17.949	17.765
40	18.708	18.238	17.741	17.210	16.870	16.632	16.457
50	16.993	16.563	16.105	15.612	15.294	15.071	14.905
60	16.011	15.604	15.169	14.698	14.393	14.177	14.016
70	15.375	14.983	14.563	14.107	13.810	13.599	13.441
80	14.930	14.548	14.139	13.693	13.401	13.194	13.038
100	14.348	13.981	13.585	13.152	12.868	12.664	12.511
200	13.310	12.968	12.597	12.188	11.915	11.719	11.569
500	12.756	12.427	12.070	11.673	11.407	11.214	11.066
1000	12.581	12.257	11.904	11.511	11.247	11.054	10.907
∞	12.412	12.092	11.743	11.353	11.091	10.899	10.752

$d = 10$

m_E \ m_H	10	12	15	20	25	30	35
5% 14	98.999	98.013	97.002	95.963	95.326	94.9	94.6
16	58.554	57.814	57.050	56.260	55.772	55.44	55.20
18	43.061	42.454	41.824	41.169	40.762	40.485	40.284
20	35.146	34.620	34.071	33.497	33.140	32.895	32.716
25	26.080	25.660	25.219	24.753	24.458	24.255	24.107
30	22.140	21.773	21.384	20.970	20.706	20.523	20.388
35	19.955	19.618	19.260	18.876	18.630	18.458	18.331
40	18.569	18.252	17.914	17.550	17.316	17.151	17.029
50	16.913	16.622	16.309	15.969	15.748	15.592	15.476
60	15.960	15.684	15.385	15.059	14.847	14.695	14.582
70	15.341	15.074	14.786	14.469	14.261	14.113	14.002
80	14.907	14.647	14.365	14.055	13.851	13.705	13.595
100	14.338	14.087	13.814	13.513	13.313	13.170	13.061
200	13.319	13.085	12.828	12.542	12.351	12.212	12.106
500	12.774	12.548	12.301	12.023	11.836	11.699	11.594
1000	12.602	12.379	12.134	11.859	11.674	11.538	11.432
∞	12.434	12.214	11.972	11.700	11.515	11.380	11.275
1% 14	180.90	178.28	175.62	172.91	171.24	170	--
16	89.068	87.414	85.270	83.980	82.91	82.2	81.7
18	59.564	58.328	57.055	55.742	54.933	54.384	53.990
20	45.963	44.951	43.905	42.821	42.150	41.693	41.362
25	31.774	31.029	30.253	29.440	28.932	28.583	28.328
30	26.115	25.489	24.832	24.139	23.701	23.399	23.177
35	23.116	22.556	21.966	21.338	20.939	20.663	20.459
40	21.267	20.749	20.201	19.615	19.241	18.980	18.787
50	19.114	18.646	18.148	17.611	17.266	17.023	16.842
60	17.901	17.462	16.992	16.484	16.154	15.922	15.748
70	17.124	16.703	16.252	15.762	15.443	15.216	15.046
80	16.583	16.175	15.738	15.260	14.948	14.726	14.559
100	15.881	15.490	15.069	14.608	14.305	14.088	13.925
200	14.641	14.280	13.889	13.457	13.169	12.962	12.803
500	13.986	13.641	13.266	12.848	12.569	12.366	12.210
1000	13.780	13.441	13.070	12.658	12.381	12.179	12.023
∞	13.581	13.246	12.881	12.472	12.198	11.997	11.842

Reproduced, in part, with permission of John Wiley and Sons, Inc. from Seber (1984); in part, with permission of the Biometrika Trustees from Davis (1970a) and, in part, with permission of Marcel Dekker from Davis (1970b, 1980).

APPENDIX G.6

Tables of the Extreme Roots of a Covariance Matrix

	Smallest root, l_p				Largest root, l_1			
α $\backslash$ v	0·005	0·010	0·025	0·050	0·005	0·010	0·025	0·050
				$p = 2$				
2	0·0⁴1518	0·0⁴6287	0·0³3858	0·0²1500	13·66	12·16	10·15	8·594
3	·0²5012	·0²1005	·02532	·05129	16·16	14·57	12·42	10·74
4	·04047	·06477	·1216	·1980	18·40	16·73	14·46	12·68
5	·1264	·1812	·2948	·4314	20·48	18·73	16·36	14·49
6	·2659	·3573	·5340	·7333	22·45	20·64	18·17	16·21
7	·4550	·5858	·8278	1·090	24·33	22·47	19·91	17·88
8	·6880	·8595	1·167	1·489	26·15	24·23	21·59	19·49
9	·9597	1·172	1·544	1·926	27·92	25·95	23·24	21·06
10	1·265	1·518	1·953	2·392	29·65	27·63	24·84	22·60
15	3·184	3·629	4·358	5·059	37·83	35·59	32·48	29·96
20	5·558	6·177	7·166	8·094	45·51	43·08	39·69	36·94
25	8·233	9·009	10·23	11·37	52·86	50·27	46·63	43·67
30	11·13	12·05	13·49	14·80	59·99	57·24	53·39	50·24
40	17·38	18·56	20·38	22·03	73·76	70·75	66·50	63·02
50	24·07	25·48	27·65	29·60	87·08	83·84	79·24	75·46
60	31·07	32·70	35·18	37·39	100·1	96·72	91·72	87·66
70	38·31	40·14	42·91	45·37	112·9	109·2	104·0	99·70
80	45·75	47·76	50·79	53·48	125·4	121·6	116·1	111·6
90	53·34	55·52	58·81	61·71	137·8	133·8	128·1	123·4
100	61·06	63·40	66·93	70·04	150·1	145·9	140·0	135·0
				$p = 3$				
3	0·0⁵9820	0·0⁴3927	0·0³2454	0·0³9817	18·96	17·18	14·90	13·11
4	0·0³3342	0·0³6701	·01688	·03420	21·26	19·50	17·12	15·24
5	0·02844	·04550	·08538	·1390	23·45	21·66	19·18	17·22
6	·09224	·1322	·2149	·3142	25·55	23·69	21·13	19·09
7	·1997	·2682	·4004	·5492	27·56	25·64	22·99	20·88
8	·3495	·4497	·6346	·8339	29·49	27·52	24·80	22·62
9	·5383	·6719	·9106	1·160	31·37	29·34	26·55	24·31
10	·7625	·9300	1·223	1·522	33·19	31·12	28·26	25·96
15	2·301	2·638	3·191	3·724	41·79	39·52	36·36	33·80
20	4·338	4·836	5·623	6·364	49·82	47·37	43·95	41·18
25	6·710	7·350	8·356	9·285	57·49	54·89	51·24	48·27
30	9·326	10·10	11·31	12·41	64·90	62·15	58·30	55·15
40	15·08	16·10	17·66	19·07	79·18	76·18	71·96	68·50
50	21·33	22·57	24·45	26·14	92·95	89·73	85·18	81·44
60	27·93	29·37	31·55	33·48	106·4	102·9	98·09	94·09
70	34·81	36·43	38·88	41·04	119·5	115·9	110·8	106·5
80	41·90	43·69	46·39	48·77	132·4	128·6	123·2	118·8
90	49·16	51·12	54·06	56·64	145·2	141·2	135·6	130·9
100	56·57	58·69	61·85	64·63	157·7	153·6	147·8	143·0
				$p = 4$				
4	0·0⁵7074	0·0⁴2830	0·0³1769	0·0³7085	23·78	21·97	19·49	17·52
5	·0²2506	·0²5025	·01266	·02565	26·11	24·24	21·67	19·63
6	·02197	03514	·06595	·1073	28·31	26·39	23·74	21·62
7	·07289	·1045	·1698	·2481	30·41	28·43	25·71	23·53
8	·1607	·2158	·3220	·4414	32·43	30·41	27·61	25·37
9	·2854	·3671	·5177	·6798	34·39	32·32	29·45	27·15
10	·4451	·5552	·7519	·9574	36·29	34·18	31·25	28·90
15	1·675	1·935	2·365	2·781	45·25	42·94	39·73	37·13
20	3·435	3·839	4·488	5·096	53·58	51·10	47·64	44·84
25	5·555	6·095	6·946	7·730	61·51	58·88	55·21	52·22
30	7·939	8·607	9·645	10·59	69·16	66·40	62·53	59·37

v \ α	\multicolumn{4}{c}{Smallest root, l_p}	\multicolumn{4}{c}{Largest root, l_1}
	0·005	0·010	0·025	0·050	0·005	0·010	0·025	0·050

$p = 4$ (cont.)

v	0·005	0·010	0·025	0·050	0·005	0·010	0·025	0·050
40	13·28	14·18	15·56	16·79	83·86	80·86	76·64	73·18
50	19·16	20·27	21·95	23·45	98·00	94·79	90·25	86·53
60	25·43	26·73	28·69	30·43	111·8	108·3	103·5	99·55
70	31·99	33·47	35·69	37·65	125·2	121·6	116·5	112·3
80	38·80	40·44	42·90	45·06	138·4	134·7	129·3	124·9
90	45·79	47·59	50·27	52·63	151·4	147·5	142·0	137·4
100	52·94	54·89	57·79	60·33	164·3	160·2	154·4	149·7

$p = 5$

v	0·005	0·010	0·025	0·050	0·005	0·010	0·025	0·050
5	$0{\cdot}0^5 5521$	$0{\cdot}0^4 2209$	$0{\cdot}0^3 1381$	$0{\cdot}0^3 5527$	28·85	26·62	23·97	21·85
6	$\cdot 0^2 2005$	$\cdot 0^2 4020$	·01013	·02052	31·01	28·86	26·13	23·95
7	·01791	·02865	·05377	·08750	33·11	31·00	28·21	25·96
8	·06035	·08648	·1405	·2054	35·17	33·05	30·19	27·88
9	·1347	·1809	·2698	·3698	37·17	35·04	32·11	29·75
10	·2418	·3110	·4383	·5754	39·14	36·98	33·98	31·57
15	1·210	1·411	1·746	2·073	48·40	46·05	42·79	40·15
20	2·728	3·063	3·602	4·109	56·99	54·49	50·99	48·14
25	4·629	5·092	5·820	6·493	65·15	62·51	58·80	55·78
30	6·811	7·394	8·301	9·128	73·01	70·23	66·34	63·16
40	11·79	12·59	13·82	14·93	88·00	85·08	80·84	77·37
50	17·34	18·35	19·87	21·23	102·6	99·34	94·81	91·08
60	23·32	24·51	26·30	27·88	116·6	113·2	108·4	104·4
70	29·61	30·97	33·01	34·81	130·3	126·8	121·7	117·5
80	36·16	37·68	39·95	41·94	143·8	140·1	134·8	130·4
90	42·91	44·58	47·08	49·25	157·0	153·2	147·6	143·1
100	49·84	51·66	54·36	56·71	170·1	166·1	160·4	155·6

$p = 6$

v	0·005	0·010	0·025	0·050	0·005	0·010	0·025	0·050
6	$0{\cdot}0^5 4590$	$0{\cdot}0^4 1835$	$0{\cdot}0^3 1148$	$0{\cdot}0^3 4596$	33·22	31·19	28·39	26·14
7	$\cdot 0^2 1671$	$\cdot 0^3 3350$	$\cdot 0^2 8440$	·01710	35·48	33·40	30·54	28·23
8	·01512	·02420	·04540	·07389	37·65	35·53	32·60	30·24
9	·05153	·07383	·1200	·1753	39·75	37·59	34·60	32·19
10	·1161	·1558	·2325	·3185	41·79	39·59	36·54	34·08
15	·8580	1·012	1·272	1·529	51·33	48·96	45·64	42·96
20	2·162	2·440	2·889	3·313	60·16	57·63	54·09	51·21
25	3·865	4·264	4·893	5·475	68·53	65·87	62·13	59·07
30	5·866	6·379	7·178	7·907	76·58	73·79	69·86	66·66
40	10·51	11·24	12·35	13·25	92·00	88·98	84·72	81·24
50	15·78	16·69	18·09	19·33	106·8	103·6	99·00	95·27
60	21·49	22·58	24·23	25·69	121·1	117·7	112·9	108·9
70	27·53	28·79	30·69	32·35	135·1	131·5	126·4	122·3
80	33·85	35·27	37·39	39·24	148·7	145·0	139·7	135·4
90	40·39	41·95	44·28	46·32	162·2	158·3	152·8	148·3
100	47·12	48·82	51·35	53·56	175·5	171·5	165·8	161·1

$p = 7$

v	0·005	0·010	0·025	0·050	0·005	0·010	0·025	0·050
7	$0{\cdot}0^5 3835$	$0{\cdot}0^4 1534$	$0{\cdot}0^4 9592$	$0{\cdot}0^3 3841$	37·82	35·70	32·76	30·40
8	$\cdot 0^2 1432$	$\cdot 0^2 2872$	$\cdot 0^2 7234$	·01466	40·05	37·89	34·90	32·48
9	·01309	·02095	·03930	·06395	42·22	40·02	36·96	34·49
10	·04498	·06444	·1047	·1530	44·31	42·07	38·96	36·45
15	·5909	·7071	·9057	1·105	54·11	51·70	48·34	45·61
20	1·701	1·931	2·306	2·662	63·16	60·60	57·02	54·10

	Smallest root, l_p				Largest root, l_1			
α / v	0·005	0·010	0·025	0·050	0·005	0·010	0·025	0·050
				$p = 7$ (*cont.*)				
25	3·224	3·569	4·114	4·620	71·72	69·03	65·26	62·17
30	5·058	5·513	6·220	6·867	79·95	77·13	73·18	69·95
40	9·403	10·06	11·07	11·98	95·67	92·64	88·36	84·86
50	14·40	15·24	16·53	17·66	110·7	107·5	102·9	99·18
60	19·86	20·88	22·41	23·77	125·3	121·9	117·1	113·1
70	25·69	26·36	28·63	30·18	139·5	135·9	130·9	126·7
80	31·79	33·12	35·11	36·84	153·4	149·7	144·4	140·0
90	38·13	39·60·	41·80	43·71	167·1	163·2	157·7	153·2
100	44·67	46·28	48·67	50·74	180·5	176·5	170·8	166·1
				$p = 8$				
8	$0.0^{5}3327$	$0.0^{4}1331$	$0.0^{4}8318$	$0.0^{3}3332$	42·35	40·15	37·10	34·63
9	$\cdot0^{5}1253$	$\cdot0^{5}2513$	$\cdot0^{5}6330$	·01283	44·57	42·33	39·22	36·71
10	·01154	·01847	·03465	·05638	46·72	44·45	41·28	38·72
15	·3902	·4753	·6235	·7744	56·76	54·32	50·91	48·15
20	1·323	1·513	1·824	2·122	66·01	63·43	59·81	56·86
25	2·681	2·979	3·452	3·892	74·76	72·04	68·23	65·12
30	4·361	4·764	5·392	5·968	83·15	80·31	76·33	73·07
40	8·426	9·026	9·945	10·77	99·16	96·11	91·81	88·29
50	13·17	13·95	15·14	16·19	114·5	111·2	106·7	102·9
60	18·41	19·36	20·78	22·04	129·3	125·8	121·0	117·0
70	24·02	25·12	26·78	28·23	143·7	140·1	135·0	130·9
80	29·93	31·18	33·05	34·69	157·8	154·0	148·8	144·4
90	36·09	37·48	39·55	41·35	171·6	167·8	162·3	157·8
100	42·45	43·97	46·23	48·20	185·3	181·3	175·6	170·9
				$p = 9$				
9	$0.0^{5}2936$	$0.0^{4}1175$	$0.0^{4}7343$	$0.0^{3}2941$	46·84	44·57	41·40	38·84
10	$\cdot0^{5}1114$	$\cdot0^{5}2234$	$\cdot0^{5}5626$	·01140	49·05	46·74	43·52	40·91
15	·2426	·3024	·4090	·5202	59·32	56·85	53·39	50·58
20	1·014	1·169	1·425	1·672	68·76	66·14	62·48	59·50
25	2·218	2·475	2·884	3·267	77·67	74·93	71·09	67·94
30	3·753	4·111	4·670	5·183	86·21	83·35	79·34	76·05
40	7·557	8·104	8·944	9·698	102·5	99·43	95·11	91·57
50	12·07	12·79	13·89	14·86	118·0	114·8	110·2	106·4
60	17·09	17·98	19·31	20·49	133·1	129·6	124·8	120·8
70	22·50	23·54	25·10	26·47	147·7	144·1	139·0	134·8
80	28·25	29·41	31·18	32·72	162·0	158·2	152·9	148·6
90	34·21	35·53	37·49	39·20	176·0	172·1	166·6	162·1
100	40·41	41·86	44·01	45·88	189·8	185·8	180·1	175·4
				$p = 10$				
10	$0.0^{5}2628$	$0.0^{4}1051$	$0.0^{4}6573$	$0.0^{3}2632$	51·32	48·98	45·73	43·12
15	·1382	·1777	·2503	·3284	61·78	59·28	55·78	52·94
20	·7608	·8863	1·095	1·298	71·41	68·77	65·07	62·05
25	1·821	2·042	2·396	2·728	80·48	77·72	73·84	70·67
30	3·221	3·538	4·035	4·493	89·17	86·29	82·24	78·93
40	6·777	7·278	8·047	8·738	105·7	102·6	98·28	94·72
50	11·07	11·74	12·76	13·66	121·5	118·2	113·6	109·8
60	15·89	16·72	17·97	19·07	136·7	133·3	128·4	124·4
70	21·11	22·09	23·56	24·85	151·5	147·9	142·8	138·7
80	26·66	27·78	29·46	30·92	166·0	162·2	157·0	152·6
90	32·48	33·74	35·60	37·23	180·2	176·3	170·8	166·3
100	38·52	39·90	41·95	43·74	194·1.	190·1	184·5	179·8

Bibliography

Aastveit, A.H. and Martens, H. (1986) ANOVA interactions interpreted by partial least squares regression, *Biometrics*, **42**, 829–844.

Abbott, E.A. (1884) *Flatland*, 2d Ed., Barnes and Noble, New York.

Achrol, R.S. and Stern, L.W. (1988) Environmental determinants of decision making uncertainty in marketing channels, *J. Mark. Res.*, **25**, 36–50.

Acito, F. and Anderson, R.D. (1980) A monte carlo comparison of factor analytic methods, *J. Mark. Res.*, **17**, 228–236.

Acito, F. and Anderson, R.D. (1986) A simulation study of factor score indeterminancy, *J. Mark. Res.*, **23**, 111–118.

Adcock, R.J. (1878) A problem in least squares, *Analyst*, **5**, 53–54.

Afifi, A.A. and Clark, V. (1984) *Computer-aided Multivariate Analysis*, Lifetime Learning Publications, Belmont, Calif.

Ahamad, B. (1967) An analysis of crimes by the method of principal components, *Appl. Stat.*, **16**, 17–35.

Ahamad, B. (1968) A note on the interrelationships of crimes—a reply to Mrs. Walker, *Appl. Stat.*, **17**, 49–51.

Aitchison, J. (1982) The statistical analysis of compositional data (with discussion). *J. Roy. Stat. Soc. B*, **44**, 139–177.

Aitchison, J. (1983) Principal component analysis of compositional data, *Biometrika*, **70**, 57–65.

Akaike, H. (1974) A new look at the statistical model identification, *IEEE Trans. Automatic Control*, **AC-19**, 716–773.

Akaike, H. (1977) On entropy maximization principle, *Applications of Statistics* (P.R. Krishnaiah, editor), North-Holland, New York, pp. 27–41.

Akaike, H. (1987) Factor analysis and AIC, *Psychometrika*, **52**, 317–332.

Aldenderfer, M.S. and Blashfield, R.K. (1984) *Cluster Analysis*, Sage Publications, Inc., Beverly Hills, Calif.

Algina, J. (1980) A note on identification in the oblique and orthogonal factor analysis models, *Psychometrika*, **45**, 393–396.

Allen, S.J. and Hubbard, R. (1986) Regression equations for latent roots of random data correlation matrices with unities on the diagonal, *Mult. Behav. Res.*, **21**, 393–398.

Alt, F.B. (1985) Multivariate control charts, *Encyclopedia of Statistical Sciences*, **6** (S. Kotz and N.L. Johnson, editors), John Wiley and Sons, Inc., New York, pp. 110–122.

Ammann, L. and Van Ness, J. (1989) Standard and robust orthogonal regression, *Comm. Stat.—Theor. Meth.*, **18**, 145–162.

Anderson, G.A. (1965) An asymptotic expansion for the distribution of the latent roots of an estimated covariance matrix, *Ann. Math. Stat.*, **36**, 1153–1173.

Anderson, T.W. (1963) Asymptotic theory for principal component analysis, *Ann. Math. Stat.*, **34**, 122–148.

Anderson, T.W. (1965) Some optimum confidence limits for roots of determinental equations, *Ann. Math. Stat.*, **36**, 468–488.

Anderson, T.W. (1971) *Statistical Analysis of Time Series*, John Wiley and Sons, Inc., New York.

Anderson, T.W. (1980) Recent results on the estimation of a linear functional relationship, *Multivariate Analysis*, **V** (P.R. Krishnaiah, editor), North-Holland, New York, pp. 23–34.

Anderson, T.W. (1984a) *An Introduction to Multivariate Analysis*, 2d Ed., John Wiley and Sons, Inc., New York.

Anderson, T.W. (1984b) Estimating linear statistical relationships, *Ann. Stat.*, **12**, 1–45.

Anderson, T.W. and Amemiya, Y. (1988) The asymptotic normal distribution of estimators in factor analysis under general conditions, *Ann. Stat.*, **16**, 759–771.

Anderson, T.W. and Rubin, H. (1956) Statistical inference in factor analysis, *Proc. 3rd Berkeley Symp. on Math. Stat. and Prob.*, **V**, University of California Press, Berkeley, pp. 111–150.

Andrews, D.F. (1972) Plots of high-dimensional data, *Biometrics*, **28**, 125–136.

Andrews, D.F. (1974) A robust method for multiple linear regression, *Technometrics*, **16**, 523–531.

Andrews, D.F. (1979) The robustness of residual displays, *Robustness in Statistics* (R.L. Launer and G.N. Wilkinson, editors), Academic Press, Inc., New York, pp. 19–32.

ANSI (1969) *Specifications for Audiometers*, **Standard S3.6-1969**, American National Standards Institute, New York.

Apel, H. and Wold, H. (1982) Soft modeling with latent variables in two or more dimensions: PLS estimation and testing for predictive relevance, *Systems Under Indirect Observation*, **II** (K.G. Jöreskog and H. Wold, editors), North-Holland, New York, pp. 209–247.

Arbuckle, J. and Friendly, M.L. (1977) On rotating to smooth functions, *Psychometrika*, **42**, 127–140.

Archer, C.O. and Jennrich, R.I. (1973) Standard errors for rotated factor loadings, *Psychometrika*, **38**, 581–592.

Archer, C.O. and Jennrich, R.I. (1976) A look, by simulation, at the validity of some asymptotic distribution results for rotated loadings, *Psychometrika*, **41**, 537–541.

Areskoug, B. (1982) The first canonical correlation: Theoretical PLS analysis and simulation experiments. *Systems Under Indirect Observation*, **II** (K.G. Jöreskog and H. Wold, editors), North-Holland, New York, pp. 95–117.

Atchley, W.R. and Bryant, E.H., editors (1975) *Multivariate Statistical Methods: Among-groups Covariation*, Dowden, Hutchinson and Ross, Inc., Stroudsburg, Pa.

Bagozzi, R.P. (1977) Structural equation models in experimental research, *J. Mark. Res.*, 14, 209–226.

Bagozzi, R.P., Fornell, C., and Larcker, D.F. (1981) Canonical correlation analysis as a special case of structural relations model, *Mult. Behav. Res.*, 16, 437–454.

Baker, G.A. (1954) Organoleptic ratings and analytical data for wines analyzed into orthogonal factors, *Food Research*, 19, 575–580.

Bargmann, R.E. and Baker, F.D. (1977) A minimax approach to component analysis, *Appl. Stat.* (P.R. Krishnaiah, editor), North-Holland, New York, pp. 55–69.

Barker, F., Soh, Y.C., and Evans, R.J. (1988) Properties of the geometric mean functional relationship, *Biometrics*, 44, 279–281.

Bartholomew, D.J. (1980) Factor analysis for categorical data (with discussion), *J. Roy. Stat. Soc. B*, 42, 293–321.

Bartholomew, D.J. (1984) Scaling binary data using a factor model, *J. Roy. Stat. Soc. B*, 46, 120–123.

Bartholomew, D.J. (1987) *Latent Variable Models and Factor Analysis*, Oxford University Press, New York.

Bartlett, M.S. (1937) The statistical conception of mental factors, *Br. J. Psych.*, 28, 97–104.

Bartlett, M.S. (1948) Statistical estimation of supply and demand relations, *Econometrica*, 16, 323–329.

Bartlett, M.S. (1950) Tests of significance in factor analysis, *Br. J. Psych. Stat. Sec.*, 3, 77–85.

Bartlett, M.S. (1951a) The effect of standardization on a χ^2 approximation in factor analysis, *Biometrika*, 38, 337–344.

Bartlett, M.S. (1951b) A further note on tests of significance in factor analysis, *Brit. J. Psych. Stat. Sec.*, 4, 1–2.

Bartlett, M.S. (1954) A note on the multiplying factors for various χ^2 approximations, *J. Roy. Stat. Soc. B*, 16, 296–298.

Basilevsky, A. (1981) Factor analysis regression, *Can. J. Stat.*, 9, 109–117.

Bauer, K.W., Jr. (1981) *A Monte Carlo Study of Dimensionality Assessment and Factor Interpretation in Principal Component Analysis*, Air Force Inst. of Tech., Wright-Patterson AFB, Ohio.

Baye, M.R. and Parker, D.F. (1984) Combining ridge and principal component regression: A money demand illustration, *Comm. Stat. Theor. Meth.*, 13, 197–205.

Beale, E.M.L., Kendall, M.G., and Mann, D.W. (1967) The discarding of variables in multivariate analysis, *Biometrika*, 54, 357–366.

Beale, E.M.L. and Little, R.J.A. (1975) Missing values in multivariate analysis, *J. Roy. Stat. Soc. B*, 37, 129–145.

Beaton, A. and Tukey, J.W. (1974) The fitting of power series, meaning polynomials, illustrated in band-spectroscopic data (with discussion), *Technometrics*, 16, 147–192.

Bebbington, A.C. and Smith, T.M.F. (1977) The effect of survey design in multivariate analysis, *The Analysis of Survey Data*, II: *Model Fitting* (C.A. O'Muircheartaigh and C. Payne, editors) John Wiley and Sons, Inc., New York, pp. 175–192.

Bechtel, G.G., Tucker, L.R., and Chang, W.-C. (1971) A scalar product model for the multidimensional scaling of choice, *Psychometrika*, **36**, 369–388.

Becker, R.A. and Cleveland, W.S. (1987) Brushing scatterplots, *Technometrics*, **29**, 127–142.

Bekker, P. and de Leeuw, J. (1988) Relations between variants of non-linear principal component analysis, *Component and Correspondence Analysis*, (J.L.A. van Rijckevorsel and J. de Leeuw, editors) John Wiley and Sons, Inc., New York, pp. 1–31.

Belinfante, A. and Coxe, K.L. (1986) Principal components regression—Selection rules and application, *Proc. Bus. & Econ. Sec., Amer. Stat. Assoc.*, 429–431.

Bennett, J.F. (1956) Determination of the number of independent parameters of a score matrix from the examination of rank orders, *Psychometrika*, **21**, 383–393.

Bennett, J.F. and Hays, W.L. (1960) Multidimensional unfolding: Determining the dimensionality of ranked preference data, *Psychometrika*, **25**, 27–43.

Bentler, P.M. (1982a) Confirmatory factor analysis via noniterative estimation: A fast, inexpensive method, *J. Mark. Res.*, **19**, 417–424.

Bentler, P.M. (1982b) Linear systems with multiple levels and types of latent variables, *Systems Under Indirect Measurement*, I (K.G. Jöreskog and H. Wold, editors), North-Holland, New York, pp. 101–130.

Bentler, P.M. (1986) Structural modeling and Psychometrika: An historical perspective on growth and achievements, *Psychometrika*, **51**, 35–51.

Bentler, P.M. and Lee, S.-Y. (1978) Statistical aspects of a three-mode factor analysis model, *Psychometrika*, **43**, 343–352.

Benzecri, J.P. (1973) *L'Analyse des Données*, II: *L'Analyse des Correspondances*, Dunod, Paris.

Benzecri, J.P. (1977) Histoire et prehistoire de l'analyse des données, V. L'analyse des correspondances, *Cahiers de l'Analyse des Données*, **2**, 9–40.

Beran, R. and Srivastava, M.S. (1985) Bootstrap tests and confidence regions for functions of a covariance matrix, *Ann. Stat.*, **13**, 95–115.

Bernstein, I.H. and Teng, G. (1989) Factoring items and factoring scales are different: Spurious evidence for multidimensionality due to item categorization, *Psych. Bull.*, **105**, 467–477.

Bertholet, J.L. and Wold, H. (1985) Recent developments in categorical data analysis by PLS, *Measuring the Unmeasurable* (P. Nijkamp, editor), Martinus Nijhoff Publishers, Boston, pp. 253–286.

Besse, P. (1988) Spline functions and optimal metric in linear principal component analysis, *Component and Correspondence Analysis* (J.L.A. van Rijckevorsel and J. de Leeuw, editors), John Wiley and Sons, Inc., New York, pp. 81–101.

Besse, P. and Ramsay, J.O. (1986) Principal components analysis of sampled functions, *Psychometrika*, **51**, 285–311.

Best, R.J. (1976) The predictive aspects of a joint-space theory of stochastic choice, *J. Mark. Res.*, **13**, 198–204.

Bibby, J. (1980) Some effects of rounding optimal estimates, *Sankhyā B*, **42**, 165–178.

Bishop, Y.M.M., Fienberg, S.E., and Holland, P.W. (1975) *Discrete Multivariate Analysis*, MIT Press, Cambridge, Mass.

Blackman, A.W.Jr., Seligman, E.J. and Sogliero, G.C. (1973) An innovation index based on factor analysis, *Tech. Forc. and Soc. Change*, **4**, 301–316.

Bloomfield, P. (1974) Linear transformations for multivariate binary data, *Biometrics*, **30**, 609-617.

Bloxom, B. (1968) A note on invariance in three-mode factor analysis, *Psychometrika*, **33**, 347-350.

Blum, E.K. (1972) *Numerical Analysis and Computations: Theory and Practice*, Addison-Wesley Publishing Co., Reading, Mass.

Bock, R.D. and Bargmann, R.E. (1966) Analysis of covariance structures, *Psychometrika*, **31**, 507-534.

Bock, R.D. and Jones, L.V. (1968) *The Measurement and Prediction of Judgment and Choice*, Holden-Day, San Francisco.

Boeckenholt, I. and Gaul, W. (1986) Analysis of choice behavior via probabilistic ideal point and vector models, *Appl. Stoch. Models and Data Anal.*, **2**, 209-226.

Boente, G. (1987) Asymptotic theory for robust principal components, *J. Mult. Anal.*, **21**, 67-78.

Boik, R.J. (1986) Testing the rank of a matrix with applications to the analysis of interaction in ANOVA, *J. Amer. Stat. Assoc.*, **81**, 243-248.

Bookstein, F.L. (1982) The geometric meaning of soft modeling, with some generalizations, *Systems Under Indirect Observation*, **II** (K.G. Jöreskog and H. Wold, editors), North-Holland, New York, pp. 55-74.

Boomsma, A. (1982) The robustness of LISREL against small sample sizes in factor analysis models, *Systems Under Indirect Measurement*, **I** (K.G. Jöreskog and H. Wold, editors), North-Holland, New York, pp. 149-173.

Borg, I. (1978) Procrustean analysis of matrices with different row order, *Psychometrika*, **43**, 277-278.

Borg, I. (1979) Geometric representation of individual differences, *Geometric Representations of Relational Data* (J.C. Lingoes *et al.* editors), Mathesis Press, Ann Arbor, Mich., pp. 609-656.

Borg, I. and Lingoes, J.C. (1978) What weight should weights have in individual differences scaling?, *Quality and Quantity*, **12**, 223-237. (Also in Davies and Coxon, 1982.)

Box, G.E.P. (1949) A general distribution theory for a class of likelihood criteria, *Biometrika*, **36**, 317-346.

Box, G.E.P., Hunter, W.G. and Hunter, J.S. (1978) *Statistics for Experimenters; An Introduction to Design, Data Analysis and Model Building*, John Wiley and Sons, Inc., New York.

Box, G.E.P., Hunter, W.G., MacGregor, J.F., and Erjavac, J. (1973) Some problems associated with the analysis of multiresponse data, *Technometrics*, **15**, 33-51.

Box, G.E.P. and Jenkins, G.M. (1970) *Time Series Analysis, Forecasting and Control*, Holden-Day, San Francisco.

Boynton, R.M. and Gordon, J. (1965) Bezold-Bruke hue shift measured by color-naming technique, *J. Optical. Soc. Amer.*, **55**, 78-86.

Bozdogan, H. and Ramirez, D.E. (1988) FACAIC: Model selection algorithm for the orthogonal factor model using AIC and CAIC, *Psychometrika*, **53**, 407-415.

Bozik, J.E. and Bell, W.R. (1987) Forecasting age specific fertility using principal components, *Proc. Soc. Stat. Sec. Amer. Stat. Assoc.*, 396-401.

Bradley, R.A. and Terry, M. (1952) Rank analysis of incomplete block designs, I: The method of paired comparisons, *Biometrika*, **39**, 324-345.

Bradu, D. and Gabriel, K.R. (1974) Simultaneous statistical inference on interactions in two-way analysis of variance, *J. Amer. Stat. Assoc.*, **69**, 428–436.

Bradu, D. and Gabriel, K.R. (1978) The biplot as a diagnostic tool for models of two-way tables, *Technometrics*, **20**, 47–68.

Brady, H.E. (1989) Factor ideal point analysis for interpersonally incomparable data, *Psychometrika*, **54**, 181–202.

Brakstad, F., Karstang, T.V., Sørensen, J., and Steen, A. (1988) Prediction of molecular weight and density of distillation fractions from gas chromatographic—mass spectrometric detection and multivariate calibration, *Chemom. and Intel. Lab. Syst.*, **3**, 321–328.

Brambilla, F. (1976) Stability of distance between structured groups in social organism: Empirical research, *Contributions to Appl. Stat.* (W.J. Ziegler, editor), *Experimentia Supplementum*, **22**, Birkhauser Verlag, Basel, 213–220.

Bratchell, N. (1989) Multivariate response surface modeling by principal components analysis, *J. Chemom.*, **3**, 579–588.

Brauer, A. (1953) Bounds for characteristic roots of matrices, *Simultaneous Linear Equations and the Determination of Eigenvalues*, Appl. Math. Series #29, National Bureau of Standards, U.S. Govt. Printing Office Washington, D.C., pp. 101–106.

Breckenbach, E.F. and Bellman, R. (1965) *Inequalities*, Springer-Verlag, New York.

Breckler, S.J. (1990) Applications of covariance structure modeling in psychology: Cause for concern? *Psych. Bull.*, **107**, 260–273.

Bretaudiere, J., Dumont, G., Rej, R. and Bailly, M. (1981) Suitability of control materials. General principles and methods of investigation, *Clinical Chem.*, **27**, 798–805.

Brewer, W.L. and Williams, F.C. (1954) An objective method for determination of equivalent neutral densities of color film images, I. Definitions and basic concepts, *J. Optical Soc. Amer.*, **44**, 460–464.

Brillinger, D.R. (1969) The canonical analysis of stationary time series, *Multivariate Analysis*, II (P.R. Krishnaiah, editor), Academic Press, New York, pp. 331–350.

Brillinger, D.R. (1975) *Time Series, Data Analysis and Theory*, Holt, Rinehart and Winston, Inc., New York.

Brokken, F.B. (1983) Orthogonal procrustes rotation maximizing congruence, *Psychometrika*, **48**, 343–352.

Brokken, F.B. (1985) The simultaneous maximization of congruence for two or more matrices under orthogonal rotation, *Psychometrika*, **50**, 51–56.

Brown, C.H. (1983) Asymptotic comparison of missing data procedures for estimating factor loadings, *Psychometrika*, **48**, 269–291.

Brown, M.B. (1975) A method for combining non-independent one-sided tests of significance, *Biometrics*, **31**, 987–992.

Brown, M.L. (1982) Robust line estimation with errors in both variables, *J. Amer. Stat. Assoc.*, **77**, 71–79.

Brown, P.J. (1982) Multivariate calibration (with discussion), *J. Roy. Stat. Soc. B*, **44**, 287–321.

Brown, R.C. (1963) *Smoothing, Forecasting and Prediction*, Prentice-Hall, Englewood Cliffs, NJ.

Brown, S.D., Barker, T.Q., Larivef, R.J., Monfre, S.L., and Wilk, H.R. (1988) Chemometrics, *Anal. Chem.*, **60**, 252R–273R.

Brown, W.R.J., Howe, W.G., Jackson, J.E., and Morris, R.H. (1956) Multivariate normality of the color-matching process, *J. Optical Soc. Amer.*, **46**, 46–49.

Browne, M.W. (1968) A comparison of factor analytic techniques, *Psychometrika*, **33**, 267–334.

Browne, M.W. (1972) Oblique rotation to a partially specified target, *Br. J. Math. Stat. Psychol.*, **25**, 207–212.

Browne, M.W. (1988) Properties of the maximum likelihood solution in factor analysis regression, *Psychometrika*, **53**, 585–590.

Bru, M.F. (1989) Diffusions of perturbed principal component analysis, *J. Mult. Anal.*, **29**, 127–136.

Bryant, E.H. and Atchley, W.R., editors (1975) *Multivariate Statistical Methods: Within-groups Covariation*, Dowden, Hutchinson and Ross, Inc., Stroudsburg, Pa.

Bureau of the Census (Various) *Statistical Abstract of the United States*, U.S. Govt. Printing Office, Washington, D.C. for years 1967, 1973, 1976, 1979, 1986.

Burt, C. (1917) *The Distribution and Relation of Educational Abilities*, P.S. King & Son, London.

Burt, C. (1950) The factorial analysis of qualitative data, *Br. J. Stat. Psych. (Stat. Sec.)*, **3**, 166–185.

Buzas, T.E., Fornell, C., and Rhee, B.-D. (1989) Conditions under which canonical correlation and redundancy maximization produce identical results, *Biometrika*, **76**, 618–621.

Büyükkurt, B.K. and Büyükkurt, M.D. (1990) Robustness and small-sample properties of the estimators of probabilistics multidimensional scaling (PROSCAL), *J. Mark. Res.*, **27**, 139–149.

Cadotte, E.R., Woodruff, R.B., and Jenkins, R.L. (1987) Expectations and norms on models of consumer satisfaction, *J. Mark. Res.*, **24**, 305–314.

Calder, P. (1986) *Influence Functions in Multivariate Analysis*, Unpublished Ph.D. Thesis, University of Kent, England.

Campbell, N.A. (1980a) Shrunken estimators in discriminant and canonical variate analysis, *Appl. Stat.*, **29**, 5–14.

Campbell, N.A. (1980b) Robust procedures in multivariate analysis I: Robust covariance estimation, *Appl. Stat.*, **29**, 231–237.

Campbell, N.A. and Atchley, W.R. (1981) The geometry of canonical variate analysis, *Syst. Zool.*, **30**, 268–280.

Carey, R.N., Wold, S., and Westgard, J.O. (1975) Principal component analysis: An alternative to "referee" methods in method comparison studies, *Anal. Chem.*, **47**, 1824–1829.

Carlson, P.G. (1956) A least squares interpretation of the bivariate line of organic correlation, *Skand. Aktuar.*, **39**, 7–10.

Carmone, F.J., Green, P.E., and Robinson, P.J. (1968) TRICON—an IBM 360/65 FORTRAN IV program for the triangularization of conjoint data, *J. Mark. Res.*, **5**, 219–220.

Carroll, B.H., Higgins, G.C., and James, T.H. (1980) *Introduction to Photographic Theory*, John Wiley and Sons, Inc., New York.

Carroll, J.B. (1953) An analytical solution for approximating simple structure in factor analysis, *Psychometrika*, **18**, 23–38.

Carroll, J.B. (1957) Biquartimin criterion for rotation to oblique simple structure in factor analysis, *Science*, **126**, 1114–1115.

Carroll, J.D. (1972) Individual differences in multidimensional scaling, *Multidimensional Scaling: Theory and Application in the Behavioral Sciences*, I (R.H. Shepard *et al.*, editors), Academic Press, New York, pp. 105–155. (Also, in part, in Davies and Coxon, 1982).

Carroll, J.D. (1980) Models and methods for multidimensional analysis of preferential choice (or other dominance) data, *Proc. of Aachen Symp. on Decision Making and Multidimensional Scaling* (E.D. Lautermann and H. Ferger, editors), Springer-Verlag, New York, pp. 234–289.

Carroll, J.D. and Chang, J.-J. (1970a) Analysis of individual differences in multidimensional scaling via an N-way generalization of the "Eckart-Young" decomposition, *Psychometrika*, **35**, 283–319. (Also, in part, in Davies and Coxon, 1982.)

Carroll, J.D. and Chang, J.-J. (1970b) Reanalysis of some color data of Helm's by the INDSCAL procedure for individual differences multidimensional scaling, *Proc. 78th APA Convention*, pp. 137–138.

Carroll, J.D. and Green, P.E. (1988) An INDSCAL-based approach to multiple correspondence analysis, *J. Mark. Res.*, **25**, 193–203.

Carroll, J.D., Green, P.E., and Schaffer, C.M. (1986) Interpoint distance comparisons in correspondence analysis, *J. Mark. Res.*, **23**, 271–280.

Carroll, J.D., Greene, P.E., and Schaffer, C.M. (1987) Comparing interpoint distances in correspondence analysis: A clarification, *J. Mark. Res.*, **24**, 445–450.

Carroll, J.D., Green, P.E., and Schaffer, C.M. (1989) Reply to Greenacre's commentary on the Carroll–Green–Schaffer scaling of two-way correspondence analysis solutions, *J. Mark. Res.*, **26**, 366–368.

Carter, E.M. (1980) Tests of location based on principal components, *Multivariate Statistical Analysis* (R.P. Gupta, editor), North-Holland, New York, pp. 47–52.

Carter, E.M. and Srivastava, M.S. (1980) Asymptotic distribution of the latent roots of the noncentral Wishart distribution and the power of the likelihood ratio test for nonadditivity *Can. J. Stat.*, **8**, 119–134.

Cartier, A. and Rivail, J.-L. (1987) Electronic descriptors in quantitative structure–activity relationships, *Chemom. and Intel. Lab. Syst.*, **1**, 335–347.

Cartwright, H. (1987) Determination of the dimensionality of spectroscopic data by submatrix analysis, *J. Chemom.*, **1**, 111–120.

Castro, P.E., Lawton, W.H., and Sylvestre, E.A. (1986) Principal modes of variations for processes with continuous sample curves, *Technometrics*, **28**, 329–337.

Cattell, R.B. (1966) The scree test for the number of factors, *Mult. Behav. Res.*, **1**, 245–276.

Cattell, R.B. and Jaspers, J. (1967) A general plasmode (No. 30-10-5-2) for factor analytic exercises and research, *Mult. Behav. Res. Monographs*, **67–3**, 1–212.

Cattell, R.B. and Khanna, D.K. (1977) Principles and procedures for unique rotation in factor analysis, *Statistical Methods for Digital Computers*, III (K. Enslein *et al.*, editors), John Wiley and Sons, Inc., New York, pp. 166–202.

Cattell, R.B. and Muerle, J.L. (1960) The MAXPLANE program for factor rotation to obtain simple structure, *Educ. & Psych. Meas.*, **20**, 569–590. (Also in Bryant and Atchley, 1975.)

Cattell, R.B. and Vogelmann, S. (1977) A comprehensive trial of the SCREE and KG criteria for determining the number of factors, *Mult. Behav. Res.*, **12**, 289–325.

Cerdan, S.V. (1989) A note on the behaviour of augmented principal-component plots in regression, *Comm. Stat.—Theor. Meth.*, **18**, 331–342.

Chalmers, C.P. (1975) Generation of correlation matrices with a given eigen-structure, *J. Stat. Comp. & Simul.*, **4**, 133–139.

Chang, J.-J. and Carroll, J.D. (1969) How to use MDPREF, a computer program for multidimensional analysis of preference data. Unpublished report, Bell Telephone Laboratories, Murray Hill, N.J.

Chatelin, F. and Belaid, D. (1987) Numerical analysis for factorial data analysis I. Numerical software—the package INDA for microcomputers, *Appl. Stoch. Models and Data Anal.*, **3**, 193–206.

Chatfield, C. and Collins, J. (1980) *Introduction to Multivariate Analysis*, Chapman and Hall, New York,

Chen, C.W. (1974) An optimal property of principal components, *Comm. Stat.—Theor. Meth.*, **3**, 979–983.

Choulakian, V. (1988) Exploratory analysis of contingency tables by loglinear formulation and generalizations of correspondence analysis, *Psychometrika*, **53**, 235–250. Errata, **53**, 593.

Chow, G.C. (1975) *Analysis and Control of Dynamic Economic Systems*, John Wiley and Sons, Inc., New York.

Christoffersson, A. (1975) Factor analysis of dichotomized variables, *Psychometrika*, **40**, 5–32.

Christoffersson, A. (1977) Two step weighted least squares factor analysis of dichotomized variables, *Psychometrika*, **42**, 433–438.

Church, A., Jr. (1966) Analysis of data when the response is a curve, *Techometrics*, **8**, 229–246.

Clark, D., Davies, W.K.D., and Johnston, R.J. (1974) The application of factor analysis in human geography, *Statistician*, **23**, 259–281.

Clarke, M.R.B. (1980) The reduced major axis of a bivariate sample, *Biometrika*, **67**, 441–446.

Clarkson, D.B. (1979) Estimating the standard errors of rotated factor loadings by jackknifing, *Psychometrika*, **44**, 297–314.

Clarkson, D.B. and Jennrich, R.J. (1988) Quartic rotation criteria and algorithms, *Psychometrika*, **53**, 251–259.

Clemm, D.S., Krishnaiah, P.R., and Waikar, V.B. (1973) Tables of the extreme roots of a Wishart matrix, *J. Stat. Comp. and Simul.*, **2**, 65–92.

Cleveland, W.S. and Guarino, R. (1976) Some robust statistical procedures and their application to air pollution data, *Technometrics*, **18**, 401–409.

Cliff, N. (1966) Orthogonal rotation to congruence, *Psychometrika*, **31**, 33–42.

Cliff, N. (1983) Some cautions concerning the application of causal modeling methods, *Mult. Behav. Res.*, **18**, 115–126.

Cliff, N. and Krus, D.J. (1976) Interpretation of canonical analysis: Rotated vs. unrotated solutions, *Psychometrika*, **41**, 35–42.

Clogg, C.G. and Goodman, L.A. (1984) Latent structure analysis of a set of multidimensional contingency tables, *J. Amer. Stat. Assoc.*, **79**, 762–771.

Cochran, R.N. and Horne, F.H. (1977) Statistically weighted principal component analysis of rapid scanning wavelength kinetics experiments, *Anal. Chem.*, **49**, 846–853.

Cochran, W.G. (1977) *Sampling Techniques*, 3d Ed., John Wiley and Sons, Inc., New York.

Cohen, A. and Jones, R.H. (1969) Regression on a random field, *J. Amer. Stat. Assoc.*, **64**, 1172–1182.

Coleman, D. (1986) Hotelling's T^2, robust principal components and graphics for SPC, *Proc. 18th Symp. on the Interface* (T.J. Boardman, ed.), pp. 312–326.

Comrey, A.L. (1950) A proposed method for absolute ratio scaling, *Psychometrika*, **15**, 317–325.

Constantine, A.G. and Gower, J.C. (1978) Graphical representation of asymmetric matrices, *Appl. Stat.*, **27**, 297–304.

Cooley, W.W. and Lohnes, P.R. (1971) *Multivariate Data Analysis*, John Wiley and Sons, Inc., New York.

Coombs, C.H. (1950) Psychological scaling without a unit of measurement, *Psychol. Rev.*, **57**, 148–158.

Coombs, C.H. (1964) *A Theory of Data*, John Wiley and Sons, Inc., New York. (Reprinted in 1976 by Mathesis Press, Ann Arbor, Mich.)

Coombs, C.H. (1975) A note on the relation between the vector model and the unfolding model for preferences, *Psychometrika*, **40**, 115–116.

Coombs, C.H. and Kao, R.C. (1960) On a connection between factor analysis and multidimensional scaling, *Psychometrika*, **25**, 219–231.

Cooper, L.G. (1972) COSCAL: Program for metric multidimensional scaling, *J. Mark. Res.*, **9**, 201–202.

Cooper, L.G. and Nakanishi, M. (1983) Two logit models for external analysis of preferences, *Psychometrika*, **48**, 607–620.

Corballis, M.C. (1971) Comparison of ranks of cross-product and covariance solutions in component analysis, *Psychometrika*, **36**, 243–249.

Corballis, M.C. and Traub, R.E. (1970) Longitudinal factor analysis, *Psychometrika*, **35**, 79–98.

Cornelius, P.L. (1980) Functions approximating Mandel's tables for the means and standard deviations of the first three roots of a Wishart matrix. *Technometrics*, **22**, 613–616.

Cornelius, P.L., Templeton, W.C., Jr., and Taylor, T.H. (1979) Curve fitting by regression on smoothed singular vectors, *Biometrics*, **35**, 849–859.

Corsten, L.C.A. and Gabrial, K.R. (1976) Graphical exploration in comparing variance matrices, *Biometrics*, **32**, 851–863.

Corsten, L.C.A. and van Eijbergen, A.C. (1972) Multiplicative effects in two-way analysis of variance, *Stat. Neer.*, **26**, 61–68.

Cox, D.R. (1972) The analysis of multivariate binary data, *Appl. Stat.*, **21**, 113–120.

Coxe, K.L. (1982) Selection rules for principal components regression: Comparison with latent root regression, *Proc. Bus. Econ. Sec., Amer. Stat. Assoc.*, 222–227.

Coxon, A.P.M. (1974) The mapping of family composition preferences: A scaling analysis, *Social Sci. Res.*, **3**, 191–210. (Also in Davies and Coxon, 1982.)

Coxon, A.P.M. (1982) *A User's Guide to Multidimensional Scaling*, Heinemann Educational Books, Exeter, N.H.

Craddock, J.M. (1966) A metrological application of principal component analysis, *Statistician*, **15**, 143–156.

Craddock, J.M. (1973a) Problems and prospects for eigenvector analysis of meterology (with discussion), *Statistician*, **22**, 133–145.

Craddock, J.M. (1973b) A reply to Professor G.M. Jenkins, *Statistician*, **22**, 233–235.

Creasy, M.A. (1956) Confidence limits for the gradient in the linear functional relationship, *J. Roy. Stat. Soc. B*, **18**, 65–69.

Critchley, F. (1985) Influence in principal components analysis, *Biometrika*, **72**, 627–636.

Crosby, L.A. and Stephens, N. (1987) Effects of relationship marketing on satisfaction, retention and prices in the life insurance industry, *J. Mark. Res.*, **24**, 404–411.

Crosier, R.B. (1988) Multivariate generalizations of cumulative sum quality-control schemes, *Technometrics*, **30**, 291–303.

Cudeck, R. and Browne, M.W. (1983) Cross-validation of covariance matrices, *Mult. Behav. Res.*, **18**, 147–167.

Cureton, E.E. and D'Agostino, R.B. (1983) Factor Analysis: An Applied Approach, Lawrence Erlbaum Assoc., Hillsdale, N.J.

Cureton, E.E. and Mulaik, S.A. (1975) The weighted varimax rotation and the promax rotation, *Psychometrika*, **40**, 183–195.

Daling, J.R. and Tamura, H. (1970) Use of orthogonal factors for selection of variables in a regression equation—an illustration, *Appl. Stat.*, **19**, 260–268.

Dallal, G.E. (1988) Statistical microcomputing—like it is, *Amer. Stat.*, **42**, 212–216.

Daniel, C. and Wood, F.S. (1971) *Fitting Equations to Data*, John Wiley and Sons, Inc., New York.

Danielyan, S.A., Zharinov, G.M., and Osipova, T.T. (1986) Application of the principal component method and the proportional hazards regression model to analysis of survival data, *Biometrical J.*, **28**, 73–79.

Darroch, J.N. (1965) An optimal property of principal components, *Ann. Math. Stat.*, **36**, 1579–1582.

Daudin, J.J., Duby, C., and Trecourt, P. (1988) Stability of principal component analysis studied by the bootstrap method, *Statistics*, **19**, 241–258.

Davenport, E.C., Jr. (1990) Significance testing of congruence coefficients: A good idea? *Educ. & Psych. Meas.*, **50**, 289–296.

Davidson, J. (1972) A geometrical analysis of the unfolding model: Nondegenerate solutions, *Psychometrika*, **37**, 193–216.

Davidson, J. (1973) A geometrical analysis of the unfolding model: General solutions, *Psychometrika*, **38**, 305–336.

Davidson, M.L. (1988) A reformulation of the general Euclidean model for the external analysis of preference data, *Psychometrika*, **53**, 305–320.

Davies, P.M. and Coxon, A.P.M., editors (1982) *Key Texts in Multidimensional Scaling*, Heinemann Educational Books, Exeter, N.H.

Davis, A.W. (1970a) Exact distributions of Hotelling's generalized T_0^2, *Biometrika*, **57**, 187–191. Correction, (1972), **59**, 498.

Davis, A.W. (1970b) Further applications of a differential equation for Hotelling's generalized T_0^2, *Ann. Inst. Stat. Math.*, **22**, 77–87.

Davis, A.W. (1977) Asymptotic theory for principal component analysis: non-normal case, *Austral. J. Stat.*, **19**, 206–212.

Davis, A.W. (1980) Further tabulation of Hotelling's generalized T_0^2, *Comm. Stat.—Simul. Comput. B*, **9**, 321–336.

Dawkins, B. (1989) Multivariate analysis of national track records, *Amer. Stat.*, **43**, 110–115.

Dawson-Saunders, B.K. (1982) Correcting for bias in the canonical redundancy statistic, *Educ. & Psych. Meas.*, **42**, 131–143.

Delaney, M.F. (1988) Multivariate detection limits for selected ion monitoring gas chromatography—mass spectrometry, *Chemom. and Intel. Lab. Syst.*, **3**, 45–51.

DeLeeuw, J. (1982) Generalized eigenvalue problems with positive semi-definite matrices, *Psychometrika*, **47**, 87–93.

DeLeeuw, J. (1983) On the prehistory of correspondence analysis, *Stat. Neer.*, **37**, 161–164.

DeLeeuw, J. and Pruzansky, S. (1978) A new computational method to fit the weighted euclidean distance model, *Psychometrika*, **43**, 479–490.

DeLeeuw, J. and Van der Heijden, P.G.M. (1988) Correspondence analysis of incomplete contingency tables, *Psychometrika*, **53**, 223–233.

DeLeeuw, J. and Van Rijckevorsel, J.L.A. (1988) Beyond homogeneity analysis, *Component and Correspondence Analysis* (J.L.A. Van Rijckevorsel and J. DeLeeuw, editors), John Wiley and Sons, Inc., New York, pp. 55–80.

DeLeeuw, J., Young, F.W., and Takane, Y. (1976) Additive structure in qualitative data: An alternating least squares method with optimal scaling features, *Psychometrika*, **41**, 471–503.

De Ligny, C.L., and Nieuwdorp, G.H.E., Brederode, W.K., Hammers, W.E., and Van Houwelingen, J.C. (1981) An application of factor analysis with missing data, *Technometrics*, **23**, 91–95.

Dempster, A.P. (1963) Stepwise multivariate analysis of variance based on principal variables, *Biometrics*, **19**, 478–490.

Derde, M.P. and Massart, D.L. (1988) Comparison of the performance of the class modelling techniques UNEQ, SIMCA and PRIMA, *Chemom. Intel. Lab. Syst.*, **4**, 65–93.

Derflinger, G. (1984) A loss function for alpha factor analysis, *Psychometrika*, **49**, 325–330.

DeSarbo, W.S. (1981) Canonical/redundancy factoring analysis, *Psychometrika*, **46**, 307–329.

DeSarbo, W.S. and Carroll, J.D. (1985) Three-way unfolding via alternating least squares, *Psychometrika*, **50**, 275–300.

DeSarbo, W.S., Hausman, R.E., Lin, S., and Thompson, W. (1982) Constrained canonical correlation, *Psychometrika*, **47**, 489–516.

DeSarbo, W.S. and Hoffman, D.L. (1987) Constructing MDS joint spaces from binary choice data: A multivariate unfolding threshold model for marketing research, *J. Mark. Res.*, **24**, 40–54.

Devaux, M.-F., Bertrand, D., Robert, P., and Morat, J.-L. (1987) Extraction of near infra-red spectral information by fast fourier transform and principal component analysis. Application to the discrimination of baking quality of wheat flours, *J. Chemom.*, **1**, 103–110.

Devlin, S.J., Gnanadesikan, R., and Kettenring, J.R. (1975) Robust estimation and outlier detection with correlation coefficients, *Biometrika*, **62**, 531–545.

Devlin, S.J., Gnanadesikan, R., and Kettenring, J.R. (1981) Robust estimation of dispersion matrices and principal components, *J. Amer. Stat. Assoc.*, **76**, 354–362.

Dey, D.K. (1988) Simultaneous estimation of eigenvalues, *Ann. Inst. Stat. Math.*, **40**, 137–147.

Didow, N.M., Keller, K.L., Barksdale, H.C., Jr., and Franke, G.R. (1985) Improving measure quality by alternating least squares optimal scaling, *J. Mark. Res.*, **22**, 30–40.

Dietz, P.O., Fogler, H.R., and Smith, M. (1985) Factor analysis of portfolio styles, *Interfaces*, **15**, #2, 50–62.

Dillon, W.R. and Goldstein, M. (1984) *Multivariate Analysis*, John Wiley and Sons, Inc., New York.

Dillon, W.R., Madden, T.J., and Mulani, N. (1983) Scaling models for categorical variables: An application of latent structure models, *J. Cons. Res.*, **10**, 209–224.

Dillon, W.R., Mulani, N., and Frederick, D.G. (1989) On the use of component scores in the presence of group structure, *J. Cons. Res.*, **16**, 106–112.

Doran, H.E. (1976) A spectral principal components estimator of the distributed lag model, *Int. Econ. Rev.*, **17**, 8–25.

Draper, N.R. and Smith, H. (1980). *Applied Regression Analysis*, 2d. Ed., John Wiley and Sons, Inc., New York.

Droge, J.B.M., Rinsma, W.J., Van't Klooster, H.A., Tas, A.C., and Van Der Greef, J. (1987) An evaluation of SIMCA, Part 2. Classification of pyrolysis mass spectra of pseudomonas and serratia bacteria by pattern recognition using the SIMCA classifier, *J. Chemom.*, **1**, 231–241.

Droge, J.B.M. and Van't Klooster, H.A. (1987) An evaluation of SIMCA, Part 1. The reliability of the SIMCA pattern recognition method for a varying number of objects and features, *J. Chemom.*, **1**, 221–230.

Dudzinski, M. L. (1975) Principal components analysis and its use in hypothesis generation and multiple regression, *Developments in Field Experiment Design and Analysis* (V.J. Bofinger and J.L. Wheeler, editors) Bulletin #50, Commonwealth Bureau of Pastures and Field Crops, Commonwealth Agricultural Bureaux, Farnham Royal, Slough, U.K., pp. 85–105.

Dudzinski, M.L., Norris, J.M., Chmura, J.T., and Edwards, C.B.H. (1975) Repeatability of principal components in samples: normal and non-normal data sets compared, *Mult. Behav. Res.*, **10**, 109–118.

Duncan, O.D., Sloane, D.M., and Brody, C. (1982) Latent classes inferred from response-consistency effects, *Systems Under Indirect Observation*, I (K.G. Jöreskog and H. Wold, editors), North-Holland, New York, pp. 19–64.

Dunn, T.R. and Harshman, R.A. (1982) A multidimensional scaling model for the size-weight illusion, *Psychometrika*, **47**, 25–45.

Dunn, W.J. III and Wold, S. (1978) A structure–carcinogenicity study of 4-nitroquinoline 1-oxides using the SIMCA method of pattern recognition, *J. Med. Chem.*, **21**, 1001–1007.

Dunn, W.J. III, Wold, S. and Martin, Y.C. (1978) Structure–activity study of β-adrenergic agents using the SIMCA method of pattern recognition, *J. Med. Chem.*, **21**, 922–930.

Dunteman, G.H. (1989) *Principal Component Analysis*, Sage Publications, Beverly Hills, Calif.

Dupačová, J. and Wold, H. (1982) On some identification problems in ML modeling of systems with indirect observations, *Systems Under Indirect Observations*, II (K.G. Jöreskog and H. Wold, editors), North-Holland, New York, pp. 293–315.

Durban, J. and Knott, M. (1972) Components of Cramer–von Mises statistics, I, *J. Roy. Stat. Soc. B*, **34**, 290–307. Correction (1975), **37**, 237.

Durban, J., Knott, M., and Taylor, C.C. (1975) Components of Cramer—von Mises statistics, II, *J. Roy. Stat. Soc. B*, **37**, 216–237.

Dwyer, F.R. and Oh, S. (1987) Output sector munificence effects on the internal political economy of marketing channels, *J. Mark. Res.*, **24**, 347–358.

Eastment, H.T. and Krzanowski, W.J. (1982) Cross-validatory choice of the number of components from a principal component analysis, *Technometrics*, **24**, 73–78.

Eckart, C. and Young, G. (1936) The approximation of one matrix by another of lower rank, *Psychometrika*, **1**, 211–218.

Efron, B. and Tibshirani, R. (1986) Bootstrap methods for standard errors, confidence intervals and other measures of statistical accuracy, *Stat. Sci.*, **1**, 54–75.

Efroymson, M.A. (1960) Multiple regression analysis, *Mathematical Methods for Digital Computers* (Ralston, A. and Wilf, H.S., editors), John Wiley and Sons, Inc., New York, pp. 191–203.

Emerson, J.D., Hoaglin, D.C., and Kempthorne, P.J. (1984) Leverage in least squares additive-plus-multiplicative fits for two-way tables, *J. Amer. Stat. Assoc.*, **79**, 329–335.

Eplett, W.J.R. (1978) A note about the multipliers in latent root regression, *J. Roy. Stat. Soc. B*, **40**, 184–185.

Escoufier, Y. (1973) Le traitment des variables vectorielles, *Biometrics*, **29**, 751–760.

Escoufier, Y. and Grorud, A. (1980) Analysis factorielle des matrices carrees non symetriques, *Data Analysis and Informatics* (E. Diday et al., editors), North-Holland, New York, pp. 263–276.

Espeland, M.A. and Handelman, S.L. (1989) Using latent class models to characterize and assess relative error in discrete measurements, *Biometrics*, **45**, 587–599.

Eubank, R.L. and Webster, J.T. (1985) The singular-value decomposition as a tool for solving estimability problems, *Amer. Stat.*, **39**, 64–66.

Evans, G.T. (1971) Transformation of factor matrices to achieve conguence, *Br. J. Math. Stat. Psychol.*, **24**, 22–48.

Evans, M.J., Gilula, Z., and Guttman, I. (1989) Latent class analysis of two-way contingency tables by Bayesian methods, *Biometrika*, **76**, 557–563.

Everitt, B.S. (1980) *Cluster Analysis*, 2d Ed., Heinemann Educational Books, Inc., Exeter, N.H.

Everitt, B.S. (1984) *An Introduction to Latent Variable Models*, Chapman and Hall, New York.

Everitt, B.S. and Dunn, G. (1983) *Advanced Methods of Data Exploration and Modeling*, Heinemann Educational Books, Inc., Exeter, N.H.

Exton, H. (1976) *Multiple Hypergeometric Functions and Applications*, John Wiley and Sons, Inc., New York.

Exton, H. (1978) *Handbook of Hypergeometric Integrals*, John Wiley and Sons, Inc., New York.

Fang, C. and Krishnaiah, P.R. (1982) Asymptotic distributions of functions of the eigenvalues of some random matrices for nonnormal populations, *J. Mult. Anal.*, **12**, 39–63.

Fang, C. and Krishnaiah, P.R. (1986) On asymptotic distribution of the test statistic for the mean of the non-isotropic principal component, *Comm. Stat.—Simul. Comp.*, **15**, 1163–1168.

Farmer, S.A. (1971) An investigation into the results of principal component analysis of data derived from random numbers, *Statistician*, **20**, 63–72.

Fellegi, I.P. (1975) Automatic editing and imputing of quantitative data, *Bull. Int. Stat. Inst.*, **46**, 249–253.

Fielding, A. (1977) Latent structure models, *The Analysis of Data* I: *Exploring Data Structures* (C.A. O'Muircheartaigh and C. Payne, editors), John Wiley and Sons, Inc., New York, pp. 125–157.

Finkbeiner, C. (1979) Estimation for the multiple factor model when data are missing, *Psychometrika*, **44**, 409–420.

Finn, A. (1988) Print ad recognition readership scores: An information processing perspective, *J. Mark. Res.*, **25**, 168–177.

Finn, J.D. (1977) Multivariate analysis of variance and covariance, *Statistical Methods for Digital Computers* III (K. Enslein *et al.*, editors), John Wiley and Sons, Inc., New York, pp. 203–264.

Finney, D.J. (1956) Multivariate analysis and agricultural experiments, *Biometrics*, **12**, 67–71.

Fisher, M.T., Lee, J., and Mara, M.K. (1986) Techniques for evaluating control of automated multi-determinant analytical instruments by computer, *Analyst*, **111**, 1225–1229.

Fisher, R.A. (1932) *Statistical Methods for Research Workers*, 4th Ed., Oliver and Boyd, Edinburgh.

Fisher, R.A. (1936) The use of multiple measurements in taxonomic problems, *Ann. Eugenics*, **7**, 179–188.

Fisher, R.A. (1939) The sampling distribution of some statistics obtained from non-linear equations, *Ann. Eugenics*, **9**, 238–249.

Fisher, R.A. and MacKenzie, W.A. (1923) Studies in crop variation II. The manurial response of different potato variates, *J. Agr. Sci.*, **13**, 311–320.

Fisher, R.A. and Yates, F. (1974) *Statistical Tables for Biological Agricultural and Medical Research*, 6th Edition, Longman Group Ltd, London.

Fletcher, R. and Powell, M.J.D. (1963) A rapidly convergent descent method for minimization, *Comp. J.*, **6**, 163–168.

Flury, B.N. (1983) Some relations between the comparisons of covariance matrices and principal component analysis, *Comp. Stat. and Data Anal.*, **1**, 97–109.

Flury, B.N. (1984) Common principal components in *k* groups, *J. Amer. Stat. Assoc.*, **79**, 892–898.

Flury, B.N. (1986) Asymptotic theory for common principal component analysis, *Ann. Stat.*, **14**, 418–430.

Flury, B.N. (1987) Two generalizations of the common principal component model, *Biometrika*, **74**, 59–69.

Flury, B.N. (1988) *Common Principal Components and Related Multivariate Models*, John Wiley and Sons, Inc., New York.

Flury, B.N. and Constantine, G. (1985) Algorithm AS211: The FG diagonalization algorithm, *Appl. Stat.*, **34**, 177–183.

Flury, B.N. and Gautschi, W. (1986) An algorithm for simultaneous orthogonal transformation of several positive definite symmetric matrices to nearly diagonal form, *SIAM J. Sci. Stat. Comp.*, **7**, 169–184.

Fomby, T.B., Hill, R.C., and Johnson, S.R. (1978) An optimal property of principal components in the context of restricted least squares, *J. Amer. Stat. Assoc.*, **73**, 191–193.

Forgas, J.P. (1979) Multidimensional scaling: a discovery method in social psychology, *Emerging Strategies in Social Psychological Research* (G.P. Ginsburg, editor), John Wiley and Sons, Inc., New York, pp. 253–288. (Partially reprinted in Davies and Coxon, 1982.)

Formann, A.K. (1988) Latent class models for nonmonotone dichotomous items, *Psychometrika*, **53**, 45–62.

Fornell, C. (1979) Improving methodology for the analysis of multidimensional phenomena, *Educator's Conference AMA*, 9–13.

Fornell, C., Barclay, D.W., and Rhee, B.-D. (1988) A model and simple algorithm for redundancy analysis, *Mult. Behav. Res.*, **23**, 349–360.

Fornell, C. and Rust, R.T. (1989) Incorporating prior theory in covariance structure analysis: A Bayesian approach, *Psychometrika*, **54**, 249–259.

Fortier, J.J. (1966) Simultaneous linear prediction, *Psychometrika*, **31**, 369–381.

Frane, J.W. (1975) A new BMDP program for the description and estimation of missing data, *Proc. Stat. Comp. Sec., Amer. Stat. Assoc.*, 110–113.

Frane, J.W. (1976) Some simple procedures for handling missing data in multivariate analysis, *Psychometrika*, **41**, 409–415.

Frank, I.E., and Feikema, J., Constantine, N., and Kowalski, B.R. (1984) Prediction of product quality from spectral data using the partial least squares method, *J. Chem. Inf. Comp. Sci.*, **24**, 20–24.

Frankel, S., and Chan, R., and Lewandowski, M.E. (1984) Demonstration of multidimensional scaling as an evaluation technique: Evaluation of the vertical team concept in an evaluation training program, *J. Appl. Behav. Res.*, **20**, 193–201.

Freeman, G.H. (1975) Analysis of interactions in incomplete two-way tables, *Appl. Stat.*, **24**, 46–55.

Freund, R.A. (1960) A reconsideration of the variables control chart with special reference to the chemical industries, *Ind. Qual. Control*, **16**, No. 11, 35–41.

Freund, R.J. and Minton, P.D. (1979) *Regression Methods*, Marcel Dekker, Inc., New York.

Frisch, R. (1929) Correlation and scatter in statistical variables, *Nord. Stat. Tid.*, **8**, 36–102.

Fujikoshi, Y. (1977) Asymptotic expansions for distributions of some multivariate tests, *Multivariate Analysis*, IV (P.R. Krishnaiah, editor), North-Holland, New York, pp. 55–71.

Fujikoshi, Y. (1978) Asymptotic expansions for the distribution of some sample functions of the latent roots of matrices in three situations, *J. Mult. Anal.*, **8**, 63–72. Correction (1980), **10**, 140.

Fujikoshi, Y. (1980) Asymptotic expansions for the distributions of the sample roots under normality, *Biometrika*, **67**, 45–51.

Gabriel, K.R. (1968) Simultaneous test procedures in multivariate analysis of variance, *Biometrika*, **55**, 489–504.

Gabriel, K.R. (1969) A comparison of some methods of simultaneous inference in MANOVA, *Multivariate Analysis*, II (P.R. Krishnaiah, editor), Academic Press, New York, pp. 67–86. (Also in Atchley and Bryant, 1975.)

Gabriel, K.R. (1971) The biplot-graphic display of matrices with application to principal component analysis, *Biometrika*, **58**, 453–467.

Gabriel, K.R. (1978) Least squares approximation of matrices by additive and multiplicative models, *J. Roy. Stat. Soc. B*, **40**, 186–196.

Gabriel, K.R. (1981) Biplot display of multivariate matrices for inspection of data and diagnosis, *Interpreting Multivariate Data* (V. Barnett, editor) John Wiley and Sons, Inc., New York, pp. 147–173.

Gabriel, K.R. and Haber, M. (1973) The Moore–Penrose inverse of a data matrix—a statistical tool with some meteorological applications, *Int. Symp. on Prob. and Stat. in the Atmospheric Sciences*, Boulder, Colo.

Gabriel, K.R. and Odoroff, C.L. (1984) Resistant lower rank approximation of matrices, *Data Analysis and Informatics*, **III**, (E. Diday *et al.*, editors), North-Holland, New York, pp. 22–30.

Gabriel, K.R. and Odoroff, C.L. (1986a) Some diagnoses of models by 3-D biplots, *Proc. of Multidimensional Data Analysis Workshop*, (J. DeLeeuw *et al.*, editors), DSWO-Press, Leiden).

Gabriel, K.R. and Odoroff, C.L. (1986b) Illustrations of model diagnosis by means of three-dimensional biplots, *Statistical Image Processing and Graphics* (E.J. Wegman and D.J. DePriest, editors), Marcel Dekker, Inc., New York, pp. 257–274.

Gabriel, K.R. and Odoroff, C.L. (1986c) *The Biplot for Exploration and Diagnosis: Examples and Software*, Dept. of Stat. Rep. #86/03, University of Rochester, Rochester, New York.

Gabriel, K.R. and Odoroff, C.L. (1990) Biplots in biomedical research, *Stat. in Medicine*, **9**, 469–483.

Gabriel, K.R. and Zamir, S. (1979) Lower rank approximation of matrices by least squares with any choice of weights, *Technometrics*, **21**, 489–498.

Galpin, J.S. and Hawkins, D.M. (1987) Methods of L_1 estimation of a covariance matrix, *Comp. Stat. & Data Anal.*, **5**, 305–319.

Gaski, J.F. (1986) Interrelations among a channel entity's power sources: Impact of the exercise of reward and coercion on expert, referent and legitimate power sources, *J. Mark. Res.*, **23**, 62–77.

Gaski, J.F. (1987) Commentary on Howell's observations, *J. Mark. Res.*, **24**, 127–129.

Gebhardt, F. (1968) Counterexample to two-dimensional varimax-rotation, *Psychometrika*, **33**, 35–36.

Geisser, S. (1965) Bayesian estimation in multivariate analysis, *Ann. Math. Stat.*, **36**, 150–159.

Geladi, P. (1988) Notes on the history and nature of partial least squares modeling, *J. Chemom.*, **2**, 231–246.

Geladi, P. and Kowalski, B. (1986) Partial least squares regression: a tutorial, *Analytica Chimica Acta*, **185**, 1–17.

Gemperline, P.J. (1989) Mixture analysis using factor analysis I: Calibration and quantitation, *J. Chemom.*, **3**, 549–568.

Geweke, J.F. and Singleton, K.J. (1980) Interpreting the likelihood ratio statistic in factor models when sample size is small, *J. Amer. Stat. Assoc.*, **75**, 133–137.

Gheva, D. (1988) Diagnosing departures from stability in a multidimensional multivariate system, *Appl. Stoch. Models & Data Anal.*, **4**, 55–64.

Girshick, M.A. (1936) Principal components, *J. Amer. Stat. Assoc.*, **31**, 519–528.

Girshick, M.A. (1939) On the sampling theory of the roots of determinental equations, *Ann. Math. Stat.*, **10**, 203–224.

Gittins, R. (1969) The application of ordination techniques, *Ecological Aspects of the Mineral Nutrition of Plants* (I.H. Rorison, editor), Blackwell Scientific Publications, Inc., Oxford, U.K., pp. 37–66. (Also in Bryant and Atchley, 1975.)

Glahn, H.R. (1968) Canonical correlation and its relationship to discriminant analysis and multiple regression, *J. Atmospheric Sci.*, **25**, 23–31. (Also in Bryant and Atchley, 1975.)

Glasser, M. (1964) Linear regression analysis with missing observations among the independent variables, *J. Amer. Stat. Assoc.*, **59**, 834–844.

Gleason, T.C. (1976) On redundancy in canonical analysis, *Psych. Bull.*, **83**, 1004–1006.

Gleason, T.C. and Staelin, R. (1973) Improving the metric quality of questionnaire data, *Psychometrika*, **38**, 393–410.

Gleason, T. C. and Staelin, R. (1975) A proposal for handling missing data, *Psychometrika*, **40**, 229–252.

Gnanadesikan, R. (1977) *Methods for Statistical Data Analysis of Multivariate Observations*, John Wiley and Sons, Inc., New York.

Gnanadesikan, R. and Kettenring, J.R. (1972) Robust estimates, residuals, and outlier detection with multiresponse data, *Biometrics*, **28**, 81–124.

Gnanadesikan, R. and Lee, E.T. (1970) Graphical techniques for internal comparisons amongst equal degree of freedom groupings in multiresponse experiments, *Biometrika*, **57**, 229–237.

Gnanadesikan, R. and Wilk, M.B. (1969) Data analytic methods in multivariate statistical analysis, *Multivariate Analysis*, II (P.R. Krishnaiah, editor), Academic Press, New York, pp. 593–638.

Gold, E.M. (1973) Metric unfolding: Data requirement for unique solution and clarification of Schönemann's algorithm, *Psychometrika*, **38**, 555–569.

Goldstein, H. and McDonald, R.P. (1988) A general model for the analysis of multilevel data, *Psychometrika*, **53**, 455–467.

Gollob, H.F. (1968a) A statistical model which combines features of factor analytic and analysis of variance techniques, *Psychometrika*, **33**, 73–115.

Gollob, H.F. (1968b) Confounding of sources of variation in factor-analytic techniques, *Psych. Bull.*, **70**, 330–344.

Gondran, M. (1977) Eigenvalues and eigenvectors in hierarchical classification, *Recent Developments in Statistics* (J.R. Barra *et al.*, editors), North-Holland, New York, pp. 775–781.

Good, I.J. (1950) On the inversion of circulant matrices, *Biometrika*, **37**, 185–186.

Good, I.J. (1969) Some applications of the singular decomposition of a matrix, *Technometrics*, **11**, 823–831. Errata, **12**, 722, (1970).

Goodman, L.A. (1986) Some useful extensions of the usual correspondence analysis approach and the usual log-linear models approach in the analysis of contingency tables (with discussion), *Int. Stat. Rev.*, **54**, 243–309.

Goodman, N.R. (1963) Statistical analysis based on a certain multivariate complex Gaussian distribution, *Ann. Math. Stat.*, **34**, 152–177.

Gower, J.C. (1966a) Some distance properties of latent root and vector methods used in multivariate analysis, *Biometrika*, **53**, 325–338.

Gower, J.C. (1966b) A Q-technique for the calculation of canonical variates, *Biometrika*, **53**, 588–590.

Gower, J.C. (1968) Adding a point to vector diagrams in multivariate analysis, *Biometrika*, **55**, 582–585.

Gower, J.C. (1975) Generalized Procrustes analysis, *Psychometrika*, **40**, 33–51.

Gower, J.C. (1977) The analysis of asymmetry and orthogonality, *Recent Developments in Statistics* (J.R. Barra et al., editors), North-Holland, New York, pp. 109–123.

Gower, J.C. and Harding, S.A. (1988) Nonlinear biplots, *Biometrika*, **75**, 445–455.

Graff, J. and Schmidt, P. (1982) A general model for decomposition of effects, *Systems under Indirect Observation*, I (K.G. Jöreskog and H. Wold, editors), North-Holland, New York, pp. 131–148.

Grant, E.L. and Leavenworth, R.S. (1988) *Statistical Quality Control*, 6th Ed., McGraw-Hill, New York.

Green, B.F. (1952) The orthogonal approximation of an oblique structure in factor analysis, *Psychometrika*, **17**, 429–440.

Green, B.F., Jr. (1976) On the factor score controversy, *Psychometrika*, **41**, 263–266.

Green, B.F., Jr. (1977) Parameter sensitivity in multivariate methods, *Mult. Behav. Res.*, **12**, 263–287.

Green, P.E. (1978) *Analyzing Multivariate Data* (with contributions by J.D. Carroll), Dryden Press, Hinsdale, Ill.

Green, P.E., Carmone, F.J., Jr., and Smith, S.M. (1989) *Multidimensional Scaling: Concepts and Applications*, Allyn and Bacon, Needham Heights, Mass.

Green, P.E. and Rao, V.R. (1972) *Applied Multidimensional Scaling*, Dryden Press, Hinsdale, Ill.

Green, P.E. and Wind, Y. (1973) *Multiattribute Decisions in Marketing: A Measurement Approach*, The Dryden Press, Hinsdale, Ill.

Green, P.E., Wind, Y. and Jain, A.K. (1972) A note on measurement of social-psychological belief systems, *J. Mark. Res.*, **9**, 204–208.

Green, P.J. (1984) Iteratively reweighted least squares for maximum likelihood estimation and some robust and resistant alternatives (with discussion), *J. Roy. Stat. Soc. B*, **46**, 149–192.

Green, R.F., Guilford, J.P., Christensen, P.R., and Comrey, A.L. (1953) A factor-analytic study of reasoning abilities, *Psychometrika*, **18**, 135–160.

Greenacre, M.J. (1984) *Theory and Applications of Correspondence Analysis*, Academic Press, New York.

Greenacre, M.J. (1988) Correspondence analysis of multivariate categorical data by weighted least squares, *Biometrika*, **75**, 457–467.

Greenacre, M.J. (1989) The Carroll–Green–Schaffer scaling in correspondence analysis: A theoretical and empirical appraisal, *J. Mark. Res.*, **26**, 358–365.

Greenacre, M.J. and Browne, M.W. (1986) An efficient alternating least-squares algorithm to perform multidimensional scaling, *Psychometrika*, **51**, 241–250.

Greenacre, M.J. and Hastie, T. (1987) The geometric interpretation of correspondence analysis, *J. Amer. Stat. Assoc.*, **82**, 437–447.

Greenberg, E. (1975) Minimum variance properties of principal component regression, *J. Amer. Stat. Assoc.*, **70**, 194–197.

Grubbs, F.E. (1948) On estimating precision of measuring instruments and product variability, *J. Amer. Stat. Assoc.*, **43**, 243–264.

Grubbs, F.E. (1973) Errors of measurement, precision, accuracy and the statistical comparison of measuring instruments, *Technometrics*, **15**, 53–66.

Grum, F. and Wightman, T. (1960) Measure of the contribution of fluorescence to the brightness of papers treated with whitening agents, *TAPPI*, **43**, 400–405.

Guilford, J.P. and Michael, W.B. (1948) Approaches to univocal factor scores, *Psychometrika*, **13**, 1–22.

Gulliksen, H. (1954) A least squares solution for successive intervals assuming unequal standard deviations, *Psychometrika*, **19**, 117–139.

Gulliksen, H. (1975) Characteristic roots and vectors indicating agreement of data with different scaling laws, *Sankhyā B*, **37**, 363–384.

Gunst, R.F. (1983) Latent root regression, *Ency. of Stat. Sci.*, **4** (S. Kotz and N.L. Johnson, editors), John Wiley and Sons, Inc., New York, pp. 495–497.

Gunst, R.F. and Mason, R.L. (1977a) Advantages of examining multicollinearities in regression analysis, *Biometrics*, **33**, 249–260.

Gunst, R.F. and Mason, R.L. (1977b) Biased estimation in regression; An evaluation using mean squared error, *J. Amer. Stat. Assoc.*, **72**, 616–628.

Gunst, R.F., Webster, J.T., and Mason, R.L. (1976) A comparison of least squares and latent root regression estimators, *Technometrics*, **18**, 75–83.

Gupta, R.D. (1973) On testing latent vectors of a normal covariance matrix, *Can. J. Stat.*, **1**, 255–260.

Gupta, R.P. (1967) Latent roots and vectors of a Wishart matrix, *Ann. Inst. Stat. Math.*, **19**, 157–165.

Gupta, R.P. (1972) Principal components in the complex case, *Metrika*, **19**, 150–155.

Guttman, L. (1953) Image theory for the structure of quantitative variates, *Psychometrika*, **18**, 277–296.

Guttman, L. (1954) Some necessary conditions for common-factor analysis, *Psychometrika*, **19**, 149–161.

Guttman, L. (1956) Best "possible" systematic estimates of communalities, *Psychometrika*, **21**, 273–285.

Guttman, L. (1968) A general nonmetric technique for finding the smallest coordinate space for a configuration of points, *Psychometrika*, **33**, 469–506.

Hadi, A.S. (1987) The influence of single rows on the eigenstructure of a matrix, *Proc. Stat. Comp. Sec. Amer. Stat. Assoc.*, 85–90.

Hadi, A.S. and Wells, M.T. (1990) Assessing the effects of multiple rows on the condition number of a matrix, *J. Amer. Stat. Assoc.*, **85**, 786–792.

Hafner, R. (1981) An improvement of the Harman–Fukuda method for the Minres solution in factor analysis, *Psychometrika*, **46**, 347–349.

Hägglund, G. (1982) Factor analysis by instrumental variables, *Psychometrika*, **47**, 209–222.

Hakstian, A.R. (1971) A comparative evaluation of several prominent methods of oblique factor transformation, *Psychometrika*, **36**, 175–193.

Hakstian, A.R. (1976) Two-matrix orthogonal rotation procedures, *Psychometrika*, **41**, 267–272.

Hakstian, A.R. and Abell, R.A. (1974) A further comparison of oblique factor transformation methods, *Psychometrika*, **39**, 429–444.

Hampel, F.R. (1974) The influence curve and its role in robust estimation, *J. Amer. Stat. Assoc.*, **69**, 383–393.

Hannan, E.L. (1983) An eigenvalue method of evaluating contestants, *Computers and Oper. Res.*, **10**, 41–46.

Hanumara, R.C. and Thompson, W.A., Jr. (1968) Percentage points of the extreme roots of a Wishart matrix, *Biometrika*, **55**, 505–512.

Harmon, H.H. (1976) *Modern Factor Analysis*, 3d Ed., University of Chicago Press, Chicago.

Harmon, H.H. (1977) Minres method of factor analysis, *Statistical Methods for Digital Computers*, III (K. Enslein *et al.*, editors), John Wiley and Sons, Inc., pp. 154–165.

Harmon, H.H. and Fukuda, Y. (1966) Resolution of the Heywood case in the minres solution, *Psychometrika*, **31**, 563–571.

Harmon, H.H. and Jones, W.H. (1966) Factor analysis by minimizing residuals (Minres), *Psychometrika*, **31**, 351–368.

Harris, C.W. (1967) On factors and factor scores, *Psychometrika*, **32**, 363–379.

Harris, C.W. and Kaiser, H.F. (1964) Oblique factor analytic solutions by orthogonal transformations, *Psychometrika*, **29**, 347–362.

Harris, P. (1985) Testing for variance homogeneity of correlated variables, *Biometrika*, **72**, 103–107.

Harris, R.J. (1975) *A Primer of Multivariate Statistics*, Academic Press, New York.

Harshman, R.A., Green, P.E., Wind, Y., and Lundy, M.E. (1982) A model for the analysis of asymmetric data in marketing research, *Mark. Sci.*, **1**, 205–242.

Hartigan, J.A. (1975) *Clustering Algorithms*, John Wiley and Sons, Inc., New York.

Hashiguchi, S. and Morishima, H. (1969) Estimation of genetic contribution of principal components to individual variates concerned, *Biometrics*, **25**, 9–15.

Hastie, T. and Stuetzle, W. (1989) Principal curves. *J. Amer. Stat. Assoc.*, **84**, 502–516.

Hauser, J.R. and Urban, G.L. (1986) The value priority hypothesis for consumer budget plans, *J. Cons. Res.*, **12**, 446–462.

Hawkins, D.M. (1973) On the investigation of alternative regressions by principal component analysis, *Appl. Stat.*, **22**, 275–286.

Hawkins, D.M. (1974) The detection of errors in multivariate data using principal components, *J. Amer. Stat. Assoc.*, **69**, 340–344.

Hawkins, D.M. (1975) Relations between ridge regression and eigenanalysis of the augmented correlation matrix, *Technometrics*, **17**, 477–480.

Hawkins, D.M. (1980) *Identification of Outliers*, Chapman and Hall, London.

Hawkins, D.M. and Eplett, W.J.R. (1982) The Cholesky factorization of the inverse correlation or covariance matrix in multiple regression, *Technometrics*, **24**, 191–198.

Hawkins, D.M. and Fatti, L.P. (1984) Exploring multivariate data using the minor principal components, *Statistician*, **33**, 325–338.

Hayashi, C. (1980) Data analysis in a comparative study, *Data Analysis and Informatics* (E. Diday *et al.*, editors), North-Holland, New York, pp. 31–51.

Hayduk, L.A. (1987) *Structural Equation Modeling with LISREL*, Johns Hopkins University Press, Baltimore, Md.

Hays, W.L. and Bennett, J.F. (1961) Multidimensional unfolding: Determining configurations from complete rank order preference data, *Psychometrika*, **26**, 221–238.

Heath, H. (1952) A factor analysis of women's measurements taken for garment and pattern consideration, *Psychometrika*, **17**, 87–100.

Heermann, E.F. (1963) Univocal or orthogonal estimators of orthogonal factors, *Psychometrika*, **28**, 161–172.

Heermann, E.F. (1964) The geometry of factorial indeterminancy, *Psychometrika*, **29**, 371–381.

Heermann, E.F. (1966) The algebra of factorial indeterminancy, *Psychometrika*, **31**, 539–543.

Heesacker, M. and Heppner, P.P. (1983) Using real-client perceptions to examine psychometric properties of the counselor rating form, *J. Counseling Psych.*, **30**, 180–187.

Hegemann, V. and Johnson, D.E. (1976a) The power of two tests for nonadditivity, *J. Amer. Stat. Assoc.*, **71**, 945–948.

Hegemann, V. and Johnson, D.E. (1976b) On analyzing two-way AoV data with interaction, *Technometrics*, **18**, 273–281.

Heinlein, R.A. (1943) —And he built a crooked house, *The Pocket Book of Science Fiction* (D.A. Wollheim, editor), Pocket Books, Inc., New York, pp. 284–310.

Heiser, W.J. (1987) Correspondence analysis with least absolute residuals, *Comp. Stat. & Data Anal.*, **5**, 337–356.

Heiser, W.J. and Meulman, J. (1983) Analyzing rectangular tables by joint and constrained multidimensional scaling, *J. Econometrics*, **22**, 139–167.

Helland, I.S. (1988) On the structure of partial least squares regression, *Comm. Stat.—Simul. Comp.*, **17**, 581–607.

Helm, C.E. (1964) Multidimensional ratio scaling analysis of perceived color relations, *J. Optical Soc. Amer.*, **54**, 256–262.

Hendrickson, A.E. and White, P.O. (1964) PROMAX: A quick method for rotation to oblique simple structure, *Br. J. Math. Stat. Psychol.*, **17**, 65–70.

Hill, M.O. (1974) Correspondence analysis: A neglected multivariate method, *Appl. Stat.*, **23**, 340–354.

Hill, R.C., Fomby, T.B., and Johnson, S.R. (1977) Component selection norms for principal components regression, *Comm. Stat.—Theor. Meth.*, **A6**, 309–334.

Hirsch, R.F., Wu, G.L., and Tway, P.C. (1987) Reliability of factor analysis in the presence of random noise or outlying data, *Chemom. and Intel. Lab. Syst.*, **1**, 265–272.

Hocking, R.R. (1976) The analysis and selection of variables in linear regression, *Biometrics*, **32**, 1–49.

Hocking, R.R. (1984) Discussion of "K-clustering as a detection tool for influential subsets in regression" by J.B. Gray and R.F. Ling, *Technometrics*, **26**, 321–334.

Hocking, R.R., Speed, F.M., and Lynn, M.J. (1976) A class of biased estimators in linear regression, *Technometrics*, **18**, 425–437.

Hoel, P.G. (1937) A significance test for component analysis, *Ann. Math. Stat.*, **8**, 149–158.

Hoerl, A.E. and Kennard, R.W. (1970) Ridge regression: Biased estimation for nonorthogonal problems, *Technometrics*, **12**, 55–67.

Hoerl, R.W. (1985) Ridge analysis 25 years later, *Amer. Stat.*, **39**, 186–192.

Hoerl, R.W., Schuenemeyer, J.H., and Hoerl, A.E. (1986) A simulation of biased estimation and subset selection regression techniques, *Technometrics*, **28**, 369–380.

Hoffman, D.L. and Franke, G.R. (1986) Correspondence analysis: Graphical representation of categorical data in marketing research, *J. Mark. Res.*, **23**, 213–227.

Hogg, R.V. (1979a) Statistical robustness: one view of its use in applications today, *Amer. Stat.*, **33**, 108–115.

Hogg, R.V. (1979b) An introduction to robust estimation, *Robustness in Statistics* (R.L. Launer and G.N. Wilkinson, editors), Academic Press, Inc., New York, pp. 1–17.

Holland, D.A. (1969) Component analysis—an approach to the interpretation of soil data, *J. Sci. Food Agric.*, **20**, 26–31.

Hooley, G.J. (1984) Modelling product positions through the use of multidimensional scaling techniques: An empirical investigation, *Eur. J. Op. Res.*, **16**, 34–41.

Hooper, J.W. (1959) Simultaneous equations and canonical correlation theory, *Econometrica*, **27**, 245–256.

Hopkins, J.W. (1966) Some considerations in multivariate allometry, *Biometrics*, **22**, 747–760.

Horan, C.B. (1969) Multidimensional scaling: Combining observations when individuals have different perceptual structures, *Psychometrika*, **34**, 139–165.

Horn, J.L. (1965) A rationale and test for the number of factors in factor analysis, *Psychometrika*, **30**, 179–185.

Horst, P. (1961) Relations among *m* sets of measures, *Psychometrika*, **26**, 129–149.

Horst, P. (1965) *Factor Analysis of Data Matrices*, Holt, Rinehart & Winston, New York.

Horvitz, D.G. and Thompson, D.J. (1952) A generalization of sampling without replacement from a finite universe, *J. Amer. Stat. Assoc.*, **47**, 663–685.

Höskuldsson, A. (1988) PLS regression methods, *J. Chemom.*, **2**, 211–228.

Hotelling, H. (1931) A generalization of Student's ratio, *Ann. Math. Stat.*, **2**, 360–378. (Also in Atchley and Bryant, 1975.)

Hotelling, H. (1933) Analysis of a complex of statistical variables into principal components, *J. Educ. Psychol.*, **24**, 417–441, 498–520. (pp. 417–441 are also in Bryant and Atchley, 1975.)

Hotelling, H. (1935) The most predictable criterion, *J. Educ. Psychol.*, **26**, 139–142. (Also in Atchley and Bryant, 1975.)

Hotelling, H. (1936a) Simplified calculation of principal components, *Psychometrika*, **1**, 27–35.

Hotelling, H. (1936b) Relations between two sets of variates, *Biometrika*, **28**, 321–377.

Hotelling, H. (1942) Rotation in psychology and the statistical revolution. (A review of "Factor Analysis. A Synthesis of Factorial Methods" by K.J. Holzinger and H.H. Harmon.) *Science*, **95**, 504–507.

Hotelling, H. (1943a) Some new methods in matrix calculation, *Ann. Math. Stat.*, **14**, 1–34.

Hotelling, H. (1943b) Further points on matrix calculation and simultaneous equations, *Ann. Math. Stat.*, **14**, 440–441.

Hotelling, H. (1947) Multivariate quality control, illustrated by the air testing of sample bombsites, *Selected Techniques of Statistical Analysis* (C. Eisenhart *et al.*, editors), McGraw-Hill, New York, pp. 111–184.

Hotelling, H. (1949) Practical problems of matrix calculation, *Proc. Berkeley Symp. on Math. Stat. & Prob.*, **1**, University of California Press, Berkeley, pp. 275–293.

Hotelling, H. (1951) A generalized *T* test and measure of multivariate dispersion, *Proc. 2nd Berkeley Symp. on Math. Stat. and Prob.*, **1**, University of California Press, Berkeley, pp. 23–41.

Hotelling, H. (1954) *Multivariate Methods in Testing Complex Equipment*, University of North Carolina Mimeo Series #106, Chapel Hill, N.C.

Hotelling, H. (1957) The relations of newer multivariate statistical methods to factor analysis, *Br. J. of Stat. Psych.*, **10**, 69–79.

Householder, A.S. and Young, G. (1938) Matrix approximations and latent roots, *Amer. Math. Monthly*, **45**, 165–171.

Howe, W.G. (1955) *Some Contributions to Factor Analysis*, Oak Ridge National Laboratory Report ORNL-1919, Oak Ridge, Tenn.

Howell, R.D. (1987) Covariance structure modeling and measurement issue: A note on "Interrelations among a channel entity's power sources", *J. Mark. Res.*, **24**, 119–126.

Howery, D.G. (1977) Factor analysis, *Statistics: 1977 Eastern Analytical Symposium* (R.F. Hirsch, editor), The Franklin Institute Press, Philadelphia, pp. 185–213.

Howery, D.G. and Soroka, J.M. (1987) Target factor analysis of parameters influencing solute-monomeric stationary phase interactions, *J. Chemom.*, **1**, 91–101.

Hsu, P.L. (1939) On the distribution of roots of certain determinental equations, *Ann. Eugenics*, **9**, 250–258.

Hsu, P.L. (1940) On generalized analysis of variance, *Biometrika*, **31**, 221–237.

Hsuan, F.C. (1981) Ridge regression from principal component point of view, *Commun. Stat.—Theor. Meth.*, **A10**, 1981–1995.

Huber, J. (1976) Ideal point models of preference, *Advances in Consumer Research*, **III**, 138–142.

Huber, P.J. (1964) Robust estimation of a location parameter, *Ann. Math. Stat.*, **35**, 73–101.

Huber, P.J. (1981) *Robust Statistics*, John Wiley and Sons, Inc., New York.

Hudlet, R. and Johnson, R.A. (1982) An extension of some optimal properties of principal components, *Ann. Inst. Stat. Math.*, **34**, 105–110.

Hui, B.S. and Wold, H. (1982) Consistency and consistency at large of partial least square estimates, *Systems Under Indirect Observation*, **II** (K.G. Jöreskog and H. Wold, editors), North-Holland, New York, pp. 119–130.

Huitson, A. (1989) Problems with Procrustes analysis, *J. Appl. Stat.*, **16**, 39–45.

Hurley, J.R. and Cattell, R.B. (1962) The PROCRUSTES program: Producing direct rotation to test a hypothesized factor structure, *Behav. Sci.*, **7**, 258–262.

Hutchinson, J.W. (1989) NETSCALE: A network scaling algorithm for nonsymmetric proximity data. *Psychometrika*, **54**, 25–51.

Iglarsh, H.J. and Cheng, D.C. (1980) Weighted estimators in regression with multicollinearity, *J. Stat. Comp. Simul.*, **10**, 103–112.

Indow, T. and Kanazawa, K. (1960) Multidimensional mapping of Munsell colors varying in hue, chroma and value, *J. Exp. Psych.*, **59**, 330–336.

Indow, T. and Uchizono, T. (1960) Multidimensional mapping of Munsell colors varying in hue and chroma, *J. Exp. Psych.*, **59**, 321–329.

Isogawa, Y. and Okamoto, M. (1980) Linear prediction on the factor analysis model, *Biometrika*, **67**, 482–484.

Israels, A.Z. (1984) Redundancy analysis for qualitative variables, *Psychometrika*, **49**, 331–346.

Israels, A.Z. (1986) Interpretation of redundancy analysis: Rotated vs. unrotated solutions, *Appl. Stoch. Models & Data Anal.*, **2**, 121–130.

Izenman, A.J. and Williams, J.S. (1989) A class of linear spectral models and analyses for the study of longitudinal data, *Biometrics*, **45**, 831–849.

Jackson, J.E. (1956) Quality control methods for two related variables, *Ind. Qual. Control*, **12**, No. 7, 4–8.

Jackson, J.E. (1959) Quality control methods for several related variables, *Technometrics*, **1**, 359–377.

Jackson, J.E. (1960) Multivariate analysis illustrated by Nike-Hercules, *Proc. 6th Conf. on the Des. of Exp. in Army Res. Dev. and Testing*, U.S. Army Research Office, Durham, N.C., pp. 307–327.

Jackson, J.E. (1978) Multidimensional Scaling, *Photo. Sci. & Eng.*, **22**, 97–101.

Jackson, J.E. (1980) Principal components and factor analysis I: Principal components, *J. Qual. Tech.*, **12**, 201–213.

Jackson, J.E. (1981a) Principal components and factor analysis II: Additional topics related to principal components, *J. Qual. Tech.*, **13**, 46–58.

Jackson, J.E. (1981b) Principal components and factor analysis III: What is factor analysis?, *J. Qual. Tech.*, **13**, 125–130.

Jackson, J.E. (1985) Multivariate quality control, *Comm. Stat.—Theor. Meth.*, **14**, 2657–2688.

Jackson, J.E. and Bradley, R.A. (1961a) Sequential χ^2-and T^2-tests and their application to an acceptance sampling problem, *Technometrics*, **3**, 519–534.

Jackson, J.E. and Bradley, R.A. (1961b) Sequential χ^2- and T^2-tests, *Ann. Math. Stat.*, **32**, 1063–1077.

Jackson, J.E. and Bradley, R.A. (1966) Sequential multivariate procedures for means with quality control applications, *Multivariate Analysis*, I (P.R. Krishnaiah, editor), Academic Press, Inc., New York, pp. 507–519.

Jackson, J.E., Fassett, D.W., Riley, E.C., and Sutton, W.L. (1962) Evaluation of the variability in audiometric procedures, *J. Acoustical Soc. Amer.*, **34**, 218–222.

Jackson, J.E. and Hearne, F.T. (1973) Relationships among coefficients of vectors used in principal components, *Technometrics*, **15**, 601–610.

Jackson, J.E. and Hearne, F.T. (1978) Allowance of age in the multivariate analysis of hearing loss, *Biometrie-Praximetrie*, **18**, 83–104.

Jackson, J.E. and Hearne, F.T. (1979a) Hotelling's T_M^2 for principal components—what about absolute values? *Technometrics*, **21**, 253–255.

Jackson, J.E. and Hearne, F.T. (1979b) PRINCOM—A principal components program, *Proc. Stat. Comp. Sec., Amer. Stat. Assoc.*, 344–347.

Jackson, J.E. and Lawton, W.H. (1967) Regression residual analysis: Query 22, *Technometrics*, **9**, 339–340.

Jackson, J.E. and Lawton, W.H. (1969) Comparison of ANOVA and harmonic components of variance, *Technometrics*, **11**, 75–90.

Jackson, J.E. and Lawton, W.H. (1976) Some probability problems associated with cross-impact analysis, *Tech. Forecast. and Soc. Change*, **8**, 263–273.

Jackson, J.E. and Morris, R.H. (1957) An application of multivariate quality control to photographic processing, *J. Amer. Stat. Assoc.*, **52**, 186–199.

Jackson, J.E. and Mudholkar, G.S. (1979) Control procedures for residuals associated with principal component analysis, *Technometrics*, **21**, 341–349; Addendum, **22**, 136 (1980).

James, A.T. (1960) The distribution of the latent roots of the covariance matrix, *Ann. Math. Stat.*, **31**, 151–158.

James, A.T. (1964) Distributions of matrix variates and latent roots derived from normal samples, *Ann. Math. Stat.*, **35**, 475–501.

James, A.T. (1966) Inference on latent roots by calculation of hypergeometric functions of matrix argument, *Multivariate Analysis*, I (P.R. Krishnaiah, editor), Academic Press, Inc., New York, pp. 209–235.

James, A.T. (1969) Tests of equality of latent roots of the covariance matrix, *Multivariate Analysis*, II (P.R. Krishnaiah, editor), Academic Press, Inc., New York, pp. 205–218.

James, A.T. (1977) Tests for a prescribed subspace of principal components, *Multivariate Analysis*, IV (P.R. Krishnaiah, editor), North-Holland, New York, pp. 73–77.

James, A.T. and Venables, W. (1980) Interval estimates for bivariate principal axis, *Multivariate Analysis*, V (P.R. Krishnaiah, editor), North-Holland, New York, pp. 399–411.

Jeffers, J.N.R. (1962) Principal component analysis of designed experiment, *Statistician*, **12**, 230–242.

Jeffers, J.N.R. (1967) Two case studies on the application of principal component analysis, *Appl. Stat.*, **16**, 225–236.

Jeffers, J.N.R. (1981) Investigation of some alternative regressions: Some practical examples, *Statistician*, **30**, 79–88.

Jenkins, G.M. (1973) Comments on papers of Cook and Craddock, *Statistician*, **22**, 227–232.

Jenkins, G.M. and Watts, D.G. (1968) *Spectrum Analysis and its Applications*, Holden-Day, San Francisco.

Jennrich, R.I. (1973a) Standard errors for obliquely rotated factor loadings, *Psychometrika*, **38**, 593–604.

Jennrich, R.I. (1973b) On the stability of rotated loadings. The Wexler phenomenon. *Br. J. Math. Stat. Psychol.*, **26**, 167–176.

Jennrich, R.I. (1974) Simplified formula for standard errors in maximum-likelihood factor analysis, *Br. J. Math. Stat. Psychol.*, **27**, 122–131.

Jennrich, R.I. (1979) Admissible values of γ in direct oblimin rotation, *Psychometrika*, **44**, 173–177.

Jennrich, R.I. (1987) Tableau algorithms for factor analysis by instrumental variable methods, *Psychometrika*, **52**, 469–476.

Jennrich, R.I. and Sampson, P.F. (1966) Rotation for simple loadings, *Psychometrika*, **31**, 313–323. (Also in Bryant and Atchley, 1975.)

Jennrich, R.I. and Thayer, D.T. (1973) A note on Lawley's formulas for standard errors in maximum likelihood factor analysis, *Psychometrika*, **38**, 571–580.

Jensen, D.R. (1986) The structure of ellipsoidal distributions, II: Principal components, *Biom. J.*, **28**, 363–369.

Jensen, D.R. (1988) *Conditional Properties of Principal Components*, Tech. Rep., Dept. of Stat., Virginia Poly. Inst. & State. Univ., Blacksburg, Va.

Jensen, R.E. (1981) Scenario probability scaling: an eigenvector analysis of elicited scenario odds ratios, *Futures*, **13**, 489–498.

Joe, G.W. and Woodward, J.A. (1976) Some developments in multivariate generalizability, *Psychometrika*, **41**, 205–217.

Johansson, J.K. (1981) An extension of Wollenberg's redundancy analysis, *Psychometrika*, **46**, 93–103.

Johnson, D.E. (1976) Some new multiple comparison procedures for the two-way AoV model with interaction, *Biometrics*, **32**, 929–934.

Johnson, D.E. and Graybill, F.A. (1972) An analysis of a two-way model with interaction and no replication, *J. Amer. Stat. Assoc.*, **67**, 862–868.

Johnson, R.A. and Wichern, D.W. (1982) *Applied Multivariate Statistical Analysis*, Prentice-Hall, Inc., Englewood Cliffs, N.J.

Johnson, R.M. (1963) On a theorem stated by Eckart and Young, *Psychometrika*, **28**, 259–263.

Johnson, R.M. (1971) Market segmentation: A strategic management tool, *J. Mark. Res.*, **8**, 13–18.

Jolicoeur, P. (1963) The multivariate generalization of the allometry equation, *Biometrics*, **19**, 479–499.

Jolicoeur, P. (1968) Interval estimation of the slope of the major axis of a bivariate normal distribution in the case of a small sample, *Biometrics*, **24**, 679–682.

Jolicoeur, P. (1973) Imaginary confidence limits of the slope of the major axis of a bivariate normal distribution; A sampling experiment, *J. Amer. Stat. Assoc.*, **68**, 866–871.

Jolicoeur, P. (1984) Principal components, factor analysis and multivariate allometry: A small-sample direction test, *Biometrics*, **40**, 685–690.

Jolicoeur, P. and Mosimann, J.E. (1960) Size and shape variation in the painted turtle. A principal component analysis, *Growth*, **24**, 339–354. (Also in Bryant and Atchley, 1975.)

Jolliffe, I.T. (1972) Discarding variables in principle component analysis. I: Artificial data, *Appl. Stat.*, **21**, 160–173.

Jolliffe, I.T. (1973) Discarding variables in principal component analysis. II: Real data, *Appl. Stat.*, **22**, 21–31.

Jolliffe, I.T. (1982) A note on the use of principal components in regression, *Appl. Stat.*, **31**, 300–303.

Jolliffe, I.T. (1986) *Principal Component Analysis*, Springer-Verlag, New York.

Jolliffe, I.T. (1989) Rotation of ill-defined components, *Appl. Stat.*, **38**, 139–147.

Jones, C.L. (1983) A note on the use of directional statistics in weighted euclidean distances multidimensional scaling models, *Psychometrika*, **48**, 473–476.

Jöreskog, K.G. (1963) *Statistical Estimation in Factor Analysis*, Almqvist and Wiksells, Uppsala, Sweden.

Jöreskog, K.G. (1966) Testing a simple structure hypothesis in factor analysis, *Psychometrika*, **31**, 165–178.

Jöreskog, K.G. (1967) Some contributions to maximum likelihood factor analysis, *Psychometrika*, **32**, 443–482.

Jöreskog, K.G. (1969a) Efficient estimation in image factor analysis, *Psychometrika*, **34**, 51–75.

Jöreskog, K.G. (1969b) A general approach to confirmatory maximum likelihood factor analysis, *Psychometrika*, **34**, 182–202. (Also in Magidson, 1979 with addendum.)

Jöreskog, K.G. (1970) A general method for analysis of covariance structures, *Biometrika*, **57**, 239–251.

Jöreskog, K.G. (1971) Simultaneous factor analysis in several populations, *Psychometrika*, **36**, 409–426. (Also in Magidson, 1979.)

Jöreskog, K.G. (1974) Analyzing psychological data by structural analysis of covariance matrices, *Contemporary Developments in Math. Psych.*, II (D.H. Krantz *et al.*, editors), W.H. Freeman and Co., San Francisco, Calif. (Also in Magidson, 1979.)

Jöreskog, K.G. (1977a) Factor analysis by least-squares and maximum-likelihood methods, *Statistical Methods for Digital Computers*, III (K. Enslein *et al.*, editors), John Wiley and Sons, Inc., New York, pp. 125–135.

Jöreskog, K.G. (1977b) Structural equation models in the social sciences: Specification, estimation and testing, *Applications of Statistics* (P.R. Krishnaiah, editor), North-Holland, New York, pp. 265–287.

Jöreskog, K.G. (1978) Structural analysis of covariance and correlation matrices, *Psychometrika*, **43**, 443–477.

Jöreskog, K.G. (1982) The LISREL approach to causal model-building in the social sciences, *Systems Under Indirect Observation*, I (K.G. Jöreskog and H. Wold, editors), North-Holland, New York, pp. 81–99.

Jöreskog, K.G. and Goldberger, A.S. (1972) Factor analysis by generalized least squares. *Psychometrika*, **37**, 243–260.

Jöreskog, K.G. and Lawley, D.N. (1968) New methods in maximum likelihood factor analysis, *Br. J. Math. Stat. Psychol.*, **21**, 85–96.

Jöreskog, K.G. and Sörbom, D. (1984) *LISREL VI*; *Analysis of Linear Structural Relationships by the Method of Maximum Likelihood*, Scientific Software, Inc., Mooresville, Ind.

Jöreskog, K.G. and Sörbom, D. (1989) *LISREL 7, A Guide to the Program and Applications*, SPSS, Inc., Chicago.

Jöreskog, K.G. and Wold, H. (1982) The ML and PLS techniques for modeling with latent variables; Historical and comparative aspects, *Systems Under Indirect Observation*, I (K.G. Jöreskog and H. Wold, editors), North-Holland, New York, pp. 263–270.

Kabe, D.G. and Gupta, A.K. (1990) On a multiple correlation ratio, *Stat. & Prob. Lett.*, **9**, 449–451.

Kaiser, H.F. (1958) The varimax criterion for analytic rotation in factor analysis, *Psychometrika*, **23**, 187–200. (Also in Bryant and Atchley, 1975.)

Kaiser, H.F. (1959) Computer program for varimax rotation in factor analysis, *Educ. & Psych. Meas.*, **19**, 413–420.

Kaiser, H.F. (1960) The application of electronic computers to factor analysis, *Educ. & Psych. Meas.*, **20**, 141–151.

Kaiser, H.F. (1963) Image analysis, *Problems in Measuring Change* (C.W. Harris, editor), University of Wisconsin Press, Madison, Wis.

Kaiser, H.F. (1967) Uncorrelated linear composites maximally related to a complex of correlated observations, *Educ. & Psych. Meas.,* **27,** 3–6.

Kaiser, H.F. (1970) A second generation Little Jiffy, *Psychometrika,* **35,** 401–415.

Kaiser, H.F. and Caffrey, J. (1965) Alpha factor analysis, *Psychometrika,* **30,** 1–14.

Kaiser, H.F. and Rice, J. (1974) Little Jiffy, Mark IV, *Educ. & Psych. Meas.,* **34,** 111–117.

Kane, V.E., Baer, T., and Begovich, C.L. (1977) Principal component testing for outliers, *Oak Ridge Tech. Rep.* **K/UR-7,** Oak Ridge, Tenn., July 1977.

Kane, V.E., Ward, R.C., and Davis, G.J. (1985) Assessment of linear departures in multivariate data, *SIAM J. Sci. & Stat. Comp.,* **6,** 1022–1032.

Kapteyn, A., Neudecker, H., and Wansbeek, T. (1986) An approach to *n*-mode components analysis, *Psychometrika,* **51,** 269–275.

Karson, M.J. (1982) *Multivariate Statistical Methods,* Iowa State University Press, Ames, Iowa.

Katz, J.O. and Rohlf, F.J. (1974) Functionplane—A new approach to simple structure rotation, *Psychometrika,* **39,** 37–51.

Kell, R.L., Pearson, J.C.G., and Taylor, W. (1970) Hearing thresholds of an island population in North Scotland, *Int. Audiol.,* **9,** 334–349.

Kelley, T.L. (1928) *Crossroads in the Mind of Man,* Stanford University Press, Palo Alto, Calif.

Kelley, T.L. (1935) Essential traits of mental life, *Harvard Studies in Education,* **26,** Harvard University Press, Cambridge, Mass.

Kendall, M.G. (1939) The geographical distribution of crop productivity in England, *J. Roy. Stat. Soc. A,* **102,** 21–62.

Kendall, M.G. (1957) *A Course in Multivariate Analysis,* Charles Griffin & Co., Ltd., London.

Keramidas, E.M., Devlin, S.J., and Gnanadesikan, R. (1987) A graphical procedure for comparing the principal components of several covariance matrices, *Comm. Stat. Simul. Comp.,* **16,** 161–191.

Kermack, K.A. and Haldane, J.B.S. (1950) Organic correlations and allometry, *Biometrika,* **37,** 30–41.

Kettenring, J.R. (1971) Canonical analysis of several sets of variables, *Biometrika,* **58,** 433–451.

Kiers, H.A.L. (1989) An alternating least squares algorithm for fitting the two- and three-way DEDICOM model and the IDIOSCAL model, *Psychometrika,* **54,** 515–521.

Kiers, H.A.L. and TenBerge, J.M.F. (1989) Alternating least squares algorithms for simultaneous components analysis with equal component weight matrices in two or more populations, *Psychometrika,* **54,** 467–473.

Klahr, D.A. (1969) A Monte Carlo investigation of the statistical significance of Kruskal's nonmetric scaling procedure, *Psychometrika,* **34,** 319–330.

Kloek, T. and Mennes, L.B.M. (1960) Simultaneous equations estimation based on principal components of predetermined variables, *Econometrica,* **28,** 45–61.

Knol, D.L. and TenBerge, J.M.F. (1989) Least-squares approximation of an improper correlation matrix by a proper one, *Psychometrika,* **54,** 53–61.

Konishi, S. (1978) Asymptotic expansions for the distributions of statistics based on a correlation matrix, *Can. J. Stat.*, **6**, 49–56.

Konishi, S. (1979) Asymptotic expansions for the distributions of statistics based on the sample correlation matrix in principal component analysis, *Hiroshima Math. J.*, **9**, 647–700.

Koopman, R.F. (1978) On Bayesian estimation in unrestricted factor analysis, *Psychometrika*, **43**, 109–110.

Korhonen, P.J. (1984) Subjective principal component analysis, *Comp. Stat. & Data Anal.*, **2**, 243–255.

Korhonen, P.J. and Lasko, J. (1986) A visual interactive method for solving the multiple criteria problem, *Eur. J. Op. Res.*, **24**, 277–287.

Korth, B. and Tucker, L.R. (1975) The distribution of chance congruence coefficients from simulated data, *Psychometrika*, **40**, 361–372.

Koziol, J.A. (1986) Assessing multivariate normality: A compendium, *Comm. Stat.— Theor. Meth.*, **15**, 2763–2783.

Krishnaiah, P.R. (1969) Simultaneous test procedures under general ANOVA models, *Multivariate Analysis*, II (P.R. Krishnaiah, editor), Academic Press, Inc., New York, pp. 121–143.

Krishnaiah, P.R. (1976) Some recent developments on complex multivariate distributions, *J. Mult. Anal.*, **6**, 1–30.

Krishnaiah, P.R. (1978) Some recent developments on real multivariate distributions, *Developments in Statistics*, **1** (P.R. Krishnaiah, editor), Academic Press, Inc., New York, pp. 135–169.

Krishnaiah, P.R. (1979) Some developments on simultaneous test procedures, *Developments in Statistics*, **2** (P.R. Krishnaiah, editor), Academic Press, Inc., New York, pp. 157–201.

Krishnaiah, P.R. (1980) Computation of some multivariate distributions, *Handbook of Statistics*, I (P.R. Krishnaiah, editor), North-Holland, New York, pp. 745–971.

Krishnaiah, P.R. and Lee, J.C. (1977) Inference on the eigenvalues of the covariance matrices of real and complex multivariate normal populations, *Multivariate Analysis*, IV (P.R. Krishnaiah, editor), North-Holland, New York, pp. 95–103.

Krishnaiah, P.R. and Pathak, P.K. (1968) A note on confidence bounds for certain ratios of characteristic roots of covariance matrices, *Austral. J. Stat.*, **10**, 116–119.

Krishnaiah, P.R. and Schuurmann, F.J. (1974) On the evaluation of some distributions that arise in simultaneous tests for the equality of the latent roots of the covariance matrix, *J. Mult. Anal.*, **4**, 265–281.

Krishnaiah, P.R. and Waikar, V.B. (1971) Exact joint distributions of any few ordered roots of a class of random matrices, *J. Mult. Anal.*, **1**, 308–315.

Kroonenberg, P.M. (1983) *Three-mode Principal Component Analysis*, DSWO-Press, Leyden.

Kroonenberg, P.M. and DeLeeuw, J. (1980) Principal component analysis of three-mode data by means of alternating least squares algorithms, *Psychometrika*, **45**, 69–97.

Kruskal, J.B. (1964a) Multidimensional scaling by optimizing goodness-of-fit to a non-metric hypothesis, *Psychometrika*, **29**, 1–27. (Also in Atchley and Bryant, 1975, and Davies and Coxon, 1982.)

Kruskal, J.B. (1964b) Nonmetric multidimensional scaling; A numerical method, *Psychometrika*, **29**, 115–129. (Also in Davies and Coxon, 1982.)

Kruskal, J.B. (1977) Multidimensional scaling and other methods for discovering structure, *Statistical Methods for Digital Computers*, **III** (K. Enslein *et al.*, editors), John Wiley and Sons, Inc., New York, pp. 296–339.

Kruskal, J.B. and Carroll, J.D. (1969) Geometric models and badness-of-fit functions, *Multivariate Analysis*, **II** (P.R. Krishnaiah, editor), Academic Press, New York, pp. 639–671.

Kruskal, J.B. and Shepard, R.N. (1974) A nonmetric variety of linear factor analysis, *Psychometrika*, **39**, 123–157.

Kruskal, J.B. and Wish, M. (1978) *Multidimensional Scaling*, Sage Publications, Beverly Hills, Calif.

Kruskal, W.H. (1953) On the uniqueness of the line of organic correlation, *Biometrics*, **9**, 47–58.

Kryter, K.D. (1973) Impairment to hearing from exposure to noise, *J. Acoustical Soc. Amer.*, **53**, 1211–1234.

Krzanowski, W.J. (1979a) Some exact percentage points of a statistic useful in analysis of variance and principal component analysis, *Technometrics*, **21**, 261–263.

Krzanowski, W.J. (1979b) Between-groups comparison of principal components, *J. Amer. Stat. Assoc.*, **74**, 703–707. Correction: **76**, 1022 (1981).

Krzanowski, W.J. (1982) Between-group comparison of some principal components— Some sampling results, *J. Stat. Comp. Simul.*, **15**, 141–154.

Krzanowski, W.J. (1983) Cross-validatory choice in principal component analysis; Some sampling results, *J. Stat. Comp. Simul.*, **18**, 299–314.

Krzanowski, W.J. (1984a) Principal component analysis in the presence of group structure, *Appl. Stat.*, **33**, 164–168.

Krzanowski, W.J. (1984b) Sensitivity of principal components, *J. Roy. Stat. Soc. B*, **46**, 558–563.

Krzanowski, W.J. (1987a) Selection of variables to preserve multivariate data structure, using principal components, *Appl. Stat.*, **36**, 22–33.

Krzanowski, W.J. (1987b) Cross-validation in principal component analysis, *Biometrics*, **43**, 575–584.

Krzanowski, W.J. (1988) *Principles of Multivariate Analysis*, Oxford University Press, New York.

Krzanowski, W.J. (1989) On confidence regions in canonical variate analysis, *Biometrika*, **76**, 107–116.

Krzanowski, W.J. (1990) Between-group analysis with heterogeneous covariance matrices: The common principal component model, *J. Classifi.*, **7**, 81–98.

Kshirsagar, A.M. (1961) The goodness-of-fit of a single (non-isotropic) hypothetical principal component, *Biometrika*, **48**, 397–407.

Kshirsagar, A.M. (1966) The non-null distribution of a statistic in principal components analysis, *Biometrika*, **53**, 590–594.

Kshirsagar, A.M. (1972) *Multivariate Analysis*, Marcel Dekker, Inc., New York.

Kshirsagar, A.M. and Gupta, R.P. (1965) The goodness of fit of two (or more) hypothetical principal components, *Ann. Inst. Stat. Math.*, **17**, 347–356.

Kshirsagar, A.M., Kocherlakota, S. and Kocherlakota, K. (1990) Classification procedures using principal component analysis and stepwise discriminant function, *Comm. Stat.—Theor. Meth.*, **19**, 91–109.

Kulkarni, S.R. and Paranjape, S.R. (1984) Use of Andrews' function plot technique to construct control curves for multivariate process, *Comm. Stat.—Theor. Meth.*, **13**, 2511–2533.

Kullback, S. (1959) *Information Theory and Statistics*, John Wiley and Sons, Inc., New York.

Lambert, Z.V., Wildt, A.R., and Durand, R.M. (1990) Assessing sampling variation relative to number-of-factors criteria, *Educ. & Psych. Meas.*, **50**, 33–48.

Langeheine, R. (1982) Statistical evaluation of measures of fit in the Lingoes–Borg procrustean individual differences scaling, *Psychometrika*, **47**, 427–442.

Lastovicka, J.L. (1981) The extension of component analysis to four-mode matrices, *Psychometrika*, **46**, 47–57.

Lautenschlager, G.J. (1989) A comparison of alternatives to conducting Monte Carlo analyses for determining parallel analysis criteria, *Mult. Behav. Res.*, **24**, 365–395.

Lautenschlager, G.J., Lance, C.E. and Flaherty, V.L. (1989) Parallel analysis criteria: revised regression equations for estimating the latent roots of random data correlation matrices, *Educ. & Psych. Meas.*, **49**, 339–345.

Lawley, D.N. (1938) A generalization of Fisher's z-test, *Biometrika*, **30**, 180–187. Correction, **30**, 467–469.

Lawley, D.N. (1940) The estimation of factor loadings by the method of maximum likelihood, *Proc. Roy. Soc. Edin.* **60**, 64–82.

Lawley, D.N. (1942) Further investigations in factor estimation, *Proc. Roy. Soc. Edin.*, **61**, 176–185.

Lawley, D.N. (1956) Tests of significance for the latent roots of covariance and correlation matrices, *Biometrika*, **43**, 128–136.

Lawley, D.N. (1958) Estimation in factor analysis under various initial assumptions, *Br. J. Stat. Psych.*, **11**, 1–12.

Lawley, D.N. (1963) On testing a set of correlation coefficients for equality. *Ann. Math. Stat.*, **34**, 149–151.

Lawley, D.N. and Maxwell, A.E. (1971) *Factor Analysis as a Statistical Method*, 2d Ed., Butterworths, London.

Lawley, D.N. and Maxwell, A.E. (1973) Regression and factor analysis, *Biometrika*, **60**, 331–338.

Lawton, W.H. and Sylvestre, E.A. (1971) Self modeling curve resolution, *Technometrics*, **13**, 617–633.

Lazersfeld, P.F. and Henry, N.W. (1968) *Latent Structure Analysis*, Houghton Mifflin, Boston.

Leamer, E.E. and Chamberlain, G. (1976) A Bayesian interpretation of pretesting, *J. Roy. Stat. Soc. B*, **38**, 85–94.

Lebart, L., Morineau, A., and Tabard, N. (1977) *Techniques de la Description Statistique*, Dunod, Paris.

Lebart, L., Morineau, A., and Warwick, K.M. (1984) *Multivariate Descriptive Statistical Analysis. Correspondence Analysis and Related Techniques for Large Matrices*, John Wiley and Sons, Inc., New York.

Ledermann, W. (1937) On the rank of the reduced correlation matrix in multiple-factor analysis, *Psychometrika*, **2**, 85–93.

Lee, S.-Y. (1981) A Bayesian approach to confirmatory factor analysis, *Psychometrika*, **46**, 153–160.

Lee, S.-Y. and Jennrich, R.I. (1984) The analysis of structural equation models by means of derivative-free nonlinear least squares, *Psychometrika*, **49**, 521–528.

Lee, W. and Birch, J.B. (1988) Fractional principal components regression: A general approach to biased estimators, *Comm. Stat.—Simul. Comp.*, **17**, 713–727.

Leffingwell, A. (1892) *The Influence of Seasons Upon Conduct*, Charles Scribner's Sons, Inc., New York.

Lehman, E.L. (1975). *Nonparametrics*, Holden-Day, Inc., San Francisco.

Levin, J. (1966) Simultaneous factor analysis of several Gramian matrices, *Psychometrika*, **31**, 413–419.

Levine, M.S. (1977) *Canonical Analysis and Factor Comparison*, Sage Publications, Beverly Hills, Calif.

Lewi, P.J. (1982) *Multivariate Data Analysis in Industrial Practice*, John Wiley and Sons, Inc., New York.

Li, G. and Chen, Z. (1985) Projection-pursuit approach to robust dispersion matrices and principal components: Primary theory and Monte Carlo, *J. Amer. Stat. Assoc.*, **80**, 759–766. Corrigenda: **80**, 1084.

Lin, S.P. and Bendel, R.B. (1985) Generation of population correlation matrices with specified eigenvalues (AS213), *Appl. Stat.*, **34**, 193–198.

Lincoln, S.V., Piepe, A., and Prior, R. (1971) An application of principal components analysis to voting in the Scottish municipal elections 1967–9, *Statistician*, **20**, 73–88.

Lindberg, W., Öhman, J., and Wold, S. (1986) Multivariate resolution of overlapping peaks in liquid chromatography using diode array detection, *Anal. Chem.*, **58**, 299–303.

Lindberg, W., Persson, J.-A., and Wold, S. (1983) Partial least-squares method for spectrofluorimetric analysis of mixtures of humic acid and ligninsulfonate, *Anal. Chem.*, **55**, 643–648.

Lingoes, J.C. (1968) The multivariate analysis of qualitative data, *Mult. Behav. Res.*, **3**, 61–94.

Lingoes, J.C. (1971) Some boundary conditions for a monotone analysis of symmetric matrices, *Psychometrika*, **36**, 195–203. (Also in Lingoes, *et al.*, 1979.)

Lingoes, J.C. (1972) A general survey of the Guttman–Lingoes nonmetric program series, *Multidimensional Scaling* I: *Theory* (Shephard, R.N. *et al.*, editors), Seminar Press, New York, pp. 49–68. (Also in Lingoes, *et al.*, 1979.)

Lingoes, J.C. (1973) *The Guttman–Lingoes Nonmetric Program Series*, Mathesis Press, Ann Arbor, Mich.

Lingoes, J.C. (1979) Additional programs in the Guttman–Lingoes nonmetric program series: an update, *Geometric Representations of Relational Data* (J.C. Lingoes *et al.*, editors), Mathesis Press, Ann Arbor, Mich., pp. 283–287.

Lingoes, J.C. (1980) Testing regional hypothesis in multidimensional scaling, *Data Analysis and Informatics* (E. Diday *et al.*, editors), North-Holland, New York, pp. 191–207.

Lingoes, J.C. and Borg, I. (1978) A direct approach to individual differences scaling using increasingly complex transformations, *Psychometrika*, **43**, 491–520.

Lingoes, J.C. and Borg, I. (1979) Identifying spatial manifolds for interpretation, *Geometric Representations of Relational Data* (J.C. Lingoes et al., editors), Mathesis Press, Ann Arbor, Mich., pp. 127–148.

Lingoes, J.C., Roskam, E.E., and Borg, I., editors (1979) *Geometric Representations of Relational Data*, 2d Ed., Mathesis Press, Ann Arbor, Mich.

Lingoes, J.C. and Schönemann, P.H. (1974) Alternative measures of fit for the Schönemann–Carroll matrix fitting algorithm, *Psychometrika*, **39**, 423–427.

Littell, R.C. and Folks, J.L. (1971) Asymptotic optimality of Fisher's method of combining independent tests, *J. Amer. Stat. Assoc.*, **66**, 802–806.

Littell, R.C. and Folks, J.L. (1973) Asymptotic optimality of Fisher's method of combining independent tests II, *J. Amer. Stat. Assoc.*, **68**, 193–194.

Little, R.J.A. (1988a) Robust estimation of the mean and covariance matrix from data with missing values, *Appl. Stat.*, **37**, 23–38.

Little, R.J.A. (1988b) A test of missing completely at random for multivariate data with missing values, *J. Amer. Stat. Assoc.*, **83**, 1198–1202.

Little, R.J.A. and Rubin, D.B. (1987) *Statistical Analysis with Missing Data*, John Wiley and Sons, Inc., New York.

Loh, W.-Y. and Vanichsetakul, N. (1988) Tree-structured classification via generalized discriminant analysis (with discussion), *J. Amer. Stat. Assoc.*, **83**, 715–728.

Lohmöller, J.-B. (1989) *Latent Variable Path Modeling with Partial Least Squares*, Physica-Verlag, Heidelberg.

Longley, J.W. (1967) An appraisal of least squares programs for the electronic computer from the point of view of the user, *J. Amer. Stat. Assoc.*, **62**, 819–841.

Longman, R.S., Cota, A.A., Holden, R.R., and Fekken, G.C. (1989) A regression equation for the parallel analysis criterion in principal component analysis: Mean and 95th percentile eigenvalues, *Mult. Behav. Res.*, **24**, 59–69.

Lott, W.F. (1973) The optimal set of regression components restrictions on a least squares regression, *Comm. Stat.*, **2**, 449–464.

Luce, R.D. (1959) *Individual Choice Behavior*, John Wiley and Sons, Inc., New York.

Luotamo, M., Aitio, A., and Wold, S. (1988) Serum polychlorinated biphenyls: Quantitation and identification of source of exposure by the SIMCA pattern recognition method, *Chemom. and Intel Lab. Sys.*, **4**, 171–181.

MacCallum, R.C. (1976a) Effects of INDSCAL of non-orthogonal perceptions of object space dimensions, *Psychometrika*, **41**, 177–188.

MacCallum, R.C. (1976b) Transformation of a three-mode multidimensional scaling solution to INDSCAL form, *Psychometrika*, **41**, 385–400.

MacCallum, R.C. (1977) Effects of conditionality on INDSCAL and ALSCAL weights, *Psychometrika*, **42**, 297–305.

MacCallum, R.C. (1983) A comparison of factor analysis programs in SPSS, BMDP and SAS, *Psychometrika*, **48**, 223–231.

MacCallum, R.C. and Cornelius, E.T.A., III (1977) A Monte Carlo investigation of recovery of structure by ALSCAL, *Psychometrika*, **42**, 401–428.

MacKay, D.B. and Zinnes, J.L. (1986) A probabilistic model for the multidimensional scaling of proximity and preference data (with discussion), *Mark. Sci.*, **5**, 325–349.

MacKenzie, S.B., Lutz, R.J., and Belch, G.E. (1986) The role of attitude toward the ad as a mediator of advertising effectiveness: A test of competing explanations, *J. Mark. Res.*, **23**, 130–143.

Madansky, A. (1964) Instrumental variables in factor analysis, *Psychometrika*, **29**, 105–113.

Madow, W.G., Nisselson, J., and Olkin, I., editors. (1983) *Incomplete Data in Sample Surveys* I: *Report and Case Studies*, Academic Press, Inc., New York.

Madow, W.G. and Olkin, I., editors (1983) *Incomplete Data in Sample Surveys* III: *Proceedings of the Symposium*, Academic Press, Inc., New York.

Madow, W.G., Olkin, I., and Rubin, D.B., editors. (1983) *Incomplete Data in Sample Surveys* II: *Theory and Bibliographies*, Academic Press, Inc., New York.

Mager, P.P. (1980) Principal components regression analysis applied to structure–activity relationships, 2. Flexible opioids with unusually high safety margin, *Biom. J.*, **22**, 535–543.

Magidson, J., editor. (1979) *Advances in Factor Analysis and Structural Models*, Abt Books, Cambridge, Mass.

Malhorta, N.K. (1986) An approach to measurement of consumer preferences using limited information, *J. Mark. Res.*, **23**, 33–40.

Malinowski, E.R. (1977) Theory of error in factor analysis, *Anal. Chem.*, **49**, 606–612.

Malinowski, E.R. (1987) Theory of distribution of error eigenvalues resulting from principal component analysis with applications to spectroscopic data, *J. Chemom.*, **1**, 33–40.

Malinowski, E.R. and Howery, D.G. (1980) *Factor Analysis in Chemistry*, John Wiley and Sons, Inc., New York.

Malinowski, E.R. and McCue, M. (1977) Qualitative and quantitative determination of suspected components in mixtures by target transformation factor analysis of their mass spectra, *Anal. Chem.*, **49**, 284–287.

Mallows, C.L. (1961) Latent vectors of random symmetric matrices, *Biometrika*, **48**, 133–149.

Mallows, C.L. (1973) Some comments on C_p, *Technometrics*, **15**, 661–675.

Mandel, J. (1961) Non-additivity in two-way analysis of variance, *J. Amer. Stat. Assoc.*, **56**, 878–888.

Mandel, J. (1969) The partitioning of interaction in analysis of variance, *J. Res. NBS, Math. Sci.*, **73B**, 309–328.

Mandel, J. (1970) Distribution of eigenvalues of covariance matrices of residuals in analysis of variance, *J. Res. NBS, Math. Sci.*, **74B**, 149–154.

Mandel, J. (1971) A new analysis of variance model for non-additive data, *Technometrics*, **13**, 1–18.

Mandel, J. (1972) Principal components, analysis of variance and data structure, *Stat. Neer.*, **26**, 119–129.

Mandel, J. (1982) Use of singular value decomposition in regression analysis, *Amer. Stat.*, **36**, 15–24.

Mandel, J. (1989) The nature of collinearity, *J. Qual. Tech.*, **21**, 268–276.

Mandel, J. and Lashof, T.W. (1974) Interpretation and generalization of Youden's two-sample diagram, *J. Qual. Tech.*, **6**, 22–36.

Mansfield, E.R., Webster, J.T., and Gunst, R.F. (1977) An analytic variable selection technique for principal component regression, *Appl. Stat.*, **26**, 34–40.

Marascuilo, L.A. and Levin, J.R. (1983) *Multivariate Statistics in the Social Sciences*, Brooks/Cole Publishing Co., Monterey, Calif.

Marasinghe, M.G. and Johnson, D.E. (1981) Testing for subhypotheses in the multiplicative interaction model, *Technometrics*, **23**, 385–393.

Marasinghe, M.G. and Johnson, D.E. (1982a) Estimation of σ^2 on the multiplicative interaction model, *Comm. Stat.—Theor. Meth.*, **11**, 315–324.

Marasinghe, M.G. and Johnson, D.E. (1982b) A test of incomplete additivity in the multiplicative interaction model, *J. Amer. Stat. Assoc.*, **77**, 869–877.

Mardia, K.V., Kent, J.T., and Bibby, J.M. (1979) *Multivariate Analysis*, Academic Press, Inc., New York.

Maronna, R.A. (1976) Robust M-estimators of multivariate location and scatter, *Ann. Stat.*, **4**, 51–67.

Marquardt, D.W. and Snee, R.D. (1975) Ridge regression in practice, *Amer. Stat.*, **29**, 3–19.

Marshall, A.W. and Olkin, I. (1979) *Inequalities: Theory of Majorization and its Applications*, Academic Press, Inc., New York.

Martens, H. and Naes, T. (1985) Multivariate calibration by data compression. *Near Infrared Reflection Spectroscopy* (P. Williams, editor), Amer. Assoc. Cereal Chem., St. Paul, Minn.

Martens, H., Naes, T., and Jensen, S.A. (1984) Computer-aided spectrophotometric analysis of cereals. From dirty data to clean information, *Proc. Nordic Cerealist Meeting*, Willehammas, Norway.

Martens, H., Paulsen, F., Spjøtvoll, E., and Volden, R. (1980) Regression on disjoint factor analysis models, *Data Analysis and Informetics* (Diday et al., editors), North-Holland, New York, pp. 101–107.

Martin, J.K. and McDonald, R.P. (1975) Bayesian estimation in unrestricted factor analysis: A treatment for Heywood cases, *Psychometrika*, **40**, 505–517.

Mason, R.L. (1986) Latent root regression: A biased regression method for use with collinear prediction variables, *Comm. Stat.—Theor. Math.*, **15**, 2651–2678.

Mason, R.L. and Gunst, R.F. (1985) Outlier-induced collinearities, *Technometrics*, **27**, 401–407.

Massey, M.F. (1965) Principal components regression in exploratory statistical research, *J. Amer. Stat. Assoc.*, **60**, 234–256.

Mathai, A.M. and Saxena, R.K. (1978) *The H-function with Applications in Statistics and Other Disciplines*, John Wiley and Sons, Inc., New York.

Matthews, J.N.S. (1984) Robust methods in the assessment of multivariate normality, *Appl. Stat.*, **33**, 272–277.

Mauchley, J.W. (1940) Significance tests for sphericity of a normal n-variate distribution, *Ann. Math. Stat.*, **11**, 204–209.

Maxwell, A.E. (1977) *Multivariate Analysis in Behavioural Research*, John Wiley and Sons, Inc., New York.

Maxwell, A.E. (1978) Factoring correlation matrices. *International Encyclopedia of Statistics* (W.H. Kruskal and J.M. Tanur, editors), Macmillan Publishing Co., Inc., New York, pp. 330–337.

McCabe, G.P. (1984) Principal variables, *Technometrics*, **26**, 137–144.

McClelland, G. and Coombs, C.H. (1975) ORDMET: A general algorithm for constructing all numerical solutions to ordered metric data, *Psychometrika*, **40**, 269–290.

McCloy, C.H., Metheny, E., and Knott, V. (1938) A comparison of the Thurstone method of multiple factors with the Hotelling method of principal components, *Psychometrika*, **3**, 61–67.

McDonald, G.C. (1980) Some algebraic properties of ridge estimators, *J. Roy. Stat. Soc. B*, **42**, 31–34.

McDonald, G.C. and Galarneau, D.I. (1975) A Monte Carlo evaluation of some Ridge-type estimators, *J. Amer. Stat. Assoc.*, **70**, 407–416.

McDonald, R.P. (1968) A unified treatment of the weighting problem, *Psychometrika*, **33**, 351–381.

McDonald, R.P. (1970a) The theoretical foundations of principal factor analysis, canonical factor analysis and alpha factor analysis, *Br. J. Math. & Stat. Psych.*, **23**, 1–21. (Also in Bryant and Atchley, 1975.)

McDonald, R.P. (1970b) Three common factor models for groups of variables, *Psychometrika*, **35**, 111–128.

McDonald, R.P. (1975) Descriptive axioms for common factor theory, image theory and component theory, *Psychometrika*, **40**, 137–152.

McDonald, R.P. (1979) The simultaneous estimation of factor loadings and scores, *Br. J. Math. & Stat. Psych.*, **32**, 212–228.

McDonald, R.P. and Burr, E.J. (1967) A comparison of four methods of constructing factor scores, *Psychometrika*, **32**, 381–401. (Also in Bryant and Atchley, 1975.)

McGee, V.C. (1968) Multidimensional scaling of *n* sets of similarity measures: A nonmetric individual differences approach, *Mult. Behav. Res.*, **3**, 233–248.

Menozzi, P., Piazza, A., and Cavalli-Sforza, L. (1978) Synthetic maps of human gene frequencies in Europeans, *Science*, **201**, 786–791.

Meredith, W. (1964a) Notes on factorial invariance, *Psychometrika*, **29**, 177–185.

Meredith, W. (1964b) Rotation to achieve factorial invariance, *Psychometrika*, **29**, 187–206.

Meredith, W. (1977) On weighted Procrustes and hyperplane fitting in factor analytic rotation, *Psychometrika*, **42**, 491–522.

Meredith, W. and Millsap, R.E. (1985) On component analysis, *Psychometrika*, **50**, 495–507.

Meredith, W. and Tisak, J. (1982) Canonical measures of longitudinal and repeated measures data with stationary weights, *Psychometrika*, **47**, 47–67.

Meredith, W. and Tisak, J. (1990) Latent curve analysis, *Psychometrika*, **55**, 107–123.

Merrington, M. and Thompson, C.M. (1943) Tables of percentage points of the inverted beta (F) distribution, *Biometrika*, **33**, 73–88.

Mijares, T.A. (1990) On the normal approximation to the Lawley–Hotelling trace criterion, *Biometrika*, **77**, 443.

Miller, K.E. (1976) An investigation of the inclusion of the explicit ideal point in the multi-attribute attitude model, *Advances in Consumer Research*, **III**, 110–113.

Mills, J.H. (1978) Effect of noise on young and old people. *Noise and Audiology* (D.M. Lipscomb, editor), University Park Press, Baltimore, Md., pp. 229–241.

Millsap, R.E. and Meredith, W. (1988) Component analysis in cross-sectional and longitudinal data, *Psychometrika*, **53**, 123–134.

Montanelli, R.G., Jr. and Humphreys, L.G. (1976) Latent roots of random data correlation matrices with squared multiple correlations on the diagonal: A monte carlo study, *Psychometrika*, **41**, 341–348.

Mood, A.M. (1951) On the distribution of the characteristic roots of normal second-moment matrices, *Ann. Math. Stat.*, **22**, 266–273.

Mood, A.M. (1990) Miscellaneous reminiscences, *Stat. Sci.*, **5**, 35–43.

Mooijaart, A. (1982) Latent structure analysis for categorical variables, *Systems under Indirect Observation*, I (K.G. Jöreskog and H. Wold, editors), North-Holland, New York, pp. 1–18.

Moore, W.L., Pessemier, E.A., and Little, T.E. (1979) Predicting brand purchase behavior: Marketing application to the Schönemann and Wang unfolding model, *J. Mark. Res.*, **16**, 203–210.

Morris, R.H. and Morrissey, J.H. (1954) An objective method for the determination of equivalent neutral densities of color images, II. Determination of primary equivalent neutral densities, *J. Optical Soc. Amer.*, **44**, 530–534.

Morrison, D.F. (1976) *Multivariate Statistical Methods*, 2d Ed., McGraw-Hill, New York.

Mosier, C.I. (1939) Determining a simple structure when loadings for certain tests are known, *Psychometrika*, **4**, 149–162.

Mosteller, F. and Tukey, J.W. (1977) *Data Analysis and Regression; A Second Course in Statistics*, Addison-Wesley, Reading, Mass.

Mosteller, F. and Wallace, D.L. (1963) Inference in an authorship problem, *J. Amer. Stat. Assoc.*, **58**, 275–309.

Mudholkar, G.S., Davidson, M.L., and Subbaiah, P. (1974) Extended linear hypothesis and simultaneous tests in multivariate analysis of variance, *Biometrika*, **61**, 467–477.

Mudholkar, G.S. and George, E.O. (1979) The logit statistic for combining probabilities—an overview, *Optimizing Methods in Statistics* (J.S. Rustagi, editor), Academic Press, Inc., New York, pp. 345–365.

Mudholkar, G.S. and Subbaiah, P. (1980) A review of step-down procedures for multivariate analysis of variance, *Multivariate Statistical Analysis* (R.P. Gupta, editor), North-Holland, New York, pp. 161–178.

Muirhead, R.J. (1974) Bounds for distribution functions of the extreme latent roots of a sample covariance matrix, *Biometrika*, **61**, 641–642.

Muirhead, R.J. (1975) Expressions for some hypergeometric functions of matrix argument with applications, *J. Mult. Anal.*, **5**, 283–292.

Muirhead, R.J. (1978) Latent roots and matrix variates: a review of some asymptotic results, *Ann. Stat.*, **6**, 5–33.

Muirhead, R.J. (1982) *Aspects of Multivariate Statistical Theory*, John Wiley and Sons, Inc., New York.

Mukherjee, B.N. and Maity, S.S. (1988) On conditions for equality of OLSE, GLSE and MLE in analysis of covariance structures, *Bull. Calcutta Stat. Assoc.*, **37**, 171–191.

Mulaik, S.A. (1986) Factor analysis and Psychometrika: Major developments, *Psychometrika*, **51**, 23–33.

Muller, K.E. (1981) Relationships between redundancy analysis, canonical correlation, and multivariate regression, *Psychometrika*, **46**, 139–142.

Muller, K.E. (1982) Understanding canonical correlation through the general linear model and principal components, *Amer. Stat.*, **36**, 342–354.

Muller, K.E. and Peterson, B.L. (1984) Practical methods for computing power in testing the multivariate general linear hypothesis, *Comp. Stat. & Data Anal.*, **2**, 143–158.

Muthén, B. (1978) Contributions to factor analysis of dichotomous variables, *Psychometrika*, **43**, 551–560.

Muthén, B. (1982) Some categorical response models with continuous latent variables, *Systems Under Indirect Observation*, I (K.G. Jöreskog and H. Wold, editors), North-Holland, New York, pp. 65–79.

Muthén, B. (1989) Latent variable modeling in heterogeneous populations, *Psychometrika*, **54**, 557–585.

Muthén, B. and Christoffersson, A. (1981) Simultaneous factor analysis of dichotomous variables in several groups. *Psychometrika*, **46**, 407–419.

Muthén, B. and Hofacker, C. (1988) Testing the assumptions underlying tetrachoric correlations, *Psychometrika*, **53**, 563–578.

Naes, T. (1985) Multivariate calibration when the error covariance matrix is structured, *Technometrics*, **27**, 301–311.

Naes, T. (1989) Leverage and influence measures for principal component regression, *Chemom. & Int. Lab. Syst.*, **5**, 155–168.

Naes, T., Irgens, C., and Martens, H. (1986) Comparison of linear statistical methods for calibration of NIR instruments, *Appl. Stat.*, **35**, 195–206.

Naes, T. and Martens, H. (1985) Comparison of prediction methods for multicollinear data, *Comm. Stat.—Simul. Comp.*, **14**, 545–576.

Naes, T. and Martens, H. (1988) Principal components regression in NIR analysis: Viewpoints, background details and selection of components, *J. Chemom.*, **2**, 155–167.

Nagao, H. (1988) The jackknife statistics for eigenvalues and eigenvectors of a correlation matrix, *Ann. Inst. Stat. Math.*, **40**, 477–489.

Nash, J.C. (1979) *Compact Numerical Methods for Computers: Linear Algebra and Functional Minimization*, John Wiley and Sons, Inc., New York.

Nathan, G. and Holt, D. (1980) The effect of survey design on regression analysis, *J. Roy. Stat. Soc. B*, **42**, 377–386.

Nayatani, Y., Takahana, K., and Sobagaki, H. (1983) A proposal of new standard deviate observers, *Color Res. and Applic.*, **8**, 47–56.

Nelder, J.A. (1985) An alternative interpretation of the singular value decomposition in regression, *Amer. Stat.*, **39**, 63–64.

Neudecker, H. and Wansbeek, T. (1983) Some results on commutation matrices with statistical applications, *Can. J. Stat.*, **11**, 221–231.

Neuhaus, J.O. and Wrigley, C. (1954) The quartimax method: An analytical approach to orthogonal simple structure, *Br. J. Stat. Psychol.*, **7**, 81–91.

Nevels, K. (1979) On Meridith's solution for weighted procrustes rotation, *Psychometrika*, **44**, 121–122.

Newcomb, R.W. (1961) On the simultaneous diagonalization of two semi-definite matrices, *Q. Appl. Math.*, **19**, 144–146.

Newman, R.I. and Sheth, J.N. (1985) A model of primary voter behavior, *J. Cons. Res.*, **12**, 178–187.

Nicewander, W.A., Littlejohn, R.L., and Sarle, W. (1984) Pattern and structure in image analysis. Presented at the Psychometrics Society Meeting, Santa Barbara, Calif.

Nishisato, S. (1980) *Analysis of Categorical Data: Dual Scaling and Applications*, University of Toronto Press, Toronto.

Noma, E. and Smith, D.R. (1985) Scaling sociomatrices by optimizing an explicit function: correspondence analysis of binary single response sociomatrices, *Mult. Behav. Res.*, **20**, 179–197.

Nomakuchi, K. and Sakata, T. (1984) The union–intersection principle applied to the test of dimensionality, *Comm. Stat.—Theor. Meth.*, **13**, 753–760.

Noonan, R. and Wold, H. (1982) PLS path modeling with indirectly observed variables: A comparison of alternative estimates for the latent variable, *Systems Under Indirect Observation*, II (K.G. Jöreskog and H. Wold, editors), North-Holland, New York, pp. 75–94.

Obenchain, R.L. (1972) Regression optimality of principal components, *Ann. Math. Stat.*, **43**, 1317–1319.

O'Brien, P.N., Parente, F.J., and Schmitt, C.J. (1982) A Monte Carlo study on the robustness of four MANOVA criterion tests, *J. Stat. Comp. Sim.*, **15**, 183–192.

Ohta, N. (1973) Estimating absorption bands of component dyes by means of principal component analysis, *Anal. Chem.*, **45**, 553–557.

Okamoto, M. (1969) Optimality of principal components, *Multivariate Analysis* II (P.R. Krishnaiah, editor), Academic Press, Inc., New York, pp. 673–685.

Okamoto, M. (1972) Four techniques of principal component analysis, *J. Japan Stat. Soc.*, **2**, 63–69.

Okamoto, M. (1973) Distinctness of the eiganvalues of a quadratic form in a multivariate sample, *Ann. Stat.*, **1**, 763–765.

Okamoto, M. and Kanazawa, M. (1968) Minimization of eigenvalues of a matrix and optimality of principal components, *Ann. Math. Stat.*, **39**, 859–863.

Oliver, R.L. and Bearden, N.O. (1985) Crossover effects in the theory of reasoned action: A moderating influence attempt, *J. Cons. Res.*, **12**, 324–340.

Olson, C.L. (1974) Comparative robustness of six tests in multivariate analysis of variance, *J. Amer. Stat. Assoc.*, **69**, 894–908.

Olsson, U. (1979) On the robustness of factor analysis against crude classification of the observations, *Mult. Behav. Res.*, **14**, 485–500.

Oman, S.D. (1978) A Bayesian comparison of some estimators used in linear regression with multicolinear data, *Comm. Stat.-Theor. Meth.*, **A7**, 517–534.

Osmond, C. (1985) Biplot models applied to cancer mortality rates, *Appl. Stat.*, **34**, 63–70.

Ottestad, P. (1975) Component analysis: an alternative system, *Int. Stat. Rev.*, **43**, 83–108.

Pack, P., Jolliffe, I.T., and Morgan, B.J.T. (1988) Influential observations in principal component analysis; A case study, *J. Appl. Stat.*, **15**, 39–52.

Panel on Discriminant Analysis, Classification and Clustering (1989) Discriminant analysis and clustering, *Stat. Sci.*, **4**, 34–69.

Paulson, A.S., Roohan, P. and Sullo, P. (1987) Some empirical distribution function tests for multivariate normality, *J. Stat. Comp. Simul.*, **28**, 15–30.

Pearce, S.C. and Holland, D.A. (1960) Some applications of multivariate methods in botany, *Appl. Stat.*, **9**, 1–7.

Pearson, E.S., and Hartley, H.O., editors (1966) *Biometrika Tables for Statisticians,* **I**, 3rd Edition, Cambridge University Press, London.

Pearson, E.S. and Hartley, H.O., editors (1972) *Biometrika Tables for Statisticians,* **II**, Cambridge University Press, London.

Pearson, K. (1901) On lines and planes of closest fit to systems of points in space, *Phil. Mag., Ser. B*, **2**, 559–572. (Also in Bryant and Atchley, 1975.)

Pearson, K. (1904) Mathematical contributions to the theory of evolution XIII. On the theory of contingency and its relation to association and normal correlation, *Drapers Co. Res. Mem.*, Biometric Series I, Cambridge University Press, London.

Pearson, K. (1906) On certain points connected with scale order in the case of the correlation of two characters for which some arrangements give a linear regression line, *Biometrika*, **5**, 176–178.

Peay, E.R. (1988) Multidimensional rotation and scaling of configurations to optimal agreement, *Psychometrika*, **53**, 199–208.

Perry, J.M. (1977) AMDAHL speaks: Carter really won the election, *The National Observer*, Feb. 12, 1977, p. 5.

Persson, J.-A., Johansson, E., and Albano, C. (1986) Quantitative thermogravimetry on peat. A multivariate approach, *Anal. Chem.*, **58**, 1173–1178.

Phelan, M.K., Barlow, C.H., Kelly, J.J., Jinguji, T.M., and Callis, J.B. (1988) Measurement of caustic and caustic brine by spectroscopy detection of the hyrdoxide ion in wavelength range 700–1150 nm, Univ. Washington, Dept. Chemistry Report bg-10, July 29, 1988, Seattle, Wash.

Pillai, K.C.S. (1976) Distributions of characteristic roots in multivariate analysis, Part I: Null distributions, *Can. J. Stat.*, **4**, 157–184.

Pillai, K.C.S. (1977) Distributions of characteristic roots in multivariate analysis, Part II: Non-null distributions, *Can. J. Stat.*, **5**, 1–62.

Pillai, K.C.S., Al-Ani, S., and Jouris, G.M. (1969) On the distribution of the ratios of the roots of a covariance matrix and Wilk's criterion for the tests of three hypotheses, *Ann. Math. Stat.*, **40**, 2033–2040.

Pillai, K.C.S. and Chang, T.C. (1970) An approximation to the C.D.F. of the largest root of a covariance matrix, *Ann. Inst. Stat. Math. Suppl. No. 6*, 115–124.

Pillai, K.C.S. and Young, D.L. (1971) An approximation to the distribution of the largest root of a complex Wishart matrix, *Ann. Inst. Stat. Math.*, **23**, 89–96.

Press, S.J. (1972) *Applied Multivariate Analysis*, Holt, Rinehart and Winston, Inc., New York.

Price, J.M. and Nicewander, W.A. (1977) Maximally correlated orthogonal composites and oblique factor analytic solutions, *Psychometrika*, **42**, 439–442.

Pudney, S.E. (1982) Estimating latent variable systems when specification is uncertain: Generalized component analysis and the eliminant method, *J. Amer. Stat. Assoc.*, **77**, 883–889.

Qualls, W.J. (1987) Household decision behavior: The impact of husband's and wive's sex role orientation, *J. Cons. Res.*, **14**, 264–279.

Radcliff, J. (1964) The construction of a matrix used in deriving tests of significance in multivariate analysis, *Biometrika*, **51**, 503–504.

Radhakrishnan, R. and Kshirsagar, A.M. (1981) Influence functions for certain parameters in multivariate analysis, *Comm. Stat.-Theor. Meth.*, **A10**, 519–529.

Ralston, A. and Rabinowitz, P. (1978) *A First Course in Numerical Analysis*, 2d Ed., McGraw-Hill, New York.

Ramsay, J.O. (1977) Maximum likelihood estimation in multidimensional scaling, *Psychometrika*, **42**, 241–266. (Also in Davies and Coxon, 1982.)

Ramsay, J.O. (1978a) Confidence regions for multidimensional scaling analysis, *Psychometrika*, **43**, 145–160.

Ramsay, J.O. (1978b) *MULTISCALE: Four Programs for Multidimensional Scaling by the Method of Maximum Likelihood*, National Educational Services, Chicago.

Ramsay, J.O. (1980a) Some small sample results for maximum likelihood estimation in multidimensional scaling, *Psychometrika*, **45**, 139–144.

Ramsay, J.O. (1980b) The joint analysis of direct ratings, pairwise preferences and dissimilarities, *Psychometrika*, **45**, 149–165.

Ramsay, J.O. (1982) When data are functions, *Psychometrika*, **47**, 379–396.

Ramsay, J.O. (1984) Some metrics for robust multivariate analysis, given at Psychometric Society meeting, Santa Barbara, Calif.

Ramsay, J.O. (1988) Monotone regression splines in action (with discussion), *Stat. Sci.*, **3**, 425–461.

Ramsay, J.O. (1990) MAPFIT: A FORTRAN subroutine for comparing two matrices in a subspace, *Psychometrika*, **55**, 351–353.

Ramsay, J.O. and Abrahamowicz, M. (1989) Binomial regression with monotone splines: A psychometric application, *J. Amer. Stat. Assoc.*, **84**, 906–915.

Ramsey, F.L. (1986) A fable of PCA, *Amer. Stat.*, **40**, 323–324. (Also two letters to the editor, **41**, 341, 1987.)

Rao, C.R. (1952) *Advanced Statistical Methods in Biometric Research*, John Wiley and Sons, Inc., New York.

Rao, C.R. (1955) Estimation and tests of significance in factor analysis, *Psychometrika*, **20**, 93–111. (Also in Bryant and Atchley, 1975.)

Rao, C.R. (1958) Some statistical methods for comparison of growth curves, *Biometrics*, **14**, 1–17.

Rao, C.R. (1964) The use and interpretation of principal component analysis in applied research, *Sankhyā A*, **26**, 329–358. (Also in Bryant and Atchley, 1975.)

Rao, C.R. (1979) Separation theorems for singular values of matrices and their applications in multivariate analysis, *J. Mult. Anal.*, **9**, 362–377.

Rao, C.R. (1980) Matrix approximations and reduction of dimensions in multivariate statistical analysis, *Multivariate Analysis*, V (P.R. Krishnaiah, editor), North-Holland, New York, pp. 3–22.

Rao, C.R. (1987) Prediction of future observations in growth curve models (with discussion), *Stat. Sci.*, **2**, 434–471.

Rao, J.N.K. and Scott, A.J. (1981) The analysis of categorical data from complex sample surveys: Chi-squared tests for goodness of fit and independence in two-way tables, *J. Amer. Stat. Assoc.*, **76**, 221–230.

Rao, V.R. (1972) Changes in explicit information and brand perceptions, *J. Mark. Res.*, **9**, 209–213.

Rao, V.R. and Katz, R. (1971) Alternative multidimensional scaling methods, *J. Mark Res.*, **8**, 488–494.

Reddon, J.R. (1985) Monte carlo type I error rates for Velicer's partial correlation test for the number of principal components, *Criminometrica*, **1**, 13–23.

Reyment, R.A. (1963) Multivariate analytical treatment of quantitative species association. An example from palgeoecology, *J. Animal Ecol.*, **32**, 535–547. (Also in Bryant and Atchley, 1975.)

Riley, E.C., Sterner, J.H., Fassett, D.W., and Sutton, W.L. (1961) Ten years' experience with industrial audiometry, *Am. Ind. Hygiene Assoc. J.*, **22**, 151–159.

Ritter, G.L., Lowry, S.R., Isenhour, T.L., and Wilkens, C.L. (1976) Factor analysis of mass spectra of mixtures, *Anal. Chem.*, **48**, 591–595.

Robert, P., Cleroux, R., and Ranger, N. (1985) Some results on vector correlation, *Comp. Stat. & Data Anal.*, **3**, 25–32.

Robert, P. and Escoufier, Y. (1976) A unifying tool for linear multivariate statistical methods: The RV-coefficient, *Appl. Stat.*, **25**, 257–265.

Roff, M. (1936) Some properties of the communality in multiple factor theory, *Psychometrika*, **1**, No. 2, 1–6.

Röhr, M. (1985) On the rotation of canonical solutions, *Biom. J.*, **27**, 89–96.

Roskam, E.E. (1979a) The nature of data: Interpretation and representation, *Geometric Representations of Relational Data*, 2d Ed. (J.C. Lingoes *et al.*, editors) Mathesis Press, Ann Arbor, Mich., pp. 149–235.

Roskam, E.E. (1979b) A survey of the Michigan–Israel–Netherlands Integrated Series, *Geometrical Representations of Relation Data*, 2d Ed. (J.C. Lingoes *et al.*, editors), Mathesis Press, Ann Arbor, Mich., pp. 289–312.

Ross, J. (1964) Mean performance and the factor analysis of learning data, *Psychometrika*, **29**, 67–73.

Ross, J. and Cliff, N. (1964) A generalization of the interpoint distance method, *Psychometrika*, **29**, 167–176.

Rossa, P.J. (1982) Explaining international political behavior and conflict through partial least squares modeling, *Systems Under Indirect Observation*, II (K.G. Jöreskog and H. Wold, editors), North-Holland, New York, pp. 131–159.

Rossi, T.M. and Warner, I.M. (1986) Rank estimation of excitation-emission matrices using frequency analysis of eigenvectors, *Anal. Chem.*, **58**, 810–815.

Rost, J. (1988) Rating scale analysis with latent class models, *Psychometrika*, **53**, 327–348.

Rothkopf, E.Z. (1957) A measure of stimulus similarity and errors in some pair-associated learning tasks, *J. Exp. Psych.*, **53**, 94–101.

Rouvier, R. (1966) L'Analyse en composantes principales: son utilisation en genetique et ses rapports avec l'analyse discriminatoire, *Biometrics*, **22**, 343–357.

Rouvier, R. (1969) Ponderation des valeurs genotypiques dans la selection par index sur plusieurs caracters, *Biometrics*, **25**, 295–307.

Roy, J. (1958) Step-down procedure in multivariate analysis, *Ann. Math. Stat.*, **29**, 1177–1187.

Roy, S.N. (1939) *P*-Statistics or some generalizations in analysis of variance appropriate to multivariate problems, *Sankhyā*, **4**, 381–396.

Roy, S.N. (1953) On a heuristic method of test construction and its use in multivariate analysis, *Ann. Math. Stat.*, **24**, 220–238.

Roy, S.N. (1954) Some further results in simultaneous confidence interval estimation, *Ann. Math. Stat.*, **25**, 752–761.

Roy, S.N. (1957) *Some Aspects of Multivariate Analysis*, John Wiley and Sons, Inc., New York.

Royston, J.P. (1983) Some techniques for assessing multivariate normality based on the Shapiro-Wilk W, *Appl. Stat.*, **32**, 121-133.

Rozett, R.W. and Petersen, E.M. (1975a) Methods of factor analysis of mass spectra, *Anal. Chem.*, **47**, 1301-1308.

Rozett, R.W. and Petersen, E.M. (1975b) Factor analysis of the mass spectra of the isomers of $C_{10}H_{14}$, *Anal. Chem.*, **47**, 2377-2384.

Russell, T.S. and Bradley, R.A. (1958) One-way variances in a two-way classification, *Biometrika*, **45**, 111-129.

Ruymgaart, F.H. (1981) A robust principal component analysis, *J. Mult. Anal.*, **11**, 485-497.

Sampson, A.R. (1984) A multivariate correlation ratio, *Stat. & Prob. Lett.*, **2**, 77-81.

Sands, R. and Young, F.W. (1980) Component models for three-way data: ALSCOMP3, an alternating least squares algorithm with optimal scaling features, *Psychometrika*, **45**, 39-67.

SAS Institute, Inc. (1986) *SUGI Supplemental Library User's Guide*, Version 5 Edition, SAS Institute, Inc., Cary, N.C.

SAS Institute, Inc. (1989) *SAS/SAT User's Guide*, Version 6, 4th Edition, SAS Institute, Inc., Cary, N.C.

Saunders, D.R. (1961) The rationale for an "oblimax" method of transformation in factor analysis, *Psychometrika*, **26**, 317-324.

Schall, S. and Chandra, M.J. (1987) Multivariate quality control using principal components, *Int. J. Production. Res.*, **25**, 571-588.

Schiffman, S.S., Reynolds, M.L., and Young, F.W. (1981) *Introduction to Multidimensional Scaling: Theory, Method and Applications*, Academic Press, Inc., New York.

Schilling, E.G. (1982) *Acceptance Sampling in Quality Control*, Marcel Dekker, Inc., New York.

Schmidt, P. (1976) *Econometrics*, Marcel Dekker, Inc., New York.

Schönemann, P.H. (1966) A generalized solution of the orthogonal procrustes problem, *Psychometrika*, **31**, 1-10.

Schönemann, P.H. (1968) One two-sided orthogonal procrustes problems, *Psychometrika*, **33**, 19-33.

Schönemann, P.H. (1970) On metric multidimensional scaling, *Psychometrika*, **35**, 349-366.

Schönemann, P.H. and Carroll, R.M. (1970) Fitting one matrix to another under choice of a central dilation and a rigid motion, *Psychometrika*, **35**, 245-255.

Schönemann, P.H., James, W.L., and Carter, F.S. (1978) COSPA: Common space analysis—A Program for fitting and testing Horan's subjective metrics model, *J. Mark. Res.*, **15**, 268-270.

Schönemann, P.H., James, W.L., and Carter, F.S. (1979) Statistical inference in multidimensional scaling: A method for fitting and testing Horan's model, *Geometric Representations of Relational Data* (J.C. Lingoes *et al.*, editors), Mathesis Press, Ann Arbor, Mich., pp. 791-826.

Schönemann, P.H. and Wang, M.-M. (1972) An individual difference model for the multidimensional analysis of preference data, *Psychometrika*, **37**, 275-309.

Schott, J.R. (1986) Tests concerning two non-isotropic principal components, *Comm. Stat.—Theor. Meth.*, **15**, 3307-3320.

Schott, J.R. (1987a) An improved chi-squared test for a principal component, *Stat. & Prob. Lett.*, **5**, 361–365.

Schott, J.R. (1987b) Two-stage tests of hypotheses for principal components, *Sankyhā B*, **49**, 105–112.

Schott, J.R. (1988) Testing the equality of the smallest latent roots of a correlation matrix, *Biometrika*, **75**, 794–796.

Schott, J.R. (1989) An adjustment for a test concerning a principal component subspace, *Stat. & Prob. Lett.*, **7**, 425–430.

Schuurmann, F.J., Krishnaiah, P.R., and Chattopadhyay, A.K. (1973) On the distribution of the ratios of the extreme roots to the trace of the Wishart matrix, *J. Mult. Anal.*, **3**, 445–453.

Schuurmann, F.J. and Waikar, V.B. (1974) Upper percentage points of the individual roots of the complex Wishart matrix, *Sankyhā B*, **36**, 299–305.

Scott, J.T., Jr. (1966) Factor analysis and regression, *Econometrica*, **34**, 552–562.

Seber, G.A.F. (1984) *Multivariate Observations*, John Wiley and Sons, Inc., New York.

Shapiro, S.S. and Wilk, M.B. (1965) An analysis of variance test for normality (Complete samples), *Biometrika*, **52**, 591–611.

Sharma, A.S., Durvasula, S., and Dillon, W.R. (1989) Some results on the behavior of alternate covariance structure estimation procedures in the presence of non-normal data, *J. Mark. Res.*, **26**, 214–221.

Shepard, R.N. (1962) The analysis of proximities: multidimensional scaling with an unknown distance function, *Psychometrika*, **27**; **I**: 125–140; **II**: 219–246.

Shepard, R.N. (1966) Metric structures in ordinal data, *J. Math. Psych.*, **3**, 287–315. (Also in Davies and Coxon, 1982.)

Shepard, R.N. and Carroll, J.D. (1966) Parametric representation of nonlinear data structures, *Multivariate Analysis*, **I** (P.R. Krishnaiah, editor), Academic Press, Inc., New York, pp. 561–592.

Shewhart, W.A. (1931) *Economic Control of Quality of Manufactured Product*, D. Van Nostrand Co., Inc., New York.

Shine, L.C., II (1972) A note on McDonald's generalization of principal components analysis, *Psychometrika*, **37**, 99–101.

Shocker, A.D. and Srinivasan, V. (1975) LINMAP: Linear programming techniques for the multidimensional analysis of preferences, *J. Mark. Res.*, **12**, 214–215.

Sibson, R. (1978) Studies in the robustness of multidimensional scaling: Procrustes statistics, *J. Roy. Stat. Soc. B*, **40**, 234–238.

Sibson, R. (1979) Studies in the robustness of multidimensional scaling: Perturbational analysis of classical scaling, *J. Roy. Stat. Soc. B*, **41**, 217–229.

Sibson, R., Bowyer, A., and Osmond, C. (1981) Studies in the robustness of multidimensional scaling: Euclidean models and simulation studies, *J. Stat. Comp. Simul.*, **13**, 273–296.

Silverstein, J.W. (1984) Comments on a result of Yin, Bai and Krishnaiah for large dimensional multivariate *F*-matrices, *J. Mult. Anal.*, **15**, 408–409.

Simonds, J.L. (1958a) Analysis of the variability among density–log exposure curves of black-and-white negative films by the method of principal components, *Photo Sci. and Eng.*, **2**, 205–209.

Simonds, J.L. (1958b) The use of principal-component description of D–log E curves in a study of the ASA–DIN speed relationships, *Photo Sci. and Eng.*, **2**, 210–212.

Simonds, J.L. (1963) Application of characteristic vector analysis to photographic and optical response data, *J. Optical Soc. Amer.*, **53**, 968–974.

Skinner, C.J., Holmes, D.J., and Smith, T.M.F. (1986) The effect of sample design on principal component analysis, *J. Amer. Stat. Assoc.*, **81**, 789–798.

Smith, D.M. and Bremner, J.M. (1989) All possible subset regressions using the QR decomposition, *Comp. Stat. & Data Anal.*, **7**, 217–235.

Smith, R.E. and Lusch, R.F. (1976) How advertising can position a brand, *J. Adv. Res.*, **16** (No. 1, Feb. 1976), 37–43.

Snee, R.D. (1972a) On the analysis of response curve data, *Technometrics*, **14**, 47–62.

Snee, R.D. (1972b) A useful method for conducting carrot shape studies, *J. Hortic. Sci.*, **47**, 267–277.

Snee, R.D., Acuff, S.K., and Gibson, J.R. (1979) A useful method for the analysis of growth studies, *Biometrics*, **35**, 835–848.

Snook, S.C. and Gorsuch, R.L. (1989) Component analysis versus common factor analysis: A Monte Carlo study, *Psych. Bull.*, **106**, 148–154.

Sörbom, D. (1974) A general method for studying differences in factor means and factor structure among groups, *Br. J. Math. & Stat. Psych.*, **27**, 229–239. (Also in Magidson, 1979.)

Sörbom, D. (1978) An alternative to the methodology for analysis of covariance, *Psychometrika*, **43**, 381–396. (Also in Magidson, 1979.)

Spearman, C. (1904) General intelligence, objectively determined and measured, *Amer. J. Psychol.*, **15**, 201–293.

Spence, I. (1972) A Monte Carlo evaluation of three nonmetric multidimensional scaling algorithms, *Psychometrika*, **37**, 461–486.

Spence, I. (1979) A simple approximation for random ranking STRESS values, *Mult. Behav. Res.*, **14**, 355–365. (Also in Davies and Coxon, 1982.)

Spence, I. and Lewandowsky, S. (1989) Robust multidimensional scaling, *Psychometrika*, **54**, 501–513.

Spence, I. and Ogilvie, S.C. (1973) A table of expected stress values for random rankings in nonmetric multidimensional scaling, *Mult. Behav. Res.*, **8**, 511–517.

Spindler, G.A. (1987) *A Comparison of Three Residual Statistics used in Principal Component Analysis*, Master's Thesis, Rochester Inst. of Technology, Rochester, N.Y.

Spjøtvoll, E., Martens, H., and Volden, R. (1982) Restricted least squares estimation of spectra and concentration of two unknown constituents available in mixtures, *Technometrics*, **24**, 173–180.

Sprent, P. (1968) Linear relationships in growth and size studies, *Biometrics*, **24**, 639–656.

Sprent, P. (1969) *Models in Regression and Related Topics*, Methuen, London.

Srinivasan, V. and Shocker, A.D. (1973a) Linear programming techniques for multidimensional analysis of preferences, *Psychometrika*, **38**, 337–369.

Srinivasan, V. and Shocker, A.D. (1973b) Estimating weights for multiple attributes in a composite criterion using pairwise judgments, *Psychometrika*, **38**, 473–493.

Srivastava, M.S. and Carter, E.M. (1980) Asymptotic distribution of latent roots and applications, *Multivariate Statistical Analysis* (R.P. Gupta, editor), North-Holland, New York, pp. 219–236.

Srivastava, M.S. and Carter, E.M. (1983) *An Introduction to Applied Multivariate Statistics*, North-Holland, New York, pp. 219–236.

Stahle, L. and Wold, S. (1987) Partial least squares analysis with cross-validation for the two-class problem: A Monte-Carlo Study, *J. Chemom.*, 1, 185–196.

Steece, B.M. (1986) Regressor space outliers in ridge regression, *Comm. Stat.—Theor. Meth.*, 15, 3599–3605.

Steiger, J.H. (1979) Factor indeterminancy in the 1930's and the 1970's: Some interesting parallels, *Psychometrika*, 44, 157–167.

Steiger, J.H., Shapiro, A., and Browne, M.W. (1985) On the multivariate asymptotic distribution of sequential chi-square statistics, *Psychometrika*, 50, 253–264.

Stewart, D. and Love, W. (1968) A general canonical correlation index, *Psych. Bull.*, 70, 160–163.

Stewart, G.W. (1973) *Introduction to Matrix Computations*, Academic Press, Inc., New York.

Stewart, G.W. (1987) Collinearity and least squares regression (with discussion), *Stat. Sci.*, 2, 68–100.

Stone, M. (1974) Cross-validatory choice and assessment of statistical predictions, *J. Roy. Stat. Soc. B*, 36, 111–133.

Stone, M. and Brooks, R.J. (1990) Continuum regression: Cross-validated sequentially constructed prediction embracing ordinary least squares, partial least squares and principal components regression (with discussion), *J. Roy. Stat. Soc. B*, 52, 237–269.

Stone, R. (1947) On the interdependence of blocks of transactions (with discussion), *J. Roy. Stat. Soc. B*, 9, 1–45.

Stroebel, L., Compton, J., Current, I. and Zakia, R. (1986) *Photographic Materials and Processes*, Focal Press, Boston.

Stroud, C.P. (1953) An application of factor analysis to systematics of Kalotermes, *Syst. Zool.*, 2, 76–92. (Also in Bryant and Atchley, 1975.)

Sugiyama, T. (1970) Joint distribution of the extreme roots of a covariance matrix, *Ann. Math. Stat.*, 41, 655–657.

Sugiyama, T. and Tong, H. (1976) On a statistic useful in dimensionality reduction in multivariable linear stochastic system, *Comm. Stat.—Theor. Meth.*, A5, 711–721.

Sujan, H. (1986) Smarter vs. harder: An exploratory attributional analysis of salespeople's motivation, *J. Mark. Res.*, 23, 41–49.

Sylvestre, E.A., Lawton, W.H., and Maggio, M.S. (1974) Curve resolution using a postulated chemical reaction, *Technometrics*, 16, 353–368.

Takane, Y. (1987) Analysis of contingency tables by ideal point discriminant analysis, *Psychometrika*, 52, 493–513.

Takane, Y. and DeLeeuw, J. (1987) On the relationship between item response theory and factor analysis of discretized variables, *Psychometrika*, 52, 393–408.

Takane, Y., Young, F.W., and DeLeeuw, J. (1977) Nonmetric individual differences multidimensional scaling: An alternating least squares method with optimum scaling features, *Psychometrika*, 42, 7–67.

Takane, Y., Young, F.W., and DeLeeuw, J. (1980) An individual differences additive model: An alternating least squares method with optimal scaling features, *Psychometrika*, **45**, 183–209.

Tanaka, Y. (1988) Sensitivity analysis in principal component analysis: Influence on the subspace spanned by principal components, *Comm. Stat.—Theor. Meth.*, **17**, 3157–3175.

Tanaka, Y. and Odaka, Y. (1989) Influential observations in principal factor analysis, *Psychometrika*, **54**, 475–485.

Taniguchi, M. and Krishnaiah, P.R. (1987) Asymptotic distributions of the functions of the eigenvalues of sample covariance matrix and canonical correlation matrix in multivariate time series, *J. Mult. Anal.*, **22**, 156–175.

Tarumi, T. (1986) Sensitivity analysis of descriptive multivariate methods formulated by the generalized singular value decomposition, *Math. Japon.*, **31**, 957–977.

Taylor, C.C. (1977) Principal components and factor analysis, *Exploring Data Structures* (C.A. O'Muircheartaigh and C. Payne, editors), John Wiley and Sons, Inc., New York, pp. 89–124.

Taylor, W., Pearson, J., Maier, A., and Burns, W. (1965) Study of noise and hearing in jute weaving, *J. Acoustical Soc. Amer.*, **38**, 113–120.

TenBerge, J.M.F. (1977) Orthogonal Procrustes rotation for two or more matrices, *Psychometrika*, **42**, 267–276.

TenBerge, J.M.F. (1979) On the equivalence of two oblique congruence rotation methods and orthogonal approximations, *Psychometrika*, **44**, 359–364.

TenBerge, J.M.F. (1984) A joint treatment of varimax rotation and the problem of diagonalizing symmetric matrices simultaneously in the least squares sense, *Psychometrika*, **49**, 347–358.

TenBerge, J.M.F. (1985) On the relationship between Fortier's simultaneous linear prediction and van den Wollenberg's redundancy analysis, *Psychometrika*, **50**, 121–122.

TenBerge, J.M.F. (1986) Rotation to perfect congruence and the cross-validation of component weights across populations, *Mult. Behav. Res.*, **21**, 41–64.

TenBerge, J.M.F. (1988) Generalized approaches to the Maxbet problem and the Maxdiff problem, with applications to canonical correlations, *Psychometrika*, **53**, 487–494.

TenBerge, J.M.F., DeLeeuw, J., and Kroonenberg, P.M. (1987) Some additional results on principal component analysis of three-mode data by means of alternating least squares algorithms, *Psychometrika*, **52**, 183–191.

TenBerge, J.M.F. and Kiers, H.A.L. (1989) Fitting the off-diagonal DEDICOM model in the least-squares sense by a generalization of the Harmon and Jones MINRES procedure of factor analysis, *Psychometrika*, **54**, 333–337.

TenBerge, J.M.F., Kiers, H.A.L., and DeLeeuw, J. (1988) Explicit Candecomp/Parafac solutions for a contrived 2 × 2 × 2 array of rank three, *Psychometrika*, **53**, 579–584.

TenBerge, J.M.F. and Knol, D.L. (1984) Orthogonal rotations to maximal agreement for two or more matrices of different column order, *Psychometrika*, **49**, 49–55.

TenBerge, J.M.F. and Nevels, K. (1977) A general solution to Mosier's oblique Procrustes problem, *Psychometrika*, **42**, 593–600.

Tenenhaus, M. and Young, F.W. (1985) An analysis and synthesis of multiple correspondence analysis, optimal scaling, dual scaling, homogeneity analysis and other methods for quantifying categorical multivariate data, *Psychometrika*, **50**, 91–119.

TerBraak, C.J.F. (1983) Principal components biplots and alpha and beta diversity, *Ecology*, **64**, 454–462.

Thielemans, A., Lewi, P.J., and Massart, D.L. (1988) Similarities and differences among multivariate display techniques illustrated by Belgian cancer mortality distribution data, *Chemom. and Intel. Lab. Syst.*, **3**, 277–300.

Thomas, B.A.M. (1961) Some industrial applications of multivariate analysis, *Appl. Stat.*, **10**, 3–8.

Thompson, B. (1990) Finding a connection for the sampling error in multivariate measures of relationship: A Monte Carlo study, *Educ. & Psych. Meas.*, **50**, 13–51.

Thompson, G.H. (1934) Hotelling's method modified to give Spearman's g, *J. Educ. Psych.*, **25**, 366–374.

Thurstone, L.L. (1931) Multiple factor analysis, *Psychol. Rev.*, **38**, 406–427. (Also in Bryant and Atchley, 1975.)

Thurstone, L.L. (1932) *Theory of Multiple Factors*, Edwards Bros., Ann Arbor, Mich.

Thurstone, L.L. (1935) *The Vectors of the Mind*, University of Chicago Press, Chicago.

Thurstone, L.L. (1947) *Multiple Factor Analysis*, University of Chicago Press, Chicago.

Tiao, G.C. and Fienberg, S. (1969) Bayesian estimation of latent roots and vectors with special reference to the bivariate normal distribution, *Biometrika*, **56**, 97–108.

Tiao, G.C. and Tsay, R.S. (1989) Model specification in multivariate time series (with discussion), *J. Roy. Stat. Soc. B*, **51**, 157–213.

Timm, N.H. (1975) *Multivariate Analysis; With Applications in Education and Psychology*, Wadsworth Publ. Co., Monterey, Calif.

Tisak, J. and Meredith, W. (1989) Exploratory longitudinal factor analysis in multiple populations, *Psychometrika*, **54**, 261–281.

Torgerson, W.S. (1952) Multidimensional scaling I: Theory and method, *Psychometrika*, **17**, 401–419.

Torgerson, W.S. (1958) *Theory and Methods of Scaling*, John Wiley and Sons, Inc., New York.

Tortora, R.D. (1980) The effect of a disproportionate stratified design on principal component analysis used for variable elimination, *Proc. Survey Sec., Amer. Stat. Assoc.*, 746–750.

Trenkler, D. and Trenkler, G. (1984) On the Euclidean distance between biased estimators, *Comm. Stat.—Theor. Meth.*, **13**, 273–284.

Tucker, L.R. (1958a) Determination of parameters of a functional relation by factor analysis, *Psychometrika*, **23**, 19–23.

Tucker, L.R. (1958b) An interbattery method of factor analysis, *Psychometrika*, **23**, 111–136.

Tucker, L.R. (1966) Some mathematical notes on three-mode factor analysis, *Psychometrika*, **31**, 279–311.

Tucker, L.R. (1972) Relations between multidimensional scaling and three-mode factor analysis, *Psychometrika*, **37**, 3–27.

Tucker, L.R. and Messick, S. (1963) An individual difference model for multidimensional scaling, *Psychometrika*, **28**, 333–367.

Tukey, J.W. (1949) One degree of freedom for non-additivity, *Biometrics*, **5**, 232–242.

Tukey, J.W. (1951) Components in regression, *Biometrics*, **7**, 33–69.

Tukey, J.W. (1962) The future of data analysis, *Ann. Math. Stat.*, **33**, 1–67. Correction, **33**, 812 (1964).

Tukey, J.W. (1977) *Exploratory Data Analysis*, Addison-Wesley, Reading, Mass.

Tyler, D.E. (1981) Asymptotic inference for eigenvectors, *Ann. Stat.*, **9**, 725–736.

Tyler, D.E. (1982) On the optimality of the simultaneous redundancy transformations, *Psychometrika*, **47**, 77–86.

Tyler, D.E. (1983a) The asymptotic distribution of principal component roots under local alternatives to multiple roots, *Ann. Stat.*, **11**, 1232–1242.

Tyler, D.E. (1983b) A class of asymptotic tests for principal component vectors, *Ann. Stat.*, **11**, 1243–1250.

Urban, G.L. and Hauser, J.R. (1980) *Design and Marketing of New Products*, Prentice-Hall, Inc., Englewood Cliffs, N.J.

Van de Geer, J.P. (1984) Linear relations among *k* sets of variables, *Psychometrika*, **49**, 79–94.

Van den Wollenberg, A.L. (1977) Redundancy analysis: An alternative for canonical correlation analysis, *Psychometrika*, **42**, 207–219.

Van der Burg, E., DeLeeuw, J., and Verdegaal, R. (1988) Homogeneity analysis with *k* sets of variables: An alternating least squares method with optimum scaling features, *Psychometrika*, **53**, 177–197.

Van der Heijden, P.G.M. and DeLeeuw, J. (1985) Correspondence analysis used complementary to loglinear analysis, *Psychometrika*, **50**, 429–447.

Van der Heijden, P.G.M., de Falguerolles, A., and DeLeeuw, J. (1989) A combined approach to contingency table analysis using correspondence analysis and log–linear analysis (with discussion), *Appl. Stat.*, **38**, 249–292.

Van der Heijden, P.G.M. and Worsley, K.J. (1988) Comment on "Correspondence analysis used complimentary to loglinear analysis", *Psychometrika*, **53**, 287–291.

Van der Kloot, W.A. and Kroonenberg, P.M. (1985) External analysis with three-mode principal component models, *Psychometrika*, **50**, 479–494.

Van der Linde, A. (1988) Rethinking factor analysis as an interpolation problem, *Statistics*, **19**, 359–367.

Van Gerven, D.P. and Oakland, G.B. (1973) Univariate and multivariate statistical model in the analysis of human sexual dimorphism, *Statistician*, **22**, 256–268.

Vani, V. and Raghavachari, M. (1985) A note on the determination of configuration and weights for a class of individual scaling models, *Psychometrika*, **50**, 539–542.

Van Pelt, W. and Van Rijckevorsel, J. (1986) Non-linear principal component analysis of maximum expiratory flow-volume curves, *Appl. Stoch. Models and Data Anal.*, **2**, 1–12.

Van Rijckevorsel, J.L.A. (1988) Fuzzy coding and B-splines, *Component and Correspondence Analysis* (J.L.A. Van Rijckevorsel and J. DeLeeuw, editors), John Wiley and Sons, Inc., New York, pp. 33–54.

Vavra, T.G. (1972) Factor analysis of perceptual change, *J. Mark. Res.*, **9**, 193–199.

Velicer, W.F. (1974) An empirical comparison of the stability of factor analysis, principal component analysis and rescaled image analysis, *Educ. & Psych. Meas.*, **34**, 563–572.

Velicer, W.F. (1976a) The relation between factor score estimates, image scores and principal component scores, *Educ. & Psych. Meas.*, **36**, 149–159.

Velicer, W.F. (1976b) Determining the number of components from the matrix of partial correlations, *Psychometrika*, **41**, 321–327.

Velicer, W.F. (1977) An empirical comparison of the similarity of principal component, image and factor patterns, *Mult. Behav. Res.*, **12**, 3–22.

Velicer, W.F. and Jackson, D.N. (1990) Component analysis versus common factor analysis: Some issues on selecting an appropriate procedure (with discussion), *Mult. Behav. Res.*, **25**, 1–114.

Vinod, H.D. (1976) Application of new ridge regression methods to a study of Bell System scale economies, *J. Amer. Stat. Assoc.*, **71**, 835–841.

Vittadini, G. (1988) On the validity of the indeterminate latent variables in the LISREL model, *Comm. Stat.—Theor. Meth.*, **17**, 861–874.

Vogt, N.B. (1988) Principal component variable discriminant plots: A novel approach for interpretation and analysis of multi-class data, *J. Chemom.*, **2**, 81–84.

Vong, R., Geladi, P., Wold, S., and Esbensen, K. (1988) Source contributions to ambient aerosol calculated by discriminant partial least squares regression (PLS), *J. Chemom.*, **2**, 281–296.

Wahba, G. (1968) On the distribution of some statistics useful in the analysis of jointly stationary time series, *Ann. Math. Stat.*, **39**, 1849–1862.

Waikar, V.B. (1973) On the joint distribution of the largest and smallest latent roots of two random matrices (noncentral case), *S. Afr. Stat. J.*, **7**, 103–108.

Waikar, V.B. and Schuurmann, F.J. (1973) Exact joint density of the largest and smallest roots of the Wishart and Manova matrices, *Utilitas Math.*, **4**, 253–260.

Walker, E. (1989) Detection of collinearity–influential observations, *Comm. Stat.—Theor. Meth.*, **18**, 1675–1690.

Walker, E. and Birch, J.B. (1988) Influence measures in ridge regression, *Technometrics*, **30**, 221–227.

Walker, M.A. (1967) Some critical comments on "An analysis of crimes by the method of principal components" by B. Ahmad, *Appl. Stat.*, **16**, 36–39.

Wallis, J.R. (1965) Multivariate statistical methods in hydrology—a comparison using data of known functional relationship, *Water Resources Res.*, **1**, 447–461.

Walters, R.G. and MacKenzie, S.B. (1988) A structural equations analysis of the impact of price promotions on store performance, *J. Mark. Res.*, **25**, 51–63.

Wang, S.-C. and Chow, S.-C. (1990) A note on adaptive generalized ridge regression estimator, *Stat. & Prob. Lett.*, **10**, 17–21.

Waternaux, C.M. (1976) Asymptotic distribution of the sample roots for a nonnormal population, *Biometrika*, **63**, 639–645.

Waternaux, C.M. (1984) Principal components in the nonnormal case: The test of equality of Q roots, *J. Mult. Anal.*, **14**, 323–335.

Watson, G.S. (1984) The calculation of confidence regions for eigenvectors, *Austral. J. Stat.*, **26**, 272–276.

Webster, J.T., Gunst, R.F., and Mason, R.L. (1974) Latent root regression analysis, *Technometrics*, **16**, 513–522.

Weeks, D.G. and Bentler, P.M. (1982) Restricted multidimensional scaling models for asymmetric proximities, *Psychometrika*, **47**, 201–208.

Weihs, C. and Schmidli, H. (1990) OMEGA (On line multivariate exploratory graphical analysis): Routine searching for structure (with discussion), *Stat. Sci.*, **5**, 175–226.

Weinberg, S.L., Carroll, J.D., and Cohen, H.S. (1984) Confidence regions for INDSCAL using jackknife and bootstrap techniques, *Psychometrika*, **49**, 475–491.

Weiner, P.H. (1973) The estimation of relative gas phase acidities of substituted benzoic acids from solution measurements by factor analysis, *J. Amer. Chem. Soc.*, **95**, 5845–5851.

Weitzman, R.A. (1963) A factor analytic method for investigating differences between groups of individual learning curves, *Psychometrika*, **28**, 69–80.

Wernimont, G. (1967) Evaluating laboratory performance of spectrophotometers, *Anal. Chem.*, **39**, 554–562.

White, J.W. and Gunst, R.F. (1979) Latent root regression: Large sample analysis, *Technometrics*, **21**, 481–488. Corrigendum **22**, 452 (1980).

Whittle, P. (1952) On principal components and least square methods of factor analysis, *Skand. Aktuar.*, **35**, 223–239.

Whittle, P. and Adelman, I., appendix by T.K. Dijkstra (1982) The fitting of restricted rank regression with prior information, *Systems Under Indirect Observation*, II (K.G. Jöreskog and H. Wold, editors), North-Holland, New York, pp. 273–291.

Wilk, M.B. and Gnanadesikan, R. (1964) Graphical methods for internal comparisons in multiresponse experiments, *Ann. Math. Stat.*, **35**, 613–631.

Wilk, M.B. and Gnanadesikan, R. (1968) Probability plotting methods for the analysis of data, *Biometrika*, **55**, 1–17.

Wilk, M.B., Gnanadesikan, R., and Huyett, M.J. (1962) Probability plots for the gamma distribution, *Technometrics*, **4**, 1–20.

Wilkie, W.L. and Pessemier, E.A. (1973) Issues in marketing's use of multiattribute attitude models, *J. Mark. Res.*, **10**, 428–441.

Wilks, S.S. (1932a) Moments and distributions of estimates of population parameters for fragmentary samples, *Ann. Math. Stat.*, **3**, 165–195.

Wilks, S.S. (1932b) Certain generalizations in the analysis of variance, *Biometrika*, **24**, 471–494.

Williams, E.J. (1952) The interpretation of interactions in factorial experiments, *Biometrika*, **39**, 65–81.

Williams, E.J. (1959) *Regression Analysis*, John Wiley and Sons, Inc., New York.

Williams, J.S. (1978) A definition for the common-factor analysis model and the elimination of problems of factor score indeterminacy, *Psychometrika*, **43**, 293–306.

Windig, W. (1988) Mixture analysis of spectral data by multivariate methods, *Chemon. and Intel Lab. Syst.*, **4**, 201–213.

Windig, W., McClennen, W.H., and Meuzelaar, H.L.C. (1987) Determination of fractional concentrations and exact component spectra by factor analysis of pyrolysis mass spectra of mixtures, *Chemom. and Intel. Lab. Syst.*, **1**, 151–165.

Winsberg, S. (1988) Two techniques: Monotone spline transformations for dimension reduction in PCA and easy-to-generate metrics for PCA of sampled functions, *Component and Correspondence Analysis* (J.L.A. Van Rijckevorsel and J. DeLeeuw, editors), John Wiley and Sons, Inc., New York, pp. 115–135.

Winsberg, S. and Ramsay, J.O. (1983) Monotone spline transformations for dimension reduction, *Psychometrika*, **48**, 575–595.

Wish, M. and Carroll, J.D. (1974) Application of individual differences scaling to studies of human perception and judgment, *Handbook of Perception*, **2** (E.C. Carterette and M.P. Friedman, editors), Academic Press, Inc., New York, pp. 449–491.

Wold, H. (1966a) Nonlinear estimation by iterative least square procedures, *Research Papers in Statistics* (F.N. David, editor), John Wiley and Sons, Inc., New York, pp. 411–444.

Wold, H. (1966b) Estimation of principal components and related models by iterative least squares, *Multivariate Analysis*, **I** (P.K. Krishnaiah, editor), Academic Press, Inc., New York, pp. 391–420.

Wold, H. (1982) Soft modeling, *Systems Under Indirect Observation*, **II** (K.G. Jöreskog and H. Wold, editors), North-Holland, New York, pp. 1–54.

Wold, H. (1985) Partial least squares, *Encyc. Stat. Sci.*, **6** (S. Kotz and N. Johnson, editors), John Wiley and Sons, Inc., New York, pp. 581–591.

Wold, S. (1976) Pattern recognition by means of disjoint principal components models, *Pattern Recognition*, **8**, 127–139.

Wold, S. (1978) Cross-validatory estimation of the number of components in factor and principal component analysis, *Technometrics*, **20**, 397–405.

Wold, S., Geladi, P., Esbensen, K., and Öhman, J. (1987) Multi-way principal components- and PLS-analysis, *J. Chemom.*, **1**, 41–56.

Wold, S., Martens, H., and Wold, H. (1983) The multivariate calibration problem in chemistry solved by the PLS method, *Proc. Conf. on Matrix Pencils*, March 1982 (A. Ruhe and B. Kagström, editors), Springer-Verlag, Heidelberg, pp. 286–293.

Wold, S. and Sjöström, M. (1987) Comments on a recent evaluation of the SIMCA method, *J. Chemom.*, **1**, 243–245.

Woodhall, W.H. and Ncube, M.M. (1985) Multivariate cusum quality control procedures, *Technometrics*, **27**, 285–292.

Yanai, H. and Ichikawa, M. (1990) New upper and lower bounds for communality in factor analysis, *Psychometrika*, **55**, 405–410.

Yanai, H. and Mukherjee, B.N. (1987) A generalized method of image analysis from an intercorrelation matrix which may be singular, *Psychometrika*, **52**, 555–564.

Yendle, P.W. and MacFie, H.J.H. (1989) Discriminant principal component analysis, *J. Chemom.*, **3**, 589–600.

Yeomans, K.A. and Golder, P.A. (1982) The Guttman–Kaiser criterion as a predictor of the number of common factors, *Statistician*, **31**, 221–229.

Yin, Y.Q., Bai, Z.D., and Krishnaiah, P.R. (1983) Limiting behaviour of the eigenvalues of a multivariate *F* matrix, *J. Mult. Anal.*, **13**, 508–516.

Yochmowitz, M.G. and Cornell, R.G. (1978) Stepwise tests for multiplicative components of interaction, *Technometrics*, **20**, 79–84.

Young, F.W. (1970) Nonmetric multidimensional scaling: Recovery of metric information, *Psychometrika*, **35**, 455–473.

Young, F.W. (1981) Quantitative analysis of qualitative data, *Psychometrika*, **46**, 357–388.

Young, F.W. (1987) *Multidimensional Scaling: History, Theory and Applications* (R.M. Hamer, editor), Lawrence Erlbaum Assoc., Hillsdale, N.J.

Young, F.W. and Lewyckyj, R. (1979) *ALSCAL User's Guide*, 3d Ed., Data Analysis and Theory Associates, Chapel Hill, N.C.

Young, F.W. and Sarle, W.S. (1983) *SAS Views*®: *Exploratory Data Analysis*, 1982 Edition, SAS Institute, Inc., Cary, N.C.

Young, F.W., Takane, Y., and DeLeeuw, J. (1978) The principal components of mixed measurements level multivariate data: An alternating least squares method with optimum scaling features, *Psychometrika*, **43**, 279–281.

Young, G. and Householder, A.S. (1938) Discussion of a set of points in terms of their mutual distances, *Psychometrika*, **3**, 19–22.

Zachert, V. (1951) A factor analysis of vision tests, *Am. J. Optometry*, **28**, 405–416.

Zamar, R.H. (1989) Robust estimation in the error-in-variables model, *Biometrika*, **76**, 149–160.

Zani, S. (1980) On Hubert's clustering method applied to principal components, *Data Analysis and Informetrics* (E. Diday *et al.*, editors), North-Holland, New York, pp. 593–599.

Zinnes, J.L. and MacKay, D.B. (1983) Probabilistic multidimensional scaling: complete and incomplete data, *Psychometrika*, **48**, 27–48.

Zwick, W.R. and Velicer, W.F. (1986) Comparison of five rules for determining the number of components to retain, *Psych. Bull.*, **99**, 432–442.

Author Index

Subject Index